MATHEMATICAL PHYSICS,
AN ADVANCED COURSE

NORTH-HOLLAND SERIES IN
APPLIED MATHEMATICS AND MECHANICS

EDITORS:

H. A. LAUWERIER
Institute of Applied Mathematics
University of Amsterdam

W. T. KOITER
Laboratory of Applied Mechanics
Technical University, Delft

VOLUME 11

NORTH-HOLLAND PUBLISHING COMPANY – AMSTERDAM · LONDON

MATHEMATICAL PHYSICS, AN ADVANCED COURSE

BY

S. G. MIKHLIN

Faculty of Mathematical Physics,
Leningrad University

1970

NORTH-HOLLAND PUBLISHING COMPANY — AMSTERDAM · LONDON
AMERICAN ELSEVIER PUBLISHING COMPANY, INC. – NEW YORK

© NORTH-HOLLAND PUBLISHING COMPANY - 1970

All Rights Reserved. No part of this publication may be reproduced, stored in a retrieval system, or transmitted, in any form or by any means, electronic, mechanical, photocopying, recording or otherwise, without the prior permission of the Copyright owner.

Library of Congress Catalog Card Number 71-126505
North-Holland I.S.B.N. 0 7204 2361 9
American Elsevier I.S.B.N. 0 444 100 70 9

PUBLISHERS:

NORTH-HOLLAND PUBLISHING COMPANY - AMSTERDAM
NORTH-HOLLAND PUBLISHING COMPANY, LTD. - LONDON

SOLE DISTRIBUTORS FOR THE U.S.A. AND CANADA:

AMERICAN ELSEVIER PUBLISHING COMPANY, INC.
52 VANDERBILT AVENUE
NEW YORK, N.Y. 10017

This book was originally published in Russian under the title "Kuhrs Matematicheskoy Fiziki" by Izdatelstvo "Nauka", Glavnaya Redaktsiya, Fiziko-Matematicheskoy Interatuhry", Moskva 1968.
Translated by Multilingua (Scientific Translations), London.

PRINTED IN THE NETHERLANDS

Contents

FOREWORD xiii

FOREWORD TO THE ENGLISH EDITION xv

INTRODUCTION 1

PART I. Mean Functions and Generalized Derivatives 7

Chapter 1. Mean Functions 7

1. Averaging Kernel 9
2. Mean Functions 11
3. Convergence of Mean Functions 13
Exercises 16

Chapter 2. Generalized Derivatives 18

1. The Concept of a Generalized Derivative 18
2. Basic Properties of the Generalized Derivative 23
3. Limit Properties of Generalized Derivatives 25
4. The Case of One Independent Variable 27
5. Sobolev Spaces and Embedding Theorems 30
Exercises 31

PART II. Elements of the Calculus of Variations 33

Chapter 3. Basic Concepts 35

1. Examples of the Extremum of a Functional 35

2. Statement of the Problem of the Variational Calculus 37
3. Variation and Gradient of a Functional 40
4. Euler's Equation 50
5. Second Variation. Sufficient Condition for an Extremum 54
6. The Isoperimetric Problem 56
7. Minimizing Sequence 60
Exercises 61

Chapter 4. Functionals Depending on Numerical Functions of Real Variables 63

1. The Simplest Problem of the Variational Calculus (Problem S) 63
2. Discussion of the Second Variation 65
3. Case of Several Independent Variables 67
4. Functionals Depending on Higher-Order Derivatives 72
5. Functionals Depending on Several Functions 75
6. Natural Boundary Conditions 76

Chapter 5. Minimum of a Quadratic Functional 84

1. The Concept of a Quadratic Functional 84
2. Positive Definite Operators 85
3. Energy Space 91
4. Problem of the Minimum of a Quadratic Functional 100
5. Generalized Solution 103
6. The Separability of Energy Space 105
7. Extension of a Positive-Definite Operator 108
8. The Simplest Boundary-Value Problem for an Ordinary Linear Differential Equation 113
9. The Minimum of a Quadratic Functional: A More General Problem 118
10. The Case of a Merely Positive Operator 121
Exercises 122

Chapter 6. Eigenvalue Spectrum of a Positive Definite Operator 124

1. The Concept of the Eigenvalue Spectrum of an Operator 124
2. Eigenvalues and Eigenelements of a Symmetric Operator 126
3. Generalized Eigenvalue Spectrum of a Positive Definite Operator 127
4. Variational Formulation of the Problem of the Eigenvalue Spectrum 129

5. Theorem on the Smallest Eigenvalue 132
6. Discrete Spectrum 135
7. The Sturm-Liouville Problem 138
8. Elementary Cases 144
9. The Minimax Principle 145
10. Magnitude of the Eigenvalues in the Sturm-Liouville Problem 148
Exercises 150

PART III. Elements of the Theory of Integral Equations 151

Chapter 7. Completely Continuous Operators 153

1. Necessary Results from Functional Analysis 153
2. Fredholm Operator 155
3. Integral Operator with a Weak Singularity 158
4. Operators with a Weak Singularity in the Space of Continuous Functions 162
Exercises 165

Chapter 8. Fredholm Theory 166

1. Equation with Completely Continuous Operator. Integral Equations 166
2. Reduction to a Finite-Dimensional Equation. Proofs of Fredholm's First and Second Theorems 168
3. Proof of Fredholm's Third Theorem 172
4. Proof of Fredholm's Fourth Theorem 173
5. Fredholm's Alternative 176
6. Continuity of Solutions of the Equation with a Weak Singularity 177

PART IV. General Aspects of Partial Differential Equations 181

Chapter 9. Equations and Boundary-Value Problems 183

1. Differential Expression and Differential Equation 183
2. Classification of Second-Order Equations 185
3. Boundary Conditions and Boundary-Value Problems 188

4. Cauchy's Problem 192
5. Questions of Existence, Uniqueness and Proper Posing of Boundary-Value Problems 194

Chapter 10. Characteristics. Canonical Form. Green's Formulae 200

1. Transformation of the Independent Variables 200
2. Characteristics. Relation Between Cauchy Data on Characteristics 202
3. Reduction of a Second-Order Equation to Canonical Form 205
4. Case of Two Independent Variables 206
5. Formally Adjoint Differential Expressions 209
6. Green's Formulae 210

PART V. Equations of Elliptic Type 215

Chapter 11. Laplace's Equation and Harmonic Functions 217

1. Basic Concepts 217
2. Singular Solution of Laplace's Equation 219
3. Integral Representation of Functions of Class $C^{(2)}$ 221
4. Integral Representation of a Harmonic Function 224
5. The Concept of a Potential 225
6. Properties of the Newtonian Potential 228
7. Mean Value Theorem 237
8. Maximum Principle 240
9. Convergence of Sequences of Harmonic Functions 243
10. Extension to Equations with Variable Coefficients 246

Chapter 12. Dirichlet and Neumann Problems 254

1. Formulation 254
2. Uniqueness Theorems for Laplace's Equation 256
3. Solution of the Dirichlet Problem for a Sphere 260
4. Liouville's Theorem 267
5. Dirichlet Problem for the Exterior of a Sphere 268
6. Derivatives of Harmonic Functions at Infinity 269
7. Uniqueness Theorem for the Exterior Neumann Problem 270

Chapter 13. Elementary Solutions of the Dirichlet and Neumann Problems 273

1. Dirichlet and Neumann Problems for a Circle 273
2. Dirichlet Problem for a Circular Annulus 277
3. Application of Conformal Mapping 278
4. Spherical Harmonics and their Properties 282
5. Dirichlet and Neumann Problems: Solutions in Spherical Harmonics 285

Exercises 289

Chapter 14. Variational Method for the Dirichlet Problem. Other Positive Definite Problems 290

1. Friedrichs's Inequality 290
2. Operator of the Dirichlet Problem 292
3. Energy Space for the Dirichlet Problem 296
4. Generalized Solution of the Dirichlet Problem 299
5. Dirichlet Problem for the Homogeneous Equation 301
6. On the Existence of Second Derivatives in the Solution of the Dirichlet Problem 304
7. Elliptic Equations of Higher Order and Systems of Equations 306
8. Dirichlet Problem for an Infinite Domain 309

Exercises 311

Chapter 15. Spectrum of the Dirichlet Problem 313

1. Integral Representation of Functions which Vanish on the Boundary of a Finite Domain 313
2. Spectrum of the Dirichlet Problem for a Finite Domain 315
3. Elementary Cases 316
4. Estimate of the Magnitude of the Eigenvalues 320

Chapter 16. The Neumann Problem 325

1. The Case of Positive $C(x)$ 325
2. The Case $C(x) \equiv 0$ 327
3. Integral Representation of S. L. Sobolev 329
4. Investigation of the Operator $\mathfrak{N}_0$ 331
5. Generalized Solution of the Neumann Problem 335

Exercises 337

Chapter 17. Non-Self-Adjoint Elliptic Equations 338

1. Generalized Solution 338
2. Fredholm's Theorems 340
Exercise 343

Chapter 18. Method of Potentials for the Homogeneous Laplace Equation 344

1. Lyapunov Surfaces 344
2. Solid Angle 349
3. The Double-Layer Potential and its Direct Value 356
4. Gauss's Integral 357
5. Limiting Values of the Double-Layer Potential 360
6. Continuity of the Single-Layer Potential 363
7. Normal Derivative of the Single-Layer Potential 365
8. Reduction of the Dirichlet and Neumann Problems to Integral Equations 370
9. Dirichlet and Neumann Problems in a Half-Space 373
10. The First Pair of Adjoint Equations 374
11. The Second Pair of Adjoint Equations 376
12. Solution of the Exterior Dirichlet Problem 379
13. Case of Two Independent Variables 381
14. Equations of Potential Theory for the Circle 387

Chapter 19. The Oblique-Derivative Problem 391

1. Formulation of the Problem 391
2. Hilbert Operator 392
3. Equations with the Hilbert Operator 398
4. Number of Solutions and Index for the Oblique-Derivative Problem in a Plane 404

PART VI. Time-Dependent Equations 409

Chapter 20. The Heat-Conduction Equation 411

1. The Heat-Conduction Equation and its Characteristics 411
2. Maximum Principle 413

3. Cauchy's Problem and the Mixed Boundary-Value Problem 416
4. Uniqueness Theorems 418
5. Abstract Functions of a Real Variable 420
6. Generalized Solution of the Mixed Problem 421

Chapter 21. The Wave Equation 425

1. Basic Concepts 425
2. The Mixed Problem and its Generalized Solution 426
3. Wave Equation with Constant Coefficients. Cauchy's Problem. Characteristic Cone 430
4. Uniqueness Theorem for Cauchy's Problem. Domain of Dependence 431
5. The Propagation of Waves 434
6. Generalized Solution of the Cauchy Problem 436

Chapter 22. Fourier's Method 440

1. Fourier's Method for the Heat-Conduction Equation 440
2. Justification of the Method 442
3. On the Existence of the Classical Solution. A Special Case 445
4. Fourier's Method for the Wave Equation 448
5. Justification of the Method for the Homogeneous Equation 450
6. Justification of the Method for Homogeneous Initial Conditions 453
7. Equation of the Vibrating String. Conditions for Existence of the Classical Solution 455

Chapter 23. Cauchy's Problem for the Heat-Conduction Equation 459

1. Some Properties of the Fourier Transform 459
2. Derivation of Poisson's Formula 463
3. Justification of Poisson's Formula 467
4. Infinite Velocity of Heat-Conduction 470

Chapter 24. Cauchy's Problem for the Wave Equation 472

1. Application of the Fourier Transform 472
2. Transformation of the Solution 475
3. Case of Three-Dimensional Space 478
4. Justification of Kirchhoff's Formula 480

5. Rear Wave Front 483
6. The Case $m = 2$ (Vibrating-Membrane Equation) 484
7. Equation of the Vibrating String 485
8. Wave Equation with Variable Coefficients 487

PART VII. Properly and Improperly Posed Problems 493

Chapter 25. The Proper Posing of Problems in Mathematical Physics 495

1. The Fundamental Theorem 495
2. Positive Definite Problems 497
3. Dirichlet Problem for the Homogeneous Laplace Equation 498
4. Exterior Neumann Problem 499
5. Interior Neumann Problem 502
6. Heat-Conduction Problems 504
7. Problems Connected with the Wave Equation 507
8. Improperly Posed Problems in Mathematical Physics 508

Appendices 511

Appendix 1. Elliptic Systems 513
Appendix 2. Cauchy's Problem for Hyperbolic Equations — V. M. Babich 521
Appendix 3. Some Questions in the Theory of General Differential Operators — V. G. Maz'ya 532
Appendix 4. Non-Linear, Second-Order Elliptic Equations — I. Ya. Bakel'man 541

Bibliography 553

Subject Index 558

Foreword

This book represents a somewhat expanded version of a course of lectures in mathematical physics which I have given to students of mathematics at the University of Leningrad over the past few years.

As is usual, the course encompasses only the theory of linear partial differential equations, almost exclusively of second order. Naturally enough, a principal role in the book is played by the three classical types of equations which have been most intensively studied and which are most important in application: elliptic, parabolic, and hyperbolic.

Equations of the latter two types can be treated, at least locally, as abstract, ordinary differential equations in which the unknown function is acted on by an elliptic operator. Hence it can be inferred that the elliptic type is basic for classical mathematical physics, and that the study of the subject must begin with this type.

It soon becomes evident that positive definite problems (i.e. problems with positive definite energy) play a special role, and are specially amenable to treatment. Such problems are conveniently solved by the variational method, in the course of which it is natural to introduce generalized solutions. This approach enables us to solve many positive definite problems without additional expenditure of effort, and at the same time to advance far beyond the limits of a classical course.

I consider it advisable to give the theory of the eigenvalue spectrum of positive definite operators before proceeding to time-dependent problems and Fourier's method; this is easily done with the aid of variational methods. The mixed boundary-value problem for time-dependent equations can be solved on the basis of this theory: Fourier's method reduces to an expansion over the eigenfunction spectrum, and can be justified, with very little effort, in terms of generalized (and sometimes classical) solutions. Spectral decomposition can also be used for solving Cauchy's problem, but for equations with constant coefficients – wave and heat-conduction equations – this problem is solved more simply, and with sufficient generality, by a Fourier transform with respect to the space coordinates.

Potential theory also finds a place in this book. It is impossible to restrict ourselves solely to positive definite problems: this is evident even in the Dirichlet and Neumann problems for Laplace's equation in an infinite domain, or in the oblique derivative problem. The method of potentials is presented for Laplace's equation, where it is more easily applied and where it immediately gives convincing results.

All the analysis is formulated for the general case of multi-dimensional space.

This brings us to the structure of the book. The basic text is divided into seven parts, of varying lengths. The first three of these parts are auxiliary, although Part II ("Elements of the Calculus of Variations") has independent value. The rather brief Part IV contains necessary formal apparatus, as well as the formulation of the basic concepts and most important problems.

Part V is the longest in the book, and this fact is completely justified by the special role of elliptic equations. We draw attention to the last chapter of this part, which is devoted to the oblique derivative problem in a two-dimensional domain, a problem whose index can be different from zero.

In Part VI we consider the heat-conduction and wave equations, with both constant and variable coefficients. It seems expedient to treat both these equations together in the one part; in spite of their different properties, they can be solved by quite similar methods. Part VII is devoted to a brief discussion of the question of properly and improperly posed problems in mathematical physics.

In addition to the basic text, the book contains four short appendices which discuss some modern ideas and results in the theory of partial differential equations. Of these, only Appendix 1 was written by me; V. M. Babich, V. G. Maz'ya and I. Ya. Bakel'man kindly agreed to write the other appendices, and I am happy to express my sincere gratitude to them.

The general plan and basic contents of this book were frequently discussed with my colleagues of the Mathematical Physics faculty of Leningrad University, headed by Academician V. I. Smirnov. To all of them I express my deep thanks. I am especially grateful to V. M. Babich and V. G. Maz'ya whose advice I used on a number of occasions. In particular, the presentation of the question of Lyapunov surfaces, in Chapter 18, is in accordance with their suggestions. I wish also to thank my students I. N. Krol, S. M. Mineeva and K. G. Semenova for their help in the preparation of the manuscript. Finally, I am particularly indebted to the editor, V. V. Arestov, who read the manuscript with special care, and contributed substantially to its improvement.

Leningrad, January 1968   S. Mikhlin

Foreword to the English Edition

In writing this book, "Mathematical Physics, an Advanced Course" I have been guided by two basic ideas: 1) A course of this kind should be firmly based on the concepts and methods of functional analysis. 2) Priority should be given to the study of equations of elliptic type. The general structure and contents of the book have been determined by these considerations.

The English edition contains a few changes, the most important of which are the following.
    A number of formulations and proofs have been improved.
    The connection between generalized and continuous derivatives has been explained in greater detail.
    A criterion that an element should belong to the energy space of a given positive definite operator has been established. This criterion subsequently turns out to be useful in several different situations.
    A theorem about higher derivatives of the Newtonian potential has been proved under conditions that are substantially more general than in the original edition.

In other respects the contents of the book are identical with those of the Russian edition.

Leningrad, September 1969   S. Mikhlin

Introduction

Mathematical physics can be regarded as an aspect of the general theory of partial differential equations. The term "mathematical physics" is associated with the fact that this subject arose through the investigation of some simple, but important, problems in physics. We consider a few of these.

1. *Problem of the Vibrating String.* Suppose that the initial position of the string coincides with the Ox-axis, and that vibrations take place in a vertical plane. Let the string be somehow displaced from its equilibrium state, as, for example, through the effect of a blow applied to it. The string thereby changes its shape; every point of it experiences some displacement. Suppose for simplicity that the displacement is perpendicular to the Ox-axis, and is always confined to the plane (x, u) (Fig. 1). The ordinate u

Fig. 1.

gives the deflection of the string from its equilibrium position. Obviously it is a function of two variables, $u = u(x, t)$. If we assume that the string is homogeneous and of constant thickness, that no external forces act on it after the initial instant, and, finally, that the string is inextensible but does not resist bending, then we can show that the function u satisfies the linear partial differential equation

$$\frac{\partial^2 u}{\partial x^2} = \frac{1}{a^2} \frac{\partial^2 u}{\partial t^2}. \qquad (1)$$

Here a is a constant quantity which depends on the physical properties of the string.

Equation (1) is approximate; it is valid in the case of so-called small oscillations of a string. This equation is called the *wave equation with two independent variables*, or the *vibrating-string equation*.

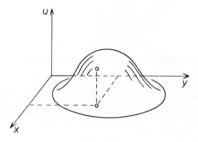

Fig. 2.

More complex physical problems lead to differential equations, similar to equation (1), but themselves more complex. Thus, the transverse vibrations of a thin membrane, whose equilibrium position lies in the (x, y)-plane (Fig. 2), are, under certain conditions, described by a second-order partial differential equation:

$$\frac{\partial^2 u}{\partial x^2} + \frac{\partial^2 u}{\partial y^2} = \frac{1}{a^2}\frac{\partial^2 u}{\partial t^2}, \qquad a = \text{const.} \tag{2}$$

Equation (2) is called the *vibrating-membrane equation*, or the *wave equation with three independent variables*. As with the string equation, it is sufficiently exact only for small oscillations of the membrane.

The *wave equation with four independent variables* has the form

$$\frac{\partial^2 u}{\partial x^2} + \frac{\partial^2 u}{\partial y^2} + \frac{\partial^2 u}{\partial z^2} = \frac{1}{a^2}\frac{\partial^2 u}{\partial t^2}. \tag{3}$$

This equation defines, for example, the velocity field of oscillations in a gas when the velocity is small and has a potential, i.e., when there exists a function u such that $\boldsymbol{v} = \text{grad } u$, where $\boldsymbol{v}$ is the velocity vector of a gas particle.

2. Consider a homogeneous body, part of whose surface is heated. In such a body a temperature field exists, and obviously the temperature changes from point to point in the body, and varies from one instant of time to another. Denoting the temperature by u, we see that u is a function

of the independent variables x, y, z, t:

$$u = u(x, y, z, t).$$

It can be shown that this function satisfies the partial differential equation

$$\frac{\partial^2 u}{\partial x^2} + \frac{\partial^2 u}{\partial y^2} + \frac{\partial^2 u}{\partial z^2} = k \frac{\partial u}{\partial t}, \qquad k = \text{const.} \tag{4}$$

We note that the expression

$$\frac{\partial^2 u}{\partial x^2} + \frac{\partial^2 u}{\partial y^2} + \frac{\partial^2 u}{\partial z^2}$$

is usually called the *Laplace operator* on the function u, and we designate it with the symbol Δ:

$$\Delta u = \frac{\partial^2 u}{\partial x^2} + \frac{\partial^2 u}{\partial y^2} + \frac{\partial^2 u}{\partial z^2};$$

consequently equation (4) can be rewritten in the form

$$\Delta u = k \frac{\partial u}{\partial t}. \tag{4'}$$

Equation (4) (or (4′)) is called the *heat-conduction equation*. It is a linear, second-order, partial differential equation. Although it was already known to Euler, it is more commonly associated with the name of Fourier.

3. Suppose now that the temperature is constant with respect to time. Then u is a function of the space coordinates and is independent of time:

$$u = u(x, y, z).$$

Equation (4) reduces to the following:

$$\frac{\partial^2 u}{\partial x^2} + \frac{\partial^2 u}{\partial y^2} + \frac{\partial^2 u}{\partial z^2} = 0 \tag{5}$$

or

$$\Delta u = 0. \tag{5'}$$

Equation (5) (or (5′)) is called *Laplace's equation*; it is a linear, second-order, partial differential equation.

In each of the foregoing examples we arrived at a linear, second-order, partial differential equation. But these are not the only equations to be found in mathematical physics.

There are many cases where linear equations of higher order are of relevance in physical applications. Again, there are often geometrical and physical problems which lead to non-linear partial differential equations, or to systems of equations; for example, the well-known systems of differential equations in the theory of elasticity, in hydrodynamics and in electrodynamics.

The equations introduced above – wave, heat-conduction and Laplace – correspond to different physical problems, but they also differ on a purely mathematical level. They are representative of the three most important types of partial differential equation: *hyperbolic*, *parabolic*, and *elliptic*. The study of these types of equation is a theme of mathematical physics, the one to which this book is devoted. It is necessary to emphasize that these three do not exhaust the possible types of partial differential equations; they have been selected because they are the best understood and have the greatest importance in application.

A few words now about the notation and scheme of numbering adopted in this book.

We denote by the symbol E_m a Euclidean space of m dimensions. If a point in this space is denoted, for example, by the letter x, then the cartesian coordinates of this point are denoted by $x_1, x_2, \ldots, x_m$.

We shall repeatedly have occasion to perform integrations with respect to sets of different dimensions in the space E_m; most often we shall need to integrate over a domain of the space or over a $(m-1)$-dimensional surface. In such cases we shall always designate the integration by a single integral sign, whatever the multiplicity. If the variable of integration is denoted, for example, by x, then the element of Lebesgue measure ("volume element") in the space E_m will be written dx. An element of measure on a surface ("element of surface area") we denote by $dS, d\Gamma, \ldots$, if the surface itself is $S, \Gamma, \ldots$. If M is a set of points of the space E_m, then the closure of this set we denote by $\overline{M}$. In particular, if Ω is some region of the space E_m, and Γ is its boundary, then $\overline{\Omega} = \Omega \cup \Gamma$.

Let G be a set in the Euclidean space E_m. The set of functions which are continuous and bounded in G will be denoted by $C(G)$; it is self-evident that a boundedness condition need not be formulated if the set G is compact.

We shall denote by $C^{(k)}(G)$ the set of functions which have all possible derivatives up to order k in G, these derivatives being continuous and bounded in G. The most frequent case we shall encounter will be that in which $G = \overline{\Omega}$, where Ω is a finite domain.

We do not exclude the case $k = \infty$, and shall denote by $C^{(\infty)}(G)$ the set of functions having continuous and bounded derivatives of all orders in G.

We shall make extensive use of the following notation: if a domain is denoted, for example, by the letter Ω, then its volume will be denoted by $|\Omega|$. Analogously, $|\Gamma|$ will signify the area of the surface Γ.

A sphere of radius R in the space E_m will be denoted by S_R. The area of its surface is

$$|S_R| = R^{m-1}|S_1|,$$

where S_1 is the sphere of unit radius.

The well-known formula

$$|S_1| = \frac{2\pi^{\frac{1}{2}m}}{\Gamma(\frac{1}{2}m)},$$

where Γ is the Gamma function, will be useful.

The chapters are numbered consecutively through the book, but sections are numbered within each chapter. When referring to a formula within a section, we use only the number of the formula; in the case of a formula in another section, but in the same chapter, we first give the section number in brackets, followed by the formula number. If we need to refer to a formula from a different chapter, we write both its section and formula number in brackets, and the chapter number outside.

At the end of the book there is a short bibliography to each part and appendix. It contains a list of basic books and monographs (and occasionally articles in journals) which bear on the particular part. References to this literature are sometimes encountered in the text, in which case the number of the cited work is given in square brackets.

PART I

*Mean Functions
and Generalized Derivatives*

Chapter 1. Mean Functions

§ 1. Averaging Kernel *

Let x and y be arbitrary points of the space E_m, and h an arbitrary positive number. The function $\omega_h(x, y)$ is called an *averaging kernel* ** if it possesses the following properties:

1. The function $\omega_h(x, y)$ depends only on h and r, where r is the distance between x and y; $r = |x-y|$. Because of this we shall usually write $\omega_h(r)$ instead of $\omega_h(x, y)$.

2. $\omega_h(r) > 0, \quad r < h,$
   $\omega_h(r) = 0, \quad r \geq h.$

3. $\int_{r<h} \omega_h(r)\,dy = \int_{r<h} \omega_h(r)\,dx = 1.$

4. $\omega_h(r)$ is differentiable infinitely-many times with respect to the cartesian coordinates of the points x and y.

We shall first show that at least one such function does exist. Let

$$\omega_h(r) = \begin{cases} c_h e^{-h^2/(h^2 - r^2)}, & r < h, \quad c_h = \text{const.} > 0; \\ 0, & r \geq h. \end{cases} \tag{1}$$

This function obviously has properties 1 and 2. It will have property 3 if we put

$$c_h = \left\{ \int_{r<h} \exp\left\{ -\frac{h^2}{h^2 - r^2} \right\} dy \right\}^{-1}. \tag{2}$$

We observe that the integral in (2) can be expressed in a more simple form. We introduce spherical coordinates with centre x, use the well-known

---

* The concept of averaging kernel and the closely-related concept of mean function (see below) were first introduced by V. A. Steklov. These concepts were further developed by S. L. Sobolev, whose ideas are expounded in this book. Sobolev also introduced and developed the notion of generalized derivatives, which will be discussed in Chapter 2.

** This is frequently called a *test function*. Ed.

formula $dy = r^{m-1} dr\, dS_1$, and effect the transformation $r = ht$. Then for c_h we obtain the expression

$$c_h = \frac{1}{h^m |S_1|} \left\{ \int_0^1 \exp\left\{ -\frac{1}{1-t^2} \right\} t^{m-1} dt \right\}^{-1}.$$

We now establish property 4. The function (1) is symmetric with respect to x and y, and therefore it is sufficient to prove that this function is differentiable infinitely-many times with respect to the coordinates $y_1, y_2, \ldots, y_m$, with x held fixed. For this we need only show that the given function has derivatives of arbitrary order with respect to r. This is obviously the case if $r < h$ or $r > h$, all derivatives being identically zero in the latter case. We therefore only have to establish that when $r = h$ the function (1) has derivatives of arbitrary order which are equal to zero, and that the derivatives evaluated for $r < h$ all tend to a zero limit when $r \to h$.

We shall give the proof for the first derivative; for higher derivatives a similar proof applies.

(a) The function $\omega_h(r)$ is continuous at $r = h$. In fact, formula (1) shows that $\omega_h(h+0) = 0 = \omega_h(h)$, and

$$\omega_h(h-0) = c_h \lim_{r \to h-0} \exp\left\{ -\frac{h^2}{h^2 - r^2} \right\} = 0,$$

since

$$-\frac{h^2}{h^2 - r^2} \xrightarrow[r \to h-0]{} -\infty.$$

(b) The derivative $\omega_h'(h)$ exists and is equal to zero. In fact,

$$\lim_{r \to h+0} \frac{\omega_h(r) - \omega_h(h)}{h - r} = \lim_{r \to h+0} 0 = 0.$$

At the same time

$$\lim_{r \to h-0} \frac{\omega_h(r) - \omega_h(h)}{r - h} = \lim_{r \to h-0} \frac{\exp\{-h^2/(r^2 - h^2)\}}{r - h} = 0,$$

which can be verified by appeal to L'Hôpital's Rule. Thus the limit

$$\lim_{r \to h} \frac{\omega_h(r) - \omega_h(h)}{r - h} = \omega_h'(h)$$

exists and is equal to zero.

(c) The relation

$$\lim_{r \to h-0} \omega'_h(r) = c_h \lim_{r \to h-0} \frac{2rh^2}{(h^2-r^2)^2} \exp\left\{-\frac{h^2}{h^2-r^2}\right\} = 0$$

holds, as can also be shown without difficulty from L'Hôpital's Rule.

Hence, the first derivative $\omega'_h(r)$ exists and is continuous for all r. In precisely the same way we can demonstrate the existence and continuity of higher derivatives. This establishes property 4.

§ 2. Mean Functions

Let Ω be a finite domain of the space E_m and let $u(y)$ be a function which is integrable* in Ω. We extend the definition of this function outside Ω by setting it equal to zero there. Let x be an arbitrary point of the space E_m. We put

$$u_h(x) = \int_\Omega \omega_h(r) u(y) \, dy, \tag{1}$$

where $\omega_h(r)$ is an arbitrary averaging kernel, having the properties 1–4, § 1. The function u_h is called a *mean function* with respect to u; the quantity h is called the *averaging radius*. There are two other forms in which the mean function can be represented:

1) bearing in mind that $u(y) = 0$, $y \bar\in \Omega$, we can extend the integral (1) over all space, so that

$$u_h(x) = \int_{E_m} \omega_h(r) u(y) \, dy; \tag{1a}$$

2) because of property 2 of the averaging kernel, the integration need not be performed over all space, but only over a sphere with centre x and radius h:

$$u_h(x) = \int_{r<h} \omega_h(r) u(y) \, dy. \tag{1b}$$

Let us note the simplest properties of mean functions:

1. A mean function has infinitely-many derivatives throughout all space; its derivatives of arbitrary order can be obtained by differentiating any of the formulae (1), (1a), (1b).

---

* Throughout this book, integrable means Lebesgue-integrable (or, summable), unless the contrary is stated. Ed.

We show, for example, that

$$\frac{\partial u_h}{\partial x_j} = \int_\Omega u(y) \frac{\partial \omega_h(r)}{\partial x_j} \, dy. \tag{A}$$

Denote by x' the point whose coordinates are $x_1, \ldots, x_{j-1}, x_j+\lambda, x_{j+1}, \ldots, x_m$; where $x_1, \ldots, x_{j-1}, x_j, x_{j+1}, \ldots, x_m$ are the coordinates of the point x. Moreover, let $r' = |x'-y|$. We form the expression

$$\frac{u_h(x')-u(x)}{\lambda} = \int_\Omega u(y) \frac{\omega_h(r')-\omega_h(r)}{\lambda} \, dy = \int_\Omega u(y) \frac{\partial \omega_h(r^*)}{\partial x_j} \, dy,$$

where $r^* = |x^*-y|$, and x^* is some point on the line-element joining x and x'. The derivative $\partial \omega_h(r)/\partial x_j$ is continuous, and therefore uniformly continuous in $\overline{\Omega}$. Hence, for a given $\varepsilon > 0$ we can find $\lambda_0(\varepsilon)$ such that, for $\lambda < \lambda_0$,

$$\left| \frac{\partial \omega_h(r^*)}{\partial x_j} - \frac{\partial \omega_h(r)}{\partial x_j} \right| < \varepsilon \left\{ \int_\Omega |u(y)| \, dy \right\}^{-1}.$$

Now

$$\left| \frac{u_h(x')-u_h(x)}{\lambda} - \int_\Omega u(y) \frac{\partial \omega_h(r)}{\partial x_j} \, dy \right| < \varepsilon,$$

and so formula (A) is proved. It is now easy to establish by induction that the mean function $u_h(x)$ has continuous derivatives of arbitrary order, which can be found by differentiating under the integral sign. The derivatives of mean functions can be obtained from any of the following formulae:

$$\frac{\partial^k u_h}{\partial x_1^{k_1} \partial x_2^{k_2} \ldots \partial x_m^{k_m}} = \int_\Omega \frac{\partial^k \omega_h(r)}{\partial x_1^{k_1} \partial x_2^{k_2} \ldots \partial x_m^{k_m}} u(y) \, dy, \tag{2}$$

$$\frac{\partial^k u_h}{\partial x_1^{k_1} \partial x_2^{k_2} \ldots \partial x_m^{k_m}} = \int_{E_m} \frac{\partial^k \omega_h(r)}{\partial x_1^{k_1} \partial x_2^{k_2} \ldots \partial x_m^{k_m}} u(y) \, dy, \tag{2a}$$

$$\frac{\partial^k u_h}{\partial x_1^{k_1} \partial x_2^{k_2} \ldots \partial x_m^{k_m}} = \int_{r<h} \frac{\partial^k \omega_h(r)}{\partial x_1^{k_1} \partial x_2^{k_2} \ldots \partial x_m^{k_m}} u(y) \, dy. \tag{2b}$$

The property 1 can be written in the concise form: $u_h \in C^{(\infty)}(E_m)$.

2. A mean function is equal to zero at all points whose distance from the domain * Ω is not less than h. Actually, the sphere $r < h$ lies entirely outside Ω in this case, so that $u(y) \equiv 0$ in the integral (1b).

Thus, a mean function can be different from zero only in a domain, denoted by $\Omega^{(h)}$, which we can construct in the following way: with each point $x \in \Omega$ as centre, draw a sphere of radius h; the union of these spheres is then $\Omega^{(h)}$. Clearly $\Omega^{(h)} \supset \Omega$; if, for example, Ω is a sphere of radius R, then $\Omega^{(h)}$ is a sphere, concentric with Ω, and of radius $R+h$.

§ 3. Convergence of Mean Functions

THEOREM 1.3.1. *If* $u \in C(\bar{\Omega})$, *then the mean function*

$$u_h(x) \underset{h \to 0}{\to} u(x)$$

uniformly in any closed, interior sub-domain ** *of the domain* Ω.

Let Ω' be an interior sub-domain of the domain Ω. We construct a domain Ω'', which is an interior sub-domain for Ω, and for which Ω' is in turn an interior sub-domain (Fig. 3).

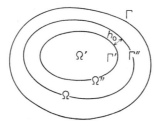

Fig. 3.

Denote by Γ' and Γ'' the boundaries of Ω' and Ω'' respectively, and let h_0 be the shortest distance between points on the boundaries Γ' and Γ''. We take $h < h_0$. From formula (2.1b) and property 3 of the averaging kernel (§ 1), we have

$$u_h(x) - u(x) = \int_{r<h} [u(y) - u(x)] \omega_h(r) \, dy. \tag{1}$$

* The distance $\rho(x, \Omega)$ from a point x to Ω is defined by the formula

$$\rho(x, \Omega) = \inf_{y \in \Omega} |x - y|.$$

Obviously $\rho(x, \Omega) = 0$ when $x \in \Omega$.

** Any domain $\Omega' \subset \Omega$ is called a sub-domain of Ω. The sub-domain Ω' is called interior if its closure $\bar{\Omega}' \subset \Omega$, that is, if Ω' together with its boundary lies within Ω.

If $x \in \bar{\Omega}'$, then in integral (1), $y \in \bar{\Omega}''$. In the closed domain $\bar{\Omega}''$, the continuous function u is uniformly continuous, and therefore, for sufficiently small h and $r \leq h$, we have $|u(y)-u(x)| < \varepsilon$, where ε is an arbitrarily small positive number. Taking account of the fact that $\omega_h(r) \geq 0$ (property 2), we obtain from equation (1),

$$|u_h(x)-u(x)| \leq \varepsilon \int_{r<h} \omega_h(r)\,dy = \varepsilon.$$

The theorem is thus proved.

THEOREM 1.3.2. *The norm in $L_2(\Omega)$ does not increase under averaging.*

Let $u \in L_2(\Omega)$. We shall show that, in the metric of $L_2(\Omega)$,

$$\|u_h\| \leq \|u\|. \tag{2}$$

We estimate the square of the mean function by use of the Schwartz-Bunyakovskii inequality *:

$$u_h^2(x) = \left\{\int_\Omega u(y)\omega_h(r)\,dy\right\}^2 = \left\{\int_\Omega u(y)\sqrt{\omega_h(r)}\sqrt{\omega_h(r)}\,dy\right\}^2$$
$$\leq \int_\Omega u^2(y)\omega_h(r)\,dy \int_\Omega \omega_h(r)\,dy \leq \int_\Omega u^2(y)\omega_h(r)\,dy, \tag{3}$$

since by property 2 of the averaging kernel

$$\int_\Omega \omega_h(r)\,dy = \int_{\Omega \cap (r<h)} \omega_h(r)\,dy \leq \int_{r<h} \omega_h(r)\,dy = 1.$$

Integrating the inequality (3) through the domain Ω, we obtain

$$\|u_h\|^2 = \int_\Omega u_h^2(x)\,dx \leq \int_\Omega u^2(y)\left[\int_\Omega \omega_h(r)\,dx\right]dy \leq \int_\Omega u^2(y)\,dy = \|u\|^2,$$

which it was required to show.

THEOREM 1.3.3. *If $u \in L_2(\Omega)$, then*

$$\|u-u_h\|_{L_2(\Omega)} \xrightarrow[h\to 0]{} 0.$$

---

* This inequality has various names. We use the term Schwartz-Bunyakovskii inequality for the integral form, and Cauchy's inequality for the algebraic form (cf. page 104 below). Ed.

It is well-known* that, for arbitrary $\varepsilon > 0$, it is possible to construct a polynomial f, such that
$$\|u-f\|_{L_2(\Omega)} < \tfrac{1}{3}\varepsilon.$$
We use the triangle inequality,
$$\|u-u_h\| \leq \|u-f\| + \|f-f_h\| + \|f_h-u_h\|.$$
By Theorem 1.3.2
$$\|f_h-u_h\| \leq \|f-u\|,$$
so that
$$\|u-u_h\| \leq 2\|f-u\| + \|f-f_h\| < \tfrac{2}{3}\varepsilon + \|f-f_h\|.$$

We now take a domain Ω_1, for which Ω is a strictly interior sub-domain. The polynomial f is continuous in Ω_1, and therefore
$$f_h \xrightarrow[h\to 0]{} f$$
uniformly in any interior, closed sub-domain of Ω_1, in particular in $\overline{\Omega}$. But uniform convergence in a closed region implies convergence in the mean, and for sufficiently small h,
$$\|f-f_h\|_{L_2(\Omega)} < \tfrac{1}{3}\varepsilon.$$
From this our proposition follows quite easily.

DEFINITION. A function ϕ is said to be *finite* in a bounded domain Ω, if it has infinitely-many derivatives in Ω, and is different from zero only in some interior sub-domain of the domain Ω.

A finite function can be defined somewhat differently. Let Γ be the boundary of the domain Ω. We give the name *boundary-strip* to the set of points of the domain Ω which have the property that their distances from Γ do not exceed a given constant δ, called the *width of the strip*. A function is called finite in Ω if it is differentiable infinitely-many times in Ω, and if it is equal to zero in some boundary-strip of the domain Ω.

This boundary strip of Ω, having width δ, we shall denote by Ω_δ.

If Ω is an unbounded domain, then we require of a finite function that it should be equal to zero outside some sphere (a particular sphere for each function).

---

* See, for example, V. I. Smirnov, A course of Higher Mathematics, Vol. 5, 1964, page 446.

The set of functions which are finite in a given domain Ω will be denoted by $\mathfrak{F}(\Omega)$.

THEOREM 1.3.4. *The set $\mathfrak{F}(\Omega)$ is dense in $L_2(\Omega)$.*

We need to show that, in the metric of $L_2(\Omega)$, any function $u \in L_2(\Omega)$ can be approximated, with arbitrary degree of exactness, by a finite function.

Choose a number δ such that the measure of the strip Ω_δ is sufficiently small, namely, take $\varepsilon > 0$ and introduce δ such that

$$\int_{\Omega_\delta} u^2 \, dx < (\tfrac{1}{2}\varepsilon)^2.$$

We consider the function $v(x)$, defined by the equation,

$$v(x) = \begin{cases} u(x), & x \in \Omega \setminus \Omega_\delta, \\ 0, & x \in \Omega_\delta. \end{cases}$$

Obviously $v \in L_2(\Omega)$; therefore

$$\|u-v\|^2 = \int_{\Omega_\delta} u^2 \, dx < (\tfrac{1}{2}\varepsilon)^2,$$

and consequently,

$$\|u-v\| < \tfrac{1}{2}\varepsilon. \tag{4}$$

Now take $h < \tfrac{1}{2}\delta$ and construct the mean function $v_h(x)$. It is finite in Ω, since it is infinitely differentiable and is equal to zero in the boundary-strip $\Omega_{\delta-h}$ (properties 1, 2 of mean functions, § 2). By Theorem 1.3.3, it is possible to choose a number h_0, such that for $h < h_0$

$$\|v-v_h\| < \tfrac{1}{2}\varepsilon. \tag{5}$$

From the triangle inequality and the relations (4) and (5), it follows that

$$\|u-v_h\| \leq \|u-v\| + \|v-v_h\| < \varepsilon, \quad h < h_0.$$

This proves the theorem.

COROLLARY 1.3.1. *If $M \subset L_2(\Omega)$ is a set which contains the set of all functions finite in Ω, then M is dense in $L_2(\Omega)$.*

Exercises

1. Show that averaging does not increase the norm in the space $L_p(\Omega)$, $1 \leq p \leq \infty$.

2. Show that if $u \in L_p(\Omega)$, $1 < p < \infty$, then
$$\|u - u_h\|_{L_p(\Omega)} \xrightarrow[h \to 0]{} 0.$$

3. It is possible to consider an averaging kernel, which has properties 1, 2 and 3, but not property 4. The simplest example of such a kernel is
$$\omega_h(r) = \begin{cases} c_h, & r < h \\ 0, & r \geq h, \end{cases}$$
where c_h is a suitably chosen constant. Find the constant c_h, and show that

a) if $u \in L(\Omega)$, then $u_h \in C(\overline{\Omega})$; if $u \in C^{(k)}(\overline{\Omega})$, then $u_h \in C^{(k+1)}(\overline{\Omega} \setminus \overline{\Omega}_h)$;

b) Theorems 1.3.1, 1.3.2 and 1.3.3 are true, and so are the statements of Exercises 1 and 2.

Chapter 2. Generalized Derivatives

§ 1. The Concept of a Generalized Derivative

To begin with, we derive an important formula of the integral calculus, known as the *integration by parts formula*.

Let Ω be a finite domain of m-dimensional Euclidean space, bounded by a piecewise-smooth surface Γ. We recall Gauss's theorem *:

$$\int_\Omega \frac{\partial P}{\partial x_k}\,dx = \int_\Gamma P \cos(v, x_k)\,d\Gamma;$$

where v is the normal to the surface Γ, in the outward direction with respect to Ω. It is sufficient that the function $P(x)$ should belong to the class $C^{(1)}(\overline{\Omega})$.

Consider the integral

$$\int_\Omega P \frac{\partial Q}{\partial x_k}\,dx = \int_\Omega \frac{\partial}{\partial x_k}(PQ)\,dx - \int_\Omega Q \frac{\partial P}{\partial x_k}\,dx,$$

where $P, Q \in C^{(1)}(\overline{\Omega})$. Substituting the first integral on the right-hand side into Gauss's theorem, we obtain the integration by parts formula:

$$\int_\Omega P \frac{\partial Q}{\partial x_k}\,dx = -\int_\Omega Q \frac{\partial P}{\partial x_k}\,dx + \int_\Gamma PQ \cos(v, x_k)\,d\Gamma.$$

We note several consequences of this formula. If one of the functions P, Q is equal to zero on Γ, then the surface integral vanishes and we obtain the simpler formula:

$$\int_\Omega P \frac{\partial Q}{\partial x_k}\,dx = -\int_\Omega Q \frac{\partial P}{\partial x_k}\,dx.$$

---

* This formula is also called Green's theorem and Ostrogradskii's formula. Ed.

Now consider a more complex form of the integral:

$$\int_\Omega P \frac{\partial^k Q}{\partial x_1^{k_1} \partial x_2^{k_2} \ldots \partial x_m^{k_m}} \, dx.$$

If the function P has the necessary continuous derivatives, then this expression can be integrated by parts k times, until finally the function Q appears undifferentiated in the integrand:

$$\int_\Omega P \frac{\partial^k Q}{\partial x_1^{k_1} \partial x_2^{k_2} \ldots \partial x_m^{k_m}} \, dx =$$

$$= -\int_\Omega \frac{\partial P}{\partial x_1} \frac{\partial^{k-1} Q}{\partial x_1^{k_1-1} \partial x_2^{k_2} \ldots \partial x_m^{k_m}} \, dx$$

$$+ \int_\Gamma P \frac{\partial^{k-1} Q}{\partial x_1^{k_1-1} \partial x_2^{k_2} \ldots \partial x_m^{k_m}} \cos(v, x_1) \, d\Gamma = \ldots$$

$$= (-1)^k \int_\Omega Q \frac{\partial^k P}{\partial x_1^{k_1} \partial x_2^{k_2} \ldots \partial x_m^{k_m}} \, dx + \int_\Gamma R(P, Q) \, d\Gamma.$$

Here $R(P, Q)$ denotes an expression dependent on the functions P, Q and their derivatives up to $(k-1)$th order.

We now introduce a set $\mathfrak{M}^{(k)}(\Omega)$ of functions, which are continuous and k-times continuously differentiable in $\bar{\Omega}$, and which are equal to zero, together with all their derivatives up to order $k-1$, on the boundary of the domain Ω. Obviously, $\mathfrak{M}^{(k+1)}(\Omega) \subset \mathfrak{M}^{(k)}(\Omega)$, and $\mathfrak{F}(\Omega) \subset \mathfrak{M}^{(k)}(\Omega)$, for arbitrary k.

Let functions $u(x)$ and $v(x)$ be integrable in Ω, and suppose that for any function $\phi \in \mathfrak{M}^{(k)}(\Omega)$ the following identity holds:

$$\int_\Omega u \frac{\partial^k \phi}{\partial x_1^{k_1} \partial x_2^{k_2} \ldots \partial x_m^{k_m}} \, dx = (-1)^k \int_\Omega v \phi \, dx. \tag{1}$$

Then v is called the *generalized derivative* of k-th order of the function u in the domain Ω. To designate a generalized derivative we use ordinary notation and write

$$v(x) = \frac{\partial^k u}{\partial x_1^{k_1} \partial x_2^{k_2} \ldots \partial x_m^{k_m}}. \tag{2}$$

THEOREM 2.1.1. *The generalized derivative in the form* (2) *is unique.*

It is necessary to show the following: if the function $u(x)$ is integrable in Ω, and if there exist two functions $v_1(x)$ and $v_2(x)$ also integrable in Ω and satisfying for arbitrary $\phi \in \mathfrak{M}^{(k)}(\Omega)$ the equations

$$\int_\Omega u \frac{\partial^k \phi}{\partial x_1^{k_1} \partial x_2^{k_2} \ldots \partial x_m^{k_m}} \, dx = (-1)^k \int_\Omega v_1 \phi \, dx,$$
$$\int_\Omega u \frac{\partial^k \phi}{\partial x_1^{k_1} \partial x_2^{k_2} \ldots \partial x_m^{k_m}} \, dx = (-1)^k \int_\Omega v_2 \phi \, dx, \tag{3}$$

then $v_1(x) \equiv v_2(x)$. As usual, we suppose two functions to be equal to each other if they are equivalent (they can differ only on a set of measure zero).

Subtracting the second of the identities (3) from the first, and putting $v_1(x) - v_2(x) = w(x)$, we obtain the identity

$$\int_\Omega w(x)\phi(x) \, dx = 0, \tag{4}$$

valid if $\phi \in \mathfrak{M}^{(k)}(\Omega)$. We show that the identity (4) is true for any bounded and measurable function $\phi(x)$ which width is equal to zero in some boundary-strip. Let $\phi(x)$ be such a function, and let it be equal to zero in a strip of width δ. We take $h < \tfrac{1}{2}\delta$, and construct the mean function $\phi_h(x)$. This function is infinitely differentiable, and is equal to zero in a boundary-strip of width $\delta - h$. Therefore $\phi_h(x) \in \mathfrak{F}(\Omega)$, and so, *a fortiori*, $\phi_h \in \mathfrak{M}^{(k)}(\Omega)$. Hence the function $\phi_h(x)$ satisfies the relation:

$$\int_\Omega w(x)\phi_h(x) \, dx = 0. \tag{5}$$

It is easy to see that, whatever the value of h, the functions $\phi_h(x)$ are bounded by one and the same constant: if $|\phi(x)| < N = \text{const.}$, then

$$|\phi_h(x)| = \left| \int_{r<h} \phi(y) \omega_h(r) \, dy \right| \leq N \int_{r<h} \omega_h(r) \, dy = N.$$

The function ϕ, bounded and measurable in Ω, is in every case square-integrable in Ω. By Theorem 1.2.3,

$$\|\phi_h - \phi\|_{L_2(\Omega)} \xrightarrow[h \to 0]{} 0.$$

According to a well-known theorem on sequences of functions which are convergent in the mean *, it is possible to choose a sequence of numbers

---

* Cf. I. P. Natanson, Theory of functions of a real variable, 1964; or any book on this subject.

$h_n \to 0$, such that
$$\phi_{h_n}(x) \to \phi(x)$$
almost everywhere in Ω.

In equation (5) we put $h = h_n$.

In the integral (5), the integrand function does not exceed the integrable function $N|w(x)|$, and as $n \to \infty$ it tends to the function $w(x)\phi(x)$ almost everywhere. The well-known theorem concerning passage to the limit in a Lebesgue integral now gives
$$\int_\Omega w(x)\phi(x)\,dx = 0,$$
which it was required to show. In equation (4) we now put
$$\phi(x) = \begin{cases} 0, & x \in \Omega_{\frac{1}{2}\delta}, \\ \operatorname{sign} w(x), & x \in \Omega' = \Omega \setminus \Omega_{\frac{1}{2}\delta}. \end{cases}$$
Then we obtain
$$\int_{\Omega'} |w(x)|\,dx = 0, \tag{6}$$
and, consequently, $w(x) \equiv 0$, $x \in \Omega'$. Since δ is an arbitrary positive number, $w(x) \equiv 0$ in Ω. Thus the theorem is proved.

If the function $u(x)$ is continuous in Ω, together with its derivatives up to the k-th order, then its generalized derivatives of order k exist and coincide with the "ordinary"* derivatives. In fact, the integral on the left-hand side of equation (1) can be integrated by parts k times; in the process the surface integrals vanish, because on the boundary of the domain Ω both the function and its derivatives up to the $(k-1)$-th order are equal to zero. As a result we obtain the equation
$$\int_\Omega u \frac{\partial^k \phi}{\partial x_1^{k_1} \partial x_2^{k_2} \ldots \partial x_m^{k_m}}\,dx = (-1)^k \int_\Omega \phi \frac{\partial^k u}{\partial x_1^{k_1} \partial x_2^{k_2} \ldots \partial x_m^{k_m}}\,dx, \tag{7}$$
where on the right-hand side we have the "ordinary" (continuous) derivative of u. Equation (7) shows that the generalized derivative does exist in this case, and is equal to the continuous derivative
$$\frac{\partial^k u}{\partial x_1^{k_1} \partial x_2^{k_2} \ldots \partial x_m^{k_m}}.$$

Let us now examine a few examples.

---

* Here "ordinary" is used in the sense of not generalized. Ed.

Example 1. Let Ω be the interval $(-1, 1)$. The function $u(x) = |x|$ has a generalized derivative $u'(x) = \text{sign } x$. In fact, let $\phi(x) \in \mathfrak{M}^{(1)}(-1, +1)$; then $\phi(x)$ is continuously differentiable on the segment $[-1, +1]$, and $\phi(-1) = \phi(1) = 0$. We have

$$\int_{-1}^{1} |x|\phi'(x)\,dx = -\int_{-1}^{0} x\phi'(x)\,dx + \int_{0}^{1} x\phi'(x)\,dx.$$

Integrating by parts, we obtain

$$\int_{-1}^{1} |x|\phi'(x)\,dx = \int_{-1}^{0} \phi(x)\,dx - \int_{0}^{1} \phi(x)\,dx = -\int_{-1}^{1} \phi(x)\,\text{sign } x\,dx,$$

and so our assertion is proved.

Example 2. The function sign x does not have a generalized first derivative in the interval $(-1, +1)$ (although, like the function $|x|$, it has a continuous derivative for $x \neq 0$). To show this, consider the integral

$$\int_{-1}^{1} \phi'(x)\,\text{sign } x\,dx = -\int_{-1}^{0} \phi'(x)\,dx + \int_{0}^{1} \phi'(x)\,dx = -2\phi(0), \tag{8}$$

where $\phi \in \mathfrak{M}^{(1)}(-1, 1)$.

There does not exist a function $v(x)$, integrable in the interval $(-1, 1)$, and satisfying, for arbitrary function $\phi(x) \in \mathfrak{M}^{(1)}(-1, +1)$, the equation

$$\int_{-1}^{+1} v(x)\phi(x)\,dx = 2\phi(0). \tag{9}$$

Suppose that in fact such a function exists. Then the function

$$V(x) = \int_{0}^{x} v(y)\,dy$$

is absolutely continuous on the segment $[-1, 1]$ and has on it an integrable derivative $v(x)$. Integrating by parts in equation (9), and using formula (8), we obtain the following equation, valid for any function $\phi(x) \in \mathfrak{M}^{(1)}(-1, +1)$:

$$\int_{-1}^{+1} \phi'(x)[\text{sign } x - V(x)]\,dx = 0.$$

Then, however [*],

$$\text{sign } x = V(x) + \text{const.}, \qquad x \in (-1, +1),$$

---

[*] Cf., for example, V. I. Smirnov, A course of Higher Mathematics, Vol. 5, 1964, page 145.

which is impossible, since at the point $x = 0$ the left-hand side is discontinuous, while the right-hand side is continuous.

Example 3. Let the functions $f(t)$ and $g(t)$ be continuous on the segment $[-1, +1]$, but not differentiable at any of its points. It is possible to show that a function

$$u(x) = u(x_1, x_2) = f(x_1) + g(x_2) \tag{10}$$

which is a continuous function of the two variables in the square $0 \leq x_1$, $x_2 \leq 1$, has no generalized first derivatives. However, this function has a generalized derivative of the second order, $\partial^2 u / \partial x_1 \partial x_2$, and this derivative is equal to zero. To establish this, it is sufficient to show that, for an arbitrary function $\phi(x) = \phi(x_1, x_2) \in \mathfrak{M}^{(2)}(\Omega)$, where Ω is the square $-1 \leq x_1$, $x_2 \leq 1$, the equation

$$\int_{-1}^{+1} \int_{-1}^{+1} u(x_1, x_2) \frac{\partial^2 \phi}{\partial x_1 \partial x_2} dx_1 dx_2 = 0$$

holds. But this equation follows at once from the following:

$$\int_{-1}^{+1} \int_{-1}^{+1} u(x_1, x_2) \frac{\partial^2 \phi}{\partial x_1 \partial x_2} dx_1 dx_2$$

$$= \int_{-1}^{+1} f(x_1) \left\{ \int_{-1}^{+1} \frac{\partial^2 \phi}{\partial x_1 \partial x_2} dx_2 \right\} dx_1 + \int_{-1}^{+1} g(x_2) \left\{ \int_{-1}^{+1} \frac{\partial^2 \phi}{\partial x_1 \partial x_2} dx_1 \right\} dx_2$$

$$= \int_{-1}^{+1} f(x_1) \frac{\partial \phi}{\partial x_1} \bigg|_{x_2 = -1}^{x_2 = +1} dx_1 + \int_{-1}^{+1} g(x_2) \frac{\partial \phi}{\partial x_2} \bigg|_{x_1 = -1}^{x_1 = +1} dx_2 = 0.$$

This example demonstrates that the existence of a generalized derivative of some order does not imply the existence of the generalized derivative of the preceding order.

§ 2. Basic Properties of the Generalized Derivative

THEOREM 2.2.1. *Let the function $u(x)$ have a generalized derivative $v(x)$, of the form (1.2), in the domain Ω. Then, in the domain $\Omega \setminus \Omega_h$, the mean function of this derivative is equal to the generalized derivative of the mean function of $u(x)$.*

We recall that Ω_h denotes the boundary-strip of the domain Ω, with width h. The set $\Omega^{(h)} = \Omega \setminus \Omega_h$ is an open set, and if $x \in \Omega \setminus \Omega_h$, then the distance from the point x to the boundary of the domain Ω is greater

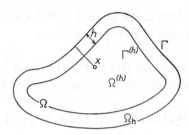

Fig. 4.

than h (Fig. 4); therefore the averaging kernel $\omega_h(r) \in \mathfrak{F}(\Omega)$. From formula (1.1),

$$\int_\Omega u(y) \frac{\partial^k \omega_h(r)}{\partial y_1^{k_1} \partial y_2^{k_2} \ldots \partial y_m^{k_m}} \, dy = (-1)^k \int_\Omega v(y) \omega_h(y) \, dy = (-1)^k v_h(x). \tag{1}$$

The averaging kernel $\omega_h(r)$ depends only on the distance $x-y$. Therefore

$$\frac{\partial^k \omega_h(r)}{\partial y_1^{k_1} \partial y_2^{k_2} \ldots \partial y_m^{k_m}} = (-1)^k \frac{\partial^k \omega_h(r)}{\partial x_1^{k_1} \partial x_2^{k_2} \ldots \partial x_m^{k_m}}.$$

Substituting this into equation (1), we obtain

$$v_h(x) = \frac{\partial^k u_h(x)}{\partial x_1^{k_1} \partial x_2^{k_2} \ldots \partial x_m^{k_m}},$$

which we set out to prove.

THEOREM 2.2.2. *Let Ω' be a sub-domain of the domain Ω. If $v(x)$ is the generalized derivative obtained from $u(x)$,*

$$v(x) = \frac{\partial^k u}{\partial x_1^{k_1} \partial x_2^{k_2} \ldots \partial x_m^{k_m}} \tag{2}$$

in the domain Ω, then $v(x)$ is also the generalized derivative of $u(x)$ in the domain Ω'.

Let $\phi(x) \in \mathfrak{M}^{(k)}(\Omega')$. We shall define the function $\phi(x)$ in $\Omega \setminus \Omega'$ by setting it equal to zero there. Obviously, we then have $\phi(x) \in \mathfrak{M}^{(k)}(\Omega)$. Now apply the formula (1.1) to the functions $u(x)$ and $v(x)$. Since they are equal to zero, the integrals over $\Omega \setminus \Omega'$ on both sides can be discarded, and we obtain the formula

$$\int_{\Omega'} u \frac{\partial^k \phi}{\partial x_1^{k_1} \partial x_2^{k_2} \ldots \partial x_m^{k_m}} \, dx = (-1)^k \int_{\Omega'} v\phi \, dx,$$

which means that $v(x)$ is a generalized derivative of $u(x)$ of the form (2), in the sub-domain Ω'.

THEOREM 2.2.3. *If in the domain Ω, the function $v(x)$ is the generalized derivative of $u(x)$ of the form (2), and $w(x)$ is the generalized derivative of $v(x)$ of the form*

$$w(x) = \frac{\partial^l v}{\partial x_1^{l_1} \partial x_2^{l_2} \ldots \partial x_m^{l_m}},$$

then $w(x)$ is the generalized derivative of $u(x)$ in the domain Ω, and has the form

$$w(x) = \frac{\partial^{k+l} u}{\partial x_1^{k_1+l_1} \partial x_2^{k_2+l_2} \ldots \partial x_m^{k_m+l_m}}.$$

Let $\phi \in \mathfrak{M}^{(k+l)}(\Omega)$. Then $\partial^l \phi / \partial x_1^{l_1} \partial x_2^{l_2} \ldots \partial x_m^{l_m} \in \mathfrak{M}^{(k)}(\Omega)$, and by formula (1.1),

$$\int_\Omega u \frac{\partial^{k+l} \phi}{\partial x_1^{k_1+l_1} \partial x_2^{k_2+l_2} \ldots \partial x_m^{k_m+l_m}} dx = (-1)^k \int_\Omega v \frac{\partial^l \phi}{\partial x_1^{l_1} \partial x_2^{l_2} \ldots \partial x_m^{l_m}} dx. \quad (3)$$

By the same formula (1.1)

$$\int_\Omega v \frac{\partial^l \phi}{\partial x_1^{l_1} \partial x_2^{l_2} \ldots \partial x_m^{l_m}} dx = (-1)^l \int_\Omega w\phi \, dx.$$

Substituting this result into equation (3), we obtain the formula

$$\int_\Omega u \frac{\partial^{k+l} \phi}{\partial x_1^{k_1+l_1} \partial x_2^{k_2+l_2} \ldots \partial x_m^{k_m+l_m}} dx = (-1)^{k+l} \int_\Omega w\phi \, dx,$$

from which Theorem 2.2.3 follows.

3. Limit Properties of Generalized Derivatives

In the present section, we shall assume that all the functions under consideration, as well as their generalized derivatives, are square-integrable in the domain Ω, which, as before, will be taken to be finite.

THEOREM 2.3.1. *Let each of the functions $u_n(x), n = 1, 2, \ldots$, have a generalized derivative in the domain Ω, of the form:*

$$v_n(x) = \frac{\partial^k u_n}{\partial x_1^{k_1} \partial x_2^{k_2} \ldots \partial x_m^{k_m}}.$$

If both the sequences $\{u_n\}$ and $\{v_n\}$ converge in the metric of $L_2(\Omega)$ to the limits $u(x)$ and $v(x)$ respectively, then in the domain Ω the function $v(x)$ is the generalized derivative of $u(x)$, of the same form.

From the definition of a generalized derivative,

$$\int_\Omega u_n \frac{\partial^k \phi}{\partial x_1^{k_1} \partial x_2^{k_2} \ldots \partial x_m^{k_m}} \, dx = (-1)^k \int_\Omega v_n \phi \, dx, \qquad \phi \in \mathfrak{M}^{(k)}(\Omega). \tag{1}$$

Each of the integrals in equation (1) is a scalar product of two functions from $L_2(\Omega)$, and within a scalar product it is permissible to carry out a limiting process. Applying this, we arrive at equation (1.1), and the theorem is proved.

THEOREM 2.3.2. *Let $v(x)$ be the generalized derivative of $u(x)$ in the domain Ω*:

$$v(x) = \frac{\partial^k u}{\partial x_1^{k_1} \partial x_2^{k_2} \ldots \partial x_m^{k_m}}.$$

In an arbitrary interior sub-domain $\Omega' \subset \Omega$, it is possible to construct a sequence of infinitely differentiable functions $\{u_n(x)\}$ such that, in the metric of the space $L_2(\Omega')$,

$$u_n \to u, \qquad \frac{\partial^k u_n}{\partial x_1^{k_1} \partial x_2^{k_2} \ldots \partial x_m^{k_m}} \to v. \tag{2}$$

The proof is very simple. We can write

$$u_n(x) = u_{h_n}(x),$$

where h_n is a sequence of positive numbers, tending to the limit zero. Then the first relation of (2) follows from Theorem 1.3.3, while the second follows from Theorems 1.3.3 and 2.2.2.

THEOREM 2.3.3. *Let Ω be a bounded domain of the space E_m. If a function of the class $L_2(\Omega)$ has all generalized derivatives of first order, continuous in Ω, then this function is continuous and continuously differentiable in Ω.*

Let $u(x)$ be the given function and $u_h(x)$ the corresponding mean function. Then $u_h(x) \xrightarrow[h \to 0]{} u(x)$ in the metric of $L_2(\Omega)$, and simultaneously $\partial u_h/\partial x_k \xrightarrow[h \to 0]{} \partial u/\partial x_k$ uniformly in $\Omega \setminus \Omega_\delta$, where δ is an arbitrary positive number. We can now introduce a sequence $h_n \to 0$, such that $u_{h_n}(x) \xrightarrow[n \to \infty]{} u(x)$ almost everywhere in Ω. Let us fix δ and construct in $\Omega \setminus \Omega_\delta$ a closed cube

Q with its sides parallel to the coordinate axes. For brevity we write $u_{h_n}(x) = u_n(x)$, and select a point x_0 of Q, at which the sequence $\{u_n(x_0)\}$ converges. Let x be an arbitrary point of the cube Q. Inside this cube we construct an open polygon L, joining the points x_0 and x, and with its sides parallel to the coordinate axes. Denoting by ρ the position vector of the variable point x, we have

$$u_n(x) = u_n(x_0) + \int_L \operatorname{grad} u_n(\xi) \, d\rho.$$

From this it can be seen that the sequence $\{u_n(x)\}$ converges uniformly in Q, and, consequently, that the function $u(x)$ is continuous (more precisely, is equivalent to a continuous function) in Q. The cube Q can be constructed in such a way as to include any prescribed point of the domain $\Omega\backslash\Omega_\delta$; therefore the function u is continuous in $\Omega\backslash\Omega_\delta$, and, because of the arbitrariness of δ, in the whole domain Ω.

In the cube Q the relations

$$u_n(x) \to u(x), \qquad \frac{\partial u_n(x)}{\partial x_k} \to \frac{\partial u(x)}{\partial x_k}, \qquad k = 1, 2, \ldots, m,$$

are satisfied; moreover, in both cases the approach to the limit is uniform. From a well-known theorem on the differentiation of sequences, the functions $\partial u/\partial x_k$, $k = 1, 2, \ldots, m$, are "ordinary" derivatives of the function u, which, consequently, is continuously differentiable in Ω.

COROLLARY 2.3.1. *If the function $u \in L_2(\Omega)$ has a generalized gradient which is identically zero in Ω, then in this domain $u(x) \equiv \mathrm{const}$.*

§ 4. The Case of One Independent Variable

In this case the class of functions having a generalized first derivative is found to be closely related to the class of absolutely continuous functions. We recall that the function $u(x)$ of the real variable x is absolutely continuous on the segment $[a, b]$, if there exists a function $v(x)$, integrable on the segment $[a, b]$ such that

$$u(x) = \int_a^x v(t) \, dt + \mathrm{const}, \qquad x \in [a, b].$$

From a well-known theorem of Lebesgue, it follows that the function

$u(x)$ has an ordinary derivative, equal to $v(x)$, almost everywhere on the segment $[a, b]$.

THEOREM 2.4.1. *Let the function $u(x)$, defined almost everywhere on the interval (a, b) and square-integrable on this interval, have on (a, b) a generalized derivative $v(x)$, which is also square-integrable. Then $u(x)$ is equivalent to a function which is absolutely continuous on the segment $[a, b]$, and has an "ordinary" derivative, equal to $v(x)$, almost everywhere in (a, b).*

Let the segment $[\alpha, \beta]$ lie inside the interval (a, b). By Theorem 2.3.2, there exists a sequence $u_n(x) \in C^{(\infty)}[\alpha, \beta]$ such that, in the metric of $L_2(\alpha, \beta)$,

$$u_n \to u, \quad u'_n \to v.$$

By the Newton-Leibnitz formula

$$u_n(x) - u_n(\alpha) = \int_\alpha^x u'_n(t)\,dt,$$

from which we have

$$u_n(\alpha) = u_n(x) - \int_\alpha^x u'_n(t)\,dt. \tag{1}$$

The right-hand side of equation (1) converges in the metric of $L_2(\alpha, \beta)$ to a limit, which is

$$u(x) - \int_\alpha^x v(t)\,dt$$

and if this is the case then the left-hand side converges in the same metric. But for functions, each of which is a constant, convergence in the mean is ordinary convergence of a numerical sequence. Hence there exists a limit – denoted by c – of the sequence $\{u_n(\alpha)\}$:

$$\lim_{n \to \infty} u_n(\alpha) = c.$$

Now letting $n \to \infty$ in equation (1), we find

$$u(x) = \int_\alpha^x v(t)\,dt + c. \tag{2}$$

Equation (2) holds almost everywhere on the segment $[\alpha, \beta]$. However, the right-hand side of this equation is defined and continuous everywhere on this segment. We now assume that equation (2) is valid everywhere on the segment $[\alpha, \beta]$; this is tantamount to replacing the given function $u(x)$

with another, its equivalent. Now the function $u(x)$ is absolutely continuous on the segment $[\alpha, \beta]$; hence it is obvious that $c = u(\alpha)$.

In formula (2), x can denote an arbitrary point of the interval (a, b), since α and β can be taken as close as we please to a and b respectively. We fix α and let $x \to b$ in the formula (2). The function $v(t)$ is integrable on the whole interval (a, b), and therefore the right-hand side of this formula has a limit, equal to

$$\int_\alpha^b v(t)\,dt + y(\alpha).$$

If we put

$$u(b) = \int_\alpha^b v(t)\,dt + u(\alpha),$$

then the function $u(x)$ is seen to be absolutely continuous on the segment $[\alpha, b]$, where α is an arbitrary point of the interval (a, b).

It is now possible to represent the function $u(x)$ in a new way:

$$u(x) = u(b) - \int_x^b v(t)\,dt. \tag{3}$$

As $x \to a$ the right-hand side of formula (3) has the limit

$$u(b) - \int_a^b v(t)\,dt.$$

Putting

$$u(a) = u(b) - \int_a^b v(t)\,dt,$$

we make the function $u(x)$ absolutely continuous on every segment $[a, b]$.

THEOREM 2.4.2. *Let the function $u(x)$ be defined almost everywhere on the interval (a, b), be square-integrable on the interval, and have a generalized k-th derivative $u^{(k)}(x) = v(x)$, also square-integrable. Then the function $u(x)$ is equivalent to a function, which is $(k-1)$-times continuously differentiable on the segment $[a, b]$ and which almost everywhere on this segment has an "ordinary" derivative of k-th order, $u^{(k)}(x) = v(x)$. Here the derivative $u^{(k-1)}(x)$ is absolutely continuous on the segment $[a, b]$.*

We shall not give the proof here – it is the same as for Theorem 2.4.1.

§ 5. Sobolev Spaces and Embedding Theorems

Let Ω be a finite domain of the space E_m. We consider the set of functions which are integrable in Ω and have in this domain all possible generalized derivatives of a given order l, integrable to some degree p, $1 < p < \infty$. The set so defined is obviously linear. It can be transformed into a Banach space, if we introduce the norm

$$\|u\| = \int_\Omega |u(x)|\,dx + \left\{\int_\Omega \sum \left|\frac{\partial^l u}{\partial x_{i_1} \partial x_{i_2} \ldots \partial x_{i_l}}\right|^p dx\right\}^{1/p}, \qquad (1)$$

the summation in the second integral extending over all values of the indices $i_1, i_2, \ldots, i_l$, each of which independently takes all values $1, 2, \ldots, m$. The space obtained by this procedure is called a *Sobolev space*, and is denoted by the symbol $W_p^{(l)}(\Omega)$.

It is not altogether necessary to introduce the norm through formula (1); in the space $W_p^{(l)}(\Omega)$ it is possible to have an arbitrary norm, equivalent to the norm (1).

In modern analysis, and especially in the theory of partial differential equations, an important role is played by so-called "embedding theorems"; these were first obtained by S. L. Sobolev, and later much strengthened and generalized. The essence of the embedding theorems is the following. If the domain Ω satisfies the so-called "cone condition" (cf. below Chapter 16, § 3) and if the function $u \in W_p^{(l)}(\Omega)$, then it has all possible generalized derivatives of all the preceding orders. The derivatives of order less than l, as well as the function itself, are integrable up to some degree greater than p. This degree increases when the order of the derivative decreases.

From the above it follows that, if $l_1 < l$, then for some $p_1 > p$ an arbitrary element of the space $W_p^{(l)}(\Omega)$ belongs to the space $W_{p_1}^{(l_1)}(\Omega)$; the space $W_p^{(l)}(\Omega)$ is "embedded" in the space $W_{p_1}^{(l_1)}(\Omega)$. We denote by E the operator which effects a correspondence between every function $u(x)$ – an element of the space $W_p^{(l)}(\Omega)$ – and the same function $u(x)$, but now regarded as an element of the space $W_{p_1}^{(l_1)}(\Omega)$. The operator E is called the *embedding operator* of the space $W_p^{(l)}(\Omega)$ into the space $W_{p_1}^{(l_1)}(\Omega)$. One of the most important embedding theorems states that the embedding operator E is bounded and (at least for less than p_1) completely continuous. As consequences of the embedding theorems we shall later have the inequalities of Friedrichs (Chapter 14) and Poincaré (Chapter 16).

Generalized derivatives of order less than l are integrable to some degree greater than p, not only in Ω, but also in piecewise-smooth manifolds of any lower dimension, which lie in Ω. The lower the order of the derivative, the lower we can take the dimension of the manifold. Derivatives of sufficiently low order can easily be shown to be continuous. There are theorems concerning the boundedness and complete continuity of the corresponding embedding operators.

The questions touched on here are amply expounded in references [1] and [2].

Exercises

1. Prove Theorem 2.3.1 for the space L_p, $1 < p < \infty$, replacing strong convergence with weak convergence.

2. Let Ω be a domain of the m-dimensional Euclidean space E_m, let $u \in L_p(\Omega)$, and let there exist a generalized derivative $\partial u/\partial x_1 \in L_p(\Omega)$. Take l to be a straight line parallel to the axis x_1, and (a, b) to be an interval on which the line l intersects the domain Ω. Prove that on almost every interval (a, b) the function u is absolutely continuous and has an "ordinary" derivative $\partial u/\partial x_1$ almost everywhere.

3. Ω is a domain of m-dimensional space, which can be enclosed in a right-angled parallelepiped with sides $a_1, a_2, \ldots, a_m$. The function $u(x)$ is continuous in the closed domain $\overline{\Omega}$, equal to zero on the boundary of the domain Ω, and has generalized first derivatives

$$\frac{\partial u}{\partial x_k} \in L_2(\Omega), \qquad k = 1, 2, \ldots, m.$$

Prove the inequality

$$\int_\Omega |u|^2 \, dx \leq \int_\Omega \sum_{k=1}^m \left|\frac{\partial u}{\partial x_k}\right|^2 dx \bigg/ \left\{\pi^2 \sum_{k=1}^n \frac{1}{a_k^2}\right\}.$$

Hint. Use an expansion in Fourier series.

PART II

Elements of the Calculus of Variations

PART II

Elements of the Oikeiōsis of Vaccinians

Chapter 3. Basic Concepts

§ 1. Examples of the Extremum of a Functional

1. The calculus of variations is usually considered to have originated with J. Bernoulli who, in the year 1696, posed the so-called *brachistochrone problem*: the points $A(0, 0)$ and $B(a, b)$ are situated in a vertical plane (x, y) (Fig. 5). What must be the curve, lying in the (x, y) plane and joining the points A and B, such that a material point, moving without friction, travels from point A to point B along this curve in the shortest time? The required curve was called a *brachistochrone*.

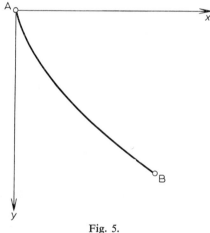

Fig. 5.

Let the equation of the curve AB be $y = u(x)$. Consider some instant of time t, and let the moving point be at a distance y from the x-axis at this instant. Then $v = \sqrt{(2gy)} = \sqrt{(2gu)}$, where v is the speed of the moving point, and g the acceleration due to gravity. At the same time,

$$v = \frac{ds}{dt} = \sqrt{1+u'^2}\,\frac{dx}{dt}.$$

Hence
$$dt = \sqrt{\frac{1+u'^2}{2gu}}\,dx.$$

We denote by T the time taken for the material point to reach the point B. Integrating, we obtain

$$T = \frac{1}{\sqrt{2g}}\int_0^a \sqrt{\frac{1+u'^2}{u}}\,dx. \qquad (1)$$

The problem reduces to the following: it is required to find a function $y = u(x)$, which satisfies the conditions

$$u(0) = 0, \qquad u(a) = b \qquad (2)$$

and which causes the integral (1) to assume its minimum value. The conditions (2) imply that the required curve must pass through the given points A and B; conditions of this type are commonly known as *boundary-* or *end-conditions*, since they refer to the end-points of the interval on which the required function is to be defined.

2. We consider another problem, which is similar to the brachistochrone problem. Light propagates in an optically non-uniform medium with velocity $v(x, y, z)$. It is required to find the path of the light-ray, which joins points $A(x_1, y_1, z_1)$ and $B(x_2, y_2, z_2)$. It is well-known from Fermat's Principle that the path of propagation is such as to enable the light to travel from point A to point B in the shortest possible time.

Let the equation of the required path be

$$y = u_1(x), \qquad z = u_2(x).$$

In accordance with Fermat's Principle, the problem reduces to the determination of two functions $u_1(x)$ and $u_2(x)$, satisfying the boundary conditions

$$u_1(x_1) = y_1, \qquad u_1(x_2) = y_2, \qquad u_2(x_1) = z_1, \qquad u_2(x_2) = z_2 \qquad (3)$$

which render the value of the integral

$$T = \int_{x_1}^{x_2} \frac{\sqrt{\{1+u_1'^2(x)+u_2'^2(x)\}}}{v(x, u_1(x), u_2(x))}\,dx \qquad (4)$$

a minimum.

3. The next problem is a little different from the preceding two. We formulate it thus: among all plane curves, whose equations are explicit, which have a given length l and which terminate in the points $A(a, 0)$ and $B(b, 0)$, find the curve such that the region bounded it and the segment $[a, b]$ of the x-axis has the largest possible area.

Let the equation of the curve be $y = u(x)$. The problem consists in finding a function $u(x)$, which satisfies the boundary conditions

$$u(a) = u(b) = 0 \tag{5}$$

and the equation

$$\int_a^b \sqrt{1 + u'^2}\, dx = l \tag{6}$$

and which renders the value of the integral

$$S = \int_a^b u\, dx \tag{7}$$

a maximum.

By comparison with the first two problems, we have here the new feature that the required function must satisfy not only the boundary conditions (5), but also the equation (6) which, obviously, does not have the character of a boundary condition. Common to all three problems is the fact that in each case we are looking for a function (or, as in problem (2), several functions – which can be regarded as a vector-function) which satisfies some prescribed conditions and renders the value of a given functional an extremum (minimum or maximum). Thus, in the brachistochrone problem, the required function must satisfy the boundary conditions (2) and minimize the value of the functional (1). The three problems discussed in the present section, together with many other problems of the same type, belong to the branch of mathematical analysis known as *calculus of variations*. A rigorous formulation of the fundamental problems of the variational calculus will be given in the next section.

§ 2. Statement of the Problem of the Variational Calculus

We recall the definition of a functional. Let $\mathfrak{M}$ be a set of elements of an arbitrary kind, and suppose that to each element $u \in \mathfrak{M}$ there corresponds one and only one number $F(u)$. In this case we say that a functional F is given on the set $\mathfrak{M}$. The set $\mathfrak{M}$ is called the *domain of definition* of the functional F and is denoted by $D(F)$; the quantity $F(u)$ is called the *value of* the functional F on the element u. The functional F is said to be real, if all of its values are real. It is called linear if its domain of definition is a linear set, and if *

$$F(\lambda u + \mu v) = \lambda F(u) + \mu F(v).$$

The problem of the variational calculus consists in the following: given a functional F with its domain of definition $D(F)$; it is required to find an element $u_0 \in D(F)$, which gives the functional either a minimum value

$$F(u_0) = \inf_{u \in D(F)} F(u), \tag{1}$$

or a maximum value

$$F(u_0) = \sup_{u \in D(F)} F(u). \tag{2}$$

The problem of maximizing the functional F is identical with the problem of minimizing the functional $-F$; therefore in what follows we shall treat only the minimum problem.

It is scarcely possible to solve the problem of the variational calculus in the general formulation just given. We therefore seek to impose some restrictions, as simple and natural as possible, on the functional F.

We shall assume that $D(F)$ is part of some Banach space X. In order to formulate the subsequent restrictions, we introduce the concept of a linear manifold. Let M be a linear set of elements of the space and $\bar{u}$ some fixed element of this space. A *linear manifold* in the space X is then the totality of elements, each of which can be expressed in the form

$$u = \bar{u} + \eta, \qquad \eta \in M. \tag{3}$$

If $\bar{u} \in M$ (for example, if $\bar{u} = 0$), then obviously the linear manifold defined in this way coincides with M.

RESTRICTION 1. *The domain of definition of the functional is a linear manifold which is dense in X.* **

We note that the linear set M will now also be dense in X. In fact, let u be an arbitrary element of the space X, and M' a manifold (dense in X) of elements of the form $\bar{u} + \eta$, where η ranges over the linear set M. The element $(u + \bar{u}) \in X$, and it is possible to find an element $\eta \in M$, such that

---

* We assume that the basic properties of linear functionals are familiar to the reader from a course in functional analysis.

** It would have been possible to consider a more general case, when $D(F)$ is the intersection of a dense linear manifold with a sphere or with some open set.

$\|(u+\bar{u})-(\bar{u}+\eta)\| < \varepsilon$, where ε is an arbitrary positive number. But then $\|u-\eta\| < \varepsilon$; the arbitrary element $u \in X$ can be approximated as closely as we please by an element of the set M. This means that the set M is dense in X. In exactly the same way it is possible to prove the converse statement: if η is an arbitrary element of a linear set dense in X, and if $\bar{u}$ is a fixed element of the space X, then the linear manifold of elements of the form $\bar{u}+\eta$ is dense in X.

Example. Consider the problem of the light-ray. For simplicity we stipulate that the path lies in the (x, y) plane, and that the velocity v is independent of z. Then the equation of the path can be written in the form $y = u(x)$, where $u(x)$ satisfies the boundary conditions

$$u(x_1) = y_1, \qquad u(x_2) = y_2, \qquad (4)$$

and the problem consists in finding a function $u(x)$ which satisfies the conditions (4) and minimizes the functional

$$T = \int_{x_1}^{x_2} \frac{\sqrt{\{1+u'^2(x)\}}}{v(x, u(x))} dx. \qquad (5)$$

It is natural to stipulate that the velocity of light is strictly positive:

$$v(x, y) \geq v_0 = \text{const.} > 0.$$

In the role of X we can take, for example, the space $L_2(x_1, x_2)$ of functions which are square-integrable on the interval (x_1, x_2). It is then natural to assign the functional (5) on the set $D(T)$ of functions, which are continuous and continuously differentiable on the segment $[x_1, x_2]$, and satisfy the conditions (4). We now show that this set is a linear manifold.

Let us write

$$\bar{u}(x) = y_1 + \frac{x-x_1}{x_2-x_1}(y_2-y_1)$$

and put $u(x) = \bar{u}(x)+\eta(x)$. If $u \in D(T)$, then $\eta(x)$ is a continuous and continuously differentiable function, satisfying the boundary conditions

$$\eta(x_1) = \eta(x_2) = 0. \qquad (6)$$

Obviously, the set M of functions $\eta(x)$ which satisfy the conditions just enumerated, is linear. This set contains as part of itself the set of functions which are finite on the segment $[x_1, x_2]$. From Corollary 1.3.1, the set M is dense in $L_2(\Omega)$.

We impose one further very important restriction on the functional under consideration. We shall assume that the space X is infinite-dimensional – if the contrary were the case, the variational problem would simply turn out to be the problem of minimizing a function having a finite number of independent variables. The linear set M, dense in X, is now also infinite-dimensional.

RESTRICTION 2. *If η ranges over an arbitrary finite-dimensional subspace, contained in M, then on this subspace the functional $F(u) = F(\bar{u} + \eta)$ is continuously differentiable a sufficient number of times.*

Let us clarify this restriction. Let η range over some n-dimensional subspace M_n. We introduce into it some basis $\eta_1, \ldots, \eta_n$. Then, if $\eta \in M_n$, η necessarily has the form

$$\eta = a_1 \eta_1 + a_2 \eta_2 + \ldots + a_n \eta_n,$$

and, consequently,

$$F(u) = F(\bar{u} + a_1 \eta_1 + \ldots + a_n \eta_n).$$

Here $\bar{u}$ is fixed, and the elements $\eta_1, \eta_2, \ldots, \eta_n$ are also fixed. Thus, F is a function of the variables $a_1, a_2, \ldots, a_n$. Because of Restriction 2, this function is differentiable with respect to $a_1, a_2, \ldots, a_n$ sufficiently many times.

We introduce the concepts of absolute and relative minimum of a functional. The functional F attains an *absolute minimum* on the element $u_0 \in D(F)$, if the inequality

$$F(u_0) \leq F(u) \tag{7}$$

holds for an arbitrary element $u \in D(F)$. Similarly the functional attains a *relative minimum* on the element u_0, if the inequality (7) holds for elements $u \in D(F)$, sufficiently close to u_0.

The concept of relative minimum depends essentially on the fact that some elements are assumed close to u_0, that is to say, on the fact that the set $D(F)$ is embedded in some space.

§ 3. Variation and Gradient of a Functional

1. We shall consider the functional F, subject to the Restrictions 1 and 2. We introduce an arbitrary element $u \in D(F)$ and an arbitrary element $\eta \in M$. Denote by α an arbitrary real number. It is not difficult to show that

§ 3] VARIATION AND GRADIENT OF A FUNCTIONAL

the element
$$u+\alpha\eta \in D(F).$$

To do this let $u = \bar{u}+\eta_0$, $\eta_0 \in M$. Then

$$u+\alpha\eta = \bar{u}+(\eta_0+\alpha\eta). \tag{1}$$

Each of the terms in the brackets belongs to the linear set M, therefore their sum $\eta_0+\alpha\eta$ also belongs to M. But then, obviously, the element $u+\alpha\eta \in D(F)$.

Let us form the expression $F(u+\alpha\eta)$. The element $\eta_0+\alpha\eta$ belongs to the two-dimensional subspace which traverses the elements η_0 and η. Because of Restriction 2, $F(u+\alpha\eta)$ is a continuously differentiable function of α. We calculate its derivative and take the value of this derivative for $\alpha = 0$:

$$\left.\frac{dF(u+\alpha\eta)}{d\alpha}\right|_{\alpha=0}. \tag{2}$$

As the result, we obtain a quantity which can be regarded as the value of the functional (2), depending on the two elements u and η.

DEFINITION. The functional

$$\left.\frac{dF(u+\alpha\eta)}{d\alpha}\right|_{\alpha=0}$$

is called the *variation* (or *first variation*) of the functional F at the point u and is designated by the symbol $\delta F(u, \eta)$:

$$\delta F(u, \eta) = \left.\frac{dF(u+\alpha\eta)}{d\alpha}\right|_{\alpha=0}. \tag{3}$$

We shall assume the element u to be fixed. Then the variation $\delta F(u, \eta)$ is a functional of η.

We shall demonstrate the following assertion: if Restrictions 1 and 2, § 2, hold, then $\delta F(u, \eta)$ is a linear functional of η; namely, if η_1, η_2 are arbitrary elements of the set M and a_1, a_2 are arbitrary numbers, then

$$\delta F(u, a_1\eta_1+a_2\eta_2) = a_1\delta F(u, \eta_1)+a_2\delta F(u, \eta_2).$$

In fact,
$$\delta F(u_1, a_1\eta_1+a_2\eta_2) = \left.\frac{d}{d\alpha} F(u+\alpha(a_1\eta_1+a_2\eta_2))\right|_{\alpha=0}.$$

If β_1, β_2 independently range over the set of real numbers, then the element $\beta_1\eta_1 + \beta_2\eta_2$ ranges over the two-dimensional subspace lying in M and spanning the elements η_1 and η_2. According to Restriction 2, § 2, the expression $F(u+\beta_1\eta_1+\beta_2\eta_2)$, regarded as a function of β_1 and β_2, is continuously differentiable at least once. Assume $\beta_1 = \alpha a_1$, $\beta_2 = \alpha a_2$, where α is an independent variable while a_1 and a_2 are arbitrary fixed numbers. Differentiating the expression $F(u+\alpha(a_1\eta_1+a_2\eta_2))$ as a function of functions we find

$$\delta F(u, a_1\eta_1+a_2\eta_2) = a_1 \frac{\partial}{\partial \beta_1} F(u+\beta_1\eta_1+\beta_2\eta_2)|_{\beta_1=\beta_2=0} +$$

$$+ a_2 \frac{\partial}{\partial \beta_2} F(u+\beta_1\eta_1+\beta_2\eta_2)|_{\beta_1=\beta_2=0} = a_1 \delta F(u, \eta_1) + a_2 \delta F(u, \eta_2),$$

which completes the proof.

For an arbitrarily-chosen $u \in D(F)$ the functional $\delta F(u, \eta)$, generally speaking, is not bounded. We select the set N of those elements $u \in D(F)$, for which the variation is a bounded functional of η.

If $u \in N$, then there exists a bounded functional g such that

$$\delta F(u, \eta) = (g, \eta), \qquad (4)$$

where (g, η) denotes the result of the action of the functional g on the element η.

The correspondence $u \to g$ defines on N an operator P such that

$$Pu = g, \qquad D(P) = N, \qquad (5)$$

and consequently

$$\delta F(u, \eta) = (Pu, \eta); \qquad u \in N, \qquad \eta \in M. \qquad (6)$$

The linear functional g, bounded in X, is an element of the space X^*, the adjoint of X. Therefore the operator P, defined by the formula (5) (or (6), which is equivalent), acts from the Banach space X into the adjoint space X^*.

DEFINITION. The operator P, defined by formula (6), is called the *gradient* of the functional $F(u)$, and is denoted by

$$P = \text{grad } F. \qquad (7)$$

If $u \in D(P)$, then the variation of the functional $F(u)$ can be written in the following form:

$$\delta F(u, \eta) = \frac{d}{d\alpha} F(u+\alpha\eta)|_{\alpha=0} = (\text{grad } F, \eta). \tag{8}$$

In general, $D(P)$ – the domain of definition of the gradient – is narrower than the domain of definition $D(F)$ of the functional F.

It is not difficult to see that the domain of definition of the gradient of a functional, and also the expression for the gradient, depend essentially on the space in which the domain of definition of the functional itself is embedded.

2. We consider the following example which is important in what follows.

Let the function $\Phi(x, y, z)$ be defined and continuous, when

$$x \in [a, b], \quad -\infty < y < +\infty, \quad -\infty < z < +\infty.$$

We shall suppose that the function $\Phi(x, y, z)$ has partial derivatives Φ_y and Φ_z, continuous in the domain of variation of the quantities x, y, z. We consider the functional

$$F(u) = \int_a^b \Phi(x, u(x), u'(x)) \, dx, \tag{9}$$

whose domain of definition $D(F)$ consists of functions satisfying the following conditions: $u \in C^{(1)}[a, b]$, and

$$u(a) = A, \quad u(b) = B, \tag{10}$$

where A and B are given constants. The condition (10) implies that the curves $y = u(x)$, where $u \in D(F)$, pass through the two fixed points (a, A) and (b, B).

We now show that the functional (9)–(10) satisfies the Restrictions 1, 2 of § 2. As X, we shall take the space $L_2(a, b)$. Obviously, $D(F)$ belongs to this space.

First of all we verify that $D(F)$ is a linear manifold. We put

$$\bar{u}(x) = A + \frac{x-a}{b-a}(B-A). \tag{11}$$

Obviously, $\bar{u} \in C^{(1)}[a, b]$, and $\bar{u}$ satisfies the conditions (10). Let $u \in D(F)$, and consider the difference $\eta(x) = u(x) - \bar{u}(x)$. Evidently, $\eta \in C^{(1)}[a, b]$, and

$$\eta(a) = \eta(b) = 0. \tag{12}$$

It is clear that the set M of functions η is linear. It contains the set of functions finite on the segment $[a, b]$. From Corollary 1.3.1, the set M is dense in $L_2(a, b)$, and then the manifold $D(F)$ is dense and linear in $L_2(a, b)$. Restriction 1 is thus fulfilled.

We now turn to Restriction 2. We have

$$F(\bar{u} + a_1\eta_1 + a_2\eta_2 + \ldots + a_n\eta_n) =$$
$$= \int_a^b \Phi\left(x, \bar{u}(x) + \sum_{k=1}^n a_k\eta_k(x), \bar{u}'(x) + \sum_{k=1}^n a_k\eta_k'(x)\right) dx. \tag{13}$$

The function (13) is continuously differentiable with respect to the variables $a_1, a_2, \ldots, a_n$. In fact, it follows from the stated properties of the function Φ that the integrand-function in (13), and its first derivatives with respect to $a_1, a_2, \ldots, a_n$, depend continuously on $x, a_1, \ldots, a_n$. From theorems on differentiating an integral depending on a parameter, it follows that the function (13) has continuous partial derivatives with respect to the variables $a_1, a_2, \ldots, a_n$, and that these derivatives can be obtained by differentiating under the integral sign.

Since Restrictions 1 and 2 are satisfied, we can form the variation of the functional (9):

$$\delta F(u, \eta) = \frac{d}{d\alpha} F(u + \alpha\eta)\big|_{\alpha=0}$$
$$= \frac{d}{d\alpha} \int_a^b \Phi(x, u + \alpha\eta, u' + \alpha\eta') dx\big|_{\alpha=0}$$
$$= \int_a^b [\Phi_u(x, u, u')\eta + \Phi_{u'}(x, u, u')\eta'] dx. \tag{14}$$

Formula (14) evidently shows that the variation $\delta F(u, \eta)$ is a linear functional of η, since the integrand function in (14) depends linearly on η and η'. Once the variation has been shown to be linear, it is possible to pose the question with regard to the gradient of the functional F. We indicate what the function $u \in D(F)$ has to be in order that the variation $\delta F(u, \eta)$ should be a bounded functional of η.

The space $L_2(a, b)$ is a Hilbert space. By the well-known Riesz's theorem, an arbitrary linear, bounded functional in this space has the form

$$(\eta, g) = \int_a^b g(x)\eta(x)\,dx, \tag{15}$$

where $g(x)$ is a completely defined function from $L_2(a, b)$.

The integral (14) divides into two parts; in the first, the quantity Φ_u is continuous, and, moreover, square-integrable on the interval (a, b); the first integral in (14) is a bounded functional of η. But, in general, this cannot be said in respect of the second integral

$$\int_a^b \Phi_{u'}\eta'\,dx \tag{16}$$

if $u(x)$ is an arbitrary function.

Let the function $u(x)$ be such that $\Phi_{u'}(x, u(x), u'(x))$ is an absolutely continuous function of x, with its derivative square-integrable on (a, b). We show that the integral (16) is then a bounded functional of η; for this we use the fact that $\eta(x)$ is continuously differentiable and satisfies the boundary conditions (12).

According to the stipulations made above with regard to the function $\Phi_{u'}$, there exist a constant c and a function $\omega \in L_2(a, b)$, such that

$$\Phi_{u'}(x, u(x), u'(x)) = c + \int_a^x \omega(t)\,dt, \tag{17}$$

from which we have that, almost everywhere in (a, b),

$$\omega(x) = \frac{d}{dx}\Phi_{u'}(x, u(x), u'(x)).$$

If this is the case, then the integral

$$\int_a^b \Phi_{u'}\eta'\,dx$$

can be integrated by parts:

$$\int_a^b \Phi_{u'}\eta'\,dx = -\int_a^b \eta \frac{d}{dx}\Phi_{u'}\,dx + \eta\Phi_{u'}\big|_a^b = -\int_a^b \eta \frac{d}{dx}\Phi_{u'}\,dx.$$

Since $(d/dx)\Phi_{u'} = \omega \in L_2(a, b)$, Riesz's theorem gives that the last integral is a bounded functional of η. But then the variation $\delta F(u, \eta)$ will also be a bounded functional of η, and it can now be written in the form

$$\delta F(u, \eta) = \int_a^b \left[\Phi_u - \frac{d}{dx} \Phi_{u'} \right] \eta \, dx. \tag{18}$$

The formula for grad F follows from relation (18):

$$\operatorname{grad} F = \Phi_u(x, u, u') - \frac{d}{dx} \Phi_{u'}(x, u, u'). \tag{19}$$

In fact, from formulae (8) and (18), it follows that

$$\left(\operatorname{grad} F - \left[\Phi_u - \frac{d}{dx} \Phi_{u'} \right], \eta \right) = 0.$$

The difference

$$\operatorname{grad} F - \left[\Phi_u - \frac{d}{dx} \Phi_{u'} \right]$$

is seen to be orthogonal to the set of functions η, dense in $L_2(a, b)$. But then

$$\operatorname{grad} F - \left[\Phi_u - \frac{d}{dx} \Phi_{u'} \right] = 0,$$

which is equivalent to formula (19).

Thus the gradient of the functional F is defined on the function $u \in D(F)$, if $\Phi_{u'}$ is an absolutely continuous function, whose derivative is square-integrable.

We show that the converse statement is also true:

If the function $u(x) \in D(F)$, and at the same time $u \in D(\operatorname{grad} F)$, then, on the segment $[a, b]$, the function $\Phi_{u'}(x, u, u')$ is absolutely continuous, the derivative $(d/dx)\Phi_{u'}(x, u, u')$ is square-integrable on the interval (a, b) and

$$(\operatorname{grad} F)(u) = \Phi_u - \frac{d}{dx} \Phi_{u'}.$$

If $u \in D(\operatorname{grad} F)$, then the variation $\delta F(u, \eta)$ is a bounded functional of η. In this case the integral (16) is also a bounded functional of η. By Riesz's theorem, there exists a function $g \in L_2(a, b)$, such that

$$\int_a^b \Phi_{u'} \eta'(x) \, dx = \int_a^b g(x) \eta(x) \, dx. \tag{20}$$

We construct the function

$$G(x) = -\int_a^x g(t)\,dt$$

and perform integration by parts on the right-hand side of (20). Since the function $\eta(x)$ satisfies the conditions (12), we have

$$\int_a^b g(x)\eta(x)dx = \int_a^b G(x)\eta'(x)dx. \tag{21}$$

Formulae (20) and (21) together give

$$\int_a^b [\Phi_{u'} - G]\eta'\,dx = 0. \tag{22}$$

The equality (22) holds for an arbitrary function $\eta(x)$, continuously differentiable on the segment $[a, b]$ and equal to zero at the end-points of this segment. Taking account of this, we substitute in equation (22),

$$\eta(x) = \sin\frac{k\pi(x-a)}{b-a}, \qquad k = 1, 2, \ldots$$

We then obtain

$$\int_a^b [\Phi_{u'} - G]\cos\frac{k\pi(x-a)}{b-a} = 0, \qquad k = 1, 2, \ldots \tag{23}$$

It is well-known that the system of functions

$$\cos k\pi \frac{x-a}{b-a}, \qquad k = 0, 1, 2, \ldots$$

is complete in $L_2(a, b)$. The function $\Phi_{u'} - G$ is orthogonal to every function of this system, with the exception of the function which is identically equal to unity. But then the function $\Phi_{u'} - G$ can differ from unity only by a constant:

$$\Phi_{u'} - G = c = \text{const.}$$

Hence

$$\Phi_{u'} = c - \int_a^x g(t)dt,$$

and, consequently, the function $\Phi_{u'}$ is absolutely continuous. Moreover, $(d/dx)\Phi_{u'} = -g(x) \in L_2(a, b)$, and our assertion is proved. The results obtained here can be formulated in the following theorem.

THEOREM 3.3.1. *Let the functional* (9) *be given on continuously differentiable functions, which satisfy the boundary conditions* (10), *and let the domain of definition of this functional be regarded as a set in the space* $L_2(a, b)$. *Then the domain of definition of the gradient of the functional* (9) *consists of those, and only those, functions, which have the following properties*: *they are continuously differentiable on the segment* $[a, b]$ *and satisfy conditions* (10); *on being substituted into the expression* $\Phi_{u'}(x, u, u')$ *they convert it into a function which is absolutely continuous on the segment* $[a, b]$, *and whose derivative on this segment is square-integrable.*

It was shown earlier that the domain of definition of the gradient (and likewise the expression for the gradient itself) depends essentially on the space in which the domain of definition of the functional is embedded. We shall illustrate this on the example of the functional (9)–(10).

We introduce a Hilbert space, which we denote by $\mathring{W}_2^{(1)}(a, b)$, and which we define in the following way: its elements are functions, absolutely continuous on the segment $[a, b]$, equal to zero at the points a and b, and having first derivatives which are square-integrable on the given segment. Scalar product $(u, v)_0$ and norm $||u||_0$ in $\mathring{W}_2^{(1)}(a, b)$ are given by the formulae

$$(u, v)_0 = \int_a^b \frac{du}{dx} \frac{dv}{dx} dx, \qquad ||u||_0^2 = \int_a^b \left(\frac{du}{dx}\right)^2 dx. \qquad (24)$$

We consider the functional (9)–(10) with the additional stipulation that in conditions (10), $A = B = 0$; the general case can be reduced to this by the substitution

$$u(x) = v(x) + A + \frac{x - a}{b - a}(B - A).$$

We embed the domain of definition of the functional (9) in the space $\mathring{W}_2^{(1)}(a, b)$; it is easy to demonstrate that the Restrictions 1, 2 (§ 2) are satisfied here, and that formula (14) holds as before.

We show that for arbitrary fixed $u \in D(F)$ the functional (14) is bounded in $\mathring{W}_2^{(1)}(a, b)$; from this it will follow that in this space $D(P) = D(F)$.

By the Schwartz-Bunyakovskii inequality

$$|\delta F(u, \eta)| \leq \left(\int_a^b \Phi_u^2 \, dx \int_a^b \eta^2 \, dx\right)^{\frac{1}{2}} + \left(\int_a^b \Phi_{u'}^2 \, dx \int_a^b \eta'^2 \, dx\right)^{\frac{1}{2}}.$$

The functions Φ_u and $\Phi_{u'}$ are continuous, and hence bounded on the segment $[a, b]$; therefore

$$|\delta F(u, \eta)| \leq C_0 \left(\int_a^b \eta^2 \, dx \right)^{\frac{1}{2}} + C_1 \left(\int_a^b \eta'^2 \, dx \right)^{\frac{1}{2}}, \quad (25)$$

where C_0 and C_1 are some constants. Furthermore,

$$\eta(x) = \eta(a) + \int_a^x \eta'(t) \, dt = \int_a^x \eta'(t) \, dt.$$

We again make use of the Schwartz-Bunyakovskii inequality, and obtain

$$\eta^2(x) \leq (x-a) \int_a^x \eta'^2(t) \, dt \leq (b-a) \int_a^b \eta'^2(t) \, dt,$$

and, consequently,

$$\int_a^b \eta^2(x) \, dx \leq (b-a)^2 \int_a^b \eta'^2(x) \, dx.$$

Substituting this in (25), we obtain

$$|\delta F(u, \eta)| \leq C \left(\int_a^b \eta'^2(x) \, dx \right)^{\frac{1}{2}} = C \|\eta\|_0; \quad C = C_0(b-a) + C_1,$$

and so our assertion is proved.

We now find the expression grad F in the space $\mathring{W}_2^{(1)}(a, b)$. Put

$$g(x) = \int_a^x \Phi_u(t, u(t), u'(t)) \, dt + C,$$

where C is a constant, as yet undefined. Integrating by parts, we obtain from formula (14)

$$\delta F(u, \eta) = \int_a^b (\Phi_{u'} - g(x)) \eta'(x) \, dx.$$

Next write

$$G(x) = \int_a^x [\Phi_{u'}(t, u(t), u'(t)) - g(t)] \, dt,$$

and choose C, such that $G(b) = 0$. Then $G \in \mathring{W}_2^{(1)}(a, b)$. At the same time

$$\delta F(u, \eta) = \int_a^b G'(x) \eta'(x) \, dx = (G, \eta)_0.$$

From this it follows that, in the space $\mathring{W}_2^{(1)}(a, b)$,

$$\text{grad } F = G(x) = \int_a^x \left[\Phi_{u'}(t, u(t), u'(t)) - \int_a^t \Phi_u(\tau, u(\tau), u'(\tau)) \, d\tau - C \right] dt. \quad (26)$$

§ 4. Euler's Equation

We consider a functional F, which satisfies the Restrictions 1, 2 (§ 2). Let this functional be defined on a linear manifold $D(F)$, whose elements have the form $u = \bar{u} + \eta$, where $\bar{u}$ is a fixed element of a given space X, and η ranges over a linear set M, dense in X.

Let the functional F attain a relative minimum at the point u_0. We take an arbitrary element $\eta \in M$, and an arbitrary real number α. Then if the absolute value of α is sufficiently small, we have that the norm of the difference

$$\|(u_0 + \alpha\eta) - u_0\| = |\alpha| \|\eta\|$$

will be as small as we please, and then, from the definition of a relative minimum,

$$F(u_0 + \alpha\eta) \geqq F(u_0). \tag{1}$$

This inequality implies that the function of one real variable α, equal to $F(u_0 + \alpha\eta)$, has a relative minimum for $\alpha = 0$. But then, necessarily,

$$\frac{d}{d\alpha} F(u_0 + \alpha\eta)\big|_{\alpha=0} = 0$$

or, equivalently,

$$\delta F(u_0, \eta) = 0. \tag{2}$$

We have stated the necessary condition for a minimum: *if a functional attains a minimum at some point, then at this point the variation of the functional is equal to zero.*

A linear functional which is identically equal to zero is obviously bounded, and so in this case $u_0 \in D$ (grad F). We write

$$\text{grad } F = Pu. \tag{3}$$

Then

$$\delta F(u_0, \eta) = (Pu_0, \eta) = 0. \tag{4}$$

The expression Pu_0 is a functional over the elements η, and is defined on that set M over which η ranges. This means that the functional Pu_0 is given on a dense set, and on this set all of its values are zero. Because of its continuity, it is possible to extend it onto all the space X, and its value will also be zero on all this space. But this implies that

$$Pu_0 = 0.$$

We have thus proved,

THEOREM 3.4.1. *If a functional F satisfying the Restrictions 1 and 2 has a relative extremum at a point u_0, then $u_0 \in D$ (grad F), and at this point the equation*

$$(\operatorname{grad} F)(u_0) = 0 \qquad (5)$$

is satisfied.

Equation (5) is called *Euler's equation*.

As an example, let us consider the so-called "simplest problem of the variational calculus", henceforth referred to as Problem S. This is the problem of the minimum of the functional

$$F(u) = \int_a^b \Phi(x, u, u') dx, \qquad (6)$$

whose domain of definition $D(F)$ consists of functions which are continuously differentiable on the segment $[a, b]$, and satisfy the boundary conditions

$$u(a) = A, \qquad u(b) = B. \qquad (7)$$

This functional was described in § 3. We suppose here that the function $\Phi(x, u, u')$ satisfies all the limitations imposed on it in § 3, and that the domain $D(F)$ is embedded in the space $L_2(a, b)$. If the function $u(x)$ attains a minimum (relative or, a fortiori, absolute), then by Euler's Theorem 3.4.1, and by formula (3.19), this function satisfies the differential equation

$$\frac{d}{dx} \Phi_{u'}(x, u, u') - \Phi_u = 0; \qquad (8)$$

being an element of the set $D(F)$, it also satisfies the conditions (7). We shall investigate this problem more extensively in Chapter 4.

As an example consider the brachistochrone problem. In this case

$$\Phi(x, u, u') = \sqrt{\frac{1+u'^2}{u}};$$

we cannot immediately use equation (8), because the function Φ has a singularity at $u = 0$, and, consequently, does not satisfy the conditions of § 3. We shall prove that equation (8) is nevertheless applicable.

Let the brachistochrone problem have a solution $u_0(x)$. It is clear from physical considerations that $u_0(x) > 0$ when $x > 0$ – otherwise the moving material point would be carried upwards on some sections of the path, and the time so expended would be superfluous.

On the interval $(0, a)$ we take an arbitrary point a'. We construct the function $\eta(x)$, which has the following properties: a) $\eta \in C^{(1)}[0, a]$; b) $\eta(x) \equiv 0, 0 \leq x \leq a'$; c) $\eta(a) = 0$; in other respects the function η is arbitrary. We put

$$F(u) = \int_0^a \sqrt{\frac{1+u'^2}{u}}\, dx,$$

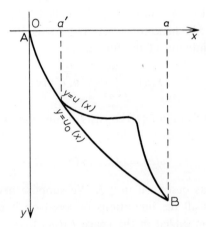

Fig. 6.

and let $u(x) = u_0(x) + \alpha\eta(x)$ (Fig. 6). If the number α is sufficiently small, then $u(x) > 0, x > 0$; in addition, $u(0) = 0, u(a) = b$. In this case $F(u) = F(u_0 + \alpha\eta) \geq F(u_0)$, and, as a function of the variable α,

$$F(u_0 + \alpha\eta) = \int_0^{a'} \sqrt{\frac{1+u_0'^2}{u_0}}\, dx + \int_{a'}^{a} \sqrt{\frac{1+(u_0' + \alpha\eta')^2}{u_0 + \alpha\eta}}\, dx \qquad (9)$$

it has a minimum for $\alpha = 0$. The integrand in the second integral, of (9) depends continuously on x and α, and can be differentiated under the integral sign:

$$\frac{d}{d\alpha} F(u_0 + \alpha\eta)|_{\alpha=0} = \int_{a'}^{a} \left[\frac{\partial}{\partial u'} \sqrt{\frac{1+u'^2}{u}}\, \eta' + \frac{\partial}{\partial u} \sqrt{\frac{1+u'^2}{u}}\, \eta \right]_{u=u_0} dx = 0. \qquad (10)$$

In order to simplify the argument, we assume – in actual fact it is not difficult to prove – that on the segment $[a', a]$ there exists a square-integrable derivative:

$$\frac{d}{dx} \left[\frac{\partial}{\partial u'} \sqrt{\frac{1+u'^2}{u}}\right]_{u=u_0}.$$

Then the first integral in (10) can be integrated by parts. It follows from the definition of the function $\eta(x)$ that $\eta(a') = \eta(a) = 0$, and so the integrated term vanishes. We arrive at the result

$$\int_{a'}^{a} \left[\frac{d}{dx} \frac{\partial}{\partial u'} \sqrt{\frac{1+u'^2}{u}} - \frac{\partial}{\partial u} \sqrt{\frac{1+u'^2}{u}} \right]_{u=u_0} \eta(x)\, dx = 0.$$

This equation is valid for an arbitrary function $\eta \in C^{(1)}[a', a]$, $\eta(a') = \eta(a) = 0$. The set of such functions is dense in $L_2(a', a)$, (cf. § 3), and hence it is necessarily the case that

$$\left[\frac{d}{dx} \frac{\partial}{\partial u'} \sqrt{\frac{1+u'^2}{u}} - \frac{\partial}{\partial u} \sqrt{\frac{1+u'^2}{u}} \right]_{u=u_0} = 0, \qquad (11)$$

which is equivalent to equation (8) for the function $\sqrt{\{(1+u'^2)/u\}}$.

Equation (11) can easily be put in the form

$$\frac{d}{dx} \sqrt{\{u_0(1+u_0'^2)\}} = 0.$$

Hence

$$u_0(1+u_0'^2) = c.$$

Let us put $u_0' = \tan \phi$. Then $u_0 = c/(1+\tan^2 \phi) = \tfrac{1}{2}c(1+\cos 2\phi)$. Differentiating, we obtain $u_0' = -c \sin 2\phi \cdot \phi'$. The substitution $u_0' = \tan \phi$ gives a differential equation involving ϕ'

$$\frac{d\phi}{dx} = -\frac{\tan \phi}{c \sin 2\phi}.$$

Now,

$$dx = -2c \cos^2 \phi\, d\phi; \qquad x = c_1 - \tfrac{1}{2}c(2\phi + \sin 2\phi).$$

Putting $2\phi = \pi + \theta$, $C_1 - \tfrac{1}{2}\pi c = C$, we obtain

$$x = C - \tfrac{1}{2}c(\theta - \sin \theta), \qquad u_0 = \tfrac{1}{2}c(1 - \cos \theta).$$

We have thus arrived at the conclusion that if the brachistochrone problem has a solution, then this solution is a cycloid.

In the preceding section it was shown that the gradient of a functional depends on that space in which the domain of definition of the functional is embedded. It follows from this that the *form* of Euler's equation depends on the space concerned, but the *solutions* of Euler's equation do not. * In fact, these solutions are essentially the elements $u \in D(F)$, for which

---

* This property is commonly called the invariance of Euler's equation.

$\delta F(u, \eta) = 0$, and the variation of a functional F depends only on the domain $D(F)$ and on the elements u, η, but not on the space in which the given domain is embedded.

We exemplify this point in the case of Problem S; let $A = B = 0$. In the space $L_2(a, b)$, Euler's equation reduces to the differential equation (8), with boundary conditions

$$u(a) = u(b) = 0. \tag{12}$$

In the space $\mathring{W}_2^{(1)}(a, b)$ (§ 3), Euler's equation reduces (because of formula (3.26)) to the equation

$$\int_a^x \left[\Phi_{u'}(t, u(t), u'(t)) - \int_a^t \Phi_u(\tau, u(\tau), u'(\tau))\,d\tau - C \right] d\tau = 0, \tag{13}$$

with the same conditions (12).

It is not difficult to see that equations (8) and (13) are equivalent: equation (8) is derived from (13) by differentiating twice; alternatively, two integrations of (8) lead to equation (13).

§ 5. Second Variation. Sufficient Condition for an Extremum

We retain the notation of the preceding sections and consider the function $F(u+\alpha\eta)$ of a real variable α; we take as fixed the elements $u \in D(F)$ and $\eta \in M$. We expand this function in a Taylor series and we put $\alpha = 1$ in the resulting expansion

$$F(u+\eta) = F(u) + \left[\frac{d}{d\alpha} F(u+\alpha\eta) \right]_{\alpha=0} + \frac{1}{2} \left[\frac{d^2}{d\alpha^2} F(u+\alpha\eta) \right]_{\alpha=\theta}$$

$$= F(u) + \delta F(u, \eta) + \frac{1}{2} \left[\frac{d^2}{d\alpha^2} F(u+\alpha\eta) \right]_{\alpha=\theta}; \quad 0 < \theta < 1. \tag{1}$$

The expression

$$\delta^2 F(u, \eta) = \frac{1}{2} \left[\frac{d^2}{d\alpha^2} F(u+\alpha\eta) \right]_{\alpha=0} \tag{2}$$

is called the *second variation* of the functional F at the point u.

We have

$$\frac{1}{2} \left[\frac{d^2}{d\alpha^2} F(u+\alpha\eta) \right]_{\alpha=0} = \frac{1}{2} \left[\frac{d^2}{d\beta^2} F(u+\theta\eta+\beta\eta) \right]_{\beta=0} = \delta^2 F(u+\theta\eta, \eta),$$

and the expansion (1) takes the following form:

$$F(u+\eta) = F(u)+\delta F(u, \eta)+\delta^2 F(u+\theta\eta, \eta), \qquad 0 < \theta < 1. \tag{3}$$

Let the functional F attain a minimum, either relative or absolute, at the point u_0. Then $\delta F(u_0, \eta) = 0$, and formula (3) gives

$$F(u_0+\eta) = F(u_0)+\delta^2 F(u_0+\theta\eta, \eta). \tag{4}$$

Formula (4) immediately enables us to formulate a necessary and sufficient condition that the element u_0, which satisfies Euler's equation, should render the value of the functional a minimum. For an absolute minimum this condition takes the form

$$\delta^2 F(u_0+\theta\eta, \eta) \geq 0, \qquad \forall \eta \in M; \tag{5}$$

for a relative minimum it consists of the requirement that the inequality (5) should be satisfied when the element η is sufficiently small in the norm. In specific problems condition (5) is difficult to establish, because the quantity θ is usually unknown, and so, as a rule, we cannot use it directly. However, it is possible to derive from (5) simpler conditions, either sufficient or necessary, which are verifiable. Specifically, the following sufficient conditions for a minimum ensue from relation (5):

THEOREM 3.5.1. *Let an element u_0 satisfy Euler's equation. If for $\eta \neq 0$ the second variation of the functional F is non-negative at any point $u \in D(F)$, then this functional has an absolute minimum at the point u_0. If for $\eta \neq 0$ the second variation is non-negative in some neighbourhood of the point u_0, then the functional has a relative minimum at this point.*

The proof is very simple. In the first case we have, for any $\eta \in M$, $\eta \neq 0$,

$$\delta^2 F(u_0+\theta\eta, \eta) \geq 0.$$

If u is an arbitrary element of $D(F)$, different from u_0, then, putting $\eta = u-u_0$, we find from (4) that $F(u) \geq F(u_0)$. Consider now the second case. There exists a number $\rho > 0$, such that, for $u \in D(F)$, $\|u-u_0\| < \rho$ and $\eta \neq 0$, we have $\delta^2 F(u, \eta) \geq 0$. We take such a value of u, $u \neq u_0$, and again put $\eta = u-u_0$. Then

$$F(u) = F(u_0)+\delta^2 F(u_0+\theta\eta, \eta).$$

We have

$$\|(u_0+\theta\eta)-u_0\| = \|\theta\eta\| = |\theta| \|\eta\| \leq \|\eta\| = \|u-u_0\| < \rho.$$

Hence it follows that $\delta^2 F(u_0+\theta\eta, \eta) \geq 0$, and, consequently, $F(u) \geq F(u_0)$.

§ 6. The Isoperimetric Problem

The isoperimetric problem is formulated in the following way: given functionals $F, G_1, G_2, \ldots, G_n$ and constants $l_1, l_2, \ldots, l_n$; from among the elements of the domain of definition $D(F)$ of the functional F, satisfying the equations

$$G_k(u) = l_k, \qquad k = 1, 2, \ldots, n, \tag{1}$$

it is required to find the element which assigns to the functional F its smallest value. The statement of the isoperimetric problem certainly requires that the intersection

$$D_0 = D(F) \cap D(G_1) \cap D(G_2) \cap \ldots \cap D(G_n) \tag{2}$$

shall not be an empty set. We shall assume below that this condition is fulfilled.

The third problem of § 1 is a special case of an isoperimetric problem. Here $n = 1$,

$$F(u) = \int_a^b u(x)\,dx, \qquad G_1(u) = \int_a^b \sqrt{(1+u'^2)}\,dx, \qquad l_1 = l.$$

As $D(F)$, we can take the set of those functions of $C[a, b]$ which are equal to zero for $x = a$ and $x = b$ (condition (5.1)), and as $D(G_1)$ we take the set of functions of $C^{(1)}[a, b]$, which also satisfy condition (5.1). Obviously $D(G_1) \subset D(G)$, and the intersection $D(G_1) \cap D(F) = D(G)$ is not an empty set.

We next assume that the functionals $F, G_1, G_2, \ldots, G_n$, satisfy the Restrictions 1 and 2 (§ 2). The intersection of linear manifolds is itself a linear manifold, since there exists an element $\bar{u} \in D_0$ and a linear set M_0, such that any element $u \in D_0$ has the form $u = \bar{u} + \eta$, $\eta \in M_0$.

We shall also assume that the set M_0 is dense in the space under consideration.

This establishes a theorem, attributed to Euler, and known as the *multiplier rule for the isoperimetric problem*.

THEOREM 3.6.1 (EULER'S THEOREM). *Let the element $u_0 \in D_0$ solve the isoperimetric problem. If there exist elements $\eta_1, \eta_2, \ldots, \eta_n \in M_0$, such that the determinant*

$$\Delta = \begin{vmatrix} \delta G_1(u_0, \eta_1) & \delta G_2(u_0, \eta_1) & \ldots & \delta G_n(u_0, \eta_1) \\ \delta G_1(u_0, \eta_2) & \delta G_2(u_0, \eta_2) & \ldots & \delta G_n(u_0, \eta_2) \\ \cdot & \cdot & & \cdot \\ \delta G_1(u_0, \eta_n) & \delta G_2(u_0, \eta_n) & \cdots & \delta G_n(u_0, \eta_n) \end{vmatrix} \tag{3}$$

is different from zero, then it is possible to find constants λ_k, $k = 1, 2, \ldots, n$, such that

$$(\mathrm{grad}\,(F + \sum_{k=1}^{n} \lambda_k G_k))\,(u_0) = 0. \tag{4}$$

Proof. We introduce the real variables $\alpha_0, \alpha_1, \ldots, \alpha_n$, and an element $\eta_0 \in M_0$. We put

$$u = u_0 + \sum_{k=0}^{n} \alpha_k \eta_k. \tag{5}$$

For arbitrary values of α_k, the element $u \in D_0$. In fact, $u_0 = \bar{u} + \eta$, $\eta \in M_0$. But then

$$u = \bar{u} + (\eta + \sum_{k=0}^{n} \alpha_k \eta_k).$$

The expression in brackets is a linear combination of elements of M_0 and consequently itself belongs to M_0. Hence $u \in D_0$.

We put

$$G_j(u_0 + \sum_{k=0}^{n} \alpha_k \eta_k) = \phi_j(\alpha_0, \alpha_1, \ldots, \alpha_n), \quad j = 1, 2, \ldots, n. \tag{6}$$

We evaluate the first partial derivatives of the function (6) for $\alpha_k = 0$, $k = 0, 1, 2, \ldots, n$. We have

$$\frac{\partial}{\partial \alpha_i} \phi_j(\alpha_0, \alpha_1, \ldots, \alpha_n)|_{\alpha_i = 0} = \frac{\partial}{\partial \alpha_i} G_j(u_0 + \sum_{\substack{k=0 \\ k \neq i}}^{n} \alpha_k \eta_k + \alpha_i \eta_i)|_{\alpha_i = 0}$$

$$= \delta G(u_0 + \sum_{\substack{k=0 \\ k \neq i}}^{n} \alpha_k \eta_k, \eta_i).$$

Further, we set $\alpha_k = 0$, $k \neq i$, and obtain

$$\frac{\partial}{\partial \alpha_i} \phi_j(\alpha_0, \alpha_1, \ldots, \alpha_n)|_{\alpha_0 = \alpha_1 = \ldots = \alpha_n = 0} = \delta G_j(u_0, \eta_i). \tag{7}$$

The elements $\eta_1, \eta_2, \ldots, \eta_n$ are selected in such a way that the determinant (3) is different from zero. Now consider the system of equations

$$\phi_j(\alpha_0, \alpha_1, \ldots, \alpha_n) - l_j = 0, \quad j = 1, 2, \ldots, n. \tag{8}$$

Systems such as (8) fulfil the conditions of theorems on implicit functions: the system (8) is satisfied at the point $\alpha_0 = \alpha_1 = \ldots = \alpha_n = 0$, and at this

point the Jacobian
$$\frac{D(\phi_1, \phi_2, \ldots, \phi_n)}{D(\alpha_1, \alpha_2, \ldots, \alpha_n)}$$

is different from zero – by virtue of relation (7), it coincides with the determinant (3). It follows from this that there exist functions $\alpha_k = \omega_k(\alpha_0)$, $k = 1, 2, \ldots, n$, which cause the equations of system (8) to be identically satisfied; these functions are continuously differentiable for values of a_0, close to zero and with $\omega_k(0) = 0$.

Equations (8) imply that
$$G_j\big(u_0 + \alpha_0 \eta_0 + \sum_{k=1}^{n} \omega_k(\alpha_0)\eta_k\big) = l_j, \qquad j = 1, 2, \ldots, n,$$

i.e., that for arbitrary α_0, sufficiently close to zero, the element
$$u(\alpha_0) = u_0 + \alpha_0 \eta_0 + \sum_{k=1}^{n} \omega_k(\alpha_0)\eta_k$$

satisfies the isoperimetric equations (1). In this case
$$F(u(\alpha_0)) \geq F(u_0).$$

For $\alpha_0 = 0$ we have $u(0) = u_0$; the function $F(u(\alpha_0))$ of the real variable α_0 has a minimum when $\alpha_0 = 0$, and therefore
$$\frac{d}{d\alpha_0} F(u(\alpha_0))\big|_{\alpha_0 = 0} = 0.$$

We have
$$\frac{d}{d\alpha_0} F(u(\alpha_0))\big|_{\alpha_0 = 0} = \frac{\partial F(u_0 + \alpha_0 \eta_0 + \sum_{k=1}^{n} \alpha_k \eta_k)}{\partial \alpha_0}\bigg|_{\alpha_0 = \alpha_1 = \ldots = \alpha_n = 0} +$$
$$+ \sum_{k=1}^{n} \frac{\partial F(u_0 + \alpha_0 \eta_0 + \sum_{k=1}^{n} \alpha_k \eta_k)}{\partial \alpha_k} \omega_k'(\alpha_0)\big|_{\alpha_0 = \alpha_1 = \ldots = \alpha_n = 0}$$

or, more concisely,
$$\frac{d}{d\alpha} F(u(\alpha_0))\big|_{\alpha_0 = 0} = \delta F(u_0, \eta_0) + \sum_{k=1}^{n} \delta F(u_0, \eta_k)\omega_k'(0). \qquad (9)$$

The value of $\omega_k'(0)$ can be determined from the system (8). In the equations of this system we put $\alpha_k = \omega_k(\alpha_0)$, differentiate the resulting expression with respect to α_0, and put $\alpha_0 = 0$. Similarly we obtain from relation (9) the equations

$$\delta G_j(u_0, \eta_0) + \sum_{k=1}^{n} \delta G_j(u_0, \eta_k)\omega_k'(0) = 0, \qquad j = 1, 2, \ldots, n. \tag{10}$$

These equations represent a linear system with unknowns $\omega_k'(0)$; its determinant, equal to the determinant (3), is different from zero. It can easily be seen that the solution of system (10) has the form

$$\omega_k'(0) = \sum_{m=1}^{n} A_{km} \delta G_m(u_0, \eta_0), \qquad k = 1, 2, \ldots, n,$$

where A_{km} are constants. Substituting this into (9), inverting the order of summation, and introducing the notation

$$\lambda_j = \sum_{k=1}^{n} A_{kj} \delta F(u_0, \eta_k), \qquad J = F + \sum_{j=1}^{n} \lambda_j G_j,$$

we obtain

$$\delta J(u_0, \eta_0) = 0.$$

Repeating the arguments used in §4 in deriving Euler's equation, we find

$$(\text{grad } J)(u_0) = (\text{grad } (F + \sum_{j=1}^{n} \lambda_j G_j))(u_0) = 0.$$

This is what we set out to show.

Note that this Theorem of Euler which we have just established only gives a necessary condition for a minimum in the isoperimetric problem.

The technique for solving isoperimetric problems is as follows: we form a functional $J = F + \sum_{k=1}^{n} \lambda_k G_k$, where the λ_k are unknown constants, and write down Euler's equation for this functional. This equation contains as unknowns the element u_0 and the constants $\lambda_1, \lambda_2, \ldots, \lambda_n$. These unknowns are then determined from Euler's equation (4) and the isoperimetric relation (1).

As an example, consider Problem 3, §1, which was also mentioned at the beginning of the present section. In accordance with Euler's theorem, we introduce constant quantities λ and construct the functional

$$J(u) = F(u) + \lambda G_1(u) = \int_a^b (u + \lambda\sqrt{(1 + u'^2)})\,dx;$$

$$u(a) = u(b) = 0.$$

The functional J is a special case of the functional for Problem S, with $\Phi(x, u, u') = u + \lambda\sqrt{(1+u'^2)}$. Euler's equation for the functional J, $(\operatorname{grad} J)(u) = 0$, together with the formula (4.8), takes the form

$$\frac{d}{dx} \frac{\lambda u'}{\sqrt{(1+u'^2)}} - 1 = 0.$$

Integration gives

$$\frac{u'}{\sqrt{(1+u'^2)}} = \frac{x-c}{\lambda}.$$

Hence

$$u' = -\frac{x-c}{\sqrt{\{\lambda^2-(x-c)^2\}}}.$$

Integrating again, we arrive at the equation of a circle of radius λ:

$$(x-c)^2 + (u-c_1)^2 = \lambda^2.$$

Thus, if the solution exists it is an arc of a circle. Its radius λ and centre (c, c_1) are defined by the three equations

$$u(a) = 0, \quad u(b) = 0, \quad \int_a^b \sqrt{(1+u'^2)}\, dx = l.$$

If we take the mid-point of the interval $[a, b]$ as the origin of coordinates, we shall have $a = -b$; our equations then take the form

$$(c+b)^2 + c_1^2 = \lambda^2, \quad (c-b)^2 + c_1^2 = \lambda^2,$$

$$\int_{-b}^{b} \frac{\lambda}{\sqrt{\{\lambda^2-(x-c)^2\}}}\, dx = l.$$

Hence $c = 0$, $c_1 = \sqrt{(\lambda^2-b^2)}$, $2\lambda \arcsin(b/\lambda) = l$. In the special case when $l = \pi b$, we have $\lambda = b$, and the solution is a semi-circle.

§ 7. Minimizing Sequence

Let F be an arbitrary functional which is bounded from below. Then there exists a lower bound to the value of F

$$\mu = \inf_{u \in D(F)} F(u).$$

The sequence $\{\mu_n\}$ of elements of $D(F)$ is called a *minimizing sequence* for the functional F, if there exists a limit $F(u_n)$ equal to μ.

THEOREM 3.7.1. *A functional which is bounded from below has at least one minimizing sequence.*

From the definition of the lower bound it follows that: 1) for an arbitrary element $u \in D(F)$, the inequality $F(u) \geq \mu$ holds; 2) for arbitrary $\varepsilon > 0$ there exists an element $u^{(\varepsilon)}$ of $D(F)$, such that $F(u^{(\varepsilon)}) < \mu + \varepsilon$. We put $\varepsilon = 1/n$ and write $u^{(1/n)} = u_n$. Then

$$\mu \leq F(u_n) < \mu + \frac{1}{n},$$

from which it follows that $\lim F(u_n) = \mu$.

THEOREM 3.7.2. *Let $D(F)$ be a linear manifold of some Banach space X. If the functional F is continuous in $D(F)$, and if the minimizing sequence has the limit $u_0 = \lim u_n$, then the element u_0 ascribes to the function F its minimum value.*

The proof is very simple: because of the continuity of the functional

$$F(u_0) = \lim F(u_n) = \mu = \inf F(u_n).$$

The theorems of the present section make it possible to by-pass Euler's equation in solving the problem of minimizing a functional. To do this it is necessary first of all to embed the set $D(F)$ in a Banach space X, such that the functional F is continuous in this space. Next we have to construct a minimizing sequence. If this converges (in the sense of convergence in the space X), then its limit solves the variational problem.

Exercises

1. Denote by $C_0^{(1)}[a, b]$ the set of functions which are continuously differentiable on the segment $[a, b]$ and equal to zero at the points a and b. The functional

$$F(u) = \int_a^b \Phi(x, u(x), u'(x)) \, dx$$

is assigned on the set $C_0^{(1)}[a, b]$, which is embedded in the Hilbert space H. This is defined in the following way: the elements of the space H are functions, absolutely continuous on $[a, b]$, equal to zero at the points a and b, and having square-integrable first derivatives. The scalar product and the norm in H are given by the formulae

$$(u, v)_H = \int_a^b u'(x)v'(x)\,dx;$$

$$\|u\|_H^2 = \int_a^b u'^2(x)\,dx.$$

The functions $\Phi(x, u, u')$ are continuously differentiable in the domain $x \in [a, b]$, $-\infty < u < +\infty$, $-\infty < u' < +\infty$. Find the expression for grad F and its domain of definition.

2. We consider an isoperimetric problem: to find the minimum of the functional

$$\Phi(u) = \int_0^1 \left[p(x) \left(\frac{du}{dx}\right)^2 + q(x)u^2 \right] dx$$

with conditions

$$u \in C_0^{(1)}[0, 1], \qquad \int_0^1 u^2\,dx = 1.$$

We stipulate that $p(x), p'(x), q(x) \in C[0, 1]$, and that $p(x) \geq p_0 = \text{const.} > 0$, and $q(x) \geq 0$. We put

$$u_n = \sum_{k=1}^n a_k \sin k\pi x$$

and define the coefficients $a_1, a_2, \ldots, a_n$ from the condition

$$\int_0^1 u_n^2(x)\,dx = \tfrac{1}{2} \sum_{k=1}^n a_k^2 = 1$$

and $\Phi(u_n) = \min$. Show that it is possible to construct the functions u_n, and that they constitute a minimizing sequence for the functional Φ.

3. In Exercise 2, take u_n to be given by

$$u_n = x(1-x) \sum_{k=0}^n a_k x^k, \qquad n = 0, 1, 2, \ldots.$$

The coefficients $a_0, a_1, a_2, \ldots, a_n$ are, as before, defined by the conditions

$$\int_0^1 u_n^2(x)\,dx = 1, \qquad \Phi(u_n) = \min.$$

Prove the same results as in Exercise 2.

Chapter 4. Functionals Depending on Numerical Functions of Real Variables

§ 1. The Simplest Problem of the Variational Calculus (Problem S)

We have already considered this problem in § 4 of Chapter 3. It is concerned with minimizing the integral

$$F(u) = \int_a^b \Phi(x, u, u')\,dx, \qquad (1)$$

where the function $u(x)$ is subject to the boundary conditions

$$u(a) = A, \qquad u(b) = B, \qquad (2)$$

and A, B are given constants. The integrand function $\Phi(x, u, u')$ was assumed to be continuous and to have continuous partial derivatives with respect to u and u' in the domain

$$a \leq x \leq b, \qquad -\infty < u < +\infty, \qquad -\infty < u' < +\infty. \qquad (3)$$

As the domain of definition $D(F)$ of the functional F we took a linear manifold of functions from $C^{(1)}[a, b]$, satisfying the conditions (2); and we took the manifold $D(F)$ to be part of the space $L_2(a, b)$. It was then shown that the gradient of the functional F was defined on those functions $u \in D(F)$, and only those, for which the function

$$\Phi_{u'}(x, u, u')$$

is absolutely continuous on the segment $[a, b]$, as well as having a square-integrable derivative; the gradient itself is defined by the formula

$$(\operatorname{grad} F)(u) = \Phi_u - \frac{d}{dx}\Phi_{u'}. \qquad (4)$$

Euler's equation for the functional (1) reduces to the differential equation

$$(\operatorname{grad} F)(u) = \Phi_u - \frac{d}{dx}\Phi_{u'} = 0 \qquad (5)$$

with boundary conditions (2). Thus, if Problem S has a solution, then this solution must satisfy the differential equation (5) and boundary conditions (2). The solutions of equation (5) are usually called *extremals* of the functional (1).

We impose some additional restrictions on the integrand function $\Phi(x, u, u')$. We shall require that in the domain (3) the function should have continuous partial derivatives of the first order with respect to all the variables x, u, u', continuous partial derivatives $\Phi_{xu'}, \Phi_{uu'}, \Phi_{u'^2}$, and that $\Phi_{u'^2} \neq 0$. We shall show that under these assumptions every extremal of the functional (1) has a continuous second derivative. First we show that, with the restrictions we have imposed on the function Φ, any function of the domain D (grad F) has a square-integrable second derivative almost everywhere.

Let $u \in D$ (grad F). Then, a fortiori, $u \in D(F)$, and, consequently, $u \in C^{(1)}[a, b]$. At the same time, $(d/dx)\Phi_{u'} \in L_2(a, b)$. We write

$$\frac{d}{dx} \Phi_{u'}(x, u, u') = \omega(x).$$

Then the limit

$$\lim_{\Delta x \to 0} \frac{\Phi_{u'}(x+\Delta x, u+\Delta u, u'+\Delta u') - \Phi_{u'}(x, u, u')}{\Delta x} = \omega(x),$$

exists almost everywhere; here $\Delta u = u(x+\Delta x) - u(x)$, $\Delta u' = u'(x+\Delta x) + -u'(x)$. By Lagrange's formula,

$$\frac{\Phi_{u'}(x+\Delta x, u+\Delta u, u'+\Delta u') - \Phi_{u'}(x, u, u')}{\Delta x} =$$

$$= \Phi_{xu'}(x+\theta\Delta x, u+\theta\Delta u, u'+\theta\Delta u')$$

$$+ \Phi_{uu'}(x+\theta\Delta x, u+\theta\Delta u, u'+\theta\Delta u') \frac{\Delta u}{\Delta x}$$

$$+ \Phi_{u'^2}(x+\theta\Delta x, u+\theta\Delta u, u'+\theta\Delta u') \frac{\Delta u'}{\Delta x}, \qquad 0 < \theta < 1.$$

From this we obtain the ratio $\Delta u'/\Delta x$; this can be done because $\Phi_{u'^2} \neq 0$. For the sake of brevity, we omit the arguments in the second derivatives:

$$\frac{\Delta u'}{\Delta x} = \frac{1}{\Phi_{u'^2}} \left\{ \frac{\Phi_{u'}(x+\Delta x, u+\Delta u, u'+\Delta u') - \Phi_{u'}(x, u, u')}{\Delta x} - \Phi_{xu'} - \Phi_{uu'} \frac{\Delta u}{\Delta x} \right\}.$$

As $\Delta x \to 0$, the right-hand side of the latter expression has a limit almost everywhere, which is equal to

$$\frac{1}{\Phi_{u'^2}(x, u, u')} \left\{ \frac{d}{dx} \Phi_{u'}(x, u, u') - \Phi_{xu'}(x, u, u') - \Phi_{uu'}(x, u, u')u' \right\} =$$

$$= \frac{1}{\Phi_{u'^2}(x, u, u')} \{\omega(x) - \Phi_{xu'}(x, u, u') - \Phi_{uu'}(x, u, u')u'\}.$$

But then the square-integrable second derivative

$$u'' = \frac{1}{\Phi_{u'^2}(x, u, u')} \{\omega(x) - \Phi_{xu'}(x, u, u') - \Phi_{uu'}(x, u, u')u'\},$$

exists almost everywhere.

If u is an extremal, then, by virtue of Euler's equation, $\omega(x) = \Phi_u(x, u, u')$, and

$$u'' = \frac{1}{\Phi_{u'^2}(x, u, u')} \{\Phi_u(x, u, u') - \Phi_{xu'}(x, u, u') - \Phi_{uu'}(x, u, u')u'\} \qquad (6)$$

from the restrictions imposed on Φ it follows that this is a function which is continuous on the segment $[a, b]$.

Equation (6) represents a modified form of Euler's equation. Thus, in the case of Problem S the extremal is defined by a second-order differential equation. Its general integral contains two arbitrary constants. We denote this general integral by $\phi(x, C_1, C_2)$. The arbitrary constants are to be determined from the equations

$$\phi(a, C_1, C_2) = A, \qquad \phi(b, C_1, C_2) = B, \qquad (7)$$

which follow from the boundary conditions (2).

§ 2. Discussion of the Second Variation

In the case of Problem S the second variation has the form

$$\delta^2 F(u, \eta) = \frac{1}{2} \left[\frac{d^2}{d\alpha^2} \int_a^b \Phi(x, u(x) + \alpha\eta(x), u'(x) + \alpha\eta'(x)) \, dx \right]_{\alpha=0}.$$

We can differentiate under the integral sign, and thus obtain

$$\delta^2 F(u, \eta) = \frac{1}{2} \int_a^b [\Phi_{u^2}(x, u, u')\eta^2(x) + 2\Phi_{uu'}(x, u, u')\eta(x)\eta'(x) +$$
$$+ \Phi_{u'^2}(x, u, u')\eta'^2(x)] \, dx. \qquad (1)$$

Let the function $u_0(x)$ satisfy equation (1.5) and the boundary conditions (1.2). It was shown in § 5, Chapter 3, that in order for this function to give the functional (1.1) a relative minimum, it is necessary (and sufficient) that the inequality $\delta^2 F(u_0 + \theta\eta, \eta) \geq 0$ should hold for elements $\eta \in M$ which are sufficiently small in the norm. We observe that in the present case M is a set of functions of $C^{(1)}[a, b]$ which are equal to zero at the points a, b. This set is denoted by $C_0^{(1)}[a, b]$.

We demonstrate that the necessary condition for a minimum is the inequality $\delta^2 F(u_0, \eta) \geq 0$, $\forall \eta \in C_0^{(1)}[a, b]$, or, writing it in expanded form,

$$\int_a^b [\Phi_{u^2}(x, u_0, u_0')\eta^2 + 2\Phi_{uu'}(x, u_0, u_0')\eta\eta' + \Phi_{u'^2}(x, u_0, u_0')\eta'^2] \, dx \geq 0,$$

$$\forall \eta \in C_0^{(1)}[a, b]. \qquad (2)$$

Suppose that in fact, for some function $\eta_0 \in C_0^{(1)}[a, b]$, we have $\delta^2 F(u_0, \eta_0) = -\gamma < 0$. Putting $\eta = \varepsilon\eta_0$, where ε is a number whose absolute value is sufficiently small, we obtain

$$\delta^2 F(u_0, \eta) = -\gamma\varepsilon^2.$$

Moreover,

$$\delta^2 F(u_0 + \theta\eta, \eta) - \delta^2 F(u_0, \eta) =$$

$$= \tfrac{1}{2}\varepsilon^2 \int_a^b \{[\Phi_{u^2}(x, u_0 + \theta\varepsilon\eta_0, u_0' + \theta\varepsilon\eta_0') - \Phi_{u^2}(x, u_0, u_0')]\eta_0^2$$

$$+ 2[\Phi_{uu'}(x, u_0 + \theta\varepsilon\eta_0, u_0' + \theta\varepsilon\eta_0') - \Phi_{uu'}(x, u_0, u_0')]\eta_0 \eta_0'$$

$$+ [\Phi_{u'^2}(x, u_0 + \theta\varepsilon\eta_0, u_0' + \theta\varepsilon\eta_0') - \Phi_{u'^2}(x, u_0, u_0')]\eta'^2\} \, dx.$$

For sufficiently small ε, this last integral will be as small as we please. Let us choose ε such that the absolute value of the integral turns out to be less than γ. Then

$$|\delta^2 F(u_0 + \theta\eta, \eta) - \delta^2 F(u_0, \eta)| < \tfrac{1}{2}\gamma\varepsilon^2.$$

Hence

$$\delta^2 F(u_0 + \theta\eta, \eta) < \delta^2 F(u_0, \eta) + \tfrac{1}{2}\gamma\varepsilon^2 = -\tfrac{1}{2}\gamma\varepsilon^2 < 0,$$

and so the necessary condition for a minimum is violated.

The necessary condition (2) can be simplified; in fact, the following theorem holds.

THEOREM 4.2.1 (LEGENDRE'S CONDITION). *In order that condition (2) be satisfied for any $\eta \in C_0^{(1)}[a, b]$, it is necessary that $\Phi_{u'^2}(x, u_0(x), u_0'(x)) \geq 0$ for any $x \in [a, b]$.*

For simplicity we put

$$\Phi_{u'^2}(x, u_0(x), u_0'(x)) = p(x),$$
$$\Phi_{uu'}(x, u_0(x), u_0'(x)) = q(x),$$
$$\Phi_{u^2}(x, u_0(x), u_0'(x)) = r(x).$$

Condition (2) takes the form

$$\int_a^b [p(x)\eta'^2 + 2q(x)\eta\eta' + r(x)\eta^2]\,dx \geq 0, \qquad \eta \in C_0^{(1)}[a, b]. \qquad (3)$$

We denote by $I(\eta)$ the integral in (3). We assume that at some point $x_0 \in [a, b]$, $p(x_0) < 0$. Then $p(x) < 0$ on some interval which contains the point x_0. Hence we can find a segment $[\alpha, \beta]$ contained in this interval, and a positive constant p_0, such that $p(x) < -p_0$, $x \in [\alpha, \beta]$. We now put

$$\eta_n(x) = \begin{cases} 0, & a \leq x \leq \alpha, \\ \sin^2 \dfrac{n\pi(x-a)}{\beta-\alpha}, & \alpha < x < \beta, \\ 0, & \beta \leq x \leq b, \end{cases}$$

where n is a positive integer. Obviously, $\eta_n \in C_0^{(1)}[a, b]$. We now have

$$I(\eta_n) = \int_a^\beta [p(x)\eta_n'^2 + 2q(x)\eta_n\eta_n' + r(x)\eta_n^2]\,dx.$$

The continuous functions $q(x)$ and $r(x)$ are bounded. Let $|q(x)| < q_0$, $|r(x)| < r_0$, where q_0 and r_0 are constants. Then

$$I(\eta_n) < -p_0 \frac{n^2\pi^2}{2(\beta-\alpha)} + q_0\pi n + r_0(\beta-\alpha);$$

the right-hand side of this inequality is negative for sufficiently large n. Thus the theorem is proved.

§ 3. Case of Several Independent Variables

Consider a finite domain Ω in m-dimensional Euclidean space. We shall assume that the boundary Γ of the domain Ω consists of a finite number of piecewise smooth, $(m-1)$-dimensional surfaces: in this case the integration by parts formula (cf. § 1, Chap. 2) is valid for the domain Ω.

Let the function
$$\Phi(x, u, z_1, z_2, \ldots, z_m) \tag{1}$$
be defined, when $x \in \overline{\Omega} = \Omega \cup \Gamma$, and $u, z_1, z_2, \ldots, z_n$ are numerical variables which can assume arbitrary finite values. We shall require that, in the indicated domain of variation of the independent variables, the function Φ should be continuous, together with its partial derivatives of the first and second order, with respect to all the variables $x_1, x_2, \ldots, x_m$, $u, z_1, z_2, \ldots, z_m$.

Consider the functional
$$F(u) = \int_\Omega \Phi\left(x, u(x), \frac{\partial u}{\partial x_1}, \ldots, \frac{\partial u}{\partial x_m}\right) dx. \tag{2}$$

We prescribe this functional on the set $D(F)$ of functions $u \in C^{(1)}(\overline{\Omega})$, satisfying the boundary condition
$$u|_\Gamma = g(x), \tag{3}$$
where $g(x)$ is a function defined and continuous on the surface Γ. We shall assume that there exists at least one function $\bar{u}(x)$ which satisfies both the restrictions
$$\bar{u} \in C^{(1)}(\overline{\Omega}), \qquad \bar{u}|_\Gamma = g(x). \tag{4}$$

Note that this last assumption is essential: in contrast with the case of one variable, it is possible here to choose a function $g(x)$, continuous on Γ, such that no function $\bar{u}(x)$, satisfying the conditions (4), will exist. In this case the domain $D(F)$ is empty, and the problem of minimizing the functional (2) loses its meaning.

If the function $\bar{u}$ exists, then the set $D(F)$ contains a linear manifold of functions of the form $u(x) = \bar{u}(x) + \eta(x)$, where the function $\eta \in \mathfrak{M}^{(1)}(\Omega)$ (cf. § 1, Chap. 2).

We shall regard the manifold $D(F)$ as a part of the space $L_2(\Omega)$. It follows at once from Corollary 1.3.1 that the set $\mathfrak{M}^{(1)}(\Omega)$ is dense in $L_2(\Omega)$. It is then not difficult to show that the functional (2) satisfies Restrictions 1, 2 of § 2, Chap. 3.

We formulate the problem of minimizing the functional (2), and derive the necessary condition. We know the most general necessary condition (§ 4, Chap. 3): if the element u_0 renders the value of the functional F a minimum, then $u_0 \in D(\mathrm{grad}\, F)$, and $(\mathrm{grad}\, F)(u_0) = 0$.

We construct the variation

$$\delta F(u, \eta) = \frac{d}{d\alpha} F(u+\alpha\eta)|_{\alpha=0}$$

$$= \frac{d}{d\alpha} \int_\Omega \Phi(x, u+\alpha\eta, u_1+\alpha\eta_1, \ldots, u_m+\alpha\eta_m)\,dx|_{\alpha=0}$$

$$= \int_\Omega \left[\frac{\partial \Phi}{\partial u}\eta + \sum_{k=1}^{m} \frac{\partial \Phi}{\partial u_k}\eta_k\right] dx. \tag{5}$$

Here we have used the notation

$$u_k = \frac{\partial u}{\partial x_k}, \qquad \eta_k = \frac{\partial \eta}{\partial x_k};$$

in the derivatives of Φ we have omitted the arguments $x, u(x), u_1(x), \ldots, u_k(x)$.

The function $u \in D(F)$ belongs to the domain $D(\operatorname{grad} F)$ if and only if the integral on the right-hand side of equation (5) is a functional bounded in $L_2(\Omega)$. The integral under consideration splits up into $(m+1)$ integrals:

$$\delta F(u, \eta) = \int_\Omega \frac{\partial \Phi}{\partial u} \eta\,dx + \sum_{k=1}^{m} \int_\Omega \frac{\partial \Phi}{\partial u_k} \eta_k\,dx$$

$$= \left(\frac{\partial \Phi}{\partial u}, \eta\right) + \sum_{k=1}^{m} \left(\frac{\partial \Phi}{\partial u_k}, \eta_k\right).$$

The functions $\partial\Phi/\partial u$, $\partial\Phi/\partial u_k$ are continuous in $\overline{\Omega}$; what is more, they belong to $L_2(\Omega)$, since the first scalar product is bounded in $L_2(\Omega)$; but in general we cannot say anything about the remaining products.

Let the function $u \in D(F)$ be such that there exist generalized derivatives

$$\frac{\partial}{\partial x_k} \frac{\partial \Phi}{\partial u_k}, \qquad k = 1, 2, \ldots, m. \tag{6}$$

Then, by formula (1.1), Chap. 2,

$$\int_\Omega \frac{\partial \Phi}{\partial u_k} \eta_k\,dx = \int_\Omega \frac{\partial \Phi}{\partial u_k} \frac{\partial \eta}{\partial x_k}\,dx = -\int_\Omega \eta \frac{\partial}{\partial x_k} \frac{\partial \Phi}{\partial u_k}\,dx,$$

and, consequently,

$$\delta F(u, \eta) = \int_\Omega \left[\frac{\partial \Phi}{\partial u} - \sum_{k=1}^{m} \frac{\partial}{\partial x_k} \frac{\partial \Phi}{\partial u_k}\right] \eta\,dx, \qquad \forall \eta \in \mathfrak{M}^{(1)}(\Omega). \tag{7}$$

Suppose also that

$$\sum_{k=1}^{m} \frac{\partial}{\partial x_k} \frac{\partial \Phi}{\partial u_k} \in L_2(\Omega). \tag{8}$$

Then

$$\delta F(u, \eta) = \left(\frac{\partial \Phi}{\partial u} - \sum_{k=1}^{m} \frac{\partial}{\partial x_k} \frac{\partial \Phi}{\partial u_k}, \eta \right)$$

is a functional of η, bounded in $L_2(\Omega)$; if so, then $u \in D(\operatorname{grad} F)$, and

$$(\operatorname{grad} F)(u) = \frac{\partial \Phi}{\partial u} - \sum_{k=1}^{m} \frac{\partial}{\partial x_k} \frac{\partial \Phi}{\partial u_k}, \qquad u|_\Gamma = g(x). \tag{9}$$

Thus, the domain of definition of the gradient of F in all cases contains functions which have the following properties: 1) they belong to $C^{(1)}(\overline{\Omega})$ and satisfy condition (3); 2) their generalized derivatives (6) exist; 3) condition (8) is fulfilled. In particular, the set $D(\operatorname{grad} F)$ contains those functions from $C^{(2)}(\overline{\Omega})$ which satisfy condition (3).

It is now easy to write down the Euler equation for our variational problem, if we assume that the function which yields a minimum of the functional (2) has the properties just enumerated. Euler's equation consists of the differential equation

$$\frac{\partial \Phi}{\partial u} - \sum_{k=1}^{m} \frac{\partial}{\partial x_k} \frac{\partial \Phi}{\partial u_k} = 0, \qquad u_k = \frac{\partial u}{\partial x_k} \tag{10}$$

and the boundary condition (3).

Example 1. Let

$$F(u) = \int_\Omega \sum_{k=1}^{m} \left(\frac{\partial u}{\partial x_k} \right)^2 dx, \qquad u|_\Gamma = g(x). \tag{11}$$

Euler's differential equation for (11) has the form

$$\sum_{k=1}^{m} \frac{\partial^2 u}{\partial x_k^2} = 0 \tag{12}$$

or, in brief, $\Delta u = 0$. This equation has to be integrated with the boundary condition $u|_\Gamma = g(x)$; here it is assumed that there exists a function $\bar{u}(x)$ satisfying the conditions (4). The integral (11) is usually called *Dirichlet's integral*.

Example 2. Let

$$F(u) = \int_\Omega \left[\sum_{k=1}^m \left(\frac{\partial u}{\partial x_k}\right)^2 - 2fu \right] dx, \qquad u|_\Gamma = 0. \tag{13}$$

Euler's differential equation for the functional (13) can easily be expressed in the form

$$-\Delta u = f(x); \tag{14}$$

this has to be integrated subject to the boundary condition $u|_\Gamma = 0$. In this case the existence of the function $\bar{u}(x)$ is trivial: we can put $\bar{u}(x) \equiv 0$.

It is possible to give a complete description of the domain of definition of the gradient of the functional (2) if we introduce the concept of generalized divergence of a vector, in conformity with the analogous concept of generalized derivative (Chap. 2). If $g(x)$ is a vector, continuously differentiable in $\overline{\Omega}$, and having components $g_1(x), g_2(x), \ldots, g_m(x)$, and $\zeta(x)$ is an arbitrary function of the class $\mathfrak{M}^{(1)}(\Omega)$ (§ 1, Chap. 2), then from a well-known formula of vector analysis,

$$\int_\Omega g(x) \cdot \operatorname{grad} \zeta(x) \, dx = - \int_\Omega \zeta(x) \operatorname{div} g(x) \, dx.$$

Now let $g(x)$ be an arbitrary vector, integrable in Ω, and let there exist a scalar function $f(x)$, integrable in Ω, such that for any function $\zeta \in \mathfrak{M}^{(1)}(\Omega)$, the equation

$$\int_\Omega g(x) \cdot \operatorname{grad} \zeta(x) \, dx = - \int_\Omega f(x) \zeta(x) \, dx \tag{15}$$

holds. Then the function $f(x)$ is called the generalized divergence of the vector $g(x)$ in the domain Ω.

Let the function $u \in D(F)$ be such that the vector whose components are $\partial\Phi/\partial u_1, \partial\Phi/\partial u_2, \ldots, \partial\Phi/\partial u_m$ has a generalized divergence which is square-integrable in Ω. We shall denote this divergence, just as in the ordinary case, by the symbol

$$\sum_{k=1}^m \frac{\partial}{\partial x_k} \frac{\partial \Phi}{\partial u_k},$$

keeping in mind, however, that the terms in this last sum have no meaning if taken separately.

Obviously, $\eta \in \mathfrak{M}^{(1)}(\Omega)$; with the aid of formula (15), we transform the variation $\delta F(u, \eta)$ to the form (7); this variation is, by virtue of our assumptions, a bounded functional of η, and, consequently, $u \in D$ (grad F).

Conversely, let $u \in D$ (grad F). Then the expression

$$\sum_{k=1}^{m} \int_{\Omega} \frac{\partial \Phi}{\partial u_k} \eta_k \, dx, \qquad \eta \in \mathfrak{M}^{(1)}(\Omega)$$

is a functional of η, bounded in $L_2(\Omega)$. By Riesz's Theorem there exists a function $v \in L_2(\Omega)$, such that

$$\sum_{k=1}^{m} \int_{\Omega} \frac{\partial \Phi}{\partial u_k} \eta_k \, dx = (v, \eta) = \int_{\Omega} v(x) \eta(x) \, dx; \qquad \forall \eta \in \mathfrak{M}^{(1)}(\Omega);$$

in keeping with the definition, the vector $(\partial \Phi / \partial u_1, \partial \Phi / \partial u_2, \ldots, \partial \Phi / \partial u_m)$ has a generalized divergence, $v \in L_2(\Omega)$.

Thus, the function $u \in D(F)$ belongs to the domain D (grad F) if and only if the vector $(\partial \Phi / \partial u_1, \partial \Phi / \partial u_2, \ldots, \partial \Phi / \partial u_m)$ has a generalized divergence which is square-integrable in Ω; in this connection it is easy to see that

$$(\text{grad } F)(u) = \frac{\partial \Phi}{\partial u} - \sum_{k=1}^{m} \frac{\partial}{\partial x_k} \frac{\partial \Phi}{\partial u_k}.$$

Euler's equation retains the form (10); it is only necessary to keep in mind that, in general, the terms in the summation do not have any meaning when considered separately.

§ 4. Functionals Depending on Higher-Order Derivatives

Consider a functional of the form

$$F(u) = \int_a^b \Phi(x, u, u', u'', \ldots, u^{(k)}) \, dx. \tag{1}$$

For simplicity we assume that the function $\Phi(x, z_0, z_1, \ldots, z_k)$ is defined in the domain of variation of the quantities

$$x \in [a, b]; \qquad -\infty < z_j < \infty, \qquad j = 0, 1, 2, \ldots, k, \tag{2}$$

and is k-times continuously differentiable in this domain. The functional (1) is prescribed on functions $u \in C^{(k)}[a, b]$, which satisfy the boundary conditions

$$\begin{aligned} u(a) &= A_0, & u'(a) &= A_1, \ldots, u^{(k-1)}(a) = A_{k-1}, \\ u(b) &= B_0, & u'(b) &= B_1, \ldots, u^{(k-1)}(b) = B_{k-1}, \end{aligned} \tag{3}$$

where A_j, B_j are given constants. In the role of $\bar{u}(x)$ we can take a polynomial of degree $2k-1$, satisfying the conditions (3); it is well-known that such a polynomial can be constructed. It is now clear that $D(F)$ is a linear manifold of functions of the form $u(x) = \bar{u}(x) + \eta(x)$, where $\eta \in \mathfrak{M}^{(k-1)}(a, b)$; if we regard $D(F)$ as part of the space $L_2(a, b)$, then it is easy to show that the functional F satisfies Restrictions 1 and 2 of §2, Chap. 3.

We calculate the variation

$$\delta F(u, \eta) = \frac{d}{d\alpha} \int_a^b \Phi(x, u+\alpha\eta, u'+\alpha\eta', \ldots, u^{(k)} + \alpha\eta^{(k)}) \, dx \bigg|_{\alpha=0}$$

$$= \int_a^b \left[\frac{\partial \Phi}{\partial u} \eta(x) + \sum_{j=1}^k \frac{\partial \Phi}{\partial u^{(j)}} \eta^{(j)}(x) \right] dx. \tag{4}$$

Let the function $u(x)$ be such that

$$\frac{\partial \Phi}{\partial u^{(j)}} (x, u(x), u'(x) \ldots u^{(k)}(x)), \qquad j = 1, 2, \ldots, k,$$

has a generalized derivative of j-th order *, and that the sum

$$\sum_{j=1}^k (-1)^j \frac{d^j}{dx^j} \frac{\partial \Phi}{\partial u^{(j)}} \in L_2(a, b). \tag{5}$$

From formula (1.1), Chap. 2, we then have

$$\int_a^b \frac{\partial \Phi}{\partial u^{(j)}} \eta^{(j)}(x) \, dx = (-1)^j \int_a^b \eta(x) \frac{d^j}{dx^j} \frac{\partial \Phi}{\partial u^{(j)}} \, dx,$$

and, consequently,

$$\delta F(u, \eta) = \int_a^b \left[\frac{\partial \Phi}{\partial u} + \sum_{j=1}^k (-1)^j \frac{d^j}{dx^j} \frac{\partial \Phi}{\partial u^{(j)}} \right] \eta(x) \, dx. \tag{6}$$

Integral (6) is a functional of η, bounded in $L_2(\Omega)$; it follows from this that a function $u(x)$ having the above-mentioned properties belongs to the domain $D(\operatorname{grad} F)$, and for such a function,

$$(\operatorname{grad} F)(u) = \frac{\partial \Phi}{\partial u} + \sum_{j=1}^k (-1)^j \frac{d^j}{dx^j} \frac{\partial \Phi}{\partial u^{(j)}}. \tag{7}$$

* By Theorem 2.4.2 this means that the function in question is continuous on the segment $[a, b]$, together with its derivatives up to and including $(j-1)$-th order, and, moreover, the derivative of $(j-1)$-th order is absolutely continuous on this segment.

Assuming that the function which minimizes the functional (1) under conditions (3) does exist and does have the above-mentioned properties, we can write down Euler's equation for this function. The equation consists of a differential equation of order $2k$,

$$\frac{\partial \Phi}{\partial u} + \sum_{j=1}^{k}(-1)^j \frac{d^j}{dx^j} \frac{\partial \Phi}{\partial u^{(j)}} = 0 \tag{8}$$

and the boundary conditions (3).

The statements of the present section can be extended in an obvious way to the case of several independent variables. For the functional

$$\int_{\Omega} \Phi\left(x, u, \frac{\partial u}{\partial x_1}, \ldots, \frac{\partial u}{\partial x_m}, \frac{\partial^2 u}{\partial x_1^2}, \ldots, \frac{\partial^k u}{\partial x_m^k}\right) dx \tag{9}$$

with boundary conditions

$$u|_\Gamma = g_0(x), \quad \frac{\partial u}{\partial \nu}\bigg|_\Gamma = g_1(x), \ldots, \frac{\partial^{k-1} u}{\partial \nu^{k-1}} = g_{k-1}(x), \tag{10}$$

where ν is the normal to Γ, Euler's equation consists of the boundary conditions (10) and a partial differential equation of order $2k$:

$$\frac{\partial \Phi}{\partial u} - \sum_{j=1}^{k} \frac{\partial}{\partial x_j} \frac{\partial \Phi}{\partial(\partial u/\partial x_j)} + \sum_{j_1, j_2=1}^{k} \frac{\partial^2}{\partial x_{j_1} \partial x_{j_2}} \frac{\partial \Phi}{\partial(\partial^2 u/\partial x_{j_1} \partial x_{j_2})} + \cdots$$

$$+(-1)^k \sum_{j_1, j_2, \ldots, j_k=1}^{k} \frac{\partial^k}{\partial x_{j_1} \partial x_{j_2} \ldots \partial x_{j_k}} \frac{\partial \Phi}{\partial(\partial^k u/\partial x_{j_1} \partial x_{j_2} \ldots \partial x_{j_k})} = 0. \tag{11}$$

Thus, for example, for the functional

$$\int_{\Omega} \sum_{j,k=1}^{m} \left(\frac{\partial^2 u}{\partial x_j \partial x_k}\right)^2 dx$$

Euler's differential equation has the form

$$2 \sum_{j,k=1}^{m} \frac{\partial^2}{\partial x_j \partial x_k} \frac{\partial^2 u}{\partial x_j \partial x_k} = 0,$$

which is easily converted into the form

$$\Delta^2 u = 0.$$

§ 5. Functionals Depending on Several Functions

For simplicity we restrict ourselves to the case of one independent variable and two functions; we shall assume, moreover, that the functional depends on derivatives of these functions of order no higher than the first. Extension to the general case is not difficult.

Thus, we consider the functional

$$F(u, v) = \int_a^b \Phi(x, u, v, u', v')\,dx. \qquad (1)$$

We prescribe this functional on the pair $\{u, v\}$ of functions of $C^{(1)}[a, b]$, which satisfy the boundary conditions

$$u(a) = A_1, \quad u(b) = B_1, \quad v(a) = A_2, \quad v(b) = B_2, \qquad (2)$$

where A_1, A_2, B_1, B_2 are constants. As usual, we denote the set of these pairs by $D(F)$. Each of the pairs we shall call a vector.

We construct the vector $\{\bar{u}, \bar{v}\}$, where

$$\bar{u}(x) = A_1 + \frac{B_1 - A_1}{b-a}(x-a), \quad \bar{v}(x) = A_2 + \frac{B_2 - A_2}{b-a}(x-a).$$

Then any vector $\{u, v\} \in D(F)$ can be expressed in the form

$$\{u, v\} = \{\bar{u}+\eta, \bar{v}+\zeta\},$$

where $\eta(x)$ and $\zeta(x)$ are functions of $C^{(1)}[a, b]$, satisfying the boundary conditions

$$\eta(a) = \eta(b) = \zeta(a) = \zeta(b) = 0. \qquad (3)$$

The set of vectors $\{\eta, \zeta\}$ is obviously linear, and $D(F)$ is a linear manifold.

We shall regard $D(F)$ as part of the space $L_2(a, b)$ of vector-functions, which are square-integrable on the interval (a, b); the scalar product and the norm in $L_2(a, b)$ are given by the relations

$$(\{u_1, v_1\}, \{u_2, v_2\}) = \int_a^b (u_1 u_2 + v_1 v_2)\,dx;$$

$$\|\{u, v\}\|^2 = \int_a^b (u^2 + v^2)\,dx.$$

It can easily be verified that the functional (1) satisfies Restrictions 1 and 2 of § 2, Chapter 3.

The variation of the functional (1) has the form

$$\delta F(u, v, \eta, \zeta) = \frac{d}{d\alpha} \int_a^b \Phi(x, u+\alpha\eta, \; v+\alpha\zeta, \; u'+\alpha\eta', \; v'+\alpha\zeta') dx|_{\alpha=0}$$

$$= \int_a^b \left(\frac{\partial \Phi}{\partial u} \eta + \frac{\partial \Phi}{\partial v} \zeta + \frac{\partial \Phi}{\partial u'} \eta' + \frac{\partial \Phi}{\partial v'} \zeta' \right) dx.$$

Repeating the arguments used in investigating Problem S, we can easily prove the following statements:

1) the domain of definition of the gradient of F consists of those, and only those, vectors $\{u, v\}$ which are contained in the domain $D(F)$, and for which $\partial \Phi / \partial u'$ and $\partial \Phi / \partial v'$ are functions absolutely continuous on the segment $[a, b]$, whose derivatives on this segment are square-integrable.

2) the gradient of the functional F is a vector, defined by the formula

$$(\text{grad } F)(\{u, v\}) = \left\{ \frac{\partial \Phi}{\partial u} - \frac{d}{dx} \frac{\partial \Phi}{\partial u'}, \; \frac{\partial \Phi}{\partial v} - \frac{d}{dx} \frac{\partial \Phi}{\partial v'} \right\}; \qquad (4)$$

3) Euler's equation for the functional F reduces to the system of differential equations

$$\frac{\partial \Phi}{\partial u} - \frac{d}{dx} \frac{\partial \Phi}{\partial u'} = 0, \quad \frac{\partial \Phi}{\partial v} - \frac{d}{dx} \frac{\partial \Phi}{\partial v'} = 0 \qquad (5)$$

and the boundary conditions (2).

§ 6. Natural Boundary Conditions

1. In order to explain the concept of natural boundary conditions, and the origin of this concept, we consider the following problem of the variational calculus.

Required to find the minimum of the functional

$$F(u) = \int_a^b \Phi(x, u, u') dx \qquad (1)$$

on the set of functions $u \in C^{(1)}[a, b]$; no boundary conditions are imposed in advance on the function $u(x)$. The function $\Phi(x, u, u')$ is subject to the same conditions as in Problem S.

The important difference between our new problem and Problem S is that, now, no boundary conditions are imposed on the function with respect to which a minimum is sought.

In the present case $D(F) = C^{(1)}[a, b]$. This set is linear, and it can be regarded as a linear manifold (we can put $\bar{u} = 0$). We shall consider it as part of the space $L_2(a, b)$. It is easily verified that the functional (1) satisfies Restrictions 1 and 2 of § 2, Chapter 2.

Let us find the gradient of the functional (1). We prove the following theorem.

THEOREM 4.6.1. *In order that the function $u(x)$ should belong to the domain of definition of the gradient of the functional* (1), *it is necessary and sufficient that it satisfy the following conditions*: 1) $u \in C^{(1)}[a, b]$; 2) *the function $\partial \Phi / \partial u' = \Phi_{u'}(x, u(x), u'(x))$ is absolutely continuous and has a square-integrable derivative on the segment $[a, b]$*; 3) *the function $u(x)$ satisfies the following boundary conditions*:

$$\left(\frac{\partial \Phi}{\partial u'}\right)_{x=a} = \Phi_{u'}(a, u(a), u'(a)) = 0;$$
$$\left(\frac{\partial \Phi}{\partial u'}\right)_{x=b} = \Phi_{u'}(b, u(b), u'(b)) = 0. \qquad (2)$$

Necessity. Let $u \in D(\operatorname{grad} F)$. Condition 1) is necessary; this follows simply from the fact that $D(\operatorname{grad} F) \subset D(F) = C^{(1)}(a, b)$. We turn to conditions 2) and 3). As in § 3, Chapter 3, the variation of the functional (1) is given by

$$\delta F(u, \eta) = \frac{d}{d\alpha} \int_a^b \Phi(x, u+\alpha\eta, u'+\alpha\eta')\, dx|_{\alpha=0} = \int_a^b \left(\frac{\partial \Phi}{\partial u}\eta + \frac{\partial \Phi}{\partial u'}\eta'\right) dx.$$

As before, this will be a bounded functional of η, if the functional

$$\int_a^b \frac{\partial \Phi}{\partial u'} \eta'\, dx \qquad (3)$$

is also bounded.

Let $u \in D(\operatorname{grad} F)$. Then the functional (3) is bounded, and there exists a function $g \in L_2(a, b)$, such that

$$\int_a^b \frac{\partial \Phi}{\partial u'} \eta'(x)\, dx = \int_a^b g(x)\eta(x)\, dx, \qquad \forall \eta \in C^{(1)}[a, b]. \qquad (4)$$

We introduce the function

$$G(x) = -\int_a^x g(t)\, dt + C, \qquad C = \text{const}.$$

Integrating by parts the right-hand side of (4), we obtain

$$\int_a^b \left[\frac{\partial \Phi}{\partial u'} - G(x)\right] \eta'(x)\,dx = G(a)\eta(a) - G(b)\eta(b). \tag{5}$$

Equation (5) holds for any function $\eta \in C^{(1)}[a, b]$. In addition we suppose that

$$\eta(a) = \eta(b) = 0. \tag{6}$$

Then

$$\int_a^b \left[\frac{\partial \Phi}{\partial u'} - G(x)\right] \eta'(x)\,dx = 0. \tag{7}$$

Equation (7) coincides with equation (3.22), Chapter 3, and from it we can draw the same conclusion:

$$\frac{\partial \Phi}{\partial u'} = G(x) + \text{const.} = -\int_a^x g(t)\,dt + \text{const.} \tag{8}$$

This proves the necessity of statement 2) of the theorem.

We can now integrate by parts in (3):

$$\int_a^b \frac{\partial \Phi}{\partial u'} \eta'\,dx = -\int_a^b \frac{d}{dx}\frac{\partial \Phi}{\partial u'} \eta\,dx + \eta(b)\left(\frac{\partial \Phi}{\partial u'}\right)_{x=b} - \eta(a)\left(\frac{\partial \Phi}{\partial u'}\right)_{x=a},$$

and the following expression is obtained for the variation:

$$\delta F(u, \eta) = \int_a^b \left[\frac{\partial \Phi}{\partial u} - \frac{d}{dx}\frac{\partial \Phi}{\partial u'}\right] \eta\,dx + \eta(b)\left(\frac{\partial \Phi}{\partial u'}\right)_{x=b} - \eta(a)\left(\frac{\partial \Phi}{\partial u'}\right)_{x=a}. \tag{9}$$

By hypothesis, the integral in (9) is a functional of η bounded in $L_2(a, b)$; the expression

$$\eta(b)\left(\frac{\partial \Phi}{\partial u'}\right)_{x=b} - \eta(a)\left(\frac{\partial \Phi}{\partial u'}\right)_{x=a},$$

$$\eta \in C^{(1)}[a, b] \tag{10}$$

must also be a functional bounded in $L_2(a, b)$. But the value of the function $\eta(x)$ at a given point is a functional of η, not bounded in $L_2(a, b)$. Hence it follows that the expression (10) will be a functional bounded in $L_2(a, b)$, only if

$$\left(\frac{\partial \Phi}{\partial u'}\right)_{x=a} = \left(\frac{\partial \Phi}{\partial u'}\right)_{x=b} = 0.$$

Thus the necessity of the condition of the theorem is proved.

From equation (9) now follows the formula for the gradient:

$$(\text{grad } F)(u) = \frac{\partial \Phi}{\partial u} - \frac{d}{dx} \frac{\partial \Phi}{\partial u'}. \tag{11}$$

Sufficiency. If the conditions 1)–3) of the theorem are fulfilled, then it is possible to integrate (3) by parts. Utilizing the boundary conditions (2), we obtain

$$\delta F(u, \eta) = \int_a^b \left[\frac{\partial \Phi}{\partial u} - \frac{d}{dx} \frac{\partial \Phi}{\partial u'} \right] \eta \, dx. \tag{12}$$

The expression in square brackets is an element of the space $L_2(a, b)$, the integral (12) is a functional bounded in this space, and, consequently, $u \in D$ (grad F). The theorem is fully established.

Thus, in the problem we have been considering, functions of the domain D (grad F) necessarily satisfy some boundary conditions (in the present case – condition (2)), which functions of the domain $D(F)$ are not obliged to satisfy. Conditions of this kind are called *natural* for the functional F.

In our case the boundary conditions (2) are natural for the functional F.

Boundary conditions which are satisfied by all functions of the domain $D(F)$ are called *principal* for the functional F. Thus, in the case of Problem S, the boundary conditions (3.10), Chapter 3, are principal.

It is now easy to write down the necessary condition for a minimum-Euler's equation – for the functional (1). Euler's equation in this case reduces to the differential equation

$$\frac{\partial \Phi}{\partial u} - \frac{d}{dx} \frac{\partial \Phi}{\partial u'} = 0 \tag{13}$$

and boundary conditions (2), under which this equation has to be solved.

Remark. Euler's equation could have been obtained without resorting to Theorem 4.6.1.

Let the function u_0 make our functional a minimum. Put $u_0(a) = A$, $u_0(b) = B$. We take a curve which passes through the points (a, A), (b, B) (Fig. 7), i.e. which has $u(a) = A$, $u(b) = B$, and compare the extremal with this curve. The latter is denoted by the broken line in Fig. 7. Then $F(u) \geq F(u_0)$, and we have again the conditions for Problem S. But then the function $u_0(x)$ has to satisfy equation (13). This equation, however, is not sufficient; we must also compare u_0 with those curves which do not pass through the points (a, A) and (b, B), for example with the curve $y = u_1(x)$ in Fig. 7.

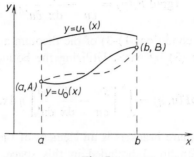

Fig. 7.

Together with u_0 we consider functions $u_0 + \alpha\eta$, where α is an arbitrary real number, and η is a continuously differentiable function, not subject to any boundary conditions. Then $F(u_0 + \alpha\eta) \geqq F(u_0)$, and, as usual,

$$\delta F(u, \eta) = \int_a^b [\Phi_u \eta + \Phi_{u'} \eta'] \, dx.$$

Integrating by parts the second term on the right-hand side, we obtain

$$\delta F(u_0, \eta) = \Phi_{u'} \eta \big|_a^b + \int_a^b \left[\Phi_u - \frac{d}{dx} \Phi_{u'} \right] \eta \, dx,$$

and then the condition for a minimum takes the form:

$$\Phi_{u'}\big|_{x=b}\, \eta(b) - \Phi_{u'}\big|_{x=a}\, \eta(a) = 0.$$

Because of the arbitrariness of η we can impose $\eta(b) = 1$, $\eta(a) = 0$. Then $\Phi_{u'}\big|_{x=b} = 0$. Similarly, we obtain $\Phi_{u'}\big|_{x=a} = 0$.

2. We consider three examples.

1. The functional

$$F_1(u) = \int_a^b [p(x)u'^2 + q(x)u^2 - 2f(x)u] \, dx \qquad (14)$$

defined on $C^{(1)}[a, b]$, is a special case of the functional (1). We assume that the functions $p(x), p'(x), q(x), f(x)$ are continuous * on the segment $[a, b]$,

---

* These requirements could be relaxed.

and that $p(x) > 0$ *. The gradient of the functional (14) is defined on functions of $C^{(1)}[a, b]$, for which the product $p(x) u'(x)$ is absolutely continuous and has a square-integrable derivative on the segment $[a, b]$ **, and which satisfy the natural boundary conditions

$$u'(a) = u'(b) = 0; \tag{15}$$

the gradient itself is equal to

$$2\left[q(x)u - \frac{d}{dx}\left(p(x)\frac{du}{dx}\right) - f(x)\right]. \tag{16}$$

Euler's relation for the functional (14) consists of the totality of boundary conditions (15) and the differential equation

$$-\frac{d}{dx}\left(p(x)\frac{du}{dx}\right) + q(x)u = f(x). \tag{17}$$

2. We consider the functional

$$F_2(u) = \int_a^b [p(x)u'^2 + q(x)u^2 - 2f(x)u]\,dx + \alpha u^2(a) + \beta u^2(b);$$

$$u \in C^{(1)}[a, b]. \tag{18}$$

Here α and β are constants, and the functions $p(x), q(x), f(x)$ are subject to the same conditions as in Example 1. Because of the presence of the last two terms, the functional F_2 is not included under the type (1); but the approach we have been using is suitable even in this case.

We find the domain D (grad F_2), and an expression for grad F_2. The variation of the functional (18) is given by

$$\delta F_2(u, \eta) = 2\left\{\int_a^b [p(x)u'\eta' + q(x)u\eta - f(x)\eta]\,dx + \right.$$

$$\left. + \alpha u(a)\eta(a) + \beta u(b)\eta(b)\right\}; \quad \eta \in C^{(1)}[a, b]. \tag{19}$$

Let $u \in D$ (grad F_2). Then the right-hand side in (19) is a functional of η bounded in $L_2(a, b)$. It retains this property, of course, if we subject η

---

* The condition $p(x) > 0$ represents a strengthened version, for the functional (14), of Legendre's condition, § 2, which in the present case has the form $p(x) \geq 0$.

** As a consequence of the properties of the function $p(x)$, it is easy to show that this condition is equivalent to the following: on the segment $[a, b]$ the derivative $u'(x)$ is absolutely continuous, and u'' is square-integrable.

to any additional restrictions. Let us suppose, for the moment, that

$$\eta(a) = \eta(b) = 0. \tag{20}$$

Then

$$\delta F_2(u, \eta) = 2\int_a^b [p(x)u'\eta' + q(x)u\eta - f(x)\eta]\,dx;$$

we have arrived at the conditions for the simplest problem of the variational calculus for the functional (14), and we can make the following assertion: if $u \in D$ (grad F_2), then $u'(x)$ is absolutely continuous, and $(p(x)u'(x))'$ is square-integrable * on the segment $[a, b]$. Taking this into account, we abandon the additional restrictions (20), and integrate by parts the right-hand side in (19):

$$\delta F_2(u, \eta) = 2\int_a^b [-(p(x)u')' + q(x)u - f(x)]\eta\,dx +$$
$$+ 2\{[p(b)u'(b) + \beta u(b)]\eta(b) - [p(a)u'(a) - \alpha u(a)]\eta(a)\}. \tag{21}$$

The integral in (19) is a functional of η bounded in $L_2(a, b)$, so that the functional in brace-brackets in (21) must also be bounded in $L_2(a, b)$. These arguments, as in the above case, lead us to the relation

$$p(a)u'(a) - \alpha u(a) = 0, \quad p(b)u'(b) + \beta u(b) = 0, \tag{22}$$

which are natural conditions for the functional (18). It is now easy to see that grad F_2 is defined by the expression (16). Euler's relation for the functional F_2 reduces to the family of differential equations (17) and boundary conditions (22).

3. Let us define the functional

$$F_3(u) = \int_0^1 (u^2 + u'^2 + u''^2 - 2f(x)u)\,dx \tag{23}$$

on a set of functions of $C^{(2)}[0, 1]$, which satisfy the boundary conditions

$$u(0) = u(1) = 0. \tag{24}$$

As usual, we shall assume that this set is part of the space $L_2(0, 1)$. The variation of the functional (23) is given by

$$\delta F_3(u, \eta) = 2\int_0^1 (u\eta + u'\eta' + u''\eta'' - f\eta)\,dx. \tag{25}$$

---

* Cf. footnote ** on the preceding page.

Here η is any function of $C^{(2)}[0, 1]$ satisfying the conditions (24), so that
$$\eta(0) = \eta(1) = 0.$$
If $\eta \in D\,(\text{grad}\,F_3)$, then the integral (25) is a functional of η, bounded in $L_2(0, 1)$; it remains bounded if η is subject to the additional conditions
$$\eta'(0) = \eta'(1) = 0. \tag{26}$$
But then we have arrived at the conditions of § 4. In our case
$$\Phi = u^2 + u'^2 + u''^2 - 2f(x)u$$
and, consequently,
$$\frac{\partial \Phi}{\partial u'} = 2u', \quad \frac{\partial \Phi}{\partial u''} = 2u''.$$

As in § 4 we suppose that $\partial \Phi/\partial u'' = 2u''$ has a generalized second derivative. This means that u''' is absolutely continuous, and that $u^{(iv)}$ is integrable. In keeping with the relation (4.5), we require that
$$[-u'' + u^{(iv)}] \in L_2(0, 1).$$
But u'' is simply continuous, so it is only necessary to ask that $u^{(iv)} \in L_2(0, 1)$. Accordingly, as in § 4,
$$(\text{grad}\,F_3)(u) = 2[u - u'' + u^{(iv)} - f(x)]. \tag{27}$$

We now dispense with the conditions (26), and integrate by parts the second and third terms on the right-hand side of (25):
$$\delta F_3(u, \eta) = 2\int_0^1 [u - u'' + u^{(iv)} - f(x)]\eta\,dx + u''(1)\eta'(1) - u''(0)\eta'(0). \tag{28}$$

The integral in (28) is a bounded functional of η, but $\eta'(1)$ and $\eta'(0)$ are easily seen to be unbounded. Therefore, if $u(x) \in D\,(\text{grad}\,F_3)$ and satisfies the differentiability conditions stated above, then $u(x)$ necessarily satisfies the following conditions:
$$u''(0) = u''(1) = 0; \tag{29}$$
and these are natural for the functional (23). Euler's equation for this functional reduces to the differential equation
$$u^{(iv)} - u'' + u = f(x) \tag{30}$$
with boundary conditions (24) and (29).

Chapter 5.
Minimum of a Quadratic Functional

§ 1. The Concept of a Quadratic Functional

In the present chapter we shall be considering functionals whose domain of definition belongs to a real Hilbert space. In some cases (which will be mentioned specifically) we shall assume that this space is separable. We recall that a Banach space is called separable if it contains a dense, enumerable set. For a Hilbert space one can give a different, though equivalent, definition: a Hilbert space is called separable if there is a complete enumerable orthonormal system within it. One of the most important separable Hilbert spaces is the space $L_2(\Omega)$, where Ω is a measurable set in a finite-dimensional space.

Let H be a given Hilbert space. In the space H we consider a *bilinear* functional $\Phi(u, v)$ – this is the name given to a functional which depends on two elements of the space H, and which possesses the following properties: for fixed v it is a linear functional of u:

$$\Phi(\alpha_1 u_1 + \alpha_2 u_2, v) = \alpha_1 \Phi(u_1, v) + \alpha_2 \Phi(u_2, v), \tag{1}$$

and for fixed u it is a linear functional of v:

$$\Phi(u, \alpha_1 v_1 + \alpha_2 v_2) = \alpha_1 \Phi(u, v_1) + \alpha_2 \Phi(u, v_2). \tag{2}$$

In equations (1) and (2), α_1 and α_2 are real numbers.

We shall consider only *symmetric* bilinear functionals, i.e., those for which

$$\Phi(u, v) = \Phi(v, u). \tag{3}$$

The simplest symmetric bilinear functional is the scalar product (u, v) of the elements u and v.

If $\Phi(u, v)$ is a symmetric bilinear functional, we shall call the expression $\Phi(u, u)$ a *homogeneous quadratic functional*, or a *quadratic form*. For brevity we shall write $\Phi(u)$ in place of $\Phi(u, u)$.

We now derive a simple, but important, relation, which is satisfied by any quadratic form. Let $\Phi(u, v)$ be a bilinear functional, and let $\alpha_1, \alpha_2, \beta_1, \beta_2$ be numbers. Applying formulae (1) and (2) successively we obtain

$$\Phi(\alpha_1 u_1 + \alpha_2 u_2, \beta_1 v_1 + \beta_2 v_2) =$$
$$= \alpha_1 \beta_1 \Phi(u_1, v_1) + \alpha_1 \beta_2 \Phi(u_1, v_2) + \alpha_2 \beta_1 \Phi(u_2, v_1) + \alpha_2 \beta_2 \Phi(u_2, v_2).$$

In particular, if the functional Φ is symmetric, then

$$\Phi(u+v, u+v) = \Phi(u, u) + 2\Phi(u, v) + \Phi(v, v)$$

or

$$\Phi(u+v) = \Phi(u) + 2\Phi(u, v) + \Phi(v). \tag{4}$$

This is the required relation.

Example. Dirichlet's integral

$$\Phi(u) = \int_\Omega \sum_{k=1}^m \left(\frac{\partial u}{\partial x_k}\right)^2 dx$$

is a quadratic form, which corresponds to the symmetric bilinear functional

$$\Phi(u, v) = \int_\Omega \sum_{k=1}^m \frac{\partial u}{\partial x_k} \frac{\partial v}{\partial x_k} dx.$$

We shall call the expression

$$F(u) = \Phi(u) - l(u), \tag{5}$$

a *quadratic functional*, when $\Phi(u)$ is a quadratic form, and $l(u)$ is a linear functional.

Example. The simplest Hilbert space is the real axis. Here, scalar multiplication is just the multiplication of numbers, and the norm of a number is its absolute value. A second-degree polynomial without the free term is a quadratic functional.

Another, more important, example, which we shall encounter later, is the quadratic functional

$$F(u) = \int_\Omega \left\{ \sum_{k=1}^m \left(\frac{\partial u}{\partial x_k}\right)^2 - 2f(x)u \right\} dx. \tag{6}$$

§ 2. Positive Definite Operators

1. Throughout our subsequent work, we shall often be considering operators which act in the Hilbert space H. If an operator is denoted by the letter A, say, then its domain of definition will be denoted by $D(A)$, and its domain of values by $R(A)$. In saying that the operator A acts in the space H, we understand, as usual, that $D(A) \subset H$ and $R(A) \subset H$.

In speaking of an operator A acting in the Hilbert space H, we shall always assume that A is a linear * operator and that its domain of definition is dense in H, i.e., $\overline{D(A)} = H$ (here the line over the symbols signifies closure in the metric of the space H).

An operator A acting in Hilbert space is called *symmetric* if $\overline{D(A)} = H$ and if for any $u, v \in D(A)$ the equation

$$(Au, v) = (u, Av) \qquad (1)$$

holds. If A is a symmetric operator then, for $u, v \in D(A)$, (Au, v) is a symmetric bilinear functional and (Au, u) is a quadratic form.

Example 1. In the space $H = L_2(\Omega)$, consider the integral operator

$$Ku = \int_\Omega K(x, y) u(y) \, dy. \qquad (2)$$

We assume that the integral having multiplicity $2m$

$$\int_\Omega \int_\Omega K^2(x, y) \, dx \, dy$$

is finite. This operator is defined on all space (cf. Theorem 7.2.1, below). If $K(x, y) = K(y, x)$, then the operator (2) is symmetric. We now show this.

We form the scalar product

$$(Ku, v) = \int_\Omega v(x) \left\{ \int_\Omega K(x, y) u(y) \, dy \right\} dx.$$

By Fubini's theorem it is permissible to change the order of integration:

$$(Ku, v) = \int_\Omega u(y) \left\{ \int_\Omega K(x, y) v(x) \, dx \right\} dy.$$

Writing x for y, and vice versa, we have

$$(Ku, v) = \int_\Omega u(x) \left\{ \int_\Omega K(y, x) v(y) \, dy \right\} dx$$

$$= \int_\Omega u(x) \left\{ \int_\Omega K(x, y) v(y) \, dy \right\} dx = (Kv, u) = (u, Kv)$$

---

* That is, additive and homogeneous, but possibly unbounded.

since, in real space, the order of multiplication in a scalar product can be reversed.

Example 2. In the space $H = L_2(0, 1)$, consider the operator

$$Au = -\frac{d^2u}{dx^2}. \tag{3}$$

Let $D(A)$ consist of functions u satisfying the following two restrictions:

$$u \in C^{(2)}[0, 1],$$
$$u(0) = u(1) = 0. \tag{4}$$

Obviously the operator A defined in this way is linear. We show that it is symmetric.

The set $D(A)$ of functions of $C^{(2)}[0, 1]$, satisfying the boundary conditions (4), contains in its turn a set, dense in $L_2(0, 1)$, of functions finite on the segment $[0, 1]$. By Corollary 1.3.1, the set $D(A)$ is itself dense in $L_2(0, 1)$.

It remains to be shown that the operator A satisfies the condition of symmetry (1). To do this, we construct the scalar product (Au, v), where $u, v \in D(A)$, i.e. $u, v \in C^{(2)}[0, 1]$, and

$$u(0) = u(1) = 0; \quad v(0) = v(1) = 0.$$

Integrating by parts, and taking into account the fact that the integrated term vanishes because of the above boundary conditions, we obtain

$$(Au, v) = -\int_0^1 v(x)u''(x)\,dx = \int_0^1 u'(x)v'(x)\,dx$$
$$= -\int_0^1 u(x)v''(x)\,dx = (u, Av).$$

2. DEFINITION 1. The symmetric operator A is called *positive* if the quadratic form $(Au, u) \geq 0$, and $(Au, u) = 0$ if and only if $u = 0$.

For example, the operator (3)–(4) is positive. In order to demonstrate this, we construct the quadratic form

$$(Au, u) = -\int_0^1 u \frac{d^2u}{dx^2}\,dx.$$

Integrating by parts, and taking account of the conditions (4), we find

$$(Au, u) = \int_0^1 u'^2(x)\,dx \geq 0. \tag{5}$$

Suppose that $(Au, u) = 0$ and, consequently, $\int_0^1 u'^2 \, dx = 0$. Then $u'(x) \equiv 0$ and $u(x) \equiv \text{const}$. It follows now from condition (4) that $u(x) \equiv 0$.

DEFINITION 2. The symmetric operator A is called *positive definite*, if

$$\inf_{\substack{u \in D(A) \\ u \neq 0}} \frac{(Au, u)}{||u||^2} > 0. \tag{6}$$

This definition is equivalent to the following: a symmetric operator A is called positive definite if there exists a constant $\gamma^2 > 0$, such that

$$(Au, u) \geq \gamma^2 ||u||^2. \tag{7}$$

The inequality (7) will be called the *positive-definiteness inequality*.

It is obvious that every positive definite operator is at the same time positive. The converse is, in general, not true.

Example. We show that the operator (3)–(4) is positive definite. We write the Newton-Leibnitz formula:

$$u(x) - u(0) = \int_0^x u'(t) \, dt. \tag{8}$$

But because of condition (4), $u(0) = 0$, and so

$$u(x) = \int_0^x u'(t) \, dt.$$

By the Schwartz-Bunyakovskii inequality

$$u^2(x) \leq \int_0^x 1 \, dt \cdot \int_0^x u'^2(t) \, dt = x \int_0^x u'^2(t) \, dt \leq \int_0^1 u'^2(t) \, dt.$$

Integrating this last inequality with respect to x between limits 0 and 1, we obtain:

$$||u||^2 = \int_0^1 u^2(x) \, dx \leq \int_0^1 u'^2(x) \, dx. \tag{9}$$

Comparing this with formula (5), we have

$$(Au, u) \geq ||u||^2.$$

The operator A is positive definite; the number γ can be taken equal to one.

3. There exist operators which are positive, but not positive definite. To convince ourselves of this, we consider the following example.

Let the operator B be defined by the formula

$$Bu = -\frac{d^2 u}{dx^2}, \quad 0 < x < \infty. \tag{10}$$

We shall regard B as an operator in the Hilbert space $L_2(0, \infty)$. As the domain of definition $D(B)$ of this operator we take the set of functions which satisfy the following restrictions: 1) $u \in C^{(2)}[0, \infty)$, 2) $u(0) = 0$, 3) for each function $u \in D(B)$ there exists a number a_u such that $u(x) \equiv 0$ for $x > a_u$. Obviously

$$D(B) \subset L_2(0, \infty).$$

We show that the operator B defined in this way is positive, but not positive definite. First of all, we prove that $\overline{D(B)} \in L_2(0, \infty)$. It is sufficient to show that for any function $\phi \in L_2(0, \infty)$, and any number $\varepsilon > 0$, we can find a function $u \in D(B)$, such that $||\phi - u|| < \varepsilon$. The integral

$$\int_0^\infty \phi^2(x) \, dx$$

is finite, since numbers $\delta > 0$ and $N > 0$ can be found such that

$$\int_0^\delta \phi^2(x) \, dx < \tfrac{1}{8}\varepsilon^2, \quad \int_N^\infty \phi^2(x) \, dx < \tfrac{1}{8}\varepsilon^2.$$

We introduce the function

$$\psi(x) = \begin{cases} 0, & 0 \leq x \leq \delta \\ \phi(x), & \delta < x < N, \\ 0, & x \geq N. \end{cases}$$

Clearly, $\psi \in L_2(0, \infty)$; thus

$$||\phi - \psi||^2 = \int_0^\infty (\phi(x) - \psi(x))^2 \, dx = \int_0^\delta \phi^2(x) \, dx + \int_N^\infty \phi^2(x) \, dx < \tfrac{1}{4}\varepsilon^2,$$

and, consequently, $||\phi - \psi|| < \tfrac{1}{2}\varepsilon$.

We now average the function ψ, taking the averaging radius to be $h < \tfrac{1}{2}\delta$, and put $u(x) = \psi_h(x)$. Obviously, $\psi_h(x) \in D(B)$: the function $\psi_h(x)$ is infinitely differentiable, and is equal to zero when $x = 0$ (in fact, for any $x < \tfrac{1}{2}\delta$); finally, the number a_u can be taken equal to $N + \tfrac{1}{2}\delta$. Then

$$||u - \psi||^2 = \int_0^\infty (u(x) - \psi(x))^2 \, dx = \int_0^{N+\frac{1}{2}\delta} (u(x) - \psi(x))^2 \, dx$$

$$= \int_0^{N+\frac{1}{2}\delta} (\psi_h(x) - \psi(x))^2 \, dx.$$

By Theorem 1.3.3 this last integral will be less than $\frac{1}{4}\varepsilon^2$ for sufficiently small h, and, consequently, $\|u-\psi\| < \frac{1}{2}\varepsilon$. Now, by the triangle inequality,

$$\|u-\phi\| \leq \|u-\psi\| + \|\psi-\phi\| < \varepsilon,$$

and our assertion is proved.

It is easy to show that the operator B is symmetric. In this connection, let $u, v \in D(B)$; then each of the functions u and v satisfies conditions 1)–3).

Construct the bilinear functional

$$(Bu, v) = -\int_0^\infty v \frac{d^2 u}{dx^2} dx = -\int_0^N v \frac{d^2 u}{dx^2} dx.$$

Here N is any number greater than a_u and a_v; for $x = N$, both functions u and v, and all their derivatives, are equal to zero.

Integration by parts gives

$$(Bu, v) = \int_0^N u'(x) v'(x) \, dx = \int_0^\infty u'(x) v'(x) \, dx. \tag{11}$$

Similarly,

$$(Bv, u) = \int_0^\infty u'(x) v'(x) \, dx,$$

and, consequently,

$$(Bu, v) = (Bv, u) = (u, Bv),$$

i.e. B is a symmetric operator.

We next show that B is a positive operator. From formula (11) we have

$$(Bu, u) = \int_0^\infty u'^2(x) \, dx \geq 0.$$

Accordingly, if $(Bu, u) = 0$, then

$$\int_0^\infty u'^2(x) \, dx = 0.$$

Since the integrand function is non-negative, we must have $u'(x) \equiv 0$ and $u(x) \equiv \text{const.}$; but $u(0) = 0$, and so, finally $u(x) \equiv 0$.

The operator B is not positive definite. To establish this, we show that the lower bound of the ratio $(Bu, u)/\|u\|^2$ is equal to zero.

We take the sequence of functions

$$u_n(x) = \begin{cases} x(n-x)^3, & \text{if } 0 \leq x \leq n, \\ 0, & \text{if } x > n. \end{cases}$$

It is easy to show that $u_n \in D(B)$. Let us find the norm of u_n. We have

$$||u_n||^2 = \int_0^\infty u_n^2(x)\,dx = \int_0^n x^2(n-x)^6\,dx.$$

Make the substitution $x = nt$:

$$||u_n||^2 = n^9 \int_0^1 t^2(1-t)^6\,dt.$$

This last integral is a positive constant, independent of n; denote it by c_1. Then $||u_n||^2 = c_1 n^9$.

Further,

$$(Bu_n, u_n) = \int_0^\infty u_n'^2(x)\,dx = \int_0^n (n-4x)^2(n-x)^4\,dx.$$

The substitution $x = nt$ gives

$$(Bu_n, u_n) = n^7 \int_0^1 (1-t)^4(1-4t)^2\,dt = c_2 n^7, \quad c_2 = \text{const}.$$

Now

$$\frac{(Bu_n, u_n)}{||u_n||^2} = \frac{c_2}{c_1 n^2} \xrightarrow[n\to\infty]{} 0$$

and, consequently,

$$\inf \frac{(Bu, u)}{||u||^2} = 0.$$

§ 3. Energy Space

1. With every positive definite operator we can associate a Hilbert space, which we shall call the *energy space* of the given operator.

Let H be a Hilbert space, and A an operator positive definite in this space. We construct a new Hilbert space as follows. The elements of the new space include all the elements of the set $D(A)$, and we define on them a new scalar product:

$$[u, v]_A = (Au, v); \quad u, v \in D(A). \tag{1}$$

It is well-known that a scalar product in Hilbert space has to satisfy three axioms:

A. Symmetry *: if (u, v) is the scalar product of elements u and v, then

$$(u, v) = (v, u).$$

---

* We are considering real Hilbert space. The symmetry axiom for a complex Hilbert space is expressed by the equation $(u, v) = \overline{(v, u)}$.

B. Linearity: if λ_1 and λ_2 are numbers, then
$$(\lambda_1 u_1 + \lambda_2 u_2, v) = \lambda_1(u_1, v) + \lambda_2(u_2, v).$$

C. Positivity:
$$(u, u) \geq 0,$$
where $(u, u) = 0$ if, and only if, $u = 0$ (i.e. if u is the zero element of the space).

We now show that the expression $[u, v]_A$, defined by equation (1), satisfies axioms A–C.

A. Symmetry. We have
$$[u, v]_A = (Au, v) = (u, Av) = (Av, u) = [v, u]_A.$$

Here we have utilized the symmetry of the operator A and the symmetry of the scalar product in the original space H.

B. Linearity. We make use of the linearity of the operator A. Then
$$[\lambda_1 u_1 + \lambda_2 u_2, v]_A = (A(\lambda_1 u_1 + \lambda_2 u_2), v)$$
$$= (\lambda_1 A u_1 + \lambda_2 A u_2, v) = \lambda_1(Au_1, v) + \lambda_2(Au_2, v)$$
$$= \lambda_1 [u_1, v]_A + \lambda_2 [u_2, v]_A.$$

C. Positivity. By the positive-definiteness inequality (2.7), we have $[u, u]_A \geq \gamma^2 \|u\|^2 \geq 0$. Moreover, if $[u, u]_A = 0$, i.e. if $(Au, u) = 0$, then it follows from the same inequality (2.7) that $u = 0$. Obviously the reverse is also true: if $u = 0$, it follows that $(Au, u) = 0$ and $[u, u]_A = 0$.

Thus, the expression (1) satisfies all the axioms for scalar product. Taking $[u, v]_A$ as a scalar product, we transform the set $D(A)$ into a Hilbert space. This space could turn out to be incomplete – in which case, we complete it in the usual way. The completed space is called an *energy space*, and will be denoted by H_A.

The new scalar product generates a new norm, which we shall designate by the symbol $\||\ \||_A$:
$$\||u\||_A = \sqrt{[u, u]_A}. \qquad (2)$$
If $u \in D(A)$, then
$$\||u\||_A = \sqrt{(Au, u)}$$
and by the positive-definiteness inequality
$$\|u\| \leq \frac{1}{\gamma} \||u\||_A. \qquad (3)$$

The quantities $[u, v]_A$ and $|||u|||_A$ will be called respectively the *energy product* of the elements u and v, and the *energy norm* of the element u.

On some occasions, when no misunderstanding can possibly arise, we shall drop the subscript A in the designation of the energy product and energy norm, and simply write $[u, v]$ and $|||u|||$.

In the energy space H_A we shall distinguish between "old" elements – the elements of the set $D(A)$, and "new", or ideal, elements, which are the result of completing the space. From well-known theorems of functional analysis we draw the following conclusions.

If u is an ideal element of the space H_A, then there exists a sequence of the old elements $\{u_n\}$ which converge to u in the energy norm:

$$|||u-u_n||| \to 0, \qquad n \to \infty.$$

In this connection the sequence $\{u_n\}$ obviously converges in itself in the energy metric. The set of the old elements is dense in the energy space.

THEOREM 5.3.1. *If the operator A is positive definite, then all the elements of its energy space can be identified with some elements of the original space.*

Proof. It is sufficient to show that a linear isomorphic correspondence can be set up between the elements of the energy space H_A and some elements of the original space H. This means the following: 1) one and only one element $u' \in H$ can correspond to each element $u \in H_A$; 2) if elements $u', v' \in H$ are brought into correspondence with elements $u, v \in H_A$, then the element $\lambda u' + \mu v' \in H$ corresponds to the linear combination $\lambda u + \mu v \in H_A$; 3) distinct elements of the space H are brought into correspondence with distinct elements of the space H_A.

For any element u of the energy space, it is possible to construct a sequence $\{u_n\}$ of old elements, such that $|||u_n-u||| \to 0$ as $n \to \infty$. In fact, the possibility of doing this for an ideal element was noted earlier, while if u is an old element, it is sufficient to put $u_n = u$. Obviously, $u_n - u_m \in D(A)$, and

$$|||u_n - u_m||| \to 0 \qquad \text{for } n, m \to \infty. \tag{4}$$

From the relation (3) between old and new norms,

$$||u_n - u_m|| \leq \frac{1}{\gamma} |||u_n - u_m|||,$$

and the sequence $\{u_n\}$ converges in itself in the sense of the old norm. By virtue of the completeness of the space H, there exists an element $u' \in H$,

such that
$$\|u' - u\| \xrightarrow[n\to\infty]{} 0.$$

This element is brought into correspondence with the element $u \in H_A$.

We now prove the uniqueness of the element u'. Suppose that, instead of the sequence $\{u_n\} \in D(A)$, we take another sequence $\{v_n\} \in D(A)$, such that $\||u - v_n\|| \xrightarrow[n\to\infty]{} 0$. Using analogous arguments we infer the existence of the element $v' \in H$, such that
$$\|v' - v_n\| \xrightarrow[n\to\infty]{} 0.$$

We next show that $u' = v'$. By the triangle inequality,
$$\||u_n - v_n\|| = \||(u_n - u) - (v_n - u)\|| \leq \||u_n - u\|| + \||v_n - u\|| \xrightarrow[n\to\infty]{} 0.$$

Since $(u_n - v_n) \in D(A)$,
$$\|u_n - v_n\| \leq \frac{1}{\gamma} \||u_n - v_n\|| \xrightarrow[n\to\infty]{} 0.$$

Proceeding to the limit as $n \to \infty$, we obtain $\|u' - v'\| = 0$, which was to be proved.

Suppose that to elements $u_1, u_2 \in H_A$, there correspond sequences of elements $u_{1n}, u_{2n} \in D(A)$, such that
$$\||u_1 - u_{1n}\|| \xrightarrow[n\to\infty]{} 0, \qquad \||u_2 - u_{2n}\|| \xrightarrow[n\to\infty]{} 0.$$

Further suppose that to the same elements u_1 and u_2 there correspond elements u'_1 and u'_2 of the space H. The elements u'_1, u'_2 are such that $\|u'_1 - u_{1n}\| \xrightarrow[n\to\infty]{} 0$ and $\|u'_2 - u_{2n}\| \xrightarrow[n\to\infty]{} 0$. But then by the triangle inequality

$$\||(\lambda_1 u_1 + \lambda_2 u_2) - (\lambda_1 u_{1n} + \lambda_2 u_{2n})\|| = \||\lambda_1(u_1 - u_{1n}) + \lambda_2(u_2 - u_{2n})\||$$
$$\leq |\lambda_1| \, \||u_1 - u_{1n}\|| + |\lambda_2| \, \||u_2 - u_{2n}\|| \xrightarrow[n\to\infty]{} 0,$$

$$\|(\lambda_1 u'_1 + \lambda_2 u'_2) - (\lambda_1 u_{1n} + \lambda_2 u_{2n})\| = \|\lambda_1(u_1 - u_{1n}) + \lambda_2(u_2 - u_{2n})\|$$
$$\leq |\lambda_1| \, \|u'_1 - u_{1n}\| + |\lambda_2| \, \|u'_2 - u_{2n}\| \xrightarrow[n\to\infty]{} 0.$$

These two relations signify that to an element $\lambda_1 u_1 + \lambda_2 u_2 \in H_A$ there corresponds an element $\lambda_1 u'_1 + \lambda_2 u'_2 \in H$; they thus establish the linearity of the correspondence.

We next show that to distinct elements $u_1, u_2 \in H_A$ there correspond elements $u'_1, u'_2 \in H$, which are also distinct.

Suppose the contrary: let $u_1' = u_2'$. We show that, if so, then $u_1 = u_2$. Introduce the difference $u_1 - u_2 = v$. Obviously $v \in H_A$, and since the correspondence is linear, then to the element v there corresponds the zero element of the space H: there exists a sequence $v_n \in D(A)$, such that

$$||v_n - 0|| = ||v_n|| \xrightarrow[n \to \infty]{} 0, \qquad |||v - v_n||| \xrightarrow[n \to \infty]{} 0.$$

Let η be an arbitrary element of the set $D(A)$. Because of the continuity of the scalar product,

$$[v_n, \eta] \xrightarrow[n \to \infty]{} [v, \eta].$$

On the other hand,

$$[v_n, \eta] = (v_n, A\eta).$$

Since $v_n \xrightarrow[n \to \infty]{} 0$, then $(v_n, A\eta) \xrightarrow[n \to \infty]{} (0, A\eta) = 0$, and, consequently, $[v, \eta] = 0$. This last equation means that the element v is orthogonal, in the metric of H_A, to the set $D(A)$, which is dense in H_A. But then v is a null element of the energy space, and so $u_1 = u_2$. Hence the theorem is finally proved.

The sets $D(A)$, H_A, H are connected by the relations

$$D(A) \subset H_A \subset H. \tag{5}$$

The inclusion $D(A) \subset H_A$ follows from the fact that H_A is obtained by completing the set $D(A)$, and the inclusion $H_A \subset H$ comes from Theorem 5.3.1.

Since the set $D(A)$ is dense in H, it can be seen from the relation (5) that the set of elements which make up the energy space of a positive definite operator is dense in the original space.

Earlier we obtained the inequality (3), establishing the relationship between the two norms of elements in the set $D(A)$:

$$||u|| \leq \frac{1}{\gamma} |||u|||, \qquad u \in D(A).$$

We show that this inequality holds for any element of the energy space. Let $u \in H_A$. There exists a sequence of elements $u_n \in D(A)$, such that

$$|||u_n - u||| \xrightarrow[n \to \infty]{} 0, \qquad ||u_n - u|| \xrightarrow[n \to \infty]{} 0.$$

The inequality (3) holds for the elements u_n:

$$||u_n|| \leq \frac{1}{\gamma} |||u_n|||.$$

Proceeding to the limit as $n \to \infty$, and making use of the continuity of the norm, we obtain

$$\|u\| \leq \frac{1}{\gamma} \|\|u\|\|, \quad u \in H_A.$$

This is what we set out to prove.

The energy space was introduced with the help of equation (1):

$$[u, v] = (Au, v), \quad u, v \in D(A).$$

We now show that this equation is valid in the more general case $u \in D(A)$, $v \in H_A$.

If $v \in H_A$, then there exists a sequence $\{v_n\}$,

$$v_n \in D(A), \quad \|\|v_n - v\|\| \xrightarrow[n \to \infty]{} 0, \quad \|v_n - v\| \xrightarrow[n \to \infty]{} 0.$$

Equation (1) is valid for the elements u and v_n:

$$[u, v_n] = (Au, v_n).$$

From the continuity of the scalar product,

$$[u, v_n] \xrightarrow[n \to \infty]{} [u, v], \quad (Au, v_n) \xrightarrow[n \to \infty]{} (Au, v).$$

Comparing right-hand sides, we find

$$[u, v] = (Au, v); \quad u \in D(A), \quad v \in H_A. \tag{6}$$

THEOREM 5.3.2. *Let A be a positive definite operator, acting in the Hilbert space H. In order that an element $u \in H$ should belong to the energy space H_A, it is necessary and sufficient that there exists a sequence $u_n \in D(A)$, $n = 1, 2, \ldots$ such that*

$$\|\|u_n - u_m\|\| \xrightarrow[n, m \to \infty]{} 0, \quad \|u_n - u\| \xrightarrow[n \to \infty]{} 0. \tag{*}$$

Necessity. Let $u \in H_A$. The set $D(A)$ is dense in H_A, so that there exists a sequence $u_n \in D(A)$, $n = 1, 2, \ldots$, such that $\|\|u_n - u\|\|_A \xrightarrow[n \to \infty]{} 0$. Being convergent, this sequence converges in itself; this leads to the first of relations (*). The second relation follows from inequality (3):

$$\|u_n - u\| \leq \frac{1}{\gamma} \|\|u_n - u\|\|_A \xrightarrow[n \to \infty]{} 0.$$

Sufficiency. Suppose the conditions (*) are fulfilled. The space H_A is complete, hence there exists an element $\hat{u} \in H_A$, such that $|||u_n - \hat{u}|||_A \xrightarrow[n \to \infty]{} 0$. Then from the isomorphic correspondence established in the course of proving Theorem 5.3.1, it follows that $u = \hat{u}$ and, consequently, $u \in H_A$.

2. As an example we shall find the energy space for the operator A of § 2. We recall that in this case $H = L_2(0, 1)$, and the operator is defined by the formula

$$Au = -\frac{d^2 u}{dx^2},$$

where the functions u of $D(A)$ satisfy the conditions

$$u \in C^{(2)}[0, 1], \qquad u(0) = u(1) = 0.$$

We show that in the present case the space H_A consists of those functions, and only those functions, which possess the following properties: 1) they are absolutely continuous on the segment [0, 1]; 2) their first derivatives are square-integrable on this segment; 3) the functions are equal to zero at the points $x = 0$ and $x = 1$.

As we saw in § 2,

$$[u, v]_A = \int_0^1 u'(x) v'(x) \, dx; \qquad u, v \in D(A).$$

Putting $v = u$ here, we obtain the formula for the norm:

$$|||u|||_A^2 = \int_0^1 u'^2(x) \, dx, \qquad u \in D(A). \tag{7}$$

1. Let u be an arbitrary element of the space H_A. By Theorem 5.3.1, $u \in L_2(0, 1)$ and there exists a sequence $\{u_n\}$, $u_n \in D(A)$, such that

$$|||u_n - u||| \xrightarrow[n \to \infty]{} 0, \qquad ||u_n - u|| \xrightarrow[n \to \infty]{} 0.$$

Since this sequence converges to the element u, it converges in itself, and so

$$|||u_n - u_m||| \xrightarrow[m, n \to \infty]{} 0.$$

But $u_n - u_m \in D(A)$, and this difference obeys equation (7). Hence

$$\int_0^1 (u_n'(x) - u_m'(x))^2 \, dx \xrightarrow[m, n \to \infty]{} 0.$$

This last relation, which can be put in the form

$$\|u'_n - u'_m\|^2 \underset{m,n\to\infty}{\to} 0$$

shows that the sequence of derivatives $\{u'_n\}$ converges in itself in the metric of $L_2(0, 1)$. The space $L_2(0, 1)$ is complete, hence there exists a function $w \in L_2(0, 1)$, such that

$$\|u'_n - w\| \underset{n\to\infty}{\to} 0.$$

The relations

$$\|u_n - u\| \underset{n\to\infty}{\to} 0, \qquad \|u'_n - w\| \underset{n\to\infty}{\to} 0$$

together with Theorem 2.3.1, lead to the conclusion that the function $u(x)$ has a generalized first derivative $u'(x) = w(x)$; being an element of the space $L_2(0, 1)$, this derivative is square-integrable on the segment $[0, 1]$. From Theorem 2.4.1 it follows that the function is absolutely continuous on the same segment.

It remains to show that $u(0) = u(1) = 0$. The functions $u_n(x)$ belong to the set $D(A)$, and therefore satisfy the analogous relations

$$u_n(0) = u_n(1) = 0.$$

By the Newton-Leibnitz formula

$$u_n(x) = u_n(0) + \int_0^x u'_n(t)\,dt = \int_0^x u'_n(t)\,dt. \tag{8}$$

If $n \to \infty$, then $u_n(x) \to u(x)$ in the metric of $L_2(0, 1)$. We show that, at the same time,

$$\int_0^x u'_n(t)\,dt \underset{n\to\infty}{\to} \int_0^x w(t)\,dt$$

uniformly on the segment $[a, b]$. In fact, by the Schwartz-Bunyakovskii inequality,

$$\left[\int_0^x u'_n(t)\,dt - \int_0^x w(t)\,dt\right]^2 = \left[\int_0^x [u'_n(t) - w(t)]\,dt\right]^2$$

$$\leq \int_0^x 1^2\,dt \int_0^x [u'_n(t) - w(t)]^2\,dt = x \int_0^x [u'_n(t) - w(t)]^2\,dt$$

$$\leq \int_0^1 [u'_n(t) - w(t)]^2\,dt = \|u'_n - w\|^2 \underset{n\to\infty}{\to} 0.$$

Proceeding to the limit in the relation (8), we obtain

$$u(x) = \int_0^x w(t)\,dt.$$

Hence $u(0) = 0$. If we had written the Newton-Leibnitz formula in the form

$$u_n(x) = u_n(1) - \int_x^1 u_n'(t)\,dt = -\int_x^1 u_n'(t)\,dt,$$

then in the same way we would have found that $u(1) = 0$.

2. Now let the function $u(x)$ satisfy the restrictions formulated above: it is absolutely continuous on the segment $[0, 1]$, it has a derivative $u' \in L_2(0, 1)$ almost everywhere, and, finally, $u(0) = u(1) = 0$. We show that $u \in H_A$.

Because of Theorem 5.3.2, it is sufficient to show that there exists a sequence $\{u_n\}$ of functions of $D(A)$, such that

$$|||u_n - u_m||| \underset{n,\,m \to \infty}{\to} 0 \quad \text{and} \quad ||u_n - u|| \underset{n \to \infty}{\to} 0.$$

The function $u(x)$ has a derivative in $L_2(0, 1)$. We expand it in a Fourier cosine series:

$$u'(x) = \sum_{k=0}^{\infty} a_k \cos k\pi x.$$

In fact, the series has no zero term since

$$a_0 = \int_0^1 u'(x)\,dx = u(1) - u(0) = 0,$$

and so

$$u'(x) = \sum_{k=1}^{\infty} a_k \cos k\pi x.$$

We integrate this last equation between the limits 0 and x; the series converges in the mean, and can be integrated term-by-term. Taking into account that $u(0) = 0$, we obtain

$$u(x) = \sum_{k=1}^{\infty} b_k \sin k\pi x,$$

where

$$b_k = \frac{a_k}{k\pi}.$$

We construct the sequence of functions

$$u_n(x) = \sum_{k=1}^{n} b_k \sin k\pi x.$$

Obviously $u_n \in D(A)$. Because of the convergence of the Fourier series,

$$\|u_n - u\| \xrightarrow[n \to \infty]{} 0.$$

It is necessary to show that $\||u_n - u_m\|| \xrightarrow[n,m \to 0]{} 0$. Without loss of generality we can assume that $n > m$. Then

$$u_n(x) - u_m(x) = \sum_{k=m+1}^{n} b_k \sin k\pi x.$$

We write the square of the energy norm of the difference $u_n - u_m$:

$$\||u_n - u_m\||^2 = \int_0^1 \left(\sum_{k=m+1}^{n} a_k \cos k\pi x \right)^2 dx = \tfrac{1}{2} \sum_{k=m+1}^{n} a_k^2 \xrightarrow[n,m \to \infty]{} 0.$$

Thus we have shown that $u \in H_A$.

§ 4. Problem of the Minimum of a Quadratic Functional

Let A be a positive definite operator in the Hilbert space H, and f a given element of this space. The quadratic functional

$$F(u) = (Au, u) - 2(u, f) \tag{1}$$

obviously has a domain of definition which coincides with the domain of definition of the operator A, so that $D(F) = D(A)$. The functional (1) will be called the *energy functional* for the operator A.

We now pose the problem of minimizing the energy functional on the set $D(A)$. We prove the following theorem:

THEOREM 5.4.1. *In order that some element $u_0 \in D(A)$ should minimize the value of the energy functional, it is necessary and sufficient that this element satisfy the equation*

$$Au_0 = f. \tag{2}$$

This element is unique.

Necessity. If the element u_0 makes the functional F a minimum, then $(\operatorname{grad} F)(u_0) = 0$. Let us find $\operatorname{grad} F$.

We have, from the definition of the variation,

$$\delta F(u, \eta) = \frac{d}{d\alpha} F(u+\alpha\eta)|_{\alpha=0} = \frac{d}{d\alpha}[(A(u+\alpha\eta), u+\alpha\eta) - 2(u+\alpha\eta, f)]_{\alpha=0}.$$

Removing the brackets, and making use of the symmetry of the operator A, we obtain

$$F(u+\alpha\eta) = F(u) + 2\alpha(Au-f, \eta) + \alpha^2(A\eta, \eta). \tag{3}$$

Hence

$$\delta F(u, \eta) = 2(Au-f, \eta). \tag{4}$$

The right-hand side of equation (4) is the scalar product of the fixed element $Au-f$ with the arbitrary element $\eta \in D(F)$. This scalar product is a functional, bounded in H. It is clear from this that

$$D(\operatorname{grad} F) = D(F) = D(A)$$

and

$$(\operatorname{grad} F)(u) = 2(Au-f); \tag{5}$$

Euler's equation $(\operatorname{grad} F)(u_0) = 0$ agrees with equation (2), which is what had to be shown.

Sufficiency. Let u_0 satisfy equation (2). If u is an arbitrary element of $D(A)$, different from u_0, then we can put $u = u_0 + \eta$, $\eta \neq 0$. In equation (3) we replace u by u_0, and put $\alpha = 1$. Taking into account equation (2), we obtain

$$F(u) = F(u_0) + (A\eta, \eta).$$

But A is a positive definite operator, and $\eta \neq 0$, so that $(A\eta, \eta) > 0$, and, consequently, $F(u) > F(u_0)$. This last relation shows that the functional F attains a minimum at the point u_0.

It remains to prove the uniqueness of the element u_0. Let the functional F attain a minimum again on the element u_1. By what we have just shown, $F(u_1) > F(u_0)$. But in exactly the same way we can show that $F(u_0) > F(u_1)$. Thus we have arrived at a contradiction, which shows that the functional (1) can attain a minimum at only one point.

Note that we have established the equivalence of the following two problems: solving the equation $Au = f$, and determining the minimum of the energy functional

$$F(u) = (Au, u) - 2(u, f);$$

once the solution to one of these problems has been found, then so has the solution to the other. However, it has not been shown that these problems, contained in Theorem 5.4.1, have a solution. In fact, a solution may indeed not exist, as the following example shows.

Let $H = L_2(0, 1)$, and let A in equation (2) denote the operator considered in § 2:

$$Au = -\frac{d^2u}{dx^2}, \tag{6}$$

where $D(A)$ consists of functions $u \in C^{(2)}[0, 1]$, satisfying the conditions

$$u(0) = u(1) = 0. \tag{7}$$

In this example, to solve the equation

$$Au = f$$

means the following: $f(x)$ is a square-integrable function; it is required to find a function $u(x)$, satisfying conditions (7), and having a continuous second derivative which differs from $f(x)$ only in sign. But this is evidently impossible if, say, $f(x)$ is discontinuous.

This example shows that the problem can admit a solution if the domain of definition of the operator A is extended in a reasonable way; in the example, it is sufficient to include in $D(A)$ functions having absolutely continuous first derivatives and square-integrable second derivatives; the conditions (7), it is understood, are retained.

If $f \in L_2(0, 1)$, then the equation $Au = f$ now has a solution. In fact, this equation implies that $u(x)$ satisfies conditions (7) and the differential equation

$$-\frac{d^2u}{dx^2} = f(x).$$

Such a function exists and is equal to

$$u(x) = x \int_x^1 (1-t)f(t)\,dt + (1-x) \int_0^x tf(t)\,dt;$$

it is easily verified that this function is contained in the extended domain of definition of our operator.

It turns out, however, that it is easier and more convenient to extend the domain of definition, not of the operator A, but of the energy functional associated with it. We shall discuss this in the following section.

§ 5. Generalized Solution

As before, let A be a positive-definite operator, acting in the Hilbert space H, f be a given element of this space, and F the corresponding energy functional

$$F(u) = (Au, u) - 2(u, f). \qquad (1)$$

Equation (1) defines the functional F on the set $D(A)$; it is easy to extend this functional on to all energy space H_A. To do this it is sufficient to observe that $(Au, u) = |||u|||_A^2$, and, consequently,

$$F(u) = |||u|||_A^2 - 2(u, f). \qquad (2)$$

In formula (2) the first term on the right is defined on elements $u \in H_A$. The second term is defined if $u \in H$, and a fortiori if $u \in H_A$. It is now clear that formula (2) allows the definition of the functional F on all energy space H_A.

Returning to our example

$$Au = -\frac{d^2 u}{dx^2}, \quad u \in C^{(2)}[0, 1], \quad u(0) = u(1) = 0,$$

we see that the functional F can be written in the form

$$F(u) = -\int_0^1 u \frac{d^2 u}{dx^2}\, dx - 2\int_0^1 fu\, dx$$

and at the same time in the form

$$F(u) = \int_0^1 u'^2\, dx - 2\int_0^1 fu\, dx,$$

where this second form of writing is suitable for all $u \in H_A$. It also permits the extension of the energy functional on all energy space.

We shall now seek to minimize the functional F, not in $D(A)$, but in H_A. We prove the following theorem:

THEOREM 5.5.1. *In energy space there exists one and only one element for which the energy functional attains a minimum.*

The proof will be based on the following theorem, due to Riesz, and well-known in functional analysis.

Let l be a linear bounded functional in some Hilbert space $\mathfrak{H}$, defined in all the space. Then there exists one and only one element $u_0 \in \mathfrak{H}$, such that

$$lu = (u, u_0)_\mathfrak{H}. \tag{3}$$

The symbol $(\ldots,\ldots)_\mathfrak{H}$ here denotes scalar product in $\mathfrak{H}$.

By Cauchy's inequality, we have

$$|(u,f)| \leq \|u\| \|f\|,$$

and by the relation between old and new norm (inequality (3.3))

$$\|u\| \leq \frac{1}{\gamma} \|\|u\|\|.$$

Hence

$$|(u,f)| \leq c\|\|u\|\|, \quad c = \frac{\|f\|}{\gamma}, \tag{4}$$

and the functional (u,f) is bounded in H_A. By Riesz's theorem, there exists one and only one element $u_0 \in H_A$, such that

$$(u,f) = [u, u_0], \quad u \in H_A. \tag{5}$$

Formula (5) enables us to rewrite the expression for the functional F in the following way:

$$F(u) = \|\|u\|\|^2 - 2[u, u_0] = [u, u] - 2[u, u_0] + [u_0, u_0] - [u_0, u_0]$$
$$= [u - u_0, u - u_0] - [u_0, u_0]$$

or, more simply,

$$F(u) = \|\|u - u_0\|\|^2 - \|\|u_0\|\|^2, \quad u \in H_A. \tag{6}$$

Formula (6) makes it perfectly apparent that the minimum of the functional F in the space H_A is attained on the element $u = u_0$, and only on this element. Hence it is obvious that

$$\min F(u) = -\|\|u_0\|\|^2. \tag{7}$$

This proves the theorem.

The element $u_0 \in H_A$, which achieves a minimum of the functional (2), will be called the *generalized solution* of the equation

$$Au = f. \tag{8}$$

It may happen that $u_0 \in D(A)$; then by Theorem 5.4.1, u_0 will be an ordinary solution of equation (8).

If the energy space is separable, it is possible to indicate a simple way which allows the generalized solution of equation (8) to be constructed. In a separable Hilbert space there exists a complete, enumerable, orthonormal system $\{\omega_n\}$:

$$[\omega_j, \omega_k] = \delta_{jk} = \begin{cases} 0, & j \neq k, \\ 1, & j = k, \end{cases} \quad j, k = 1, 2, \ldots$$

Let u_0 be the generalized solution of equation (8). We expand it in a Fourier series with respect to the system $\{\omega_k\}$:

$$u_0 = \sum_{k=1}^{\infty} [u_0, \omega_k] \omega_k. \tag{9}$$

This series converges in the energy norm: if we put

$$\phi_n = \sum_{k=1}^{n} [u_0, \omega_k] \omega_k,$$

then $|||u_0 - \phi_n||| \xrightarrow[n \to \infty]{} 0$.

The Fourier coefficients $[u_0, \omega_k]$ can easily be evaluated by formula (5): in the latter we put $u = \omega_k$, and find

$$[u_0, \omega_k] = (f, \omega_k). \tag{10}$$

Hence

$$u_0 = \sum_{k=1}^{\infty} (f, \omega_k) \omega_k. \tag{11}$$

As has been said, the series (11) converges in the energy norm of the space H_A; it is easy to see that it also converges in the norm of the original space H. In fact, if as before we denote by ϕ_n the partial sum of the series (11), we have, by inequality (3.3),

$$\|u_0 - \phi_n\| \leq \frac{1}{\gamma} |||u_0 - \phi_n||| \xrightarrow[n \to \infty]{} 0.$$

In connection with Theorem 5.5.1, there arises the question of the conditions for separability of an energy space. This question will be dealt with in the next section.

§ 6. The Separability of Energy Space

As before we shall understand by A a positive definite operator acting in Hilbert space.

LEMMA 5.6.1. *If the sequence $\{f_n\}$ is complete in the original space H, and if $\{\phi_n\}$ is the generalized solution of the equation $A\phi_n = f_n$, then the sequence $\{\phi_n\}$ is complete in the energy space H_A.*

Let $u \in D(A)$. Writing $Au = v$, we have $v \in H$. Introduce the notation

$$\sum_{k=1}^{N} a_k f_k = s_N, \qquad \sum_{k=1}^{N} a_k \phi_k = \sigma_N. \tag{1}$$

We estimate the square of the norm of the difference,

$$|||u-\sigma_N|||^2 = [u-\sigma_N, u-\sigma_N].$$

We put $u-\sigma_N = \eta$. Then

$$[u-\sigma_N, u-\sigma_N] = [u-\sigma_N, \eta] = [u, \eta] - [\sigma_N, \eta].$$

Since $u \in D(A)$, $\eta \in H_A$, therefore

$$[u, \eta] = (Au, \eta) = (v, \eta).$$

Moreover,

$$[\sigma_N, \eta] = \sum_{k=1}^{N} a_k [\phi_k, \eta],$$

but by formula (5.10),

$$[\phi_k, \eta] = (f_k, \eta);$$

and finally

$$[\sigma_N, \eta] = \sum_{k=1}^{N} a_k (f_k, \eta) = (s_N, \eta)$$

and

$$|||u-\sigma_N|||^2 = (v-s_N, \eta). \tag{2}$$

The system $\{f_n\}$ is complete in H, so that we can choose an integer N and coefficients a_k, such that the inequality

$$||v-s_N|| < \varepsilon,$$

holds, where ε is an arbitrary, given, positive number. From formula (2) we now obtain

$$|||u-\sigma_N|||^2 = (v-s_N, \eta) \leq ||v-s_N||\,||\eta|| < \frac{\varepsilon}{\gamma}|||\eta||| = \frac{\varepsilon}{\gamma}|||u-\sigma_N|||.$$

If $|||u-\sigma_N||| \neq 0$, then we now obtain the inequality

$$|||u-\sigma_N||| < \varepsilon/\gamma; \tag{3}$$

obviously this inequality also holds whenever $|||u-\sigma_N||| = 0$. Thus, if the element $u \in D(A)$, then it is possible to approximate it, to any degree of accuracy, by a linear combination of elements of the system $\{\phi_n\}$.

Now let $u \in H_A$. The set $D(A)$ is dense in H_A, so that there exists an element $u' \in D(A)$, such that

$$|||u-u'||| < \tfrac{1}{2}\varepsilon.$$

On the other hand, as we have just shown, there exists an integer N and numbers $a_1, a_2, \ldots, a_N$ such that

$$|||u' - \sum_{k=1}^{N} a_k \phi_k||| < \tfrac{1}{2}\varepsilon.$$

But by the triangle inequality,

$$|||u - \sum_{k=1}^{N} a_k \phi_k||| < \varepsilon,$$

and this proves the lemma.

THEOREM 5.6.1. *In order that the energy space of a positive definite operator should be separable, it is necessary and sufficient that the original space be separable.*

Necessity. Let A be a positive definite operator in the Hilbert space H, and let the energy space H_A be separable. Then in it there exists an enumerable, dense set $\{\psi_n\}$. We show that this set is also dense in the original space H. Let u be some element of the space H. The set of elements of the energy space is dense in the original space, and so if we are given a number $\varepsilon > 0$, we can choose an element $u' \in H_A$, such that $||u-u'|| < \tfrac{1}{2}\varepsilon$. Moreover, we can choose an element ψ_ν, such that

$$|||u' - \psi_\nu||| < \tfrac{1}{2}\varepsilon\gamma.$$

Here γ is the constant of positive-definiteness, contained in the inequality (2.7). By the relation between old and new norms (inequality (3.3)),

$$||u' - \psi_\nu|| < \tfrac{1}{2}\varepsilon,$$

and by the triangle inequality

$$||u - \psi_\nu|| \leq ||u - u'|| + ||u' - \psi_\nu|| < \varepsilon.$$

This last inequality shows that the space H contains a dense enumerable set $\{\psi_n\}$ and, consequently, this space is separable.

Sufficiency. Let the space H be separable, and let the enumerable sequence $\{f_n\}$ be complete in H. We construct elements $\phi_n \in H_A$ – generalized solutions of the equation $A\phi_n = f_n$. By Lemma 5.6.1, the sequence $\{\phi_n\}$ is complete in H_A. We now construct elements of the form

$$\sum_{k=1}^{n} \alpha_k \phi_k, \tag{4}$$

where α_k are rational numbers. The set of such numbers is enumerable. We show that it is dense in H_A.

In fact, given a quantity $\varepsilon > 0$ and an element $u \in H_A$, we can choose an integer $N > 0$ and real numbers a_k, such that

$$|||u - \sum_{k=1}^{N} a_k \phi_k||| < \tfrac{1}{2}\varepsilon.$$

We now choose rational numbers α_k, sufficiently close to the corresponding numbers a_k that

$$\sum_{k=1}^{N} |\alpha_k - a_k| \, |||\phi_k||| < \tfrac{1}{2}\varepsilon.$$

Then by the triangle inequality

$$|||u - \sum_{k=1}^{N} \alpha_k \phi_k||| < \varepsilon$$

and the set (4) is dense in the space H_A. Hence it follows that this space is separable.

§ 7. Extension of a Positive-Definite Operator

Let A be a positive-definite operator, acting in the Hilbert space H. Formula (5.5) brings into correspondence with each element $f \in H$, one and only one element $u_0 \in H_A$, which minimizes the energy functional $F(u)$. By the same token this formula defines some linear operator G:

$$u_0 = Gf, \tag{1}$$

acting in the space H. The domain of definition of this operator $D(G) = H$, and the domain of its values, $R(G)$, is part of the set of elements which constitute the energy space H_A: $R(G) \subset H_A$.

LEMMA 5.7.1. *The operator G is symmetric and bounded.*

We write the formula (5.5) in the form

$$(u,f) = [u, Gf], \quad u \in H_A. \qquad (2)$$

Take an arbitrary element $h \in H$, and put $u = Gh$; then $u \in H_A$. The formula (2) now gives

$$(Gh, f) = [Gh, Gf] = [Gf, Gh].$$

By the same formula (2)

$$[Gf, Gh] = (Gf, h) = (h, Gf).$$

Finally,

$$(Gh, f) = (h, Gf), \qquad (3)$$

and the operator G is symmetric. Further, putting $u = Gf$ in formula (2), we obtain

$$|||Gf|||_A^2 = (Gf, f).$$

Applying Cauchy's inequality to the right-hand side, and noting that the left-hand side is less than $\gamma^2 \|Gf\|^2$, we find

$$\gamma^2 \|Gf\|^2 \leq \|Gf\| \, \|f\|.$$

Hence

$$\|Gf\| \leq \frac{1}{\gamma^2} \|f\|.$$

From this inequality it follows that the operator G is bounded; thus

$$\|G\| \leq 1/\gamma^2. \qquad (4)$$

LEMMA 5.7.2. *There exists an inverse operator to the operator G.*

It is sufficient to show that the equation $Gf = 0$ has a unique solution $f = 0$. Let $Gf = 0$. Formula (2) then gives

$$(u, f) = 0, \quad u \in H_A.$$

The element f turns out to be orthogonal to a set which is dense in H, namely, the set of elements of the energy space. But then $f = 0$.

The operator G^{-1}, the inverse of G, will be denoted by $\tilde{A}$. Obviously $D(\tilde{A}) = R(G) \subset H_A$, and $R(\tilde{A}) = D(G) = H$.

THEOREM 5.7.1. *The operator $\tilde{A}$ is the positive-definite extension of the operator A. The lower bounds of the ratios*

$$\frac{(Au, u)}{\|u\|^2}, \frac{(\tilde{A}u, u)}{\|u\|^2} \tag{5}$$

are equal to each other. The equation

$$\tilde{A}u = f \tag{6}$$

has one and only one solution for any $f \in H$.

Let $u_0 \in D(A)$, and put $Au_0 = f$. By Theorem 5.4.1, the element u_0 achieves a minimum for the functional $F(u) = (Au, u) - 2(f, u)$. By formula (1), $u_0 = Gf$, and, consequently, $\tilde{A}u_0 = f$. Hence it follows that: 1) $u_0 \in D(\tilde{A})$, and, since u_0 is an arbitrary element of the set $D(A)$, so $D(A) \subset D(\tilde{A})$; 2) if $u_0 \in D(A)$, then $Au_0 = \tilde{A}u_0$. Statements 1) and 2) taken together imply that $\tilde{A}$ is the extension of the operator A.

We now show $\tilde{A}$ is a symmetric operator. First of all, its domain of definition is dense in H, since $D(\tilde{A}) \supset D(A)$. Further, we take two arbitrary elements u and v in the domain $D(\tilde{A})$, and put $\tilde{A}u = f$, $\tilde{A}v = h$. Then $u = Gf$, $v = Gh$. Substituting this into formula (3), we obtain the equation

$$(v, \tilde{A}u) = (u, \tilde{A}v),$$

which implies that the operator $\tilde{A}$ is symmetric.

The operator $\tilde{A}$ is the extension of A, and so the set of values of the ratio $(\tilde{A}u, u)/\|u\|^2$ is wider than the corresponding set for the ratio $(Au, u)/\|u\|^2$; hence in this case

$$\inf_{u \in D(\tilde{A})} \frac{(\tilde{A}u, u)}{\|u\|^2} \leq \inf_{u \in D(A)} \frac{(Au, u)}{\|u\|^2}. \tag{7}$$

On the other hand, writing

$$\inf_{u \in D(A)} \frac{(Au, u)}{\|u\|^2} = \gamma_0^2, \qquad \gamma_0^2 > 0, \tag{8}$$

we have $(Au, u) \geq \gamma_0^2 \|u\|^2$, and, consequently, $\|\|u\|\|_A \geq \gamma_0 \|u\|$, $u \in H_A$. In equation (2) we put $u = Gf$, so that $f = \tilde{A}u$:

$$(u, \tilde{A}u) = (\tilde{A}u, u) = [Gf, Gf]_A = \|\|u\|\|_A^2 \geq \gamma_0^2 \|u\|^2.$$

Hence

$$\frac{(\tilde{A}u, u)}{\|u\|^2} \geq \gamma_0^2,$$

and therefore

$$\inf_{u \in D(\tilde{A})} \frac{(\tilde{A}u, u)}{\|u\|^2} \geq \gamma_0^2 = \inf_{u \in D(A)} \frac{(Au, u)}{\|u\|^2}. \tag{9}$$

Comparison of relations (7) and (9) shows that

$$\inf_{u \in D(\tilde{A})} \frac{(\tilde{A}u, u)}{\|u\|^2} = \inf_{u \in D(A)} \frac{(Au, u)}{\|u\|^2}. \tag{10}$$

To say that equation (6) is soluble for any $f \in H$ is only a different way of stating, as we have done earlier, that $R(\tilde{A}) = H$. In fact, if $f \in H$, then $f \in R(\tilde{A})$, and this means that there exists an element u_0 such that $\tilde{A}u_0 = f$. The uniqueness of the solution is a consequence of the definition we have adopted for the operator $\tilde{A}$ (cf. Theorems 5.4.1 and 5.5.1).

From the solubility of equation (6) for any $f \in H$, it follows that for this equation the generalized solution is the ordinary solution. We note also that the generalized solution of the equation $Au = f$ is the ordinary solution of the equation $\tilde{A}u = f$.

The extension $\tilde{A}$ of the positive definite operator A, described in this section, was introduced by K. Friedrichs, and we shall henceforth call $\tilde{A}$ *Friedrich's extension of the operator A*.

Remark. For the reader who is familiar with the concept of a self-adjoint operator we may mention that Friedrichs's extension $\tilde{A}$ of the positive-definite operator A is a self-adjoint extension of this operator.

The proof of this assertion can be found in the article by K. Friedrichs [7] and in the present author's book [5], listed in the bibliography to Part II.

The quantity γ_0^2 (formula (8)) is called the *lower bound* of the positive definite operator A. Thus we arrive at Friedrichs's Theorem.

THEOREM 5.7.2. *A positive definite operator can be extended to a self-adjoint operator with the same lower bound.*

THEOREM 5.7.3. *The energy space of a positive definite operator and of its Friedrichs extension coincide.*

Let A be a positive definite operator in the Hilbert space H, and $\tilde{A}$ the Friedrichs extension of the operator A. It is necessary to show that the spaces

H_A and $H_{\tilde{A}}$ consist of exactly the same elements, and that

$$|||u_0|||_{\tilde{A}} = |||u_0|||_A, \qquad u_0 \in H_A. \tag{11}$$

1. Any element of H_A belongs to $H_{\tilde{A}}$, and its norms in the two spaces are identical. This statement is obvious for elements of the domain $D(A)$: if $u_0 \in D(A)$, then $u_0 \in D(\tilde{A}) \subset H_{\tilde{A}}$; in this regard $|||u_0|||_A^2 = (Au_0, u_0) = (\tilde{A}u_0, u_0) = |||u_0|||_{\tilde{A}}^2$. By Theorem 5.3.2, the relation $u_0 \in H_A$ implies that there exists a sequence $\{u_n\}$, $u_n \in D(A)$, with the properties

$$\|u_n - u_0\| \underset{n \to \infty}{\to} 0, \qquad |||u_n - u_m|||_A^2 \underset{n,m \to \infty}{\to} 0. \tag{12}$$

But once $u_n, u_m \in D(A)$, we have $u_n - u_m \in D(\tilde{A})$, and

$$|||u_n - u_m|||_A^2 = |||u_n - u_m|||_{\tilde{A}}^2.$$

Thus, there exists a sequence $\{u_n\}$, $u_n \in D(\tilde{A})$, with the properties

$$\|u_n - u_0\| \underset{n \to \infty}{\to} 0, \qquad |||u_n - u_m|||_{\tilde{A}} \underset{n,m \to \infty}{\to} 0; \tag{13}$$

and, again from Theorem 5.3.2. $u_0 \in H_{\tilde{A}}$. Hence, in accordance with the definition of ideal elements

$$|||u_n - u_0|||_A \underset{n \to \infty}{\to} 0, \qquad |||u_n - u_0|||_{\tilde{A}} \underset{n \to \infty}{\to} 0;$$

from which

$$|||u_0|||_A = \lim_{n \to \infty} |||u_n|||_A = \lim_{n \to \infty} |||u_n|||_{\tilde{A}} = |||u_0|||_{\tilde{A}}.$$

2. We now show that from the relation $u \in H_{\tilde{A}}$ there follow the relation $u \in H_A$ and equation (11). If $u \in H_{\tilde{A}}$, there exists a sequence $\{u_n\}$, $u_n \in D(\tilde{A})$, with the properties (13). We have seen above that $D(A) \subset H_A$; therefore $u_n \in H_A$, and from the proof in 1 above,

$$|||u_n - u_m|||_{\tilde{A}} = |||u_n - u_m|||_A,$$

the properties (13) turn into the properties (12), and then $u \in H_A$. For the elements $u \in H_A$, equation (11) was established in 1. This proves the theorem.

We have already established the formula (3.6),

$$[u, v]_A = (Au, v), \qquad u \in D(A), \qquad v \in H_A.$$

The following, somewhat more general formula, also holds:

$$[u, v]_A = (\tilde{A}u, v), \quad u \in D(\tilde{A}), \quad v \in H_A. \tag{14}$$

In fact, if $u \in D(\tilde{A})$, $v \in H_A = H_{\tilde{A}}$, then by formula (3.6),

$$[u, v]_{\tilde{A}} = (\tilde{A}u, v). \tag{15}$$

In the spaces H_A and $H_{\tilde{A}}$ the norms coincide, but then the scalar products in them also coincide:

$$[u, v]_{\tilde{A}} = \tfrac{1}{4}\{|||u+v|||_{\tilde{A}}^2 - |||u-v|||_{\tilde{A}}^2\}$$
$$= \tfrac{1}{4}\{|||u+v|||_A^2 - |||u-v|||_A^2\} = [u, v]_A.$$

Replacing $[u, v]_{\tilde{A}}$ by $[u, v]_A$ in equation (15), we again obtain formula (14).

§ 8. The Simplest Boundary-Value Problem for an Ordinary Linear Differential Equation

Consider the second-order ordinary differential equation

$$-\frac{d}{dx}\left[p(x)\frac{du}{dx}\right] + q(x)u(x) = f(x). \tag{1}$$

We pose the following problem: to find the integral of this equation on the segment $[a, b]$, subject to the boundary conditions

$$u(a) = u(b) = 0. \tag{2}$$

We shall assume that $p, p', q \in C[a, b], f \in L_2(a, b)$. Moreover, suppose that $p(x) \geq p_0 = \text{const.} > 0$, $q(x) \geq 0$. Since the functions $p(x)$, $q(x)$ are also continuous on the segment $[a, b]$, we have the inequality

$$p_0 \leq p(x) \leq p_1, \quad 0 \leq q(x) \leq q_1, \quad x \in [a, b],$$

where p_1, q_1, and also p_0, are positive constants.

As our basic space H we take $L_2(a, b)$. For the domain of definition of the operator

$$Au = -\frac{d}{dx}\left[p(x)\frac{du}{dx}\right] + q(x)u(x) \tag{3}$$

we take the set of functions $u(x)$ satisfying the following restrictions:

$$u \in C^{(2)}[a, b], \quad u(a) = u(b) = 0.$$

We now prove the positive-definiteness of the operator A. Its domain of definition is dense in $L_2(a, b)$ – this follows from Corollary 1.3.1. Next we verify that the operator A is symmetric. Let $u, v \in D(A)$; then

$$(Au, v) = -\int_a^b v \frac{d}{dx}\left[p(x)\frac{du}{dx}\right] dx + \int_a^b q(x)u(x)v(x)\,dx.$$

The first term on the left is integrated by parts. Taking into account the fact that the function $v(x)$ satisfies boundary conditions (2), we obtain

$$(Au, v) = \int_a^b \left[p(x)\frac{du}{dx}\frac{dv}{dx} + q(x)uv\right] dx. \qquad (4)$$

This expression, evidently symmetric, shows that $(Au, v) = (u, Av)$, i.e. that the operator A is symmetric.

In formula (4) we put $v = u$, and obtain

$$(Au, u) = \int_a^b \left[p(x)\left(\frac{du}{dx}\right)^2 + q(x)u^2(x)\right] dx. \qquad (5)$$

Because of the boundedness of the coefficients, we now find

$$(Au, u) \geq p_0 \int_a^b \left(\frac{du}{dx}\right)^2 dx.$$

Moreover,

$$\int_a^x u'(t)\,dt = u(x) - u(a) = u(x).$$

By the Schwartz-Bunyakovskii inequality,

$$u^2(x) \leq (x-a)\int_a^x u'^2(t)\,dt \leq (b-a)\int_a^b u'^2(t)\,dt.$$

Hence

$$\|u\|^2 = \int_a^b u^2(x)\,dx \leq (b-a)^2 \int_a^b u'^2(t)\,dt.$$

Finally

$$(Au, u) \geq \frac{p_0}{(b-a)^2}\|u\|^2,$$

which implies positive-definiteness; we can put

$$\gamma = \frac{\sqrt{p_0}}{b-a}.$$

The operator A has turned out to be positive definite, and we can in-

§ 8]   BOUNDARY-VALUE PROBLEM FOR AN EQUATION

troduce an energy space H_A. We show that H_A consists of functions which are absolutely continuous on the segment $[a, b]$, which are equal to zero at the end-points of the segment, and which have a square-integrable first derivative.

Suppose that $u \in H_A$. By Theorem 5.3.2, there exists a sequence $u_n \in D(A)$ which has the following properties:

$$|||u_n - u_m||| \underset{n, m \to \infty}{\to} 0, \qquad ||u_n - u|| \underset{n \to \infty}{\to} 0.$$

If $u \in D(A)$, then

$$|||u|||^2 = (Au, u) = \int_a^b \left[p(x) \left(\frac{du}{dx} \right)^2 + q(x) u^2(x) \right] dx.$$

Consequently,

$$|||u_n - u_m|||^2 = \int_a^b [p(x)(u_n' - u_m')^2 + q(x)(u_n - u_m)^2] \, dx \underset{n, m \to \infty}{\to} 0,$$

and since both terms under the integral sign are non-negative, we have

$$\int_a^b p(x)(u_n' - u_m')^2 \, dx \underset{n, m \to \infty}{\to} 0.$$

Recalling the restrictions on p, we obtain

$$p_0 \int_a^b (u_n' - u_m')^2 \, dx \leq \int_a^b p(x)(u_n' - u_m')^2 \, dx \leq p_1 \int_a^b (u_n' - u_m')^2 \, dx,$$

and this means that

$$\int_a^b p(x)(u_n' - u_m')^2 \, dx \underset{n, m \to \infty}{\to} 0,$$

is equivalent to

$$\int_a^b (u_n' - u_m')^2 \, dx \underset{n, m \to \infty}{\to} 0. \tag{6}$$

This last expression means, in its turn, that the sequence of derivatives $\{u_n'\}$ converges in itself in the metric of $L_2(a, b)$. The space $L_2(a, b)$ is complete, and the latter sequence converges to some function $v \in L_2(a, b)$.

Since it is possible to take the limit with respect to n in the identity

$$\int_a^x u_n'(t) \, dt = u_n(x) - u_n(a) \quad u_n(x)$$

we obtain
$$u(x) = \int_a^x v(t)\,dt.$$

This last equation implies the absolute continuity of the function $u(x)$, with $u' = v \in L_2(a, b)$. It is obvious, moreover, that $u(a) = 0$, and it remains to show that $u(b) = 0$. This can be done, for example, as follows: in the identity
$$\int_x^b u_n'(t)\,dt = u_n(b) - u_n(x) = -u_n(x)$$
we proceed to the limit
$$u(x) = -\int_x^b v(t)\,dt,$$
and it is clear that $u(b) = 0$.

We have seen above that for functions $u \in D(A)$ the following formula holds:
$$|||u|||^2 = \int_a^b \left[p(x)\left(\frac{du}{dx}\right)^2 + q(x)u^2(x) \right] dx. \tag{7}$$

We now show that this formula holds for any function from the energy space. Let $u \in H_A$. We introduce the sequence $u_n \in D(A)$, having the properties
$$|||u_n - u||| \underset{n\to\infty}{\to} 0, \qquad ||u_n - u|| \underset{n\to\infty}{\to} 0.$$

Formula (6) gives one further relation:
$$||u_n' - u'|| \underset{n\to\infty}{\to} 0.$$

The norm of the limit is equal to the limit of the norm, so that
$$||u_n||^2 \to ||u||^2, \quad |||u_n|||^2 \to |||u|||^2, \quad ||u_n'||^2 \to ||u'||^2. \tag{8}$$

For the functions u_n we have formula (7):
$$|||u_n|||^2 = \int_a^b [pu_n'^2 + qu_n^2]\,dx.$$

It follows from equation (8) that, when $n \to \infty$, the left-hand side of this last equation has the limit $|||u|||^2$. We show that the limit of the right-hand side is
$$\int_a^b [pu'^2 + qu^2]\,dx.$$

We have
$$\left|\int_a^b [pu_n'^2+qu_n^2]\,dx - \int_a^b [pu'^2+qu^2]\,dx\right|$$
$$\leq p_1 \int_a^b |u_n'^2-u'^2|\,dx + q_1 \int_a^b |u_n^2-u^2|\,dx.$$

By the Schwartz-Bunyakovskii inequality,
$$\int_a^b |u_n'^2-u'^2|\,dx \leq \left\{\int_a^b (u_n'+u')^2\,dx\right\}^{\frac{1}{2}} \left\{\int_a^b (u_n'-u')^2\,dx\right\}^{\frac{1}{2}}$$
$$= \|u_n'+u'\|\,\|u_n'-u'\|.$$

But $\|u_n'\| \to \|u'\|$, and then $\|u_n'+u'\| \leq \|u_n'\|+\|u'\|$ is a bounded quantity, so that
$$\int_a^b |u_n'^2-u'^2|\,dx \underset{n\to\infty}{\to} 0.$$

It can similarly be shown that
$$\int_a^b |u_n^2-u^2|\,dx \to 0,$$

which proves formula (7) for any function of H_A.

It is necessary now to show the converse, namely, that if the function u satisfies the three conditions formulated above, then $u \in H_A$. By Theorem 5.3.2, there exists a sequence $\{u_n\}$ with the properties
$$u_n \in D(A), \quad \|\|u_n-u_m\|\| \underset{n,m\to\infty}{\to} 0, \quad \|u_n-u\| \underset{n\to\infty}{\to} 0.$$

In order to prove the result, we expand the derivative of the function u in a Fourier cosine series:
$$u'(x) = \sum_{k=1}^{\infty} a_k \cos \frac{k\pi(x-a)}{b-a}.$$

The constant term vanishes because
$$a_0 = \frac{1}{b-a}\int_a^b u'(x)\,dx = \frac{1}{b-a}[u(b)-u(a)] = 0.$$

Integrating term by term, we obtain
$$u(x) = \sum_{k=1}^{\infty} b_k \sin \frac{k\pi(x-a)}{b-a}, \quad b_k = \frac{a_k(b-a)}{k\pi}.$$

For the function u_n it is sufficient to take the partial sum of this last series.

The generalized solution $u_0(x)$ of problem (1)–(2) exists and is unique; it is the function which achieves a minimum of the energy functional

$$F(u) = |||u|||^2 - 2(u, f)$$

in energy space. As formula (7) shows we now have

$$F(u) = \int_a^b [p(x)u'^2 + q(x)u^2 - 2f(x)u]\,dx; \qquad (9)$$

being elements of the energy space, the functions $u(x)$ must satisfy conditions (2).

The function $u_0 \in D\,(\text{grad }F)$. The variation of the functional (9) at the point u_0 has the form

$$\delta F(u_0, \eta) = 2\int_a^b [p(x)u_0'(x)\eta'(x) + q(x)u_0(x)\eta(x) - f(x)\eta(x)]\,dx,\ \eta \in H_A.$$

This will be a functional of η, bounded in $L_2(a, b)$, if, and only if, the integral

$$\int_a^b p(x)u_0'(x)\eta'(x)\,dx$$

possesses this property; just as in Problem S, it can be shown that for this to be the case it is necessary and sufficient that the function $p(x)du_0/dx$ be absolutely continuous and have a square-integrable derivative on the segment $[a, b]$. But the function $p(x)$ is strictly positive and continuously differentiable, so this last restriction is equivalent to the following: $u_0'(x)$ is absolutely continuous on the segment $[a, b]$, and $u_0'' \in L_2(a, b)$.

We have explained, among other things, that if $\tilde{A}$ denotes the Friedrichs extension of the operator A in our problem (formulae (3) and (2)), then the domain of definition $D(\tilde{A})$ of this extension consists of functions which possess the following properties: the functions themselves and their first derivatives are absolutely continuous on the segment $[a, b]$, and the second derivatives are square-integrable; and at the end-points of the segment these functions are equal to zero.

§ 9. The Minimum of a Quadratic Functional: A More General Problem

1. In § 4 we formulated the variational problem for a quadratic functional of the form

$$F(u) = (Au, u) - 2(u, f).$$

This has the important special feature that its linear part $2(u,f)$ is bounded in the original space; in § 5 we made use of this feature in proving the existence of a generalized solution to the variational problem.

We now consider the problem of minimizing a quadratic functional of the more general form

$$F(u) = (Au, u) - 2l(u), \qquad (1)$$

where A is a positive definite operator, acting in the Hilbert space H, and l is a linear (but not necessarily bounded) functional in this space; the factor 2 is introduced for convenience.

Introducing the energy space H_A of the operator A, we can write the functional (1) in the form

$$F(u) = |||u|||^2 - 2l(u) \qquad (2)$$

and regard it as a functional, given on elements (some or all) of the energy space. The interesting case arises when $D(l)$ – the domain of definition of the functional l – is dense in H_A; obviously, $D(F) = D(l)$.

Two possibilities can occur.

1. The functional l is not bounded in the energy space. In this case, the functional F is not bounded below. In fact, there now exists a sequence $\{u_n\}$ with the properties

$$|||u_n||| = 1, \qquad |l(u_n)| \xrightarrow[n \to \infty]{} \infty.$$

By changing where necessary the signs of the elements u_n, we obtain the result $l(u_n) \to +\infty$. Then

$$F(u_n) = 1 - 2l(u_n) \to -\infty.$$

The problem of minimizing the functional (2) now lacks any meaning.

2. The functional l is bounded in the energy space. Then it can be extended by continuity over all this space: in the same way the functional (2) is extended over all the space H_A. By Riesz's theorem, there exists one and only one element $u_0 \in H_A$, satisfying the equation $l(u) = [u, u_0]$. Now

$$F(u) = |||u|||^2 - 2[u, u_0].$$

If we repeat the same arguments as in § 5, we can see that the element u_0 achieves a minimum of the functional (2).

If the space H_A is separable, then it is easy to derive a formula, analogous to formula (5.11), which gives the solution to the problem of minimizing functional (2). Let ω_n, $n = 1, 2, \ldots$, be a sequence which is complete and orthonormal in the energy space. Then

$$u_0 = \sum_{n=1}^{\infty} [u_0, \omega_n] \omega_n.$$

In the formula noted above, $l(u) = [u, u_0]$, we put $u = \omega_n$. Then $[u_0, \omega_n] = [\omega_n, u_0] = l(\omega_n)$, and, consequently,

$$u_0 = \sum_{n=1}^{\infty} l(\omega_n) \omega_n. \tag{3}$$

2. Let A be the operator discussed in the preceding section (formulae (3) and (2), § 8); we retain the assumptions, introduced in that section, concerning the coefficients $p(x)$ and $q(x)$, and concerning the functions which make up the domain of definition of the operator A. We now pose the problem of minimizing the quadratic functional

$$F(u) = |||u|||^2 - 2u(c) = \int_a^b \left[p(x) \left(\frac{du}{dx} \right)^2 + q(x) u^2 \right] dx - 2u(c),$$

$$a < c < b, \tag{4}$$

in the energy space of the operator A. In particular, this implies that the function u, on which functional (4) depends, must satisfy the boundary conditions

$$u(a) = u(b) = 0. \tag{5}$$

It is easily seen that the linear functional $l(u) = u(c)$ is bounded in the energy metric. In fact, by the Schwartz-Bunyakovskii inequality,

$$|u(c)|^2 = \left| \int_a^c u'(x) dx \right|^2 \leq (c-a) \int_a^c u'^2 x(dx) \leq (c-a) \int_a^b u'^2(x) dx.$$

By formula (8.7)

$$|||u|||^2 = \int_a^b [p(x) u'^2(x) + q(x) u^2(x)] dx \geq p_0 \int_a^b u'^2(x) dx,$$

and so

$$|u(c)| \leq \sqrt{\frac{c-a}{p_0}} |||u|||. \tag{6}$$

Formula (6) shows that in the present circumstances the functional l is bounded, and also that $|||l||| \leq \sqrt{\{(c-a)/p_0\}}$. The solution of our variational problem exists; by formula (3) it can be represented by the series

$$u_0(x) = \sum_{n=1}^{\infty} \omega_n(c)\omega_n(x), \qquad (7)$$

where the system $\{\omega_n\}$ is complete and orthogonal in H_A. The series (8) converges in the metric of the space H_A, and, consequently, in the metric of $L_2(a, b)$.

Example. Consider the special case $p(x) \equiv 1$, $q(x) \equiv 0$, so that $Au = -d^2u/dx^2$. In this case, the functions

$$\omega_n(x) = \frac{\sqrt{2(b-a)}}{n\pi} \sin \frac{n\pi(x-a)}{b-a}, \qquad n = 1, 2, \ldots$$

constitute a system which is complete and orthonormal in the energy space. The proof of this is left to the reader. The minimum of the functional

$$\int_a^b u'^2 \, dx - 2u(c), \qquad u(a) = u(b) = 0 \qquad (8)$$

is achieved with the function

$$u_0(x) = \frac{2(b-a)}{\pi^2} \sum_{n=1}^{\infty} \frac{1}{n^2} \sin \frac{n\pi(c-a)}{b-a} \sin \frac{n\pi(x-a)}{b-a}.$$

This series can easily be summed if, for example, we construct and solve Euler's equation for the functional (8); this we also leave to the reader to carry through.

§ 10. The Case of a Merely Positive Operator

An operator which is positive, but not positive definite, will be called *merely positive*. For a merely positive operator it is possible to construct an energy space in just the same way as was done for a positive definite operator. But in the process one essential difference emerges; it can be shown that among the ideal elements of the energy space there will necessarily be some which do not belong to the original Hilbert space.

Consider, for example, the operator B discussed in § 2. We recall that it is defined by the formula

$$Bu = -\frac{d^2u}{dx^2}, \qquad 0 \leq x < \infty,$$

and that its domain of definition consists of functions of the class $C^{(2)}[0, \infty)$ which are equal to zero when $x = 0$ and when $x \geqq a_u$, where a_u is a constant whose value is different for different $u \in D(B)$. It is not difficult to show that the energy space H_B consists of functions with the following properties: 1) a function $u \in H_B$ is absolutely continuous on the segment $[0, a]$, where a is an arbitrary positive number: 2) $u(0) = 0$; 3) $u' \in L_2(0, \infty)$. Thus, the function $u(x) = \ln(1+x)$ belongs to the space H_B, but does not belong to the original space $L_2(0, \infty)$.

For a merely positive operator Theorem 5.6.1 remains valid: the energy space is separable if and only if the original space is separable.

If A is a merely positive operator, and l is a linear functional, then the problem of minimizing the functional

$$F(u) = (Au, u) - 2l(u), \qquad u \in D(A),$$

can be solved by exactly the same method as in the preceding section: since $(Au, u) = |||u|||^2$, we write the functional F in the form

$$F(u) = |||u|||^2 - 2l(u).$$

If l is not bounded in H_A, then our variational problem has no meaning, and if l is bounded in H_A, and defined on a set which is dense in H_A, then $l(u) = [u, u_0]$, where $u_0 \in H_A$ exists and is uniquely defined. This is the element which makes F a minimum in the energy space.

Exercises

1. The operator T_p is given by the formula

$$T_p u = -\frac{d}{dx}\left[p(x)\frac{du}{dx}\right], \qquad 0 < x \leqq 1,$$

where $p(x) \in C^{(1)}[0, 1]$: $p(0) = 0$, $p(x) > 0$, $x > 0$, and the integral

$$\int_0^1 \frac{dx}{p(x)}$$

converges. The domain $D(T_p)$ consists of functions $u(x)$ subject to the following restrictions: a) $u(x)$ and $p(x)u'(x)$ are continuous on the segment $[0, 1]$, and absolutely continuous on any segment $[\delta, 1]$, $0 < \delta < 1$; b) $T_p u \in L_2(0, 1)$; c) $u(0) = u(1) = 0$. Show that the operator T_p is positive definite in the space $L_2(0, 1)$.

2. Let $p(x) \in C^{(1)}[0, 1]$; $p(0) = 0$, $p(x) > 0$, $x > 0$,

$$\int_0^1 \frac{dx}{p(x)} = \infty, \qquad \int_0^1 \frac{x\,dx}{p(x)} < \infty.$$

We define the operator T_p by the same formulae as in Exercise 1; for the domain $D(T_p)$ we take the set of functions $u(x)$ which satisfy the same conditions as in Exercise 1, except for $u(0) = 0$. Show that the operator T_p is positive definite in the space $L_2(0, 1)$.

3. Describe the set of elements of the energy space in Exercises 1 and 2.

4. Prove the positive definiteness of the operator

$$T_{x^2} u = -\frac{d}{dx}\left(x^2 \frac{du}{dx}\right), \qquad 0 < x \leq 1,$$

defined on the set of functions in Exercise 2, with $p(x) = x^2$. Show that the lower bound of this operator is equal to $\tfrac{1}{4}$.

5. Show that for $\alpha > 2$ the operator T_{x^α} is strictly positive.

Chapter 6. Eigenvalue Spectrum of a Positieve Definite Operator

§ 1. The Concept of the Eigenvalue Spectrum of an Operator

Let A be a linear operator in a Hilbert space H. The quantity λ and the element u are called, respectively, the *eigenvalue* and *eigenelement* of the operator A, if u is a non-zero element of the space H, and if

$$Au - \lambda u = 0. \tag{1}$$

Note that if H is, for example, the space L_2, then the requirement $u \neq 0$ is equivalent to $u(x) \not\equiv 0$.

We say that the eigenelement u *corresponds* to the eigenvalue λ. Equation (1) yields a formula which enables the eigenvalue to be determined if the corresponding eigenelement is known. Specifically, on scalar-multiplying both sides of equation (1) by u, we obtain

$$(Au, u) - \lambda \|u\|^2 = 0,$$

and hence

$$\lambda = \frac{(Au, u)}{\|u\|^2}. \tag{2}$$

In a complex space eigenvalues could obviously be both real and complex. In real space multiplication of elements is defined only for real quantities; accordingly, it might be thought that only real eigenvalues should be considered in real space. But even the simplest examples show that this would be inadequate. Thus, a real, square matrix of order m generates a linear operator in m-dimensional Euclidean space; the eigenvalues of this operator are identical with the eigenvalues of the matrix, and it is well known that these can be complex. For this reason we extend somewhat the definition of eigenvalues so as to allow for their being complex.

Starting out from the given, real Hilbert space H, we construct a complex Hilbert space H^*. To do this we proceed as follows: as the set of elements of the new space H^*, we take the set of all possible formal sums of the form

$U = u' + iu''$, where $i = \sqrt{-1}$, and $u', u'' \in H$. We introduce on this new set the usual addition and multiplication rules for complex numbers; these two operations do not lead out the set H^*, which can now be assumed linear. The null element in H^* is the element $0 + i0$, where 0 means the null element of the space H; in place of $0 + i0$ we shall simply write 0; in general we shall write u and iv instead of $u + i0$ and $0 + iv$ respectively. We introduce the scalar product in H^* through the following rule: if $U = u' + iu''$, $V = v' + iv''$, where $u', u'', v', v'' \in H$, then

$$(U, V)^* = (u', v') + (u'', v'') + i[(u'', v') - (u', v'')]. \tag{3}$$

It is easy to see that with this definition all the axioms relating to scalar product are satisfied in the complex space; namely,

A. $(U, V)^* = \overline{(V, U)^*}$.

B. $(\alpha_1 U_1 + \alpha_2 U_2, V)^* = \alpha_1 (U_1, V)^* + \alpha_2 (U_2, V)^*$.

C. $(U, U)^* \geq 0$; here $(U, U)^* = 0$ if and only if $U = 0$.

The operator A can be extended to elements of the form $U = u' + iu''$, where $u', u'' \in D(A) \subset H$, by the formula

$$AU = Au' + iAu''. \tag{4}$$

With this new definition it can be shown that the operator A, originally defined in the real space H, has complex eigenvalues $\lambda = \lambda' + i\lambda''$, and corresponding eigenelements $u' + iu''$ in H^*; the equation

$$A(u' + iu'') = (\lambda' + i\lambda'')(u' + iu'')$$

is equivalent to the system of equations

$$\begin{aligned} Au' &= \lambda' u' - \lambda'' u'', \\ Au'' &= \lambda'' u' + \lambda' u''. \end{aligned} \tag{5}$$

It is possible that several eigenelements correspond to one and the same eigenvalue; if $u_1, u_2, \ldots, u_n$ are such elements, then any linear combination, not equal to the null element,

$$\sum_{k=1}^{n} c_k u_k$$

is also an eigenelement, corresponding to the same eigenvalue. This fact allows us to consider only linearly-independent eigenelements, corresponding to a given eigenvalue, and each eigenelement can be regarded as having been normalized.

The number of linearly-independent eigenelements is called the *rank* (or sometimes the *multiplicity*) of the corresponding eigenvalue. In a separable space the rank of any eigenvalue is finite or enumerable.

The totality of eigenvalues of an operator is called its *eigenvalue spectrum*.

§ 2. Eigenvalues and Eigenelements of a Symmetric Operator

THEOREM 6.2.1. *The eigenvalues of a symmetric operator are real.*

Scalar multiply the first of equations (1.5) by u'', the second by u', and then subtract the first from the second:

$$(Au'', u') - (Au', u'') = \lambda''(\|u'\|^2 + \|u''\|^2); \tag{1}$$

by virtue of the symmetry of the operator A, this is equal to zero. Since the eigenelement $u' + iu'' \neq 0$, either u' or u'' must be different from zero, and the expression in brackets in (1) is positive. Hence it follows that $\lambda'' = 0$ and the eigenvalues are real. The system (1.5) takes the form

$$Au' = \lambda' u', \quad Au'' = \lambda' u'',$$

and each of the elements u', u'' which is non-zero, is an eigenelement, corresponding to the eigenvalue λ'.

THEOREM 6.2.2. *The eigenelements of a symmetric operator corresponding to distinct eigenvalues are orthogonal.*

Let λ_1 and λ_2 be eigenvalues of the symmetric operator A, with $\lambda_1 \neq \lambda_2$. Let eigenelements u_1, u_2 correspond to the eigenvalues λ_1, λ_2 respectively.

We write down the equations

$$Au_1 = \lambda_1 u_1, \quad Au_2 = \lambda_2 u_2.$$

Scalar multiply the first equation by u_2, the second by u_1, and subtract the second from the first:

$$(Au_1, u_2) - (Au_2, u_1) = (\lambda_1 - \lambda_2)(u_1, u_2).$$

Since A is symmetric, the left-hand side is zero, and, consequently,

$$(\lambda_1 - \lambda_2)(u_1, u_2) = 0.$$

But $\lambda_1 - \lambda_2 \neq 0$. Hence

$$(u_1, u_2) = 0.$$

The theorem is proved.

COROLLARY 6.2.1. *In a separable Hilbert space a symmetric operator has not more than an enumerable set of eigenvalues.*

If to one eigenvalue there correspond several linearly-independent elements, then these elements can be subjected to a process of orthogonalization. The eigenelements can also be normalized, and this leads us to the following important conclusion:
it is always possible to assume that the eigenelements of a symmetric operator constitute an orthonormal system.

§ 3. Generalized Eigenvalue Spectrum of a Positive Definite Operator

Every positive definite operator is symmetric, and therefore all the results of the preceding section hold for positive definite operators. But for these operators it turns out to be convenient to introduce the further concept of a generalized eigenvalue spectrum – or, more precisely, of generalized eigenvalues and corresponding generalized eigenelements. These are introduced by analogy with the idea of a generalized solution.

Let A be a positive definite operator, λ one of its eigenvalues and u the eigenelement corresponding to λ. This means that $u \neq 0, u \in D(A)$, and that the equation

$$Au = \lambda u \qquad (1)$$

holds.

Introduce an arbitrary element $\eta \in H_A$. We scalar multiply both sides of equation (1) by η:

$$(Au, \eta) = \lambda(u, \eta).$$

In this expression $u \in D(A), \eta \in H_A$, and then by formula (3.6), Chapter 5, we have

$$(Au, \eta) = [u, \eta]_A.$$

Thus we have found that the eigenvalue λ and the corresponding eigenelement u satisfy the equation

$$[u, \eta]_A = \lambda(u, \eta), \qquad \forall \eta \in H_A. \qquad (2)$$

Conversely, let $u \in D(A), u \neq 0$, together with some number λ, satisfy equation (2). By formula (3.6), Chap. 5,

$$[u, \eta]_A = (Au, \eta).$$

Putting this into equation (2), we obtain

$$(Au - \lambda u, \eta) = 0, \qquad \forall \eta \in H_A.$$

Hence the completely defined element $Au - \lambda u$ of the space H is orthogonal to an arbitrary element $\eta \in H_A$. But the set of elements of the space H_A is dense in the original space H, and an element which is orthogonal to a dense set is zero. Hence it follows that

$$Au - \lambda u = 0.$$

This last equation implies that u is an eigenelement and λ an eigenvalue of the operator A.

The element $u \in H_A$, $u \neq 0$, and the number λ, will be called *generalized eigenelement* and *generalized eigenvalue* of the operator A respectively, if they satisfy equation (2).

THEOREM 6.3.1. *The generalized eigenvalues and eigenelements of a positive definite operator are the ordinary eigenvalues and eigenelements of the Friedrichs extension of this operator.*

The proof is very simple. If λ and u are a generalized eigenvalue and eigenelement of the positive definite operator A, then they satisfy equation (2). Substituting $\lambda u = f$ in the right-hand side of (2), we convert it to the form

$$[\eta, u]_A = (\eta, f), \qquad \forall \eta \in H_A. \tag{3}$$

Comparing equation (3) with equation (5.5), Chap. 5, we see that they are identical, apart from notation; hence it follows that the element u, contained in formula (3), achieves a minimum for the functional

$$F(v) = |||v|||^2 - 2(f, v), \qquad \forall v \in H_A.$$

But then $u \in D(\tilde{A})$, where $\tilde{A}$ is the Friedrichs extension of the operator A, and $\tilde{A}u = f$, or

$$\tilde{A}u = \lambda u,$$

which we set out to prove.

The operator $\tilde{A}$ is symmetric, so that Theorems 6.2.1 and 6.2.2 are valid for the generalized eigenvalues and eigenelements. We note two further properties of generalized eigenvalues and eigenelements of a positive definite operator; the word "generalized" will be omitted for the sake of brevity.

THEOREM 6.3.2. *The eigenelements of a positive definite operator are orthogonal in the energy space if they are orthogonal in the original space.*

Let u_1, u_2 be two eigenelements of the positive definite operator A and let $(u_1, u_2) = 0$. Putting $u = u_1$, $\eta = u_2$ in equation (2), we find that $[u_1, u_2]_A = 0$.

THEOREM 6.3.3. *Any eigenvalue of a positive definite operator is not less than the lower bound of this operator.*

Let γ_0^2 be the lower bound of the operator A. By Theorem 5.7.1, the operator $\tilde{A}$, obtained by Friedrichs extension of A, has the same lower bound γ_0^2. If λ and u are an eigenvalue and the corresponding eigenelement (both generalized) of the operator A, then by formula (1.2) and Theorem 6.3.1,

$$\lambda = \frac{(\tilde{A}u, u)}{\|u\|^2} \geq \gamma_0^2. \tag{4}$$

We note also that in the present case formula (1.2) can be put in the form

$$\lambda = \frac{\|\|u\|\|_A^2}{\|u\|^2}. \tag{5}$$

The right-hand side remains unchanged if u is multiplied by some non-zero constant. We choose this constant such that the eigenelement u is normalized in the metric of the original space: $\|u\| = 1$. Then we obtain a formula for the eigenvalue which is in some respects more convenient:

$$\lambda = \|\|u\|\|_A^2, \qquad \|u\|^2 = 1. \tag{6}$$

§ 4. Variational Formulation of the Problem of the Eigenvalue Spectrum

We begin with the following observation: if γ_0^2 is the lower bound of the positive definite operator A, then

$$\inf_{\substack{u \in H_A \\ u \neq 0}} \frac{\|\|u\|\|_A^2}{\|u\|^2} = \gamma_0^2. \tag{1}$$

We prove this. Since $D(A) \subset H_A$, we have

$$\inf_{\substack{u \in H_A \\ u \neq 0}} \frac{\|\|u\|\|_A^2}{\|u\|^2} \leq \inf_{\substack{u \in D(A) \\ u \neq 0}} \frac{\|\|u\|\|_A^2}{\|u\|^2} = \inf_{\substack{u \in D(A) \\ u \neq 0}} \frac{(Au, u)}{\|u\|^2} = \gamma_0^2.$$

On the other hand, $D(A)$ is dense in H_A; if $u \in H_A$, then we can choose an element $v \in D(A)$, such that $\|\|v - u\|\|_A < \varepsilon$ and $\|v - u\| < \varepsilon$, where ε is

a number as small as we please. But then $|\,|||v|||_A - |||u|||_A| < \varepsilon$, $|\,||v|| - ||u||\,|$
$< \varepsilon$; if, for some element $u \in H_A$, we had

$$\frac{|||u|||_A^2}{||u||^2} < \gamma_0^2,$$

then for sufficiently small ε we would also have

$$\frac{|||v|||_A^2}{||v||^2} = \frac{(Av, v)}{||v||^2} < \gamma_0^2,$$

which contradicts the definition of the lower bound.

THEOREM 6.4.1. *If there exists an element u_1 for which the lower bound of the ratio (1) is attained, then γ_0^2 is the smallest generalized eigenvalue of the operator A, and u_1 is the corresponding eigenelement.*

If the element u is multiplied by a constant, the ratio

$$\Psi(u) = \frac{|||u|||_A^2}{||u||^2} \tag{2}$$

is unchanged, since we can assume that the element u is normalized. Then

$$\Psi(u) = |||u|||_A^2, \qquad ||u|| = 1 \tag{3}$$

and we put $\gamma_0^2 = \lambda_1$.

The fact that the lower bound is attained for the element u_1 implies that

$$u_1 \in H_A, \qquad ||u_1||^2 = 1, \qquad |||u_1|||^2 = \lambda_1.$$

We take an arbitrary $\eta \in H_A$ and an arbitrary real α, and construct the ratio

$$\frac{|||u_1 + \alpha\eta|||^2}{||u_1 + \alpha\eta||^2}. \tag{4}$$

If η is held fixed, then the ratio (4) is a function of α, which attains a minimum for $\alpha = 0$. But then its derivative with respect to α must equal zero when $\alpha = 0$

$$\frac{d}{d\alpha} \frac{[u_1, u_1] + 2\alpha[u_1, \eta] + \alpha^2[\eta, \eta]}{(u_1, u_1) + 2\alpha(u_1, \eta) + \alpha^2(\eta, \eta)} \bigg|_{\alpha=0} = 0.$$

Carrying out the differentiation, we obtain

$$2(u_1, u_1)[u_1, \eta] - 2(u_1, \eta)[u_1, u_1] = 0. \tag{5}$$

We note that
$$(u_1, u_1) = \|u_1\|^2 = 1, \quad [u_1, u_1] = \|\|u_1\|\|^2 = \lambda_1.$$
Substituting this into expression (5), we find
$$[u_1, \eta] - \lambda_1(u_1, \eta) = 0.$$
This last equation shows that λ_1 and u_1 are an eigenvalue and an eigenelement of operator A respectively. The fact that λ_1 is the smallest eigenvalue follows from Theorem 6.3.3.

Suppose now that we have already found the smallest eigenvalue λ_1 and the corresponding element u_1 of the operator A. How do we find the next eigenvalue λ_2 and eigenelement u_2? Obviously we have to look for λ_2 among the values of the ratio (2), for functions which are orthogonal to u_1 in the metrics of both spaces H and H_A.

We denote by $H^{(1)}$ the subspace of the space H which is orthogonal to the element u_1, and by $H_A^{(1)}$ the subspace of the space H_A, which is orthogonal to u_1 in the sense of the new metric:
$$[u, u_1] = 0, \quad u \in H_A^{(1)}.$$
We show that
$$H_A^{(1)} = H_A \cap H^{(1)}.$$
Let $u \in H_A^{(1)}$. We write down the equation which defines the first eigenelement:
$$[u_1, \eta] = \lambda_1(u_1, \eta).$$
Substituting here $\eta = u$, we obtain $(u_1, u) = [u_1, u]/\lambda_1 = 0$. This implies that $u \in H^{(1)}$ and, consequently, $u \in H_A \cap H^{(1)}$.

Conversely, let $u \in H_A \cap H^{(1)}$; this implies that $u \in H_A$ and $(u, u_1) = 0$. In exactly the same way we arrive at the equation
$$[u_1, u] = \lambda_1(u_1, u) = 0,$$
from which it follows that $u \in H_A^{(1)}$.

If we know the mutually-orthogonal eigenelements $u_1, u_2, \ldots, u_n$, then we can introduce subspaces $H^{(n)}$ and $H_A^{(n)}$ of the spaces H and H_A, which are respectively orthogonal (each in its own metric) to $u_1, u_2, \ldots, u_n$. We can then similarly show that $H_A^{(n)} = H_A \cap H^{(n)}$.

THEOREM 6.4.2. *Suppose that for a positive definite operator A, the first n eigenvalues*
$$\lambda_1 \leq \lambda_2 \leq \ldots \leq \lambda_n$$

and the corresponding eigenelements

$$u_1, u_2, \ldots, u_n,$$

are known, the elements being assumed mutually orthogonal. Let λ_{n+1} be the exact lower bound of $|||u|||^2$ on the normalized elements $u \in H_A^{(n)}$. If this bound is attained, then λ_{n+1} is an eigenvalue of the operator A, and immediately follows λ_n. The element on which this lower bound is attained is the eigenelement corresponding to the eigenvalue λ_{n+1}.

By an argument similar to that used in proving the preceding theorem, we arrive at the equation

$$[u_{n+1}, \zeta] - \lambda_{n+1}(u_{n+1}, \zeta) = 0, \qquad \forall \zeta \in H_A^{(n)}. \tag{6}$$

Let η be an arbitrary element of the space H_A. We put

$$\zeta = \eta - \sum_{k=1}^{n} (\eta, u_k) u_k. \tag{7}$$

Then $\zeta \in H^{(n)}$, and, consequently, $\zeta \in H_A^{(n)}$; hence the element ζ satisfies equation (6). Substituting the expression (7) into this equation, and using the fact that $[u_{n+1}, u_k] = (u_{n+1}, u_k) = 0$, we find

$$[u_{n+1}, \eta] - \lambda_{n+1}(u_{n+1}, \eta) = 0, \qquad \forall \eta \in H_A.$$

The theorem is proved.

§ 5. Theorem on the Smallest Eigenvalue

In some respects Theorem 6.4.1 is conditional in character: it states that $\lambda_1 = \gamma_0^2$, where γ_0^2 is the lower bound of the functional

$$\Psi(u) = |||u|||_A^2, \qquad ||u|| = 1, \tag{1}$$

is the smallest eigenvalue of the operator A, *if* this lower bound is attained. In this section we establish some sufficient conditions that this should occur.

THEOREM 6.5.1. *Let $\{w_n\}$ be a minimizing sequence for the functional (1). If from this sequence it is possible to pick out a subsequence which converges in the metric of the original space H, then $\lambda_1 = \inf \Psi(u)$ is the smallest eigenvalue of the given operator, and the limit of the subsequence is the corresponding eigenelement.*

§ 5] THEOREM ON THE SMALLEST EIGENVALUE

By the conditions of the theorem, it is possible to pick out a convergent subsequence $\{w_{n_k}\}$ from the sequence $\{w_n\}$. For brevity, put $w_{n_k} = \phi_k$. It is easy to see that $\{\phi_k\}$ will also be a minimizing subsequence. Therefore we can assume that a minimizing subsequence $\{\phi_k\}$, convergent in H, is given. The elements ϕ_k possess the following properties: 1) $\phi_k \in H_A$; 2) $\|\phi_k\| = 1$; 3) $\lim \|\|\phi_k\|\|_A^2 = \lambda_1$; 4) there exists an element $u_1 \in H$, such that $\|\phi_k - u_1\| \xrightarrow[k \to \infty]{} 0$. We observe that

$$\|\phi_k - \phi_m\| \xrightarrow[k, m \to \infty]{} 0. \tag{2}$$

Our object now is to show that $u_1 \in H_A$, and that $\|\|u_1\|\|_A^2 = \lambda_1$.

Take an arbitrary element $\eta_k \in H_A$, and let t be an arbitrary real number. The element $\phi_k + t\eta_k$ belongs to the space H_A and, in general, is different from zero. Substituting it into the ratio (4.2), we obtain

$$\frac{\|\|\phi_k + t\eta_k\|\|_A^2}{\|\phi_k + t\eta_k\|^2} \geq \inf \frac{\|\|u\|\|_A^2}{\|u\|^2} = \lambda_1.$$

Cross-multiplying to remove the denominator, we find

$$[\phi_k + t\eta_k, \phi_k + t\eta_k]_A - \lambda_1(\phi_k + t\eta_k, \phi_k + t\eta_k) \geq 0,$$

from which

$$t^2\{\|\|\eta_k\|\|_A^2 - \lambda_1\|\eta_k\|^2\} + 2t\{[\phi_k, \eta_k]_A - \lambda_1(\phi_k, \eta_k)\} \\ + \|\|\phi_k\|\|_A^2 - \lambda_1\|\phi_k\|^2 \geq 0.$$

The quadratic expression on the left-hand side is non-negative for any real t, and therefore its discriminant is non-positive, and

$$|[\phi_k, \eta_k]_A - \lambda_1(\phi_k, \eta_k)| \leq \sqrt{\{\|\|\eta_k\|\|_A^2 - \lambda_1\|\eta_k\|^2\}}\sqrt{\{\|\|\phi_k\|\|_A^2 - \lambda_1\}};$$

where we have taken into account here that $\|\phi_k\| = 1$.

We can strengthen this inequality by discarding the negative term under the first radical:

$$|[\phi_k, \eta_k]_A - \lambda_1(\phi_k, \eta_k)| \leq \|\|\eta_k\|\|_A \sqrt{\{\|\|\phi_k\|\|_A^2 - \lambda_1\}}. \tag{3}$$

The elements $\eta_k \in H_A$ are arbitrary. We require that they be bounded in the aggregate, i.e. that for any k

$$\|\|\eta_k\|\|_A \leq C, \quad C = \text{const.} \tag{4}$$

Then from inequality (3) it follows that

$$|[\phi_k, \eta_k]_A - \lambda_1(\phi_k, \eta_k)| \leq C\sqrt{\{\|\|\phi_k\|\|_A^2 - \lambda_1\}}. \tag{5}$$

In inequality (5) the right-hand side tends to zero, and, consequently, the left-hand side also tends to zero; the convergence is uniform with respect to the choice of elements η_k satisfying inequality (4). With this in mind we put

$$\eta_k = \phi_k - \phi_m,$$

where the number m is arbitrary. We choose η_k according to the following considerations: the numerical sequence $|||\phi_n|||_A$ converges to a limit and is therefore bounded; there exists a constant C such that $|||\phi_n|||_A \leqq C$, and then $|||\eta_k|||_A \leqq 2C$.

It follows now from inequality (5) that

$$\lim_{k \to \infty} \{[\phi_k, \phi_k - \phi_m]_A - \lambda_1(\phi_k, \phi_k - \phi_m)\} = 0,$$

and accordingly the convergence to zero is uniform with respect to m. If so, we can take the limit $m \to \infty$, and then

$$\lim_{k, m \to \infty} \{[\phi_k, \phi_k - \phi_m]_A - \lambda_1(\phi_k, \phi_k - \phi_m)\} = 0. \qquad (6)$$

The numbers k and m here are equivalent; interchanging them, we have

$$\lim_{k, m \to \infty} \{[\phi_m, \phi_m - \phi_k]_A - \lambda_1(\phi_m, \phi_m - \phi_k)\} = 0. \qquad (7)$$

Adding equations (6) and (7), we obtain

$$\lim_{k, m \to \infty} \{|||\phi_k - \phi_m|||_A^2 - \lambda_1 ||\phi_k - \phi_m||^2\} = 0,$$

and because of relation (2),

$$|||\phi_k - \phi_m|||_A^2 \underset{k, m \to \infty}{\to} 0. \qquad (8)$$

Thus, the minimizing sequence converges in itself in the space H_A. But this space is complete; therefore the sequence $\{\phi_k\}$ converges in H_A, and converges to the same element as it does in H. Thus $u_1 \in H_A$, and

$$|||\phi_k - u_1|||_A \underset{k \to \infty}{\to} 0.$$

But then

$$|||u_1|||_A^2 = \lim_{k \to \infty} |||\phi_k|||_A^2 = \lambda_1;$$

at the same time

$$||u_1|| = \lim_{k \to \infty} ||\phi_k|| = 1.$$

Thus there exists an element $u_1 \in H_A$ such that $\|u_1\| = 1$, and $\||u_1|\|_A^2 = \lambda_1$. This implies that the functional (1) does attain its lower bound. By Theorem 6.4.1, this lower bound λ_1 is the smallest eigenvalue, and u_1 is the corresponding eigenelement of the operator A.

§ 6. Discrete Spectrum

Before proceeding to formulate and prove the basic theorem of this section, we make the following observation.

Suppose that we construct the first n eigenvalues

$$\lambda_1 \leq \lambda_2 \leq \ldots \leq \lambda_n$$

and the corresponding eigenelements

$$u_1, u_2, \ldots, u_n$$

of the operator A, these elements being orthonormal in a metric of the space H.

Consider the functional *

$$\Psi_n(u) = \||u|\|_A^2, \quad u \in H_A^{(n)}, \quad \|u\| = 1. \tag{1}$$

This is different from the functional (5.1) because it is defined on a narrower set. We write

$$\lambda_{n+1} = \inf \Psi_n(u) = \inf \||u|\|_A^2,$$

where $u \in H_A^{(n)}$, $\|u\| = 1$.

We construct a minimizing sequence for the functional (1). If we can select from this a subsequence which converges in the metric of the space H, then λ_{n+1} is the $(n+1)$th eigenvalue, and the limit of the selected sequence is the $(n+1)$th eigenelement of the operator A.

The proof of this statement is exactly the same as the proof of Theorem 6.5.1.

DEFINITION. Let the space H be infinite-dimensional. We shall say that the symmetric operator A has a *discrete spectrum* if

1) the operator A has an infinite sequence $\lambda_1, \lambda_2, \ldots, \lambda_n, \ldots$ of eigenvalues with a unique limit point at infinity;

2) the sequence $\{u_n\}$ of the corresponding eigenelements is complete in the space H.

---

* The space $H_A^{(n)}$ was defined in § 4.

The existence of a unique limit point at infinity implies that the eigenvalues can be written out in an increasing sequence of their absolute values

$$|\lambda_1| \leq |\lambda_2| \leq \ldots \leq |\lambda_n| \leq \ldots$$

where $|\lambda_n| \to \infty$. If a positive definite operator has a discrete spectrum, then its eigenvalues can simply be arranged in an increasing sequence

$$0 < \lambda_1 \leq \lambda_2 \leq \ldots \leq \lambda_n \leq \ldots, \qquad \lambda_n \to \infty.$$

THEOREM 6.6.1. *Let a positive definite operator, acting in an infinite-dimensional Hilbert space, be such that any set, bounded in the energy metric, is compact in the metric of the original space. Then the generalized spectrum of this operator is discrete.*

Proof.

1. Consider the quantity

$$\lambda_1 = \inf |||u|||^2, \qquad u \in H_A, \qquad ||u|| = 1.$$

We construct a minimizing sequence $\{\omega_k\}$. This means that

a) $\omega_k \in H_A$; b) $||\omega_k|| = 1$; c) $\lim_{k \to \infty} |||\omega_k|||^2 = \lambda_1$.

A numerical sequence having a limit is bounded, and so there exists a constant C, such that $|||\omega_k||| \leq C$. If so, then the minimizing sequence is bounded in the metric of H_A. By the conditions of the theorem, this sequence is compact in the old metric, and thus, because of Theorem 6.5.1, λ_1 is the smallest eigenvalue of the operator; we denote the corresponding eigenelement by u_1.

2. Suppose that we have already constructed the first n eigenvalues

$$\lambda_1, \lambda_2, \ldots, \lambda_n$$

and the corresponding eigenelements

$$u_1, u_2, \ldots, u_n.$$

We write $\lambda_{n+1} = \inf |||u|||^2$, $u \in H_A^{(n)}$, $||u|| = 1$, and construct a minimizing sequence $\{\omega_k^{(n)}\}$ ($k = 1, 2, \ldots$). Then $|||\omega_k^{(n)}|||^2 \xrightarrow[k \to \infty]{} \lambda_{n+1}$, and, consequently, there exists a constant C such that

$$|||\omega_k^{(n)}||| \leq C.$$

The sequence $\{\omega_k^{(n)}\}$ ($k = 1, 2, \ldots$), is compact in the old metric, and so, by virtue of the remark made at the beginning of this section, λ_{n+1} is the

($n+1$)th eigenvalue of the operator A, and there exists a corresponding eigenelement u_{n+1}.

The process comes to an end if the conditions $\|u\| = 1$ and $u \in H_A^{(n)}$ contradict each other. This can happen when the space $H_A^{(n)}$ consists of just a zero, which is possible if and only if H_A is finite-dimensional. But H_A is dense in H, and will be finite-dimensional if and only if the space H is itself finite-dimensional. Since, however, we are considering only the case of an infinite-dimensional space, the process does not terminate, and we obtain an infinite sequence of eigenvalues

$$\lambda_1 \leq \lambda_2 \leq \ldots \leq \lambda_n \leq \ldots \tag{2}$$

and a sequence of corresponding eigenelements

$$u_1, u_2, \ldots, u_n, \ldots, \tag{3}$$

orthogonal in both H and H_A, and normalized in H.

3. We show that the eigenvalues tend to infinity in the limit. Suppose the converse, and let the sequence $\{\lambda_n\}$ be bounded

$$\lambda_n \leq K = \text{const.}$$

Then

$$|\|u_n\|| = \sqrt{\lambda_n} \leq \sqrt{K};$$

the eigenelements are bounded in H_A, and therefore their sequence is compact in the metric of H. We thus find that a sequence, which is orthogonal and normalized in H, is compact in H, and this we well know to be impossible.

4. We now show that the system of eigenelements is complete in H_A. Suppose the contrary. Consider the subspace $H_A^{(\infty)}$ of the space H_A, which is orthogonal to every eigenelement u_n, $n = 1, 2, \ldots$. This subspace contains elements which are different from zero and, consequently, are normalized. We write

$$\lambda_\infty = \inf |\|u\||^2, \quad u \in H_A^{(\infty)}, \quad \|u\| = 1.$$

Repeating exactly the arguments used above, we find that λ_∞ is an eigenvalue of the operator.

Compare λ_∞ with λ_n. It is the lower bound of the same quantity $|\|u\||_A^2/\|u\|^2$, on the different sets $H_A^{(\infty)}$ and $H_A^{(n)}$. The first set is narrower than the second, and the lower bound on it is greater (at worst, not less) than on the latter. But then $\lambda_\infty \geq \lambda_n$ which is absurd because the quantities λ_n are in aggregate not bounded. From this contradiction we can infer that the sequence $\{u_n\}$ is complete in H_A.

5. We show next that the sequence of eigenelements is complete in H. We take $u \in H_A$. The system (3) is complete in H_A: for any $\varepsilon > 0$ there exists an integer N and quantities $\alpha_1, \alpha_2, \ldots, \alpha_N$, such that

$$|||u - \sum_{k=1}^{N} \alpha_k u_k|||_A < \varepsilon.$$

Then by inequality (3.3), Chapter 5,

$$||u - \sum_{k=1}^{N} \alpha_k u_k|| < \frac{\varepsilon}{\gamma}.$$

Thus, any element of the energy space can be approximated by a linear combination of elements (3) in the old metric.

Now let $u \in H$. The space H_A is dense in H, i.e. for any positive number ε there exists $u' \in H_A$, such that

$$||u - u'|| < \tfrac{1}{2}\varepsilon.$$

We choose a number N and coefficients $\alpha_1, \alpha_2, \ldots, \alpha_N$, such that the inequality

$$||u' - \sum_{k=1}^{N} \alpha_k u_k|| < \tfrac{1}{2}\varepsilon$$

is satisfied. Now by the triangle inequality

$$||u - \sum_{k=1}^{N} \alpha_k u_k|| < \varepsilon,$$

and the theorem is proved.

§ 7. The Sturm-Liouville Problem

Consider the operator

$$Au = -\frac{d}{dx}\left[p(x)\frac{du}{dx}\right] + q(x)u \tag{1}$$

on the set $D(A)$ of functions which are continuous on the segment $[a, b]$, have an absolutely continuous first derivative and a square-integrable second derivative, and are subject to the boundary conditions

$$u(a) = u(b) = 0. \tag{2}$$

We impose on $p(x)$ and $q(x)$ the same conditions as in § 8, Chap. 5, namely, the functions $p(x)$, $p'(x)$, $q(x)$ are continuous on the segment $[a, b]$; $p(x) \geq p_0$, where p_0 is a positive constant; $q(x) \geq 0$. The problem is to examine the spectrum of the operator A.

We show that this operator has a discrete spectrum in $L_2(a, b)$. We know that the operator A is positive definite, and so it is sufficient to prove that every set which is bounded in H_A is compact in

$$H = L_2(a, b).$$

By formula (8.7), Chap. 5, we have

$$|||u|||^2 = \int_a^b [p(x)u'^2 + q(x)u^2]\,dx.$$

Hence

$$|||u|||^2 \geq p_0 \int_a^b u'^2\,dx. \tag{3}$$

The rest depends on the following theorem, which for the time being we shall assume without proof. A more general theorem will be established later, in § 2, Chapter 7 (Theorem 7.2.2).

THEOREM 6.7.1. *Let the function $K(x, t)$ be defined almost everywhere in the square $a \leq x \leq b, a \leq t \leq b$; and be measurable and bounded in this square. Then the integral operator*

$$\int_a^b K(x, t)u(t)\,dt \tag{4}$$

transforms every set of functions bounded in $L_2(a, b)$ into a set of functions which is compact in this space.

For functions $u \in H_A$ we have the equation

$$u(x) = \int_a^x u'(t)\,dt. \tag{5}$$

Introduce a bounded function

$$K(x, t) = \begin{cases} 1, & a \leq t \leq x, \\ 0, & x < t \leq b \end{cases}$$

and rewrite the last equation in the form

$$u(x) = \int_a^b K(x, t)u'(t)\,dt.$$

Suppose now we are given a set $\mathfrak{M}$ of functions, bounded in energy metric

$$\forall u \in \mathfrak{M}, \qquad |||u||| \leq C = \text{const.}, \qquad \mathfrak{M} \subset H_A. \tag{6}$$

Then from inequality (3) it follows that

$$\int_a^b u'^2(t)\,dt \leq \frac{C^2}{p_0}, \qquad \|u'\| \leq \frac{C}{\sqrt{p_0}}.$$

Thus the set of derivatives u', where $u \in \mathfrak{M}$, is bounded in $L_2(a, b)$. And since the operator (5) transforms any set which is bounded in $L_2(a, b)$, into a set compact in the same space, therefore $\mathfrak{M}$ is compact in $L_2(a, b)$. By Theorem 6.6.1, the spectrum of the operator A is discrete: this operator has an infinite sequence of eigenvalues

$$0 < \lambda_1 \leq \lambda_2 \leq \ldots \leq \lambda_n \leq \ldots, \qquad \lambda_n \to \infty \tag{7}$$

and corresponding eigenelements

$$u_1(x), u_2(x), \ldots, u_n(x), \ldots, \tag{7'}$$

with respect to which we can take $\|u\| = 1$ and $(u_k, u_m) = 0$ $(k \neq n)$; the system (7') is complete in each of the spaces $L_2(a, b)$ and H_A. In the energy metric the eigenfunctions are orthogonal, as before: $[u_k, u_m] = 0$ $(k \neq m)$, but they are not normalized in this metric, since $|||u_n||| = \sqrt{\lambda_n}$.

At this point we recall that the problem being considered in this section, namely, that of finding the spectrum of the operator A, is equivalent to the following problem: find those values of the parameter λ for which there exist non-trivial (generalized) solutions to the equation

$$\frac{d}{dx}\left(p(x)\frac{du}{dx}\right) - q(x)u + \lambda u = 0,$$

satisfying boundary conditions (2).

We can pose a more general problem. Consider the equation

$$\frac{d}{dx}\left(p(x)\frac{du}{dx}\right) - q(x)u + \lambda r(x)u = 0 \tag{8}$$

with boundary conditions (2); the coefficients $p(x)$, $q(x)$ are subject to the same conditions as previously, and assume in addition that $r \in C[a, b]$ and that $r(x) \geq r_0 = \text{const.} > 0$. Analysis of the spectrum for this problem falls within the general scheme of the present chapter.

Dividing equation (8) by $r(x)$, we convert it to the form

$$\frac{1}{r(x)}\left[\frac{d}{dx}\left(p(x)\frac{du}{dx}\right) - q(x)u\right] + \lambda u = 0. \tag{9}$$

Introduce now the space $L_2(r; a, b)$ of functions, which are square-integrable on the interval (a, b) with respect to the weighting factor $r(x)$;* the norm and scalar product in this space are defined by the formulae

$$\|u\|^2 = \int_a^b r(x)u^2(x)\,dx, \quad (u, v) = \int_a^b r(x)u(x)v(x)\,dx. \tag{10}$$

Consider an operator B in this space, which acts according to the formula

$$Bu = \frac{1}{r(x)}\left[-\frac{d}{dx}\left(p(x)\frac{du}{dx}\right) + q(x)u\right]. \tag{11}$$

We define this operator on the same set of functions as the operator A considered above; thus, $D(B)$ is a set of functions which are continuous on the segment $[a, b]$ and satisfy conditions (2); the first derivatives of these functions on the segment are absolutely continuous, and the second derivatives are square-integrable.

The operator B is positive definite in the space $H = L_2(r; a, b)$. It is certainly symmetric: if $u, v \in D(B)$; then

$$(Bu, v) = \int_a^b u\left[-\frac{d}{dx}\left(p\frac{dv}{dx}\right) + qv\right]dx$$

$$= \int_a^b \left(p\frac{du}{dx}\frac{dv}{dx} + quv\right)dx = (u, Bv).$$

The fact that it is positive definite can be demonstrated as follows. First of all

$$(Bu, u) = \int_a^b (pu'^2 + qu^2)\,dx \geq p_0 \int_a^b u'^2\,dx. \tag{12}$$

The function $u(x)$ is equal to zero at the end-points of the segment $[a, b]$, therefore $u(a) = 0$, and

$$\sqrt{r(x)}\,u(x) = \sqrt{r(x)} \int_a^x u'(t)\,dt.$$

---

* Obviously the spaces $L_2(r; a, b)$ and $L_2(a, b)$ are composed of precisely the same functions.

Being continuous on the segment, the functions $r(x)$ are bounded. Let $r(x) \leq r_1$, then

$$r(x)u^2(x) \leq r_1 \left(\int_a^x u'(t)\,dt\right)^2 \leq r_1(x-a)\int_a^x u'^2(t)\,dt \leq r_1(b-a)\int_a^b u'^2(t)\,dt.$$

Integrating this with respect to x between the limits a and b, we find

$$\int_a^b u'^2(t)\,dt \geq \frac{1}{r_1(b-a)^2}\int_a^b r(x)u^2(x)\,dx = \frac{1}{r_1(b-a)^2}\|u\|^2;$$

and finally

$$(Bu, u) \geq \frac{p_0}{r_1(b-a)^2}\int_a^b r(x)u^2(x)\,dx = \frac{p_0}{r_1(b-a)^2}\|u\|^2,$$

which proves the positive definiteness of our operator.

We show finally that any set bounded in H_B is compact in $H = L_2(r; a, b)$. It is not difficult to establish that the space H_B consists of the same functions as the energy space of the operator (1)–(2); in particular, these functions satisfy conditions (2), and therefore also formula (5).

Let $\mathfrak{M} \subset H_B$ be a set of functions bounded in the norm of H_B

$$\|\|u\|\|_B \leq C = \text{const.}, \qquad u \in \mathfrak{M}.$$

From formula (12) it is easy to see that

$$\|\|u\|\|_B^2 = \int_a^b (pu'^2 + qu^2)\,dx \geq p_0 \int_a^b u'^2\,dx,$$

and, consequently,

$$\|u'\|^2 = \int_a^b u'^2\,dx \leq \frac{C}{p_0}, \qquad u \in \mathfrak{M}. \tag{13}$$

Formula (5) can be converted to the form

$$\sqrt{r(x)}\,u(x) = \int_a^b K_1(x, t)u'(t)\,dt, \tag{14}$$

where

$$K_1(x, t) = \begin{cases} \sqrt{r(x)}, & a \leq t < x, \\ 0, & x < t \leq b. \end{cases}$$

The function $K_1(x, t)$ is bounded, and then the integral operator (14) transforms the set of functions $u'(x)$, bounded in $L_2(a, b)$ (cf. inequality (13)!)

into the set of functions $\sqrt{r(x)}u(x)$, which is compact in $L_2(a, b)$. This implies the following: from the set $\mathfrak{M}$ we can select a sequence $\{u_n(x)\}$ such that

$$\|\sqrt{r}u_n - \sqrt{r}u_m\|^2_{L_2(a,\, b)} = \int_a^b r(x)[u_n(x) - u_m(x)]^2\, dx \xrightarrow[n,\, m \to 0]{} 0.$$

But the last relation simply implies that

$$\|u_n - u_m\|_{L_2(r;\, a,\, b)} \xrightarrow[n,\, m = \infty]{} 0.$$

Thus, it is possible to pick out from the set $\mathfrak{M}$ a sequence which converges in itself in the norm of $L_2(r; a, b)$; this means that the set $\mathfrak{M}$ is compact in the space $L_2(r; a, b)$. By Theorem 6.6.1, the operator B has a discrete spectrum, or, in other words, there exists an enumerable set of numbers $\lambda_n > 0$, $\lambda_n \xrightarrow[n \to \infty]{} \infty$ for which the problem defined by equations (8) and (2) has non-trivial solutions, and the totality of these solutions is complete both in $L_2(r; a, b)$ and in H_B. If, as before, these solutions are denoted by $u_n(x)$, then they are orthonormal in $L_2(r; a, b)$ and orthogonal in H_B

$$\int_a^b r(x) u_n(x) u_m(x)\, dx = \delta_{mn},$$

$$\int_a^b [p(x) u_n'(x) u_m'(x) + q(x) u_m(x) u_n(x)]\, dx = 0, \qquad m \neq n.$$

In addition

$$\int_a^b [p(x) u_n'^2(x) + q(x) u_n^2(x)]\, dx = \lambda_n.$$

The eigenvalues λ_n are all simple – this follows from the fact that the differential equation (8) is of second order. Thus, suppose that to an eigenvalue λ_n there correspond two linearly-independent eigenfunctions: $u_n(x)$ and $u_m(x)$. First of all, $u_n'(0) \neq 0$; for otherwise the function $u_n(x)$, not identically zero, would be a solution of Cauchy's problem for the homogeneous equation

$$\frac{d}{dx}\left(p \frac{du}{dx}\right) - qu + \lambda_n r u = 0 \qquad (15)$$

with the homogeneous boundary conditions

$$u_n(0) = u_n'(0) = 0,$$

and this contradicts the uniqueness theorem for Cauchy's problem. Similarly $u'_m(0) \neq 0$. For now the function

$$u(x) = \frac{u_n(x)}{u'_n(0)} - \frac{u_m(x)}{u'_m(0)},$$

not identically zero, would solve the same homogeneous Cauchy problem, which is impossible.

§ 8. Elementary Cases

The actual determination of eigenvalues and eigenfunctions on the basis of the theorems of §§ 4–7 is fraught with practical difficulties. Therefore it is interesting to study those special cases where the spectrum of the operator can be found by elementary means. We discuss two such cases below.

1. Consider the operator A (§ 7) in the simplest special case, when $p(x) \equiv 1$, $q(x) \equiv 0$. The problem now reduces to finding those values λ for which the differential equation

$$\frac{d^2 u}{dx^2} + \lambda u = 0 \tag{1}$$

has non-trivial solutions, satisfying the conditions

$$u(a) = u(b) = 0. \tag{2}$$

The general integral of equation (1) can be written in the form:

$$u(x) = C \sin \sqrt{\lambda}(x-a) + C_1 \cos \sqrt{\lambda}(x-a).$$

The condition $u(a) = 0$ gives $C_1 = 0$ and $u(x) = C \sin \sqrt{\lambda}(x-a)$. From the condition $u(b) = 0$ we find $C \sin \sqrt{\lambda}(b-a) = 0$. Here $C \neq 0$ necessarily, as otherwise we would have only the trivial solution $u = 0$. Thus $\sin \sqrt{\lambda}(b-a) = 0$, and we find the eigenvalues

$$\lambda_n = \frac{n^2 \pi^2}{(b-a)^2}, \quad n = 1, 2, \ldots, \tag{3}$$

and eigenfunctions

$$u_n(x) = C_n \sin \frac{n\pi(x-a)}{b-a}. \tag{4}$$

The constants C_n are obtained from the normalization condition

$$||u_n||^2 = C_n^2 \int_a^b \sin^2 n\pi \frac{x-a}{b-a} \, dx = 1,$$

whence $C_n = \sqrt{\{2/(b-a)\}}$, and

$$u_n(x) = \sqrt{\frac{2}{b-a}} \sin \frac{n\pi(x-a)}{b-a}. \tag{4a}$$

2. We find non-trivial solutions of equation (1) for boundary conditions

$$u'(a) = u'(b) = 0. \tag{5}$$

As before, the general solution is $u(x) = C \sin \sqrt{\lambda}(x-a) + C_1 \cos \sqrt{\lambda}(x-a)$. From the condition $u'(a) = 0$ it follows that $C = 0$, and the condition $u'(b) = 0$ gives that $\sin \sqrt{\lambda}(b-a) = 0$. Hence we have the eigenvalues

$$\lambda_n = \frac{n^2 \pi^2}{(b-a)^2}, \quad n = 0, 1, 2, \ldots,$$

and normalized eigenfunctions

$$u_n(x) = \sqrt{\frac{2}{b-a}} \cos \frac{n\pi(x-a)}{b-a}, \quad n = 0, 1, 2, \ldots$$

§ 9. The Minimax Principle

Let A be a positive definite operator satisfying the conditions of Theorem 6.6.1: any set bounded in the energy metric is compact in the metric of the original space. Then the spectrum of this operator is discrete; let λ_n and u_n, $n = 1, 2, \ldots$, be the eigenvalues and the corresponding eigenelements of the operator A, the elements being orthonormal in the original space. We now pose the following problem: to find the minimum of the functional

$$\Phi_A(u) = |||u|||_A^2 \tag{1}$$

on the set of elements of the energy space H_A, satisfying the additional conditions

$$||u||^2 = 1 \tag{2}$$

and

$$(u, v_1) = 0, \quad (u, v_2) = 0, \ldots, (u, v_{k-1}) = 0, \tag{3}$$

where $v_1, v_2, \ldots, v_{k-1}$ are fixed elements of the original space H. We shall consider the set of elements described here as the domain of definition of the functional Φ_A, and denote this domain by $D(\Phi_A)$. We show that the minimum $\Phi_A(u)$ is attained on the set $D(\Phi_A)$. In the first place we observe that the functionals (u, v_j) are bounded in H_A

$$|(u,v_j)| \leq \|u\| \cdot \|v_j\| \leq \frac{\|v_j\|}{\gamma_0} \||u\||_A,$$

where γ_0 is the lower bound of the operator A. We denote by $\mathfrak{H}_k$ a set of elements of the space H_A satisfying conditions (3). This set is obviously linear. It is also closed in H_A: if $u_n \in \mathfrak{H}_k$, $\||u_n - u\||_A \xrightarrow[n \to \infty]{} 0$ and $(u_n, v_j) = 0$, then by virtue of the boundedness of the functionals (3) in the space H_A, we have $(u, v_j) = 0$. Hence it follows that $\mathfrak{H}_k$ is a subspace of H_A.

Our variational problem can now be formulated as follows: to find the minimum of the functional (1) on the set of elements of the subspace $\mathfrak{H}_k$, satisfying the additional condition (2). It is sufficient now to repeat the arguments of Theorem 6.5.1, in order to establish that there exists in $\mathfrak{H}_k$ an element w, with $\|w\| = 1$, which achieves a minimum of our functional. This minimum we denote by $\lambda(v_1, v_2, \ldots, v_{k-1})$.

The minimax principle consists of the equation

$$\max \lambda(v_1, v_2, \ldots, v_{k-1}) = \lambda_k; \tag{4}$$

the maximum is achieved for all possible choices of the elements $v_1, v_2, \ldots, v_{k-1}$, belonging to the original space H. The proof of the minimax principle reduces to the verification of two facts: 1) $\lambda(v_1, v_2, \ldots, v_{k-1}) \leq \lambda_k$; 2) there exist elements $v_j^{(0)} \in H$, such that $\lambda(v_1^{(0)}, v_2^{(0)}, \ldots, v_{k-1}^{(0)}) = \lambda_k$. We verify these facts.

Let u be an arbitrary element of the energy space H_A. The system $\{u_n\}$ is orthonormal and complete in the space H; we expand the elements u and v_j with respect to this system

$$u = \sum_{n=1}^{\infty} a_n u_n, \tag{5}$$

$$v_j = \sum_{n=1}^{\infty} b_{jn} u_n, \quad j = 1, 2, \ldots, k.$$

The system $\{u_n\}$ is orthogonal and complete in H_A; where $\||u_n\||_A^2 = \lambda_n$. But then the system $\{u_n/\sqrt{\lambda_n}\}$ is orthonormal and complete in H_A; the expansion of the element $u \in H_A$ with respect to this system obviously has the form

$$u = \sum_{n=1}^{\infty} \sqrt{\lambda_n}\, a_n \frac{u_n}{\sqrt{\lambda_n}}. \tag{6}$$

By the closure equation

$$|||u|||_A^2 = \sum_{n=1}^{\infty} \lambda_n a_n^2. \tag{7}$$

For the element u we take the finite sum

$$\bar{u} = \sum_{n=1}^{k} a_n u_n, \tag{8}$$

where the numbers a_n are arbitrary. If we insist that the element (8) should satisfy conditions (3), then we obtain a system of $k-1$ linear, homogeneous equations in k unknowns $a_1, a_2, \ldots, a_k$

$$\sum_{n=1}^{k} b_{j_n} a_n = 0, \quad j = 1, 2, \ldots, k-1. \tag{9}$$

The number of equations is less than the number of unknowns; therefore the system (9) has an infinite set of solutions. From among these we can select at least one, such that

$$\|\bar{u}\|^2 = \sum_{n=1}^{k} a_k^2 = 1.$$

Then $\bar{u} \in D(\Phi_A)$. The formula (7) gives

$$|||\bar{u}|||_A^2 = \sum_{n=1}^{k} \lambda_n a_n^2.$$

Substituting for all the λ_n the largest of them, λ_k, we obtain

$$|||\bar{u}|||_A^2 \leq \lambda_k \sum_{n=1}^{k} a_k^2 = \lambda_k.$$

But $\bar{u}$ is one of the elements of the set $D(\Phi_A)$, and so *a fortiori*,

$$\lambda(v_1, v_2, \ldots, v_{k-1}) = \min_{u \in D(\Phi_A)} \Phi_A(u) \leq \lambda_k.$$

It can be shown quite easily that the equality sign here is attained: we have only to take

$$v_j^0 = u_j, \quad j = 1, 2, \ldots, k.$$

The validity of the minimax principle is established.

From the minimax principle there follows an important theorem which in many cases allows comparison to be made between the eigenvalues of two operators. Before formulating this theorem, we introduce one new concept.

Let A and B be positive definite operators, acting in one and the same Hilbert space H. We shall say that *the operator A is not less than the operator B*, and write this in the form $A \geqq B$ or $B \leqq A$, if 1) any element of the space H_A belongs to the space H_B; 2) for any element $u \in H_A$, the inequality

$$|||u|||_A \geqq |||u|||_B \tag{10}$$

holds.

THEOREM 6.9.1. *Let A and B be positive definite operators satisfying the conditions of Theorem 6.6.1, and let $A \geqq B$. If λ_k and μ_k are eigenvalues of the operators A and B respectively, set out in sequences of increasing magnitude, then*

$$\lambda_k \geqq \mu_k, \qquad k = 1, 2, \ldots \tag{11}$$

Denote by $\lambda(v_1, v_2, \ldots, v_{k-1})$ and $\mu(v_1, v_2, \ldots, v_{k-1})$ the minima of the functionals $|||u|||_A^2$ and $|||u|||_B^2$, subject to conditions (2) and (3). Denote by $\tilde{u}$ the element on which the first minimum is attained. By inequality (10),

$$\lambda(v_1, v_2, \ldots, v_{k-1}) = |||\tilde{u}|||_A^2 \geqq |||\tilde{u}|||_B^2 \geqq \min |||u|||_B^2 = \mu(v_1, v_2, \ldots, v_{k-1}).$$

But then

$$\max \lambda(v_1, v_2, \ldots, v_{k-1}) \geqq \max \mu(v_1, v_2, \ldots, v_{k-1}),$$

which is identical with inequality (11).

§ 10. Magnitude of the Eigenvalues in the Sturm-Liouville Problem

Denote by λ_n the eigenvalues of the operator in the Sturm-Liouville problem

$$Au = -\frac{d}{dx}\left(p\frac{du}{dx}\right) + qu, \qquad u(a) = u(b) = 0. \tag{1}$$

We impose on the coefficients $p(x)$ and $q(x)$ the same limitations as before: $p(x)$, $p'(x)$, $q(x)$ continuous, $p(x) \geqq p_0$, $q(x) \geqq 0$ on the segment $[a, b]$. In § 8, Chapter 5, we saw that the set of functions which generate the energy space of operator (1) does not depend on the coefficients $p(x)$ and $q(x)$, and that

$$|||u|||_A^2 = \int_a^b \left[p(x)\left(\frac{du}{dx}\right)^2 + q(x)u^2 \right] dx. \tag{2}$$

The functions $p(x)$ and $q(x)$, continuous on the segment, are bounded:
$$p(x) \leq p_1, \qquad q(x) \leq q_1, \qquad x \in [a, b].$$
We put
$$A_0 u = -p_0 \frac{d^2 u}{dx^2}, \qquad u(a) = u(b) = 0,$$

$$A_1 u = -p_1 \frac{d^2 u}{dx^2} + q_1 u, \qquad u(a) = u(b) = 0.$$

The operators A_0 and A_1 are special cases of the operator A, obtained by putting $p(x) \equiv p_0$, $q(x) \equiv 0$ and $p(x) \equiv p_1$, $q(x) \equiv q_0$ respectively. From formula (2) it follows that

$$|||u|||^2_{A_0} = p_0 \int_a^b \left(\frac{du}{dx}\right)^2 dx, \qquad |||u|||^2_{A_1} = p_1 \int_a^b \left(\frac{du}{dx}\right)^2 dx + q_1 \int_a^b u^2 \, dx.$$

It is clear that
$$|||u|||^2_{A_0} \leq |||u|||^2_A \leq |||u|||^2_{A_1},$$
and, consequently,
$$A_0 \leq A \leq A_1.$$

If μ_k and ν_k are eigenvalues of the operators A_0 and A_1 respectively, then by Theorem 6.9.1,
$$\mu_k \leq \lambda_k \leq \nu_k, \qquad k = 1, 2, \ldots. \tag{3}$$

The values μ_k and ν_k are easily found.

The numbers μ_k are eigenvalues for the problem
$$p_0 \frac{d^2 u}{dx^2} + \mu u = 0, \qquad u(a) = u(b) = 0.$$

Putting $\mu/p_0 = \lambda$ here, we arrive back at the problem of paragraph 1, § 8. Therefore
$$\mu_k = \frac{p_0 \pi^2 k^2}{(b-a)^2}. \tag{4}$$

In exactly the same way the quantities ν_k are eigenvalues for the problem
$$p_1 \frac{d^2 u}{dx^2} + (\nu - q_1) u = 0, \qquad u(a) = u(b) = 0,$$

and comparison with the results of § 8 gives

$$v_k = \frac{p_1 \pi^2 k^2}{(b-a)^2} + q_1. \tag{5}$$

The relations (3)–(5) give an inequality which defines the order of magnitude of the eigenvalues of the Sturm-Liouville problem:

$$\frac{p_0 \pi^2 k^2}{(b-a)^2} \leq \lambda_k \leq \frac{p_1 \pi^2 k^2}{(b-a)^2} + q_1. \tag{6}$$

Exercises

1. Show that the spectrum of the operator T_p of Exercises 1 and 2, Chapter 5, is discrete.

PART III

Elements of the Theory of Integral Equations

Chapter 7.
Completely Continuous Operators

§ 1. Necessary Results from Functional Analysis

In this section we present some concepts and facts which are needed for the study of integral equations. In the main, the results are given without proof, but an extensive discussion of the questions involved can be found, for example, in books by L. V. Kantorovich and G. P. Akilov [1] or F. Riesz and B. Sz.-Nagy [5], which are listed in the bibliography to this part.

1. A linear operator, acting from a Banach space X into another Banach space Y, and defined on a dense set in X, is called *completely continuous* if it transforms any bounded set from its domain of definition into a compact set in Y. Instead of the words „completely continuous operator", we shall often just use the initials c.c.o.

2. Every c.c.o. is bounded. The converse is true for finite-dimensional spaces, but not for infinite-dimensional spaces. For example, the identity operator in an infinite-dimensional space is not completely continuous.

A c.c.o. which is bounded and prescribed on a dense set can be extended by continuity over the whole of the space X. Henceforth we shall always assume that this extension has already been carried out.

3. The sum of a finite number of c.c.o.'s is a c.c.o. The product (irrespective of order) of two operators, one of which is completely continuous and the other bounded, is a c.c.o.

4. We call an operator of the type

$$Tu = \sum_{k=1}^{n} l_k(u)v_k, \tag{1}$$

a finite-dimensional operator, when the number n is finite and independent of u, the l_k are linear functionals bounded in X, and the v_k are fixed elements of Y.

Every finite-dimensional operator is a c.c.o.

5. Let $T_n, n = 1, 2, \ldots$ be a sequence of c.c.o.'s acting from X into Y,

and let there exist an operator T, also acting from X into Y, such that

$$\lim_{n \to \infty} \|T - T_n\| = 0.$$

Then T is a c.c.o.

6. If T is a c.c.o., then the operator T^*, the adjoint of T, will also be completely continuous.

Henceforth we shall only consider cases when the spaces X and Y coincide.

THEOREM 7.1.1. *Let T be a c.c.o. acting in the Hilbert space H. For any arbitrary number $\varepsilon > 0$, it is possible to construct a finite-dimensional operator T_ε, such that*

$$\|T - T_\varepsilon\| \leq \varepsilon. \tag{2}$$

Denote by Σ the unit sphere in H, i.e. the set of elements of the space H, whose norms are equal to unity. Denote by $T(\Sigma)$ the set into which the operator T transforms the set Σ. This latter set is bounded, and the operator T is completely continuous, therefore the set $T(\Sigma)$ is compact. It is well-known from Hausdorff's theorem that for any $\varepsilon > 0$ there exists a finite ε-net for the set $T(\Sigma)$; i.e. there exists a finite number r of elements $v_k \in H$, $k = 1, 2, \ldots, r$, having the following properties: for any arbitrary element $u \in \Sigma$ we can find an element v_j, such that

$$\|Tu - v_j\| \leq \varepsilon. \tag{3}$$

From the finite sequence $v_1, v_2, \ldots, v_r$ we can eliminate those elements which are linearly dependent on the others; the remaining elements (their number is denoted by s) are subject to the Gram-Schmidt orthogonalization procedure. As a result we obtain s elements $\phi_1, \phi_2, \ldots, \phi_s$, such that $(\phi_j, \phi_k) = \delta_{jk}$, and every element of the ε-net can be represented by a linear combination of the $\phi_1, \phi_2, \ldots, \phi_s$:

$$v_k = \sum_{l=1}^{s} a_{kl} \phi_l.$$

The inequality (3) now takes the form

$$\|Tu - \sum_{l=1}^{s} a_{jl} \phi_l\| \leq \varepsilon.$$

We now put

$$T_\varepsilon u = \sum_{l=1}^{s} (Tu, \phi_l) \phi_l = \sum_{l=1}^{s} (u, T^*\phi_l) \phi_l. \tag{4}$$

From standard theory of orthogonal series, we know that

$$\|Tu - T_\varepsilon u\| \leq \|Tu - \sum_{l=1}^{s} a_{jl}\phi_l\| \leq \varepsilon.$$

This inequality holds for any element $u \in H, \|u\| = 1$. But then $\|T - T_\varepsilon\| \leq \varepsilon$; at the same time, equation (4) shows that the operator T_ε is finite-dimensional. Hence the theorem is proved.

§ 2. Fredholm Operator

Let Ω be a measurable set in m-dimensional Euclidean space; this set may be bounded or unbounded. Let x and ξ be arbitrary points of Ω. A measurable function $K(x, \xi)$ of these points is called a *Fredholm kernel* if

$$\int_\Omega \int_\Omega K^2(x, \xi) \, dx \, d\xi < \infty \tag{1}$$

(we are restricting ourselves here to the case of a real function). The integral operator

$$(Ku)(x) = \int_\Omega K(x, \xi) u(\xi) \, d\xi, \tag{2}$$

where $K(x, \xi)$ is a Fredholm kernel, is called a *Fredholm operator*.

We note one important class of Fredholm kernels. If the set Ω has finite Lebesgue measure (for example, if it is bounded), and if the kernel $K(x, \xi)$ is bounded, then it is a Fredholm kernel. Thus, if $|K(x, \xi)| \leq C = \text{const.}$, then

$$\int_\Omega \int_\Omega |K^2(x, \xi)| \, dx \, d\xi \leq C^2 (\text{mes } \Omega)^2,$$

and inequality (1) is fulfilled.

THEOREM 7.2.1. *The Fredholm operator is defined on the whole of the space $L_2(\Omega)$, and is bounded in this space; in particular*

$$\|K\| \leq \left\{ \int_\Omega \int_\Omega K^2(x, \xi) \, dx \, d\xi \right\}^{\frac{1}{2}}. \tag{3}$$

Let $u \in L_2(\Omega)$. We show that in this case the integral (2) exists for almost all $x \in \Omega$, and represents a square-integrable function of x in Ω. We have

$$|K(x, \xi) u(\xi)| \leq \tfrac{1}{2} K^2(x, \xi) + \tfrac{1}{2} u^2(\xi). \tag{4}$$

The integral (1) converges; by Fubini's theorem, the function $K^2(x, \xi)$ is integrable with respect to ξ for almost all $x \in \Omega$; the function $u^2(\xi)$ is simply integrable with respect to ξ. Thus the right-hand side of inequality (4) is integrable with respect to ξ for almost all x. But then the left-hand side of inequality (4) also has these properties, and so integral (2) exists for almost all $x \in \Omega$.

By the Schwartz-Bunyakovskii inequality,

$$|(Ku)(x)|^2 \leq \int_\Omega K^2(x, \xi)\,d\xi \int_\Omega u^2(\xi)\,d\xi = \|u\|^2 \int_\Omega K^2(x, \xi)\,d\xi.$$

Integrating this with respect to x, we obtain the inequality

$$\|Ku\|^2 \leq \|u\|^2 \int_\Omega \int_\Omega K^2(x, \xi)\,dx\,d\xi,$$

which is identical with inequality (3). The theorem is proved.

THEOREM 7.2.2. *The Fredholm operator is completely continuous in $L_2(\Omega)$.*

Denote by $\Omega \times \Omega$ a set of points in $2m$-dimensional Euclidean space, defined in the following way: the set consists of points $(z_1, z_2, \ldots, z_{2m})$ having the property that both $(z_1, z_2, \ldots, z_m)$ and $(z_{m+1}, z_{m+2}, \ldots, z_{2m})$ belong to the set Ω. If $x \in \Omega$ and $\xi \in \Omega$, then any function of x and ξ can be regarded as a function of points of the set $\Omega \times \Omega$. Obviously the converse is also true. Inequality (1) implies that the Fredholm kernel can be regarded as an element of the space $L_2(\Omega \times \Omega)$.

The space $L_2(\Omega)$ is separable; in this space we select a complete, enumerable, orthonormal sequence $\phi_k(x)$, $k = 1, 2, \ldots$. Then the sequence

$$\phi_k(x)\phi_n(\xi); \qquad k, n = 1, 2, \ldots, \tag{5}$$

is orthonormal in $L_2(\Omega \times \Omega)$; we show that this sequence is complete in this space.

Suppose that some function $\omega(x, \xi) \in L_2(\Omega \times \Omega)$ is orthogonal to every function of the sequence (5):

$$\int_\Omega \int_\Omega \omega(x, \xi)\phi_k(x)\phi_n(\xi)\,dx\,d\xi = 0; \qquad k, n = 1, 2, \ldots.$$

Changing the order of integration in the multiple integral, we obtain

$$\int_\Omega \phi_k(x) \left\{ \int_\Omega \omega(x, \xi)\phi_n(\xi)\,d\xi \right\} dx = 0; \qquad k, n = 1, 2, \ldots. \tag{6}$$

We fix the number n and put

$$\int_\Omega \omega(x, \xi)\phi_n(\xi)\,d\xi = \omega_n(x).$$

The function $\omega(x, \xi)$ can be regarded as a Fredholm kernel: from Theorem 7.2.1 above it follows that $\omega_n \in L_2(\Omega)$.

Equation (6) takes the form

$$\int_\Omega \omega_n(x)\phi_k(x)\,dx = 0, \qquad k = 1, 2, \ldots.$$

But the sequence $\{\phi_k(x)\}$ is complete in $L_2(\Omega)$, so that

$$\omega_n(x) = \int_\Omega \omega(x, \xi)\phi_n(\xi)\,d\xi = 0, \qquad n = 1, 2, \ldots.$$

This equation holds for almost all $x \in \Omega$. We hold fixed one such value of x. Then the function of ξ, equal to $\omega(x, \xi)$, is orthogonal to the complete system $\{\phi_n(\xi)\}$. Hence it follows that $\omega(x, \xi) \equiv 0$ for almost all ξ, and the completeness of the system (5) is proven.

The function $K(x, \xi)$ can be expanded in a Fourier series with respect to the system (5). Let this series have the form

$$K(x, \xi) = \sum_{k, n=1} A_{kn}\phi_k(x)\phi_n(\xi).$$

We now put

$$K_\varepsilon(x, \xi) = \sum_{k, n=1}^N A_{kn}\phi_k(x)\phi_n(\xi),$$

and take the number N arbitrarily large, so that

$$\sum_{k>N \text{ or } n>N} A_{kn}^2 < \varepsilon^2;$$

this can be done because the series $\sum_{k, n=1}^\infty A_{kn}^2$ converges.

We put now

$$K'_\varepsilon(x, \xi) = K(x, \xi) - K_\varepsilon(x, \xi)$$

and denote by K_ε and K'_ε respectively Fredholm operators with kernels $K_\varepsilon(x, \xi)$ and $K'_\varepsilon(x, \xi)$. The operator K_ε is finite-dimensional, since

$$(K_\varepsilon u)(x) = \int_\Omega \sum_{k, n=1}^N A_{kn}\phi_k(x)\phi_n(\xi)u(\xi)\,d\xi$$

$$= \sum_{k, n=1}^N A_{kn}\phi_k(x)\int_\Omega \phi_n(\xi)u(\xi)\,d\xi = \sum_{n=1}^N (u, \phi_n)\psi_n(x),$$

where

$$\psi_n(x) = \sum_{k=1}^{N} A_{kn} \phi_k(x).$$

We estimate the norm of the operator K'_ε. We have

$$K'_\varepsilon(x, \xi) = \sum_{k>N \text{ or } n>N} A_{kn} \phi_k(x) \phi_n(\xi).$$

By the closure equation,

$$\int_\Omega \int_\Omega [K'_\varepsilon(x, \xi)]^2 \, dx \, d\xi = \sum_{k>N \text{ or } n>N} A_{kn}^2 < \varepsilon^2,$$

and from formula (3) it follows that $||K'_\varepsilon|| < \varepsilon$. Now we have $||K - K_\varepsilon|| < \varepsilon$; if we put $\varepsilon \to 0$, then $||K - K_\varepsilon|| \to 0$. Because of the statement in paragraph 5 of § 1, the Fredholm operator is completely continuous.

§ 3. Integral Operator with a Weak Singularity

Let Ω be a bounded, measurable set in m-dimensional Euclidean space, x and ξ be points of this set, and $r = |x - \xi|$ be the distance between these points. Let $A(x, \xi)$ be a function measurable and bounded on $\Omega \times \Omega$:

$$|A(x, \xi)| \leq C = \text{const.} \tag{1}$$

The function

$$K(x, \xi) = \frac{A(x, \xi)}{r^\alpha}, \qquad \alpha = \text{const.}, \qquad 0 \leq \alpha < m, \tag{2}$$

of the points x, ξ, is called a *kernel with a weak singularity*, and the operator K, defined by the formula

$$(Ku)(x) = \int_\Omega K(x, \xi) u(\xi) \, d\xi = \int_\Omega \frac{A(x, \xi)}{r^\alpha} u(\xi) \, d\xi, \tag{3}$$

is called an *integral operator with a weak singularity*.

THEOREM 7.3.1. *The integral operator with a weak singularity is defined on all the space $L_2(\Omega)$, and is bounded in this space. The norm of this operator does not exceed the value*

$$\frac{C|S_1|H^{m-\alpha}}{m - \alpha}, \tag{4}$$

where C is the constant of inequality (1), H is the upper bound of the distances between points of the set Ω (the diameter of the set).

We recall that

$$|S_1| = \frac{2\pi^{\frac{1}{2}m}}{\Gamma(\frac{1}{2}m)}. \tag{5}$$

Proof.

1. We show that the integral

$$\int_\Omega \frac{d\xi}{r^\alpha}, \quad x \in \bar{\Omega} \tag{6}$$

is bounded. Obviously, the set Ω lies in the sphere of radius H and with centre at the point x, so that

$$\int_\Omega \frac{d\xi}{r^\alpha} \leq \int_{r<H} \frac{d\xi}{r^\alpha}.$$

Introduce spherical coordinates with centre at the point x. We know that $d\xi = r^{m-1} dr\, dS_1$, where dS_1 is an element of measure on the unit sphere S_1, so that we have

$$\int_{r<H} \frac{d\xi}{r^\alpha} = \int_{S_1} \left\{ \int_0^H r^{m-\alpha-1}\, dr \right\} dS_1 = \frac{|S_1| H^{m-\alpha}}{m-\alpha}.$$

Hence

$$\int_\Omega \frac{d\xi}{r^\alpha} \leq \frac{|S_1| H^{m-\alpha}}{m-\alpha}, \quad x \in \bar{\Omega}. \tag{7}$$

We see at once that, by symmetry,

$$\int_\Omega \frac{dx}{r^\alpha} \leq \frac{|S_1| H^{m-\alpha}}{m-\alpha}, \quad \xi \in \bar{\Omega}. \tag{7'}$$

2. It is obvious that the $2m$-ple integral

$$\int_\Omega \int_\Omega \frac{u^2(\xi)}{r^\alpha}\, dx\, d\xi = \int_\Omega u^2(\xi) \left\{ \int_\Omega \frac{dx}{r^\alpha} \right\} d\xi \leq \frac{|S_1| H^{m-\alpha}}{m-\alpha} \|u\|^2$$

exists. Then by Fubini's theorem there exists at almost every $x \in \Omega$ a function of x, which is integrable in Ω, and defined by the integral

$$\int_\Omega \frac{u^2(\xi)}{r^\alpha}\, d\xi. \tag{8}$$

3. We have the inequality

$$|K(x,\xi)u(\xi)| \leq C\frac{1}{r^{\frac{1}{2}\alpha}}\frac{|u(\xi)|}{r^{\frac{1}{2}\alpha}} \leq \tfrac{1}{2}C\frac{1}{r^{\alpha}} + \tfrac{1}{2}C\frac{u^2(\xi)}{r^{\alpha}}.$$

As we have shown in paragraphs 1 and 2, the first term on the right is integrable with respect to ξ in Ω for all $x \in \Omega$, and the second for almost all $x \in \Omega$. But then the function $K(x,\xi)u(\xi)$ is integrable with respect to ξ for almost all $x \in \Omega$.

4. By the Schwartz-Bunyakovskii inequality,

$$[(Ku)(x)]^2 = \left\{\int_\Omega K(x,\xi)u(\xi)\,d\xi\right\}^2$$

$$\leq C^2\left\{\int_\Omega \frac{1}{r^{\frac{1}{2}\alpha}}\cdot\frac{u(\xi)}{r^{\frac{1}{2}\alpha}}\,d\xi\right\}^2 \leq C^2\int_\Omega \frac{d\xi}{r^\alpha}\int_\Omega \frac{u^2(\xi)}{r^\alpha}\,d\xi$$

$$\leq \frac{C^2|S_1|H^{m-\alpha}}{m-\alpha}\int_\Omega \frac{u^2(\xi)}{r^\alpha}\,d\xi.$$

Since the function (8) is integrable in Ω, the integral (3) is square-integrable in Ω. This implies that the operator K is defined on all the space $L_2(\Omega)$. Integrating the last inequality with respect to x, we obtain

$$\|Ku\|^2 \leq \frac{C^2|S_1|^2 H^{2(m-\alpha)}}{(m-\alpha)^2}\|u\|^2.$$

Hence

$$\|K\| \leq \frac{C|S_1|H^{m-\alpha}}{m-\alpha}.$$

The theorem is thus proven.

THEOREM 7.3.2. *The operator with a weak singularity is completely continuous in the space $L_2(\Omega)$.*

Let $\varepsilon > 0$ be a given number, and put

$$K_\varepsilon(x,\xi) = \begin{cases} K(x,\xi), & r \geq \varepsilon, \\ 0, & r < \varepsilon; \end{cases}$$

$$K'_\varepsilon(x,\xi) = \begin{cases} 0, & r \geq \varepsilon, \\ K(x,\xi), & r < \varepsilon, \end{cases}$$

so that

$$K(x,\xi) = K_\varepsilon(x,\xi) + K'_\varepsilon(x,\xi).$$

As in the preceding section, we denote by K_ε and K'_ε the integral operators whose kernels are $K_\varepsilon(x, \xi)$ and $K'_\varepsilon(x, \xi)$. Obviously, $K = K_\varepsilon + K'_\varepsilon$. The kernel $K_\varepsilon(x, \xi)$ is bounded:

$$|K_\varepsilon(x, \xi)| \leq \begin{cases} C/r^\alpha, & r \geq \varepsilon, \\ 0, & r < \varepsilon, \end{cases}$$

and, consequently,

$$|K_\varepsilon(x, \xi)| \leq C/\varepsilon^\alpha.$$

In this case K_ε is a Fredholm operator, and by Theorem 7.2.2 it is completely continuous in $L_2(\Omega)$. We estimate the norm of the operator K'_ε. For brevity we write $(K'_\varepsilon u)(x) = v(x)$, and then we have

$$|v(x)| = \left| \int_{\Omega \cap (r<\varepsilon)} \frac{A(x,\xi)}{r^\alpha} u(\xi)\,d\xi \right| \leq C \int_{\Omega \cap (r<\varepsilon)} \frac{|u(\xi)|}{r^{\frac{1}{2}\alpha}} \frac{1}{r^{\frac{1}{2}\alpha}}\,d\xi$$

while by the Schwartz-Bunyakovskii inequality

$$v^2(x) \leq C^2 \int_{\Omega \cap (r<\varepsilon)} \frac{u^2(\xi)}{r^\alpha}\,d\xi \int_{\Omega \cap (r<\varepsilon)} \frac{d\xi}{r^\alpha} \leq C^2 \int_\Omega \frac{u^2(\xi)}{r^\alpha}\,d\xi \int_{r<\varepsilon} \frac{d\xi}{r^\alpha}.$$

Introducing spherical coordinates with centre at the point x, we obtain

$$\int_{r<\varepsilon} \frac{d\xi}{r^\alpha} = \int_{S_1} \left\{ \int_0^\varepsilon r^{m-\alpha-1}\,dr \right\} dS_1 = \frac{|S_1|\varepsilon^{m-\alpha}}{m-\alpha}, \tag{9}$$

and, consequently,

$$v^2(x) \leq \frac{C^2|S_1|\varepsilon^{m-\alpha}}{m-\alpha} \int_\Omega \frac{u^2(\xi)}{r^\alpha}\,d\xi.$$

Integrating this last inequality with respect to x, and making use of inequality (7'), we obtain

$$\|v\|^2 = \|K'_\varepsilon u\|^2 \leq \frac{C^2|S_1|^2(\varepsilon H)^{m-\alpha}}{(m-\alpha)^2} \|u\|^2.$$

Hence

$$\|K'_\varepsilon\| \leq \frac{C|S_1|(\varepsilon H)^{\frac{1}{2}(m-\alpha)}}{m-\alpha}. \tag{10}$$

If $\varepsilon \to 0$, then $\|K'_\varepsilon\| \to 0$ and the operator K is completely continuous.

Remark. The definitions and theorems of §§ 2 and 3 for Fredholm operators and operators with weak singularities can be extended without change to the case where Ω is a smooth m-dimensional surface in a space of $(m+1)$ dimensions, with $d\xi$ now denoting an element of area on the surface.

§4. Operators with a Weak Singularity in the Space of Continuous Functions

In this section it is assumed that Ω is a bounded, closed set in m-dimensional Euclidean space, and that the function $A(x, \xi)$ in formula (3.2) is continuous in Ω with respect to the totality of points x and ξ.

THEOREM 7.4.1. *The integral operator with a weak singularity* (3.3) *is completely continuous in the space* $C(\Omega)$ *of functions which are continuous in* Ω.

Let M be a set of functions from $C(\Omega)$, such that

$$\|u\| = \max_{x \in \Omega} |u(x)| \leq c = \text{const.} \tag{1}$$

It is sufficient to show that the set $K(M)$, where K is the operator (3.3), is compact in $C(\Omega)$. This, in turn, can be shown by using Arzela's theorem and proving that the set $K(M)$ is uniformly bounded and equicontinuous.

Let $u \in M$. We put

$$v(x) = (Ku)(x) = \int_\Omega \frac{A(x, \xi)}{r^\alpha} u(\xi) \, d\xi. \tag{2}$$

Using inequalities (3.1) and (3.7), we obtain

$$|v(x)| \leq Cc \int_\Omega \frac{d\xi}{r^\alpha} \leq \frac{Cc|S_1|H^{m-\alpha}}{m-\alpha} = \text{const.} \tag{3}$$

Inequality (3) shows that the set $K(M)$ is uniformly bounded. We estimate the difference $v(x+h)-v(x)$. We have

$$v(x+h)-v(x) = \int_\Omega \left[\frac{A(x+h, \xi)}{|x+h-\xi|^\alpha} - \frac{A(x, \xi)}{|x-\xi|^\alpha}\right] u(\xi) \, d\xi. \tag{4}$$

Around the point x we describe a sphere $\Sigma_{2\delta}$, of radius 2δ, where δ is, for the time being, an arbitrary positive number. The part of the set Ω which lies outside this sphere we shall denote by Ω_1. If we stipulate that $|h| < \delta$, then the point $x+h$ will be located in the sphere $\Sigma_{2\delta}$, at a distance not less than δ

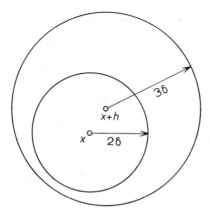

Fig. 8.

from the surface (Fig. 8). From formula (4) it follows that

$$|v(x+h)-v(x)| \leq Cc \left\{ \int_{\Sigma_{2\delta}} \frac{dx}{|x+h-\xi|^\alpha} + \int_{\Sigma_{2\delta}} \frac{d\xi}{|x-\xi|^\alpha} \right\} +$$
$$+ c \int_{\Omega_1} \left| \frac{A(x+h, \xi)}{|x+h-\xi|^\alpha} - \frac{A(x, \xi)}{|x-\xi|^\alpha} \right| d\xi. \quad (5)$$

By formula (3.9),

$$\int_{\Sigma_{2\delta}} \frac{d\xi}{|x-\xi|^\alpha} = \int_{r<2\delta} \frac{d\xi}{|x-\xi|^\alpha} = \frac{|S_1|(2\delta)^{m-\alpha}}{m-\alpha}. \quad (6)$$

If $\xi \in \Sigma_{2\delta}$, then $|x-\xi| \leq 2\delta$, and

$$|x+h-\xi| \leq |x-\xi|+|h| < 3\delta.$$

This means that the point x lies inside the sphere of radius 3δ with centre at the point $x+h$ (Fig. 8). In other words, the sphere $\Sigma_{2\delta}$ lies wholly within the sphere $r_1 < 3\delta$, $r_1 = |x+h-\xi|$. Hence

$$\int_{\Sigma_{2\delta}} \frac{d\xi}{|x+h-\xi|^\alpha} < \int_{r_1<3\delta} \frac{d\xi}{r_1^\alpha} = \frac{|S_1|(3\delta)^{m-\alpha}}{m-\alpha}. \quad (7)$$

We prescribe a number $\varepsilon > 0$ and choose δ sufficiently small, so that

$$Cc \frac{(2^{m-\alpha}+3^{m-\alpha})|S_1|\delta^{m-\alpha}}{m-\alpha} < \tfrac{1}{2}\varepsilon.$$

Then it follows from relations (5)–(7) that

$$|v(x+h)-v(x)| \leq \tfrac{1}{2}\varepsilon + c \int_{\Omega_1} \left| \frac{A(x+h,\xi)}{|x+h-\xi|^\alpha} - \frac{A(x,\xi)}{|x-\xi|^\alpha} \right| d\xi. \qquad (8)$$

We now fix the number δ. In the domain Ω_1, the inequality $|x-\xi| \geq \delta$ is satisfied, so that the function $A(x,\xi)/|x-\xi|^\alpha$ is uniformly continuous with respect to the totality of points x and ξ. We can therefore choose an arbitrarily small h_0, such that for $|h| < h_0$ we have

$$\left| \frac{A(x+h,\xi)}{|x+h-\xi|^\alpha} - \frac{A(x,\xi)}{|x-\xi|^\alpha} \right| < \frac{\varepsilon}{2c|\Omega|}.$$

Then by formula (8)

$$|v(x+h)-v(x)| < \tfrac{1}{2}\varepsilon + \frac{\varepsilon|\Omega_1|}{2|\Omega|} < \varepsilon, \qquad |h| < h_0.$$

The quantity h_0 depends only on ε; it depends neither on the point x nor the function u. It follows from this that the set $K(M)$ is uniformly continuous, and the theorem is proven.

Theorem 7.4.1 can be extended in an obvious way to the case where Ω is a smooth m-dimensional surface in an Euclidean space of $(m+1)$ dimensions, with $d\xi$ denoting an element of surface area.

In the process of proving Theorem 7.4.1, we used, not the continuity of the functions $u(x)$, but only their boundedness. As a result, we can assert the following:

If Ω is a bounded, closed set, and the function $A(x,\xi)$ is continuous in Ω, then the operator with a weak singularity

$$(Ku)(x) = \int_\Omega \frac{A(x,\xi)}{r^\alpha} u(\xi) d\xi = v(x)$$

transforms every bounded function $u(x)$ into a continuous function $v(x)$.

As before, let Ω be a bounded, closed set, and let $K(x,\xi)$ be a continuous kernel. It can be regarded as a kernel with a weak singularity for which $\alpha = 0$. Then from Theorem 7.4.1 we have the following corollary.

COROLLARY 7.4.1. *If Ω is a bounded, closed set, and if the kernel $K(x,\xi)$ is continuous in $\Omega \times \Omega$, then the Fredholm operator K, defined by the formula*

$$(Ku)(x) = \int_\Omega K(x,\xi) u(\xi) d\xi,$$

is completely continuous in $C(\Omega)$.

Exercises

1. Show that when $\alpha < \frac{1}{2}m$, the operator with a weak singularity is also a Fredholm operator in $L_2(\Omega)$.

2. Let $1 < p < \infty$ and $1/p + 1/p' = 1$. Also let $\alpha p' < m$. Show that the operator with a weak singularity is completely continuous as an operator from $L_p(\Omega)$ into $C(\bar{\Omega})$.

3. It is well-known that the so-called singular integral operator S, where

$$(Su)(x) = \int_{-1}^{1} \frac{u(\xi)}{\xi - x} \, d\xi = \lim_{\varepsilon \to 0} \left\{ \int_{-1}^{x-\varepsilon} \frac{u(\xi)}{\xi - x} \, d\xi + \int_{x+\varepsilon}^{1} \frac{u(\xi)}{\xi - x} \, d\xi \right\},$$

is bounded in $L_2(-1, 1)$. Let the function $a(x)$ be continuous on the segment $[-1, 1]$. Show that the operator T, where

$$(Tu)(x) = \int_{-1}^{1} \frac{a(\xi) - a(x)}{\xi - x} u(\xi) \, d\xi,$$

is completely continuous in $L_2(-1, 1)$.

Chapter 8. Fredholm Theory

§ 1. Equation with Completely Continuous Operator. Integral Equations

Consider the equation
$$u - \lambda Tu = f, \tag{1}$$

where T is a c.c.o. acting in the Banach space X, f is a given element and u an unknown element of the space, and λ is a numerical parameter. The equation
$$v - \bar{\lambda} T^* v = g, \tag{2}$$

is called the adjoint of equation (1). In (2), T^* is a c.c.o., the adjoint of T, g and v are respectively given and unknown elements of the space X^*, which is the adjoint (or, conjugate) of X. If f (or g) is not a zero element, then equation (1) (similarly equation (2)) is called *inhomogeneous*. Contrariwise we have the *homogeneous equations*
$$u - \lambda Tu = 0 \tag{3}$$
and
$$v - \bar{\lambda} T^* v = 0. \tag{4}$$

The values of λ for which there exist non-trivial (i.e. different from the null element) solutions of equation (3) are called *eigenvalues* of the operator T, and the non-trivial solutions themselves are called *eigenelements* corresponding to these eigenvalues. The number of linearly-independent eigenelements associated with a given eigenvalue is called the *rank* of this eigenvalue. Values of λ which are not eigenvalues are called *regular*.

Equation (1) is the subject of the following four theorems, known collectively as the *Fredholm theorems*.

THEOREM 8.1.1 (FREDHOLM'S FIRST THEOREM). *Every eigenvalue of equation* (1) *has finite rank*.

THEOREM 8.1.2 (FREDHOLM'S SECOND THEOREM). *Equation* (1) *has either a finite or an enumerable set of eigenvalues; if the set is enumerable, then it has a unique limit point at infinity.*

THEOREM 8.1.3 (FREDHOLM'S THIRD THEOREM). *If λ is an eigenvalue of equation* (1), *then $\bar{\lambda}$ is an eigenvalue of equation* (2), *and, moreover, they are of the same rank.*

THEOREM 8.1.4 (FREDHOLM'S FOURTH THEOREM). *In order that equation* (1) *should have a solution, it is necessary and sufficient that the free term f of this equation be orthogonal to every solution of the adjoint homogeneous equation* (4).

Orthogonality here is to be understood in the following sense. For a given λ, let equation (4) have some solution v. This solution belongs to the space X^* and is, consequently, a functional in the space X. In saying that f is orthogonal to v, we mean that

$$(v, f) = 0, \tag{5}$$

where (v, f) is the value of the functional v on the element f.

The most important forms of an equation with a c.c.o. are *Fredholm's equation* and the *equation with a weak singularity*; both of these we shall regard as equations in the space $X = L_2(\Omega)$; moreover, we know from Riesz's theorem that $X^* = L_2(\Omega)$ also. An equation of the form

$$u(x) - \lambda \int_\Omega K(x, \xi) u(\xi) \, d\xi = f(x) \tag{6}$$

is called a *Fredholm integral equation*, if $K(x, \xi)$ is a Fredholm kernel, and $f(x), u(x)$ belong to the space $L_2(\Omega)$.*

If $K(x, \xi)$ is a kernel with a weak singularity (in which case the set Ω is necessarily bounded), then equation (6) is called an *integral equation with a weak singularity*.

The equation adjoint to (6) in the space $L_2(\Omega)$ has the form

$$v(x) - \bar{\lambda} \int_\Omega \overline{K(\xi, x)} v(\xi) \, d\xi = g(x). \tag{7}$$

---

* The Fredholm equations can also be considered in various other functional spaces, but we shall not do this here.

To show this it is sufficient to establish that the operator K^*, the adjoint of the Fredholm operator K, where

$$(Ku)(x) = \int_\Omega K(x, \xi) u(\xi) \, \mathrm{d}\xi, \qquad (8)$$

is defined by the formula

$$(K^*v)(x) = \int_\Omega \overline{K(\xi, x)} v(\xi) \, \mathrm{d}\xi. \qquad (9)$$

A linear, bounded functional in $L_2(\Omega)$ can, by Riesz's theorem, be identified with some element of the space $L_2(\Omega)$. Let $v(x)$ be this element. Then

$$(K^*v, u) = (v, Ku) = \int_\Omega \int_\Omega \overline{K(x, \xi)} \overline{u(\xi)} v(x) \, \mathrm{d}x \, \mathrm{d}\xi$$

$$= \int_\Omega \overline{u(\xi)} \left\{ \int_\Omega \overline{K(x, \xi)} v(x) \, \mathrm{d}x \right\} \mathrm{d}\xi = \int_\Omega \overline{u(x)} \left\{ \int_\Omega \overline{K(\xi, x)} v(\xi) \, \mathrm{d}\xi \right\} \mathrm{d}x.$$

From this we see that

$$(K^*v)(x) = \int_\Omega \overline{K(\xi, x)} v(\xi) \, \mathrm{d}\xi,$$

which is what we set out to prove.

In §§ 2–4 of this chapter we shall give the proofs of Fredholm's theorems for equations with a c.c.o. in a Hilbert space. These theorems will then be demonstrated for Fredholm's equations and equations with a weak singularity. In §§ 5 and 6 we shall give the answer to one particular question – under what circumstances a square-integrable solution of the equation with a weak singularity will also be continuous.

§ 2. Reduction to a Finite-Dimensional Equation. Proofs of Fredholm's First and Second Theorems

THEOREM 8.2.1 (BANACH'S THEOREM). *Let A be a bounded, linear operator, acting in a Banach space X, and let $|\lambda| < \|A\|^{-1}$. Then the operator $(I - \lambda A)^{-1}$, where I is the unit operator, exists, is defined on the whole space X, and is bounded.*

The series

$$I + \lambda A + \lambda^2 A^2 + \ldots + \lambda^n A^n + \ldots \qquad (1)$$

§ 2] PROOFS OF FREDHOLM'S 1ST AND 2ND THEOREMS

converges in the norm, because the series of norms of the terms converges:

$$1+|\lambda|\,||A||+|\lambda|^2||A^2||+\ldots+|\lambda|^n||A^n||+\ldots$$
$$\leq 1+|\lambda|\,||A||+|\lambda|^2||A||^2+\ldots+|\lambda|^n||A||^n+\ldots$$
$$=\frac{1}{1-|\lambda|\,||A||}.$$

Thus the series (1) represents an operator which is defined through all space and is bounded; we denote the sum of series (1) by R_λ^A, and then we have

$$||R_\lambda^A|| \leq \frac{1}{1-|\lambda|\,||A||}. \tag{2}$$

By direct multiplication it can be verified that $(I-\lambda A)R_\lambda^A = R_\lambda^A(T-\lambda A) = I$ and, consequently,

$$(I-\lambda A)^{-1} = R_\lambda^A = \sum_{n=0}^{\infty} \lambda^n A^n. \tag{3}$$

The theorem is proven.

Now consider the equation

$$(I-\lambda T)u = f, \tag{4}$$

when T is a c.c.o. in a Hilbert space $\mathfrak{H}$. Given an arbitrary positive number R, we shall suppose that the parameter λ can vary in the closed circle $|\lambda| \leq R$ of the complex λ-plane. Using Theorem 7.1.1, we can construct a finite-dimensional operator – which we shall denote by T'' – such that the difference $T' = T - T''$ satisfies the inequality

$$||T'|| \leq \frac{1}{2R}. \tag{5}$$

By Banach's Theorem, the operator

$$R_\lambda' = R_\lambda^{T'} = (I-\lambda T')^{-1}, \qquad |\lambda| \leq R$$

exists, is defined on the whole of the space $\mathfrak{H}$, and is bounded. We multiply equation (4) from the left by R_λ'. Then

$$R_\lambda'(I-\lambda T) = R_\lambda'(I-\lambda T' - \lambda T'') = I - \lambda R_\lambda' T'',$$

and we obtain a new equation,

$$(I-\lambda R_\lambda' T'')u = R_\lambda' f, \tag{6}$$

which is obviously equivalent to equation (4).

We now show that the product $R'_\lambda T''$ is finite-dimensional. The finite-dimensional operator T'' is defined by an equation of the form

$$T''u = \sum_{k=1}^{n} l_k(u)v_k,$$

where the $l_k(u)$ are bounded, linear functionals, and the v_k are fixed elements of the space $\mathfrak{H}$. But then

$$R'_\lambda T''u = \sum_{k=1}^{n} l_k(u)w_k, \qquad w_k = R'_\lambda v_k,$$

and so the operator $R'_\lambda T''$ is finite-dimensional. We note that the element w_k depends on λ, so that henceforth we denote it by $w_{k,\lambda}$. Moreover, by Riesz's theorem, $l_k(u) = (u, u_k)$, where the u_k are fixed elements of the space $\mathfrak{H}$. Finally,

$$R'_\lambda T''u = \sum_{k=1}^{n} (u, u_k)w_{k,\lambda}. \tag{7}$$

Equation (6) and equation (4) can easily be reduced to the same linear algebraic system. We put

$$(u, u_k) = c_k. \tag{8}$$

Then from relations (6) and (7) we obtain

$$u = R'_\lambda f + \lambda \sum_{k=1}^{n} c_k w_{k,\lambda}. \tag{9}$$

Scalar-multiply both sides of the last equation by u_j, $1 \leq j \leq n$. This gives the above-mentioned system:

$$c_j - \lambda \sum_{k=1}^{n} \alpha_{jk}(\lambda)c_k = f_j(\lambda), \qquad j = 1, 2, \ldots, n, \tag{10}$$

where we have introduced the notation

$$\alpha_{jk}(\lambda) = (w_{k,\lambda}, u_j), \qquad f_j(\lambda) = (R'_\lambda f, u_j). \tag{11}$$

Equation (4) and the system (10) are equivalent in the following sense. By formula (9), to each solution of the system (10) there corresponds some solution of equation (4); conversely, to each solution of equation (4), formula (8) relates some solution of the system (10).

It is important to note that the $\alpha_{jk}(\lambda)$, as well as the $f_j(\lambda)$, are holomorphic functions of the complex variable λ within the closed circle $|\lambda| \leq R$. In fact,

$$R'_\lambda = \sum_{s=0}^{\infty} \lambda^s T'^s,$$

and here the series converges in the circle $|\lambda| < 2R$ – this follows from inequality (5). Now

$$\alpha_{jk}(\lambda) = \sum_{s=0}^{\infty} \lambda^s (T'^s v_k, u_j). \tag{12}$$

As before the power-series (12) converges in the circle $|\lambda| < 2R$, and therefore the function $\alpha_{jk}(\lambda)$ is holomorphic in this circle, and, *a fortiori*, in the closed circle $|\lambda| \leq R$. The determinant of the system (10),

$$D_R(\lambda) = \begin{vmatrix} 1-\lambda\alpha_{11}(\lambda) & -\lambda\alpha_{12}(\lambda) & \ldots & -\lambda\alpha_{1n}(\lambda) \\ -\lambda\alpha_{21}(\lambda) & 1-\lambda\alpha_{22}(\lambda) & \ldots & -\lambda\alpha_{2n}(\lambda) \\ \cdot & \cdot & & \cdot \\ -\lambda_1\alpha_{n1}(\lambda) & -\lambda\alpha_{n2}(\lambda) & \ldots & 1-\lambda\alpha_{nn}(\lambda) \end{vmatrix} \tag{13}$$

is also holomorphic in this circle. If $|\lambda| \leq R$ and $D_R(\lambda) \neq 0$, then the homogeneous system

$$c_j - \lambda \sum_{k=1}^{n} \alpha_{jk}(\lambda) c_k = 0, \quad j = 1, 2, \ldots, n, \tag{14}$$

and, with it, the homogeneous equation

$$(I - \lambda T)u = 0 \tag{15}$$

have only trivial solutions. If, however, $D_R(\lambda) = 0$, then (14) and (15) have a finite number of linearly-independent, non-trivial solutions. It follows from this that *the eigenvalues of the operator T which lie within the closed circle $|\lambda| \leq R$, coincide with the roots of the determinant $D_R(\lambda)$ which lie in the same circle.*

It now becomes easy to prove Fredholm's First Theorem. Let λ_0 be an eigenvalue of the operator T. Take $R > |\lambda_0|$. Then $D_R(\lambda_0) = 0$; the system (14) and equation (15) have for $\lambda = \lambda_0$ only a finite number of linearly-independent solutions, and this means that the rank of the eigenvalue λ_0 is finite.

We next prove Fredholm's Second Theorem.

The function $D_R(\lambda)$, holomorphic in the closed circle $|\lambda| \leq R$, has only a finite number of roots in this circle. Hence it follows that in any annulus

$N \leq \lambda \leq N+1$, $N = 0, 1, 2, \ldots$, there will be only a finite number of eigenvalues of the operator A. These annuli cover the whole λ-plane, and the set of all eigenvalues of the operator T represents the union of an enumerable set of finite sets; and this union is either finite or enumerable. Finally, the eigenvalues cannot have a limit point at a finite distance from the origin – for otherwise there would be a circle containing an infinite set of eigenvalues.

§ 3. Proof of Fredholm's Third Theorem

Consider the homogeneous equation

$$(I - \bar{\lambda}T^*)v = 0, \qquad |\lambda| \leq R, \tag{1}$$

which is the adjoint of equation (2.4). We reduce it to an equivalent linear algebraic system by a method much the same as that in the preceding section.

Corresponding to the expansion of $T = T' + T''$, there is an expansion of the adjoint operator $T^* = T'^* + T''^*$, where $\|T'^*\| = \|T'\| \leq 1/2R$, and where the operator T''^* is finite-dimensional; if

$$T''u = \sum_{k=1}^{n}(u, u_k)v_k,$$

then

$$T''^*v = \sum_{k=1}^{n}(v, v_k)u_k. \tag{2}$$

On the basis of Banach's theorem (Theorem 8.2.1), we conclude that the operator

$$R_\lambda'^* = (I - \bar{\lambda}T'^*)^{-1}$$

exists, is defined in all space, and is bounded. We note also that the operators $R_\lambda'^*$ and R_λ' are adjoint.

For the unknown function in equation (1) we make the substitution

$$v = R_\lambda'^* w; \tag{3}$$

equation (1) then takes the form

$$(I - \bar{\lambda}T^*)R_\lambda'^*w = 0.$$

We have

$$(I - \bar{\lambda}T^*)R_\lambda'^* = (I - \bar{\lambda}T'^* - \bar{\lambda}T''^*)R_\lambda'^* = I - \bar{\lambda}T''^*R_\lambda'^*,$$

and the new unknown function satisfies the equation

$$(I-\bar{\lambda}T''^{*}R_{\bar{\lambda}}'^{*})w = 0. \qquad (4)$$

The operator $T''^{*}R_{\bar{\lambda}}'^{*}$ is finite-dimensional:

$$T''^{*}R_{\bar{\lambda}}'^{*}w = \sum_{k=1}^{n}(R_{\bar{\lambda}}'^{*}w, v_{k})u_{k} = \sum_{k=1}^{n}(w, R_{\lambda}'v_{k})u_{k} = \sum_{k=1}^{n}(w, w_{k,\lambda})u_{k}. \qquad (5)$$

We write

$$(w, w_{k,\lambda}) = \gamma_{k}. \qquad (6)$$

From relations (4) and (5) we find

$$w - \bar{\lambda}\sum_{k=1}^{n}\gamma_{k}u_{k} = 0.$$

Scalar multiplying this by $w_{j,\lambda}$, $1 \leq j \leq n$, we obtain the homogeneous system

$$\gamma_{j} - \bar{\lambda}\sum_{k=1}^{n}\bar{\alpha}_{kj}\gamma_{k} = 0, \qquad j = 1, 2, \ldots, n, \qquad (7)$$

equivalent to equation (1); the quantity α_{kj} is defined by formula (2.11). The determinant of system (7) is equal to $\overline{D_{R}(\lambda)}$ (cf. formula (2.13)).

Let λ_{0}, $|\lambda_{0}| \leq R$, be an eigenvalue of the operator T. Then $D_{R}(\lambda_{0}) = 0$, and the matrix of system (2.10) is degenerate. Let r, $0 \leq r < n$, be the rank of this matrix. Then the homogeneous system (2.14) has exactly $n-r$ linearly-independent solutions. But equation (2.15) has just the same number of solutions. This means that the rank of the eigenvalue λ_{0} is equal to $n-r$.

The matrices of systems (2.14) and (7) are adjoint, and their ranks are equal. It follows from this that $\bar{\lambda}_{0}$ is an eigenvalue of the operator T^{*}, with rank $n-r$. This proves Fredholm's Third Theorem.

§ 4. Proof of Fredholm's Fourth Theorem

Let $\omega_{1}, \omega_{2}, \ldots, \omega_{s}$, $s \geq 0$, be linearly-independent solutions of the homogeneous equation

$$(I-\bar{\lambda}T^{*})v = 0, \qquad (1)$$

which is the adjoint of the equation

$$(I-\lambda T)u = f. \qquad (2)$$

Denote by $\mathfrak{H}_0^*$ a subspace which spans the elements $\omega_1, \omega_2, \ldots, \omega_s$; if $s = 0$ (i.e. equation (1) has no non-trivial solutions), then $\mathfrak{H}_0^*$ consists of just a zero. The orthogonal complement to the subspace $\mathfrak{H}_0^*$ we denote by $\mathfrak{H}_1^*$:

$$\mathfrak{H}_1^* = \mathfrak{H} \ominus \mathfrak{H}_0^*.$$

Further, we introduce the following notation. Linearly-independent solutions of the homogeneous equation

$$(I - \lambda T)u = 0 \tag{3}$$

will be denoted by $\phi_1, \phi_2, \ldots, \phi_s$, and the subspace which spans them by $\mathfrak{H}_0$. Also, put

$$\mathfrak{H}_1 = \mathfrak{H} \ominus \mathfrak{H}_0.$$

Finally, write $A = I - \lambda T$, and let $R(A)$ be the domain of values of the operator $A : R(A) = A(\mathfrak{H})$. Observe that the operators $\mathfrak{H}_0$ and $\mathfrak{H}_0^*$ coincide with respect to the sets of solutions of the homogeneous equations (3) and (1), or, in other words, of the equations $Au = 0$ and $A^*v = 0$.

Fredholm's Fourth Theorem is equivalent to the statement that

$$R(A) = \mathfrak{H}_1^*. \tag{4}$$

The necessity condition in the theorem implies that

$$R(A) \subset \mathfrak{H}_1^*, \tag{5}$$

while the sufficiency means

$$R(A) \supset \mathfrak{H}_1^*. \tag{6}$$

We now prove the validity of (5) and (6).

Necessity. Suppose equation (1.1), which can be written in the form

$$Au = f, \tag{1'}$$

to be soluble; i.e., there exists an element $u \in \mathfrak{H}$ which transforms equation (1') into an identity. Then $f \in R(A)$. Let ϕ be an arbitrary element of the subspace $\mathfrak{H}_0^*$. Then

$$(f, \phi) = (Au, \phi) = (u, A^*\phi) = 0.$$

Thus, f is orthogonal to $\mathfrak{H}_0^*$, and, consequently, $f \in \mathfrak{H}_1^*$. This proves the inclusion statement (5).

Sufficiency. Introduce now an operator A_1, defined only on the subspace $\mathfrak{H}_1$, and coincident there with the operator A,

$$A_1 u = Au, \qquad u \in \mathfrak{H}_1; \tag{7}$$

A_1 is called the *contraction* of the operator A on the subspace $\mathfrak{H}_1$. It is clear that * $R(A_1) \subset R(A)$. We shall show that

$$R(A_1) \supset \mathfrak{H}_1^*. \tag{8}$$

The equation $A_1 u = 0$ has only the trivial solutions $u = 0$. In fact, if $A_1 u = 0$, then $Au = 0$, and $u \in \mathfrak{H}_0$. But from the definition of A_1, $u \in \mathfrak{H}_1$. Being an element of two orthogonal spaces, u is orthogonal to itself, and, consequently, $u = 0$. Hence it follows that the operator A_1 has an inverse A_1^{-1}; its domain of definition $D(A_1^{-1}) = R(A_1) \subset R(A)$, and, once the inclusion (5) has been established, $D(A_1^{-1}) \subset \mathfrak{H}_1^*$.

We show first of all that the set $D(A_1^{-1}) = R(A_1)$ is dense in $\mathfrak{H}_1^*$. If in fact the opposite were the case, there would be an element $\omega \in \mathfrak{H}_1^*$, $\omega \neq 0$, such that $(A_1 u, \omega) = (Au, \omega) = 0$ for every $u \in \mathfrak{H}_1$. The equality $(Au, \omega) = 0$ is obviously true whenever $u \in \mathfrak{H}_0$. But then it is true everywhere in $\mathfrak{H}$. Now

$$(u, A^*\omega) = (Au, \omega) = 0, \qquad \forall u \in \mathfrak{H}.$$

The element $A^*\omega$ is orthogonal to the whole space, and so $A^*\omega = 0$ and $\omega \in \mathfrak{H}_0^*$. Thus ω belongs at the same time to two orthogonal spaces $\mathfrak{H}_1^*$ and $\mathfrak{H}_0^*$, and, consequently, $\omega = 0$.

We next show that the operator A_1^{-1} is bounded. Suppose the contrary. Then there is a sequence of elements $u_n \in D(A_1^{-1})$, such that

$$\frac{\|A_1^{-1} u_n\|}{\|u_n\|} > n, \qquad n = 1, 2, \ldots.$$

Put $A_1^{-1} u_n = \psi_n$; obviously $\psi_n \in R(A_1^{-1}) = D(A_1) = \mathfrak{H}_1$. Now we have $u_n = A_1 \psi_n = A\psi_n$, and

$$\frac{\|A\psi_n\|}{\|\psi_n\|} < \frac{1}{n}.$$

Further, put $w_n = \psi_n/\|\psi_n\|$. Then $w_n \in \mathfrak{H}_1$, $\|w_n\| = 1$, and

$$Aw_n = w_n - \lambda T w_n \underset{n \to \infty}{\to} 0. \tag{9}$$

---

* It is also easy to see here that $R(A_1) = R(A)$.

The set $\{w_n\}$ is bounded, and the operator T is completely continuous. We select a subsequence $\{w_{n_k}\}$ such that $\lambda T w_{n_k}$ tends to some limit, which we denote by w_0. From relation (9) it is clear that

$$w_{n_k} - \lambda T w_{n_k} \to 0. \tag{10}$$

Hence $w_{n_k} \to w_0$, $w_0 \in \mathfrak{H}_1$ and $\|w_0\| = \lim \|w_{n_k}\| = 1$. Proceeding to the limit in formula (10), we obtain

$$Aw_0 = w_0 - \lambda T w_0 = 0$$

and, consequently, $w_0 \in \mathfrak{H}_0$. But, as we have seen, $w_0 \in \mathfrak{H}_1$, and so necessarily $w_0 = 0$. This contradicts the equation $\|w_0\| = 1$ and hence the operator A_1^{-1} is bounded.

It is now possible to prove the validity of the inclusion statement (6). Let $f \in \mathfrak{H}_1^*$. The set $D(A_1^{-1})$ is dense in $\mathfrak{H}_1^*$, and therefore we can find a sequence $\{f_n\}$ such that $f_n \in D(A_1^{-1})$ and $f_n \to f$. Put $u_n = A_1^{-1} f_n$. The operator A_1^{-1} is bounded, and the sequence $\{f_n\}$ converges, so that the sequence $\{u_n\}$ also converges; let $u_0 = \lim u_n$. Moreover, we have $f_n = A_1 u_n$; the operator A_1 is obviously bounded, and in the last relation we can proceed to the limit: $f = A_1 u_0$. Hence it follows that $f \in R(A_1) \subset R(A)$, and the inclusion (6) is proven.

§ 5. Fredholm's Alternative

From Fredholm's Third and Fourth Theorems there ensues an important proposition, known as *Fredholm's Alternative*:

Let T be a c.c.o. in the Hilbert space $\mathfrak{H}$. Either the equation

$$(I - \lambda T)u = 0 \tag{1}$$

has only trivial solutions, so that the inhomogeneous equation

$$(I - \lambda T)u = f \tag{2}$$

is soluble for any free term $f \in \mathfrak{H}$ and the solution of this equation is unique; or equation (1) has non-trivial solutions, in which case either equation (2) has no solutions or it has infinitely many solutions.

Proof. If equation (1) has only trivial solutions then the given quantity λ is regular (not an eigenvalue) for the operator T. In this case the quantity $\bar\lambda$ is regular for the operator T^* (Fredholm's Theorem 3), and the equation

$$(I - \bar\lambda T^*)v = 0 \tag{3}$$

has only the trivial solutions $v = 0$. Any element f is orthogonal to this solution: by Fredholm's Fourth Theorem, equation (2) has a solution. This solution is unique: if u_1 and u_2 are two solutions of equation (2), then

$$(I-\lambda T)u_1 = (I-\lambda T)u_2 = f.$$

Subtracting, we find

$$(I-\lambda T)w = 0, \quad w = u_1 - u_2.$$

Since λ is regular, then $w = 0$ and $u_1 = u_2$. The first part of Fredholm's Alternative is proven.

Now let equation (1) have s linearly-independent solutions $\phi_1, \phi_2, \ldots, \phi_s$. Then the given value λ is an eigenvalue for the operator T, and of rank s. By Fredholm's Third Theorem, $\bar{\lambda}$ is an eigenvalue for the operator T^*, of the same rank s. Let $\omega_1, \omega_2, \ldots, \omega_s$ be the corresponding, linearly-independent eigenelements. By Fredholm's Fourth Theorem, equation (2) has no solution unless all the equations

$$(f, \omega_k) = 0, \quad k = 1, 2, \ldots, s \tag{4}$$

hold. If all the equations (4) do hold, then equation (2) has infinitely many solutions. Actually, Fredholm's Fourth Theorem states that in this case equation (2) has at least one solution u_0; then the general solution can be obtained, as usual, by adding to u_0 the general solution of the homogeneous equation (1)

$$u = u_0 + \sum_{k=1}^{s} c_k \phi_k, \tag{5}$$

where the c_k are arbitrary constants. Thus the second part of Fredholm's Alternative is completely established.

§ 6. Continuity of Solutions of the Equation with a Weak Singularity

Consider the integral equation with a weak singularity

$$u(x) - \int_\Omega \frac{A(x, \xi)}{r^\alpha} u(\xi) \, d\xi = f(x), \tag{1}$$

where $0 \leq \alpha < m$, $|A(x, \xi)| \leq C = $ const., and the set Ω is bounded. If $f(x) \in L_2(\Omega)$, and if the orthogonality conditions prescribed in Fredholm's Fourth Theorem are satisfied, then there exists a solution to equation (1) which also belongs to $L_2(\Omega)$.

In applications of the theory of integral equations there often occur interesting cases in which the solutions of equation (1) are continuous. The simplest case of this type is described in the following theorem:

THEOREM 8.6.1. *If Ω is a bounded, closed set, and if the functions $f(x)$ and $A(x, \xi)$ are continuous in Ω, then any solution of equation (1), belonging to the class $L_2(\Omega)$, is continuous in Ω.*

We choose an arbitrarily small number ε and a continuous function $\eta(t)$ of the real variable t, defined for $t \geq 0$ and satisfying the relations

$$\eta(t) = 1, \quad 0 \leq t \leq \tfrac{1}{2}\varepsilon, \quad 0 < \eta(t) < 1, \quad \tfrac{1}{2}\varepsilon < t < \varepsilon,$$
$$\eta(t) = 0, \quad t \geq \varepsilon.$$

Put

$$K_1(x, \xi) = \frac{A(x, \xi)\eta(r)}{r^\alpha}, \quad K_2(x, \xi) = \frac{A(x, \xi)[1-\eta(r)]}{r^\alpha}.$$

Thus

$$\frac{A(x, \xi)}{r^\alpha} = K_1(x, \xi) + K_2(x, \xi);$$

the kernel of equation (1) has been represented as the sum of two kernels, of which the first has a weak singularity but is different from zero only for $r < \varepsilon$, while the second is simply continuous.

Let $u(x)$ be any solution of equation (1), where $u \in L_2(\Omega)$. We write the equation in the form

$$u(x) - (K_1 u)(x) = g(x), \tag{2}$$

where

$$(K_1 u)(x) = \int_\Omega K_1(x, \xi) u(\xi) \, d\xi, \tag{3}$$

$$g(x) = f(x) + \int_\Omega K_2(x, \xi) u(\xi) \, d\xi. \tag{4}$$

The function $g(x)$ is continuous in Ω – this follows at once from the continuity of the functions $f(x)$ and $K_2(x, \xi)$, and from the relations

$$|(K_2 u)(x_1) - (K_2 u)(x_2)| = \left| \int_\Omega [K_2(x_1, \xi) - K_2(x_2, \xi)] u(\xi) \, d\xi \right|$$
$$\leq \left\{ \int_\Omega [K_2(x_1, \xi) - K_2(x_2, \xi)]^2 \, d\xi \right\}^{\frac{1}{2}} \|u\|_{L_2(\Omega)}.$$

If $|A(x, \xi)| \leq C$, then also $|A(x, \xi)\eta(r)| \leq C$. We choose the value of ε sufficiently small that the following two inequalities are simultaneously satisfied:

$$\frac{C|S_1|(\varepsilon H)^{\frac{1}{2}(m-\alpha)}}{m-\alpha} < 1, \tag{5}$$

$$\frac{C|S_1|\varepsilon^{m-\alpha}}{m-\alpha} < 1; \tag{6}$$

in inequality (5), H denotes the diameter of the set Ω.

By virtue of formula (3.10), Chap. 7, it follows from inequality (5) that

$$\|K_1\|_{L_2(\Omega)} < 1;$$

if we regard relation (2) as an integral equation for unknown $u(x)$ and with $g(x)$ given, then we see that Banach's Theorem, 8.2.1, is applicable to this equation, and the function $u(x)$ can be represented in series form:

$$u(x) = \sum_{n=0}^{\infty} (K_1^n g)(x), \tag{7}$$

convergent in the metric of the space $L_2(\Omega)$. We show that this series converges uniformly in Ω. The function $g(x)$ is bounded; let $|g(x)| \leq M = \mathrm{const}$. Then

$$|(K_1 g)(x)| = \left| \int_\Omega \frac{A(x, \xi)\eta(r)}{r^\alpha} g(\xi) \, d\xi \right|$$

$$= \left| \int_{\Omega \cap (r < \varepsilon)} \frac{A(x, \xi)\eta(r)}{r^\alpha} g(\xi) \, d\xi \right| \leq MC \int_{r < \varepsilon} \frac{d\xi}{r^\alpha},$$

and from formula (3.9), Chap. 7,

$$|(K_1 g)(x)| \leq \frac{MC|S_1|\varepsilon^{m-\alpha}}{m-\alpha}.$$

By induction,

$$|(K_1^n g)(x)| \leq M \left[\frac{C|S_1|\varepsilon^{m-\alpha}}{m-\alpha} \right]^n;$$

because of inequality (6), the series (7) converges uniformly in Ω. By Theorem 7.4.1, the terms of this series are continuous, and therefore the sum – which is the function $u(x)$ – is also continuous.

PART IV

General Aspects of Partial Differential Equations

Chapter 9.
Equations and Boundary-Value Problems

§ 1. Differential Expression and Differential Equation

The most general partial differential equation with m independent variables can be written in the form

$$F\left(x_1, x_2, \ldots, x_m, u, \frac{\partial u}{\partial x_1}, \ldots, \frac{\partial u}{\partial x_m}, \frac{\partial^2 u}{\partial x_1^2}, \ldots, \frac{\partial^k u}{\partial x_m^k}\right) = 0; \quad (1)$$

if k is the order of the highest derivative of the unknown function present in the equation, then the equation is said to be of *order k*. We can also write down, in a similarly general form, a *system* of partial differential equations.

In this book we consider almost exclusively *linear partial differential equations of the second order*. As in the preceding parts, the set of values of the variables $(x_1, x_2, \ldots, x_m)$ will be regarded as a point x of m-dimensional Euclidean space, with coordinates $x_1, x_2, \ldots, x_m$.

In equations connected with physical problems, the independent variables are often time and spatial coordinates; to denote these we sometimes make use of the letters t, x, y, z.

The linear equation of second order with unknown function u and independent variables $x_1, x_2, \ldots, x_m$ can be written, most generally, in the form

$$\sum_{j,k=1}^{m} A_{jk}(x) \frac{\partial^2 u}{\partial x_j \partial x_k} + \sum_{k=1}^{m} A_k(x) \frac{\partial u}{\partial x_k} + A_0(x)u = f(x), \quad (2)$$

where A_{jk}, A_k, A_0, f are given functions of x.

Note that when $j \neq k$, equation (2) contains, not the separate terms $A_{jk} \partial^2 u/\partial x_j \partial x_k$ and $A_{kj} \partial^2 u/\partial x_k \partial x_j$, but their sum

$$(A_{jk} + A_{kj}) \frac{\partial^2 u}{\partial x_j \partial x_k}.$$

The expression $A_{jk} + A_{kj}$ can be decomposed into two terms in any suitable way, and so we shall always suppose that

$$A_{kj}(x) = A_{jk}(x), \tag{3}$$

so that the matrix of the coefficients of the second derivatives ("matrix of highest coefficients") is symmetric. This matrix will play an important role in subsequent work.

The left-hand side of equation (2) will be called a *second-order differential expression*.

The function $f(x)$, on the right-hand side of equation (2), will be called its *free term*. As usual, we distinguish between *homogeneous* equations, when $f(x) \equiv 0$, and *inhomogeneous* equations, when $f(x) \not\equiv 0$.

We consider a number of examples.

1. *Vibrating-string equation*

$$\frac{\partial^2 u}{\partial t^2} - a^2 \frac{\partial^2 u}{\partial x^2} = f(x, t). \tag{4}$$

Here $m = 2$; the free term $f(x, t)$ is proportional to the external force which acts at the point x of the string at time t. The matrix of highest coefficients has the form

$$\begin{pmatrix} 1 & 0 \\ 0 & -a^2 \end{pmatrix}. \tag{5}$$

In the more complicated case when the string vibrates in a medium with resistance proportional to velocity, the vibrating-string equation takes the form:

$$\frac{\partial^2 u}{\partial t^2} - a^2 \frac{\partial^2 u}{\partial x^2} + h \frac{\partial u}{\partial t} = f(x, t), \qquad h = \text{const.} \tag{6}$$

The matrix of highest coefficients again has the form (5).

2. *Vibrating-membrane equation*

$$\frac{\partial^2 u}{\partial t^2} - a^2 \left(\frac{\partial^2 u}{\partial x^2} + \frac{\partial^2 u}{\partial y^2} \right) = f(x, y, t). \tag{7}$$

The matrix of its highest coefficients has the form

$$\begin{pmatrix} 1 & 0 & 0 \\ 0 & -a^2 & 0 \\ 0 & 0 & -a^2 \end{pmatrix}. \tag{8}$$

3. For the *heat-conduction equation*

$$k\frac{\partial u}{\partial t} - \left(\frac{\partial^2 u}{\partial x^2} + \frac{\partial^2 u}{\partial y^2} + \frac{\partial^2 u}{\partial z^2}\right) = f(x, y, z, t) \tag{9}$$

the matrix of highest coefficients has the form

$$\begin{pmatrix} 0 & 0 & 0 & 0 \\ 0 & -1 & 0 & 0 \\ 0 & 0 & -1 & 0 \\ 0 & 0 & 0 & -1 \end{pmatrix}. \tag{10}$$

4. For *Laplace's equation*

$$\Delta u = \sum_{k=1}^{m} \frac{\partial^2 u}{\partial x_k^2} = f(x) \tag{11}$$

the matrix of highest coefficients is the unit matrix of order m.

5. The equation

$$(1+y^2)\frac{\partial^2 u}{\partial x^2} - 2xy\frac{\partial^2 u}{\partial x\,\partial y} + (1+x^2)\frac{\partial^2 u}{\partial y^2} = 0 \tag{12}$$

has as its matrix of highest coefficients

$$\begin{pmatrix} 1+y^2 & -xy \\ -xy & 1+x^2 \end{pmatrix}. \tag{13}$$

§ 2. Classification of Second-Order Equations

Second-order partial differential equations are classified according to the properties of the eigenvalues of their matrix of highest coefficients.

We recall that the eigenvalues of the matrix

$$A = \begin{pmatrix} A_{11} & A_{12} & \cdots & A_{1n} \\ A_{21} & A_{22} & \cdots & A_{2n} \\ \vdots & \vdots & & \vdots \\ A_{n1} & A_{n2} & \cdots & A_{nn} \end{pmatrix}$$

are the roots of the equation

$$\mathrm{Det}\,(A-\lambda I) = \begin{vmatrix} A_{11}-\lambda & A_{12} & \cdots & A_{1n} \\ A_{21} & A_{22}-\lambda & \cdots & A_{2n} \\ \vdots & \vdots & & \vdots \\ A_{n1} & A_{n2} & \cdots & A_{nn}-\lambda \end{vmatrix} = 0$$

(I is the unit matrix), and that all eigenvalues of a symmetric matrix are real.

Consider the following differential equation, somewhat more general than equation (1.2)

$$\sum_{j,k=1}^{m} A_{jk}(x)\frac{\partial^2 u}{\partial x_j \partial x_k} + \Phi\left(x_1, x_2, \ldots, x_m, u, \frac{\partial u}{\partial x_1}, \frac{\partial u}{\partial x_2}, \ldots, \frac{\partial u}{\partial x_m}\right) = 0, \quad (1)$$

where Φ is an arbitrary function of its arguments. The matrix of highest coefficients is symmetric, so that all its eigenvalues are real. We fix some point x, at which the coefficients of equation (1) are defined, and let the matrix of highest coefficients have at this point α positive, β negative and γ zero eigenvalues; obviously

$$\alpha + \beta + \gamma = m.$$

We shall say that in this case equation (1) belongs to the type (α, β, γ) at the given point. Equation (1) belongs to type (α, β, γ) on some point-set if it belongs to type (α, β, γ) at each point of the given set. Obviously if the highest coefficients A_{jk} in equation (1) are constant, then the type of this equation is the same throughout all space. If the signs of all the terms in the differential equation are reversed, then the numbers α and β change places; hence we may assume that types (α, β, γ) and (β, α, γ) are identical.

As examples we consider the equations in paragraphs 1–5 of the preceding section. In Examples 1–4, the matrices of highest coefficients are diagonal, and their eigenvalues are just the elements of the leading diagonal. Thus we see at once that at any point of space the vibrating-string equation is of type $(1, 1, 0)$, the membrane equation is $(2, 1, 0)$, the heat-conduction equation is $(3, 0, 1)$, and Laplace's equation in m-dimensional space is of the type $(m, 0, 0)$.

The eigenvalues of the matrix (1.13) are the roots of the equation

$$\begin{vmatrix} 1+y^2-\lambda & -xy \\ -xy & 1+x^2-\lambda \end{vmatrix} = 0;$$

thus they are equal to

$$\lambda_1 = 1+x^2+y^2, \quad \lambda_2 = 1.$$

Hence we see that at a general point (x, y) equation (1.12) is of the type $(2, 0, 0)$.

It is not difficult to formulate an equation whose type is different at different points. One such equation, for example, is *Tricomi's equation*

$$y \frac{\partial^2 u}{\partial x^2} + \frac{\partial^2 u}{\partial y^2} = 0. \tag{2}$$

The matrix of its highest coefficients

$$\begin{pmatrix} y & 0 \\ 0 & 1 \end{pmatrix}$$

has eigenvalues $\lambda_1 = 1$, $\lambda_2 = y$; therefore this equation is of type $(2, 0, 0)$ for $y > 0$, type $(1, 1, 0)$ for $y < 0$, and type $(1, 0, 1)$ when $y = 0$.

Three of the types of partial differential equations considered here play a special role in mathematical physics.

A. The type $(m, 0, 0) = (0, m, 0)$ is called *elliptic*. Consequently, equation (1) is of elliptic type at a given point, if at this point all the eigenvalues of the matrix of highest coefficients are different from zero and are all of the same sign.

The most important example of an equation of elliptic type is *Laplace's equation*. Equation (1.12) is also elliptic, and so is Tricomi's equation for $y > 0$.

B. The type $(m-1, 0, 1) = (0, m-1, 1)$ is called *parabolic*. Thus, equation (1) is of parabolic type at a given point, if at this point the matrix of highest coefficients has one eigenvalue equal to zero, and all the rest different from zero and of the same sign.

The most important example of a parabolic equation is the *heat-conduction equation*, which we here write in the form

$$k \frac{\partial u}{\partial x_m} - \sum_{j=1}^{m-1} \frac{\partial^2 u}{\partial x_j^2} = f(x). \tag{3}$$

Later (c.f. Part VI) we shall express this equation, and also the wave-equation (see below), in somewhat different form. A special case of equation (3) is equation (1.9). Tricomi's equation is of parabolic type when $y = 0$.

C. The type $(m-1, 1, 0) = (1, m-1, 0)$ is called *hyperbolic*. Consequently, equation (1) will be hyperbolic at those points where the eigenvalues of the matrix of highest coefficients are all different from zero, and where one of them is different in sign from all the rest.

The most important example of a hyperbolic equation is the *wave equation*

$$\frac{\partial^2 u}{\partial x_m^2} - \sum_{k=1}^{m-1} \frac{\partial^2 u}{\partial x_k^2} = f(x); \qquad (4)$$

special cases of this are the vibrating-string and vibrating-membrane equations. Also, Tricomi's equation for $y < 0$ is of hyperbolic type.

The importance of the three types of equations – elliptic, parabolic, hyperbolic – which we have singled out here emerges from two facts. In the first place, almost all the problems of physics which have been studied up to the present day are defined in terms of equations of these three types; in the second place, the theory of these equations has been developed with much greater thoroughness than that of partial differential equations of other types.

Equations of the type $(\alpha, \beta, 0)$, where $\alpha \geq 2$ and $\beta \geq 2$, are often collected together under the general heading *ultrahyperbolic*. The simplest ultrahyperbolic equation has the form

$$\frac{\partial^2 u}{\partial x_1^2} + \frac{\partial^2 u}{\partial x_2^2} - \frac{\partial^2 u}{\partial x_3^2} - \frac{\partial^2 u}{\partial x_4^2} = 0.$$

We shall regard elliptic, parabolic and hyperbolic equations as the equations of mathematical physics.

A substantial amount of work has been done in connection with equations of "mixed" type, i.e. equations whose type can change from point to point. For details concerning these, see refs. [2] and [3]. There has also been a series of studies of equations of the type $(\alpha, 0, \gamma)$ ("elliptic-parabolic equations"). Among recent work along these lines we may mention the article [1].

§ 3. Boundary Conditions and Boundary-Value Problems

If we wish to achieve a complete description of a physical problem, we cannot limit ourselves to the differential equation alone; it is necessary to add some supplementary relations, which usually have the character of so-called boundary conditions.

We shall clarify this through simple examples. We already know that the vibrations of a string are described by the differential equation

$$\frac{\partial^2 u}{\partial t^2} - a^2 \frac{\partial^2 u}{\partial x^2} = f(x, t). \qquad (1)$$

Suppose that the string is of length l, and that in the equilibrium state it occupies the interval $[0, l]$ of the x-axis. Suppose further that at the instant of time $t = 0$ the string is disturbed from its equilibrium position and begins to vibrate. The problem consists in examining the displacement $u(x, t)$ of a point of the string, having arbitrary abscissa $x \in [0, l]$, and at an arbitrary instant after the initial instant, i.e. at an arbitrary time $t > 0$. In other words, the function $u(x, t)$, which satisfies equation (1), has to be determined in the plane of the variables x and t, in the domain shown in Fig. 9; the boundary of this domain consists of the segment $[0, l]$ of the x-axis, and of the two lines $x = 0, t > 0$ and $x = l, t > 0$.

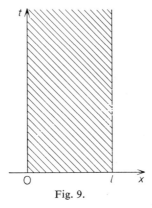

Fig. 9.

The only quantities given in the differential equation (1) are a^2, which depends in a prescribed way on the properties of the string (on its density and the tension), and the function $f(x, t)$, describing the external force, which acts at time t on the point x of the string. But equation (1) does not contain, for example, any information concerning how the string was displaced from its equilibrium position, or concerning the situation at the ends of the string; the ends might be fixed or, alternatively, free; it could be the case that although the ends are not fixed, their displacements are subject to some kind of restriction. This information has to be given separately. It could be that the string is displaced from its equilibrium state by giving the points of the string either an initial disturbance, or an initial velocity, or both of these together. Suppose the point x of the string, $0 \leq x \leq l$, is given an initial disturbance $\phi_0(x)$ and an initial velocity $\phi_1(x)$. Then the unknown function $u(x, t)$ must satisfy the conditions

$$u\bigg|_{t=0} = \phi_0(x), \quad \frac{\partial u}{\partial t}\bigg|_{t=0} = \phi_1(x), \quad 0 \leq x \leq l. \qquad (2)$$

Suppose also that the motion of the ends of the string is known: let the displacement of the left-hand end of the string be $\psi_1(t)$ for $t \geq 0$, and the displacement of the right-hand end be $\psi_2(t)$. Then the conditions

$$u|_{x=0} = \psi_1(t), \qquad u|_{x=l} = \psi_2(t) \tag{3}$$

must also be satisfied.

Conditions (3) do not have to be imposed if the string is infinite, i.e. if it occupies the whole x-axis in its equilibrium state.

The additional conditions (2) and (3) must be satisfied on the lines $t = 0, x = 0, x = l$, i.e. on the boundaries of the domain (Fig. 9), in which the function $u(x, t)$ has to be determined. Because of this these conditions are called boundary conditions.

We note that conditions (2) and (3) are not completely independent: if it is required that unknown function $u(x, t)$ be continuous not only inside, but also on the boundary of, its domain of dependence, then we must have

$$\phi_0(0) = \psi_1(0), \qquad \phi_0(l) = \psi_2(0). \tag{4}$$

The relations (4) are called *compatibility conditions*. They follow from the requirement that the displacement of the ends of the string be continuous at the initial instant of time. If in addition it should be required that on the boundary of the domain (Fig. 9) some of the derivatives of the function $u(x, t)$ be continuous, then new compatibility conditions can be generated. Thus, if we ask that the first derivative be continuous, then we must have

$$\phi_1(0) = \psi_1'(0), \qquad \phi_1(l) = \psi_2'(0); \tag{4a}$$

for continuity of the second derivatives, we shall have the conditions

$$\psi_1''(0) - a^2 \phi_0''(0) = f(0, 0),$$
$$\psi_2''(0) - a^2 \phi_0''(l) = f(l, 0). \tag{4b}$$

It will be shown later (Part VI) that, with sufficiently precise restrictions, equation (1) has one and only one solution which satisfies conditions (2) and (3). This implies that equations (1)–(3) contain all the information necessary for analysis of the vibrating string (unique solution), and does not contain excessive or contradictory information (the solution exists).

We consider one further example. A homogeneous, isotropic body occupies a domain Ω, bounded by a surface Γ, of three-dimensional space (x_1, x_2, x_3). We suppose that through the body there are distributed some sources of heat, of intensity $F(x) = F(x_1, x_2, x_3)$, independent of time.

This implies that in any sub-domain $\Omega' \subset \Omega$, and during an arbitrary time-interval of length δt, a quantity of heat equal to

$$\delta t \int_{\Omega'} F(x)\,dx$$

is released. We further suppose that the temperature-distribution in the body has achieved a steady-state, i.e. independence of time. Then at the point $x = (x_1, x_2, x_3)$ the temperature of the body satisfies Laplace's equation

$$\frac{\partial^2 u}{\partial x_1^2} + \frac{\partial^2 u}{\partial x_2^2} + \frac{\partial^2 u}{\partial x_3^2} = -f(x), \tag{5}$$

where the function $f(x)$ differs from $F(x)$ only by a constant multiplicative factor. Equation (5) alone is not sufficient completely to define the temperature distribution in the body Ω; this is apparent because, among other things, equation (5) has an infinite set of solutions. It is necessary to provide additional information. This can be done, for example, in the following way. The surface Γ of the given body is accessible to observation, and the temperature can be measured at any point of this surface. Suppose that the temperature has been measured at every point of Γ, and let it be equal to $\phi(x)$ at a point $x \in \Gamma$. Then we obtain an additional boundary condition

$$u|_\Gamma = \phi(x), \qquad x \in \Gamma. \tag{6}$$

In Part V it will be shown that, under sufficiently broad conditions, the problem (5)–(6) has one and only one solution.

If the body were not homogeneous and not isotropic, then we would obtain, not equation (5), but a more general equation of the form

$$-\sum_{j,k=1}^{3} \frac{\partial}{\partial x_j}\left(A_{jk}(x)\frac{\partial u}{\partial x_k}\right) = f(x), \tag{7}$$

which is elliptic, like (5). The problem of solving equation (7) (or, particularly, equation (5)), subject to boundary conditions (6), is called the *Dirichlet problem*.

Additional information for equation (7) can be described by boundary conditions different from (6). Thus, if it is known that at the point $x \in \Gamma$ the intensity of the flux of temperature is equal to a given function $\Psi(x)$, then

$$\left[\sum_{j,k=1}^{3} A_{jk}\frac{\partial u}{\partial x_k}\cos(v, x_j)\right]_\Gamma = \psi(x), \tag{8}$$

where $\psi(x)$ differs from $\Psi(x)$ only by some multiplicative constant, and v is the outward normal to the surface Γ. For Laplace's equation,

$$A_{jk} = \delta_{jk} = \begin{cases} 0, & j \neq k, \\ 1, & j = k, \end{cases}$$

and the boundary condition (8) takes the form

$$\left. \frac{\partial u}{\partial v} \right|_{\Gamma} = \psi(x). \tag{9}$$

The problem (7)–(8) (and in particular the problem (5), (9)) is known as the *Neumann problem*.

The Dirichlet and Neumann problems can be formulated, not only in three dimensions, but also in any m-dimensional space.

We now present a general formulation of the concepts of boundary conditions and boundary-value problems.

Suppose that we are given some partial differential equation

$$Lu = f(x). \tag{10}$$

We assume that the solution of this equation is to be determined in some domain Ω of the space E_m, where the boundary of this domain is denoted by Γ. On the whole of Γ, or on some part of it, we are given the values of one, or sometimes of several, differential expressions involving the unknown function

$$G_k u|_{\Gamma} = \phi_k(x), \quad k = 1, 2, \ldots, l. \tag{11}$$

The equations (11) are called *boundary conditions*, and the problem of integrating the differential equation (10) subject to the boundary conditions (11) is called a *boundary-value problem*.

§ 4. Cauchy's Problem

For equation (1.2) Cauchy's problem is formulated in the following way. Let Γ be some smooth surface given in the space of the variables $x_1, x_2, \ldots, x_m$. With each point $x \in \Gamma$ (x is the point whose coordinates are $x_1, x_2, \ldots, x_m$) there is associated some direction λ, not tangent to Γ. It is required to find, in some neighbourhood (either one-sided or two-sided) of the surface Γ, a solution of equation (1.2) which satisfies the so-called *Cauchy conditions*

$$u\bigg|_{\Gamma} = \phi_0(x), \qquad \frac{\partial u}{\partial \lambda}\bigg|_{\Gamma} = \phi_1(x). \qquad (1)$$

Here $\phi_0(x)$ and $\phi_1(x)$ are prescribed functions on Γ; we shall assume that the function $\phi_1(x)$ is continuous, while $\phi_0(x)$ is continuously differentiable.

The functions $\phi_0(x)$ and $\phi_1(x)$ are called *Cauchy data*, and Γ is called the *surface carrying the Cauchy data*, or, simply, the *Cauchy surface*.

Note that the boundary conditions (3.2) are Cauchy conditions for the vibrating-string equation; here the interval $[0, l]$ of the x-axis plays the role of the *Cauchy surface*.

The Cauchy problem differs from the boundary-value problems considered in § 3 in that here the domain in which the unknown solution has to be determined is not specified beforehand. Nevertheless we shall regard the Cauchy problem as a boundary-value problem.

The following observations will turn out to be useful later: if the Cauchy conditions (1) are known, it is possible to find the values of all the first derivatives of the unknown function on the Cauchy surface Γ. As proof of this, take an arbitrary point x on Γ, and construct at this point a *local coordinate system* $X_1, X_2, \ldots, X_m$, that is, a Cartesian coordinate system with original at the point x, with the axes $X_1, X_2, \ldots, X_{m-1}$ lying on a $(m-1)$-dimensional plane which is tangential to Γ at the point x, and with the axis X_m directed along the normal to Γ at this point (Fig. 10). Knowing the value of the function $u = \phi_0(x)$ on Γ, we can immediately find the

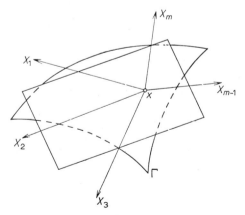

Fig. 10.

derivatives with respect to $X_1, X_2, \ldots, X_{m-1}$,

$$\left.\frac{\partial u}{\partial X_k}\right|_\Gamma = \frac{\partial \phi_0}{\partial X_k}, \quad k = 1, 2, \ldots, m-1.$$

Moreover,

$$\phi_1(x) = \frac{\partial u}{\partial \lambda} = \sum_{k=1}^{m} \frac{\partial u}{\partial X_k} \cos(\lambda, X_k).$$

The angle (λ, X_m) is not a right-angle, because the direction of λ is not tangential to Γ. But then $\cos(\lambda, X_m) \neq 0$, and the last equation yields the value of the missing derivative:

$$\left.\frac{\partial u}{\partial X_m}\right|_\Gamma = \frac{1}{\cos(\lambda, X_m)} \left[\phi_1(x) - \sum_{k=1}^{m-1} \frac{\partial \phi_0}{\partial X_k} \cos(\lambda, X_k)\right].$$

Knowing the derivatives in the local coordinate system, we can find their values in the coordinate system $x_1, x_2, \ldots, x_m$ through the formula

$$\left.\frac{\partial u}{\partial x_k}\right|_\Gamma = \sum_{j=1}^{m} \left.\frac{\partial u}{\partial X_j}\right|_\Gamma \cos(X_j, x_k).$$

§ 5. Questions of Existence, Uniqueness and Proper Posing of Boundary-Value Problems

1. Suppose that we have formulated some boundary-value problem. To find its solution means to find all the functions which satisfy the given differential equation and the given boundary conditions. Usually we subject the unknown function to some further restrictions of a general character, which often makes it possible to regard this function as an element of some functional space; we denote the latter by B_1. Thus, in posing the Dirichlet problem for Laplace's equation, we could insist that the required solution be continuous in a closed domain $\bar{\Omega} = \Omega \cup \Gamma$; in this case, the solution, if it exists, will be an element of the space $C(\bar{\Omega})$. We could impose some other restrictions on our function; for example, we could ask that the integrals

$$\int_\Omega u^2 \, dx, \quad \int_\Omega (\operatorname{grad} u)^2 \, dx = \int_\Omega \sum_{k=1}^{m} \left(\frac{\partial u}{\partial x_k}\right)^2 dx$$

be finite. In this case the required function could be regarded as an element of a Hilbert space, in which the norm was defined by the formula

$$\|u\|^2 = \int_\Omega \{u^2 + (\operatorname{grad} u)^2\} \, dx.$$

If we impose restrictions on the solution we are seeking, we are obliged to put some restrictions on the functions which are given on the right-hand sides of the differential equation and boundary conditions. It usually turns out in these circumstances that the set of these right-hand sides can also be regarded as an element of some other functional space, B_2. In many interesting cases, B_1 and B_2 are Banach spaces.

We consider the particular case when the differential expressions which occur, both in the differential equation and in the boundary conditions, are linear. The set of these differential expressions generates some linear operator $\mathfrak{A}$, which acts from the space B_1 into the space B_2, and transforms the unknown function $u(x)$ into the above-mentioned set of right-hand sides of the differential equation and boundary conditions. Denoting this set by Φ, we can write our boundary-value problem in the form

$$\mathfrak{A}u = \Phi. \tag{1}$$

The operator $\mathfrak{A}$ is called the *operator of the given boundary-value problem*.

We can now say that to solve the boundary-value problem means to find all the elements of the space B_1 which are transformed by the operator $\mathfrak{A}$ into a given element $\Phi \in B_2$.

Usually we try to pose the boundary conditions in such a way that the boundary-value problem has one and only one solution. This requires in each case proving an existence theorem and a uniqueness theorem. A uniqueness theorem is equivalent to the assertion that there exists an operator $\mathfrak{A}^{-1}$, the inverse of $\mathfrak{A}$. An existence theorem means that the domain of values of the operator $\mathfrak{A}$ coincides with the space B_2. If both theorems – existence and uniqueness – hold, then the operator $\mathfrak{A}^{-1}$ exists and is defined on the whole of the space B_2.

As well as existence and uniqueness, the solution of boundary-value problems raises a further important question. This is the so-called *proper posing* of a boundary-value problem.

This concept is best understood by approaching it from a physical viewpoint. In the final analysis, the determination of physical quantities requires the process of measurement, and this always involves a certain amount of error. In particular, the element Φ of equation (1) – the set of data in the boundary-value problem – is determinate only with a margin of error. The question then arises: to what extent is error in the data of the boundary-value problem reflected in its solution? The answer to this question can be formulated in the following definition.

A boundary-value problem is said to be *properly posed in the pair of Banach spaces B_1 and B_2* if the solution of the boundary-value problem is unique in B_1 and exists for any data from B_2 and if a sufficiently small variation of the data in the norm of B_2 results in an arbitrarily small variation of the solution in the norm of B_1.

We shall return to the question of proper posing at the end of the book, in Part VII, where it will be shown, in particular, that the proper posing of problem (1) is equivalent to the boundedness of the operator $\mathfrak{A}^{-1}$. Here we shall restrict ourselves to providing two examples of improperly-posed boundary-value problems. The first example is due to Hadamard, who was the first to introduce the concept of proper posing of boundary-value problems.

2. Consider Laplace's equation in the plane:

$$\frac{\partial^2 u}{\partial x^2} + \frac{\partial^2 u}{\partial y^2} = 0. \tag{2}$$

The role of the surface Γ will here be taken by the x-axis, on which the Cauchy data will be prescribed. The neighbourhood in which we seek the solution will be the strip $0 < y < \delta$, where δ is an arbitrary positive number; we denote this strip by Ω. The non-tangential direction λ will in this case be y. We take the Cauchy conditions to be:

$$u\Big|_{y=0} = \phi(x), \quad \frac{\partial u}{\partial y}\Big|_{y=0} = 0, \quad -\infty < x < \infty. \tag{3}$$

The set of data consists of the single function $\phi(x)$, which we shall assume to be continuous and bounded everywhere on the axis. Then this function can be regarded as an element of the space $C(-\infty, \infty)$, which in the present case plays the role of the space B_2. For B_1 we take the space $C(\Omega)$ of functions which are continuous and bounded in the strip Ω. For the domain of definition of the operator of the boundary-value problem (2)–(3), we take a set of functions from $C(\Omega)$ which have continuous second derivatives and satisfy the condition

$$\frac{\partial u}{\partial y}\Big|_{y=0} = 0.$$

We show that in the pair of spaces B_1, B_2 the problem (2)–(3) is improperly posed. It is possible to prove that the solution of this problem is unique. It obviously follows from this that to the function $\phi(x) \equiv 0$ there corresponds the solution $u \equiv 0$. Now impart a small (in the norm of the space B_2) increment to the function $\phi(x) \equiv 0$: consider the Cauchy problem for equation (2) and the Cauchy data

$$u\Big|_{y=0} = \frac{\cos nx}{n}, \quad \frac{\partial u}{\partial y}\Big|_{y=0} = 0, \tag{4}$$

where n is a sufficiently large positive integer.

The solution to our new problem is

$$u(x, y) = \frac{\cos nx \cosh ny}{n},$$

which is easily verified by substitution in equations (2) and (4). Obviously,

$$\|\phi\|_{B_2} = \left\|\frac{\cos nx}{n}\right\|_{B_2} = \max_{-\infty < x < +\infty} \left|\frac{\cos nx}{n}\right| = \frac{1}{n} \xrightarrow[n \to \infty]{} 0.$$

At the same time,

$$\|u\|_{B_1} = \max_{\substack{-\infty < x < \infty \\ 0 \leq y \leq \delta}} \left|\frac{\cos nx \cosh ny}{n}\right| = \frac{\cosh n\delta}{n} \xrightarrow[n \to \infty]{} \infty.$$

Thus, an arbitrarily small (in the norm of B_2) increment to the data can generate an arbitrarily large (in the norm of B_1) increment to the solution. This means that in the pair of spaces under consideration the Cauchy problem for Laplace's equation is not properly posed.

3. Consider the equation

$$\frac{\partial^2 u}{\partial x_1 \, \partial x_2} = 0. \tag{5}$$

This equation is of hyperbolic type. The matrix of its highest coefficients has in fact the form

$$\begin{pmatrix} 0 & \frac{1}{2} \\ \frac{1}{2} & 0 \end{pmatrix};$$

and its eigenvalues are the roots of the equation

$$\begin{pmatrix} -\lambda & \frac{1}{2} \\ \frac{1}{2} & -\lambda \end{pmatrix} = 0$$

i.e., $\lambda_1 = \frac{1}{2}, \lambda_2 = -\frac{1}{2}$. These are different from zero and have opposite signs. Note also that equation (5) is transformed into the homogeneous vibrating-string equation if we effect the following substitution of independent variables:

$$x_1 = x + at, \quad x_2 = x - at.$$

We pose the Dirichlet problem for equation (5) in the square shown in Fig. 11; denote this square by Ω, and its perimeter by Γ. Let the conditions on Γ take the form

$$\begin{aligned} u|_{x_1=0} &= \phi_1(x_2), & u|_{x_2=0} &= \psi_1(x_1), \\ u|_{x_1=1} &= \phi_2(x_2), & u|_{x_2=1} &= \psi_2(x_1). \end{aligned} \tag{6}$$

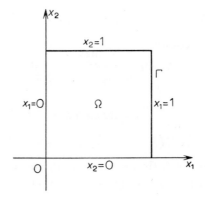

Fig. 11.

For B_1 and B_2 we take the spaces $C(\bar{\Omega})$ and $C(\Gamma)$ respectively. In order that the solution should be continuous, it is necessary that the compatibility conditions

$$\begin{aligned} \phi_1(0) &= \psi_1(0), & \phi_2(0) &= \psi_1(1), \\ \phi_1(1) &= \psi_2(0), & \phi_2(1) &= \psi_2(1) \end{aligned} \tag{7}$$

should be satisfied. Problem (5)–(6) is not soluble for arbitrary, given, continuous functions (7). To demonstrate this, we shall find the general solution of equation (5). Writing it in the form $(\partial/\partial x_1)(\partial u/\partial x_2) = 0$, we see that $\partial u/\partial x_2 = f(x_2)$, where the function f is arbitrary. Integrating now with respect to x_2 we obtain

$$u(x_1, x_2) = F_1(x_1) + F_2(x_2), \quad F_2'(x_2) = f(x_2),$$

where F_1 and F_2 are arbitrary functions. We can satisfy the first of the two conditions (6):
$$F_1(0)+F_2(x_2) = \phi_1(x_2),$$
$$F_1(x_1)+F_2(0) = \psi_1(x_1).$$

Clearly, one of the constants $F_1(0)$ and $F_2(0)$ remains arbitrary. Put $F_2(0) = 0$. Then
$$F_1(x_1) = \psi_1(x_1), \qquad F_2(x_2) = \phi_1(x_2)-\psi_1(0).$$

The solution is completely determined; the equality $F_2(0) = 0$ follows from the first of equations (7). It is now obviously impossible to satisfy the remaining conditions (6) if the functions ϕ_2 and ψ_2 are arbitrary.

From this it follows that problem (5)–(6) is not properly posed in the pair of spaces $C(\bar{\Omega})$ and $C(\Gamma)$.

Chapter 10. Characteristics. Canonical Form. Green's Formulae

§ 1. Transformation of the Independent Variables

Consider a second-order partial differential equation, which is linear with respect to the highest derivatives

$$\sum_{j,k=1}^{m} A_{jk}(x) \frac{\partial^2 u}{\partial x_j \partial x_k} + \Phi\left(x, u, \frac{\partial u}{\partial x_1}, \frac{\partial u}{\partial x_2}, \ldots, \frac{\partial u}{\partial x_m}\right) = 0. \quad (1)$$

In places of the independent variables $x_1, x_2, \ldots, x_m$, we introduce new independent variables

$$\xi_r = \xi_r(x_1, x_2, \ldots, x_m), \qquad r = 1, 2, \ldots, m. \quad (2)$$

We explain how our equation changes under this transformation.

For simplicity, let us agree to omit the summation sign above, according to the following rule: if in any single-term expression a certain variable index occurs twice, and if this index can take values from 1 to m, then we carry out summation with respect to this index from 1 to m. Equation (1) can now be written in the simpler form

$$A_{jk}(x) \frac{\partial^2 u}{\partial x_j \partial x_k} + \Phi\left(x, u, \frac{\partial u}{\partial x_1}, \frac{\partial u}{\partial x_2}, \ldots, \frac{\partial u}{\partial x_m}\right) = 0.$$

Suppose that in some domain of variation of the point x, the transformation (2) is one-to-one, and its Jacobian is different from zero. Then we shall call this transformation of independent variables *regular*. Assume further that the functions ξ_r have continuous second derivatives. We evaluate the derivatives occurring in equation (1)

$$\frac{\partial u}{\partial x_k} = \frac{\partial u}{\partial \xi_r} \frac{\partial \xi_r}{\partial x_k}, \qquad \frac{\partial^2 u}{\partial x_j \partial x_k} = \frac{\partial^2 u}{\partial \xi_r \partial \xi_s} \frac{\partial \xi_r}{\partial x_k} \frac{\partial \xi_s}{\partial x_j} + \frac{\partial u}{\partial \xi_r} \frac{\partial^2 \xi_r}{\partial x_j \partial x_k}.$$

Substituting these expressions into equation (1), we obtain a new equation:

$$A_{jk}\frac{\partial \xi_r}{\partial x_k}\frac{\partial \xi_s}{\partial x_j}\frac{\partial^2 u}{\partial \xi_r \partial \xi_s} + \Phi_1\left(\xi_1, \ldots, \xi_m, u, \frac{\partial u}{\partial \xi_1}, \frac{\partial u}{\partial \xi_2}, \ldots, \frac{\partial u}{\partial \xi_m}\right) = 0,$$

$$\Phi_1 = \Phi + A_{jk}\frac{\partial^2 \xi_r}{\partial x_j \partial x_k}\frac{\partial u}{\partial \xi_r}.$$

Introduce the notation

$$A_{jk}\frac{\partial \xi_r}{\partial x_k}\frac{\partial \xi_s}{\partial x_j} = \tilde{A}_{rs}. \tag{3}$$

Then equation (1) takes the form

$$\tilde{A}_{rs}\frac{\partial^2 u}{\partial \xi_r \partial \xi_s} + \Phi_1\left(\xi_1, \ldots, \xi_m, u, \frac{\partial u}{\partial \xi_1}, \ldots, \frac{\partial u}{\partial \xi_m}\right) = 0. \tag{4}$$

Thus, as a result of a transformation of the independent variables, equation (1) is converted into an equation of exactly the same form as it was originally; only its coefficients are changed. Note that the matrix of highest coefficients of equation (4) is symmetric

$$\tilde{A}_{rs} = A_{jk}\frac{\partial \xi_r}{\partial x_k}\frac{\partial \xi_s}{\partial x_j} = A_{kj}\frac{\partial \xi_s}{\partial x_j}\frac{\partial \xi_r}{\partial x_k}.$$

If in this last expression the symbols j and k change places, we obtain

$$\tilde{A}_{rs} = A_{jk}\frac{\partial \xi_s}{\partial x_k}\frac{\partial \xi_r}{\partial x_j} = \tilde{A}_{sr}.$$

THEOREM 10.1.1. *A regular transformation of the independent variables does not change the type of the partial differential equation* (1).

The following is a well-known algebraic fact. Suppose that some matrix is transformed by means of a regular transformation into diagonal form. Then the number of positive, negative and zero eigenvalues of the given matrix is respectively equal to the number of positive, negative and zero elements of the new (i.e. diagonal) matrix.

Denote by J the Jacobian of the matrix of the transformation (2). Its determinant – the Jacobian of this transformation – is different from zero, and so there exists an inverse matrix J^{-1}. Formula (3) is equivalent to the

matrix equation
$$\tilde{A} = JAJ', \tag{5}$$
in which the prime denotes a transposed matrix.

Let σ be a matrix which effects a regular, linear transformation of the matrix A into the diagonal matrix D:
$$A = \sigma D \sigma'.$$
Then by formula (5),
$$\tilde{A} = J\sigma D\sigma' J' = (J\sigma)D(J\sigma)',$$
and the matrix $\tilde{A}$ reduces by a regular transformation (through the matrix $J\sigma$) to the same diagonal matrix D. Therefore the number of positive, negative and zero eigenvalues of the respective matrices A and $\tilde{A}$ coincide. This proves the theorem.

§ 2. Characteristics. Relation between Cauchy Data on Characteristics

Consider again the second-order partial differential equation
$$A_{jk}(x)\frac{\partial^2 u}{\partial x_j \partial x_k} + \Phi\left(x, u, \frac{\partial u}{\partial x_1}, \ldots, \frac{\partial u}{\partial x_m}\right) = 0, \tag{1}$$
linear with regard to its highest derivatives. We formulate the first-order equation
$$A_{jk}(x)\frac{\partial \omega}{\partial x_j}\frac{\partial \omega}{\partial x_k} = 0. \tag{2}$$

This is called the *equation of characteristics* of the differential equation (1). If the function $\omega(x_1, x_2, \ldots, x_m)$ satisfies the equation of characteristics, then the surface (or in the two-dimensional case, the curve) defined by the equation
$$\omega(x_1, x_2, \ldots, x_m) = C, \tag{3}$$
where C is an arbitrary constant, is called the *characteristic surface* (or, correspondingly, the *characteristic curve*), or simply the *characteristic* of the given differential equation (1).

The characteristic equation can formally be constructed by writing down the quadratic form
$$(At, t) = A_{jk}t_j t_k, \tag{4}$$
corresponding to the matrix A of highest derivatives of equation (1), sub-

stituting in this form $t_k = \partial \omega / \partial x_k$, and setting the resultant expression equal to zero.

We note an important property of characteristics: they are invariant under a transformation of the independent variables. This implies the following: if $\omega(x_1, x_2, \ldots, x_m)$ is a solution of equation (2), and if the transformation (1.2) of the independent variables transforms the function $\omega(x_1, x_2, \ldots, x_m)$ into the function $\tilde{\omega}(\xi_1, \xi_2, \ldots, \xi_m)$ then this new function is a solution of the equation

$$\tilde{A}_{jk} \frac{\partial \tilde{\omega}}{\partial \xi_j} \frac{\partial \tilde{\omega}}{\partial \xi_k} = 0, \qquad (2a)$$

which is the characteristic equation for the transformed differential equation (1.4).

In fact,

$$\frac{\partial \omega}{\partial x_j} = \frac{\partial \tilde{\omega}}{\partial \xi_r} \frac{\partial \xi_r}{\partial x_j}, \qquad \frac{\partial \omega}{\partial x_k} = \frac{\partial \tilde{\omega}}{\partial \xi_s} \frac{\partial \xi_s}{\partial x_k}.$$

Substituting this into equation (2), and using formula (1.3), we find that $\tilde{\omega}$ satisfies equation (2a).

An equation of elliptic type does not have real characteristics. In fact, if equation (1) is elliptic, then the quadratic form (4) is defined and is equal to zero (for real t_k) only when $t_1 = t_2 = \ldots = t_m = 0$. In this case the characteristic equation has only the solution $\omega \equiv \text{constant}$, which does not define any surface.

We shall show that on a characteristic surface the Cauchy data are connected by certain relationships. It will follow from this that Cauchy data cannot be independently prescribed on a characteristic surface.

Let the Cauchy data be given on a sufficiently smooth surface Γ, defined by the equation

$$\xi(x_1, x_2, \ldots, x_m) = 0, \qquad (5)$$

and let them be of the following form:

$$u \bigg|_\Gamma = \phi_0(x), \qquad \frac{\partial u}{\partial \lambda} \bigg|_\Gamma = \phi_1(x), \qquad (6)$$

where λ is a direction not tangential to Γ. As was explained in § 4, Chap. 9, knowing the data (6) enables us to find the values of all the first derivatives of the function u on the Cauchy surface Γ.

We choose an arbitrary point on the surface Γ, and in the neighbourhood of this point introduce a new system of coordinates. The coordinates

$\xi_1, \xi_2, \ldots, \xi_{m-1}$ are introduced arbitrarily, and the coordinate ξ_m is put equal to ξ; in selecting the coordinates $\xi_1, \xi_2, \ldots, \xi_{m-1}$ we are only concerned that the transformation should be one-to-one, with a non-zero Jacobian, and that the functions ξ_r should have continuous second derivatives. In our new coordinate system the equation of the Cauchy surface takes the particularly simple form $\xi_m = 0$.

Suppose now that Γ is a characteristic surface, i.e. that

$$A_{jk} \frac{\partial \xi}{\partial x_j} \frac{\partial \xi}{\partial x_k} = 0.$$

In the variables ξ_r, the coefficient of the derivative $\partial^2 u / \partial \xi_m^2$ then becomes zero, and our equation with respect to the variable ξ_m is a first-order equation.

We now show that *on the surface Γ it is possible to calculate all the derivatives occurring in the transformed equation* (1), *starting from the Cauchy data alone*. Now the quantity $u|_\Gamma = \phi_0(x)$ is known. The first derivatives $(\partial u / \partial \xi_r)|_\Gamma$ can be defined in the manner described above. Those second derivatives which do not involve two differentiations with respect to ξ_m can be found by differentiating the first derivatives with respect to $\xi_1, \xi_2, \ldots, \xi_{m-1}$, i.e. along directions tangential to Γ. The one second derivative which cannot be found by starting from the Cauchy data alone is $\partial^2 u / \partial \xi_m^2$; but precisely this one is missing from the transformed equation.

The values of all the terms on the left-hand side of equation (1), after transformation to the variables $\xi_1, \xi_2, \ldots, \xi_m$ can be evaluated on the Cauchy surface Γ. Substituting these values into the equation, we find that some determined function must be identically equal to zero. This is then the relation between Cauchy data on the characteristic; if it is violated then the Cauchy problem with data on the characteristic does not have a solution.

As an example, consider the heat-conduction equation

$$\frac{\partial u}{\partial x_m} - \sum_{k=1}^{m-1} \frac{\partial^2 u}{\partial x_k^2} = 0. \tag{7}$$

Its characteristic equation is

$$-\sum_{k=1}^{m-1} \left(\frac{\partial \omega}{\partial x_k}\right)^2 = 0,$$

hence we obtain $\omega = f(x_m)$, where f is an arbitrary function. The equation of the characteristic surface has the form $f(x_m) = \text{const.}$; solving for x_m, we obtain an equation of the form $x_m = $ const. Thus, the characteristics of equation (7) are the planes $x_m = $ const. Let the Cauchy surface be the plane $x_m = 0$, and the Cauchy conditions take the form

$$u\big|_{x_m=0} = \phi_0(x_1, \ldots, x_{m-1}), \quad \frac{\partial u}{\partial x_m}\bigg|_{x_m=0} = \phi_1(x_1, \ldots, x_{m-1}). \qquad (8)$$

Putting $x_m = 0$ in equation (7), we immediately arrive at the relation

$$\phi_1 = \sum_{k=1}^{m-1} \frac{\partial^2 \phi_0}{\partial x_k^2}.$$

It is obvious from this that there is no point in prescribing the second of conditions (8) – it is enough simply to prescribe the condition

$$u|_{x_m=0} = \phi_0(x_1, x_2, \ldots, x_{m-1}).$$

§ 3. Reduction of a Second-Order Equation to Canonical Form

Consider the particular case of a regular, linear transformation of variables

$$\xi_r = j_{rk} x_k; \quad k = 1, 2, \ldots, m; \quad j_{rk} = \text{const}. \qquad (1)$$

Introduce the matrix J with elements j_{rk}. The transformation (1) can be written in the form

$$\xi = Jx. \qquad (1')$$

We fix the point x. Then the matrix A of highest coefficients will be constant. The matrix J can be chosen such that the transformed matrix of highest coefficients (formula (1.5)), $\tilde{A} = JAJ'$, is diagonal: $\tilde{A}_{jk} = 0, j \neq k$. Then at the fixed point equation (1.1) takes the form

$$\sum_{k=1}^{m} v_k \frac{\partial^2 u}{\partial \xi_k^2} + \Phi_1\left(\xi_1, \xi_2, \ldots, \xi_m, u, \frac{\partial u}{\partial \xi_1}, \frac{\partial u}{\partial \xi_2}, \ldots, \frac{\partial u}{\partial \xi_m}\right) = 0, \qquad (2)$$

where $v_k = \tilde{A}_{kk}$. This form of the equation, in which all second-order cross-derivatives are missing, is called the *canonical form* of the equation. Thus, a second-order partial differential equation, linear with regard to its highest derivatives, can be reduced at any point of space to canonical form with the aid of a linear transformation of the independent variables.

Obviously, the equation can be reduced to canonical form throughout all space if the highest coefficients A_{jk} are constants.

Laplace's equation, the heat-conduction and wave equations all have a canonical form.

The canonical form of an equation is closely connected with its type. Because of the inertia law for quadratic forms, the distribution of positive, negative and zero values among the quantities v_k is just the same as among the λ_k – the eigenvalues of the matrix of highest coefficients. For this reason we can define the type of a second-order partial differential equation, linear with regard to its highest coefficients, as follows: equation (1.1) belongs to the type (α, β, γ) if in the canonical form (2) of this equation there are α positive, β negative and γ zero values among the quantities v_k.

§ 4. Case of Two Independent Variables

A second-order partial differential equation having two independent variables is noteworthy in that it can be reduced to canonical form not only at a particular given point, but also in any domain where the type of the equation remains unchanged.

Denote the independent variables by x and y, and the highest coefficients by A, B, C. Then the equation can be written in the form

$$A\frac{\partial^2 u}{\partial x^2} + 2B\frac{\partial^2 u}{\partial x \partial y} + C\frac{\partial^2 u}{\partial y^2} + \Phi\left(x, y, u, \frac{\partial u}{\partial x}, \frac{\partial u}{\partial y}\right) = 0. \tag{1}$$

We shall suppose that at every point at least one of the coefficients A, B, C is different from zero. For simplicity we shall assume that these coefficients belong to class $C^{(1)}$ in a corresponding closed domain.

We write down the matrix of highest coefficients

$$\begin{pmatrix} A & B \\ B & C \end{pmatrix}$$

and the eigenvalue equation

$$\begin{vmatrix} A-\lambda & B \\ B & C-\lambda \end{vmatrix} = 0$$

or

$$\lambda^2 - (A+C)\lambda + AC - B^2 = 0.$$

This equation has the real roots

$$\lambda_{1,2} = \tfrac{1}{2}[A+C \pm \sqrt{\{(A-C)^2 + 4B^2\}}];$$

these have the same sign if $AC - B^2 > 0$, and different signs if $AC - B^2 < 0$; if $AC - B^2 = 0$, then one of the roots is zero, and the other is different from zero. Hence it follows that *equation* (1) *is elliptic, if* $AC - B^2 > 0$, *parabolic if* $AC-B^2 = 0$, *and hyperbolic if* $AC-B^2 < 0$; no other types are possible here.

The characteristic equation

$$A\left(\frac{\partial\omega}{\partial x}\right)^2 + 2B\frac{\partial\omega}{\partial x}\frac{\partial\omega}{\partial y} + C\left(\frac{\partial\omega}{\partial y}\right)^2 = 0 \qquad (2)$$

can easily be reduced to an ordinary differential equation by the following means. Let $\omega(x, y)$ be a solution of this equation. Consider the characteristic

$$\omega(x, y) = \text{const.}$$

Along this characteristic we have the relation

$$\frac{\partial\omega}{\partial x}dx + \frac{\partial\omega}{\partial y}dy = 0$$

or

$$\frac{\partial\omega}{\partial x} : \frac{\partial\omega}{\partial y} = dy : (-dx).$$

Equation (2) is homogeneous with respect to $\partial\omega/\partial x$ and $\partial\omega/\partial y$; substituting for these with the proportional quantities dy and $-dx$ we obtain a second-order ordinary differential equation

$$A\,dy^2 - 2B\,dx\,dy + C\,dx^2 = 0. \qquad (3)$$

Conversely, if $\omega(x, y) = \text{const.}$ is the general integral of equation (3), then it is easily seen that the function $\omega(x, y)$ satisfies the characteristic equation.

Equation (3) is equivalent to two equations

$$\begin{aligned}\frac{dy}{dx} &= \frac{B+\sqrt{(B^2-AC)}}{A}, \\ \frac{dy}{dx} &= \frac{B-\sqrt{(B^2-AC)}}{A},\end{aligned} \qquad (4)$$

which coincide when equation (1) is parabolic, but are otherwise distinct.

Consider first of all the case when equation (1) is elliptic: $AC-B^2 > 0$. Suppose that the coefficients A, B, C are continued analytically into a domain of complex values of x and y, in the neighbourhood of real values of these variables. Then equations (4) can be regarded as ordinary differential equations in a complex domain. Consider, for example, the first of these equations, and let $\omega(x, y) = \text{const.}$ be its general solution; as was shown earlier, the function $\omega(x, y)$ satisfies the characteristic equation (2).

We shall now suppose that x and y are real, and thus let $\omega(x, y) = \xi(x, y) + i\eta(x, y)$, where ξ and η are real functions. It is not difficult to show that the Jacobian $\partial(\xi, \eta)/\partial(x, y)$ is nowhere equal to zero. In fact, if it were equal to zero at some point, then at this point we would have

$$\frac{\partial \xi}{\partial x} = \lambda \frac{\partial \xi}{\partial y}, \quad \frac{\partial \eta}{\partial x} = \lambda \frac{\partial \eta}{\partial y},$$

or $\partial \omega/\partial x = \lambda \partial \omega/\partial y$, where λ is some real quantity. Substituting this into equation (2) we find that the equation $A\lambda^2 + 2B\lambda + C = 0$ has real roots; this is impossible, since $AC - B^2 > 0$. Now take ξ and η as new independent variables, and let $\tilde{A}$, $\tilde{B}$, $\tilde{C}$ be the new coefficients. It was shown in § 1 that the characteristics are invariant under a transformation of independent variables; therefore the new characteristic equation

$$\tilde{A}\left(\frac{\partial \tilde{\omega}}{\partial \xi}\right)^2 + 2\tilde{B}\frac{\partial \tilde{\omega}}{\partial \xi}\frac{\partial \tilde{\omega}}{\partial \eta} + \tilde{C}\left(\frac{\partial \tilde{\omega}}{\partial \eta}\right)^2 = 0 \tag{5}$$

must be satisfied by the function $\tilde{\omega} = \xi + i\eta$. Hence $\tilde{A} = \tilde{C}$, $\tilde{B} = 0$; dividing the transformed equation by $\tilde{A}$, we arrive at its canonical form

$$\frac{\partial^2 u}{\partial \xi^2} + \frac{\partial^2 u}{\partial \eta^2} + \Phi_1\left(\xi, \eta, u, \frac{\partial u}{\partial \xi}, \frac{\partial u}{\partial \eta}\right) = 0. \tag{6}$$

If the coefficients A, B, C are not continued analytically into the complex domain, then equation (1) can nevertheless be brought to the form (6) by making one natural further assumption, which we now formulate. Write

$$\frac{A}{\sqrt{(AC - B^2)}} = a, \quad \frac{B}{\sqrt{(AC - B^2)}} = b, \quad \frac{C}{\sqrt{(AC - B^2)}} = c,$$

so that $ac - b^2 = 1$. Construct the elliptic equation

$$\frac{\partial}{\partial x}\left(a\frac{\partial \eta}{\partial x} + b\frac{\partial \eta}{\partial y}\right) + \frac{\partial}{\partial y}\left(b\frac{\partial \eta}{\partial x} + c\frac{\partial \eta}{\partial y}\right) = 0. \tag{7}$$

Take an arbitrary point (x_0, y_0), and suppose that, in some neighbourhood of this point, equation (7) has a solution of class $C^{(2)}$, not identically constant, so that in this neighbourhood $\eta_x^2 + \eta_y^2 > 0$. By virtue of equation (7), the system

$$\frac{\partial \xi}{\partial x} = -\left(b\frac{\partial \eta}{\partial x} + c\frac{\partial \eta}{\partial y}\right), \quad \frac{\partial \xi}{\partial y} = a\frac{\partial \eta}{\partial x} + b\frac{\partial \eta}{\partial y} \tag{8}$$

is consistent, and there exists a function $\xi \in C^{(2)}$ satisfying this system. It is easy to show that $\partial(\xi, \eta)/\partial(x, y) \neq 0$, and that the function $\xi(x, y) + i\eta(x, y)$ satisfies the characteristic equation. The rest proceeds as above. Note that in this case we have no need of the equations (4).

We now turn to the parabolic case. Let $\xi(x, y) = $ const. be the general solution of both equations (4). Introduce new variables ξ and η, where $\eta = \eta(x, y)$ is any function independent of $\xi(x, y)$. Equation (5) has the solution $\tilde{\omega} = \xi$, so that $\tilde{A} = 0$. In addition, the type of the equation is not altered by the change of variables, so that $\tilde{A}\tilde{C} - \tilde{B}^2 = 0$. Hence $\tilde{B} = 0$. Dividing the transformed equation (1) by $\tilde{C}$, we obtain the canonical form

$$\frac{\partial^2 u}{\partial \eta^2} + \Phi_1\left(\xi, \eta, u, \frac{\partial u}{\partial \xi}, \frac{\partial u}{\partial \eta}\right) = 0. \tag{9}$$

In the hyperbolic case we introduce as new independent variables $\xi_1 = \xi + \eta$, $\eta_1 = \xi - \eta$, where $\xi(x, y) = $ const. and $\eta(x, y) = $ const. are the general solutions of equations (4); it is easily shown that the Jacobian $\partial(\xi, \eta)/\partial(x, y)$ does not vanish. Equation (5) has two solutions: $\tilde{\omega} = \frac{1}{2}(\xi_1 + \eta_1)$ and $\tilde{\omega} = \frac{1}{2}(\xi_1 - \eta_1)$, and so $\tilde{A} = -\tilde{C}$ and $\tilde{B} = 0$. Dividing the transformed equation (1) by $\tilde{A}$, we reduce it to its canonical form

$$\frac{\partial^2 u}{\partial \xi_1^2} - \frac{\partial^2 u}{\partial \eta_1^2} + \Phi_1\left(\xi_1, \eta_1, u, \frac{\partial u}{\partial \xi_1}, \frac{\partial u}{\partial \eta_1}\right) = 0. \tag{10}$$

§ 5. Formally Adjoint Differential Expressions

Consider the linear, second-order, differential expression

$$Lu = A_{jk}\frac{\partial^2 u}{\partial x_j \partial x_k} + A_k \frac{\partial u}{\partial x_k} + A_0 u. \tag{1}$$

In the Euclidean space of the coordinates $x_1, x_2, \ldots, x_m$ we prescribe a finite domain Ω, bounded by a piecewise-smooth surface Γ. We shall suppose that in the closed domain $\bar{\Omega} = \Omega \cup \Gamma$ the coefficients A_{jk} have continuous second derivatives, the A_k have continuous first derivatives, and the coefficient A_0 is continuous. We shall also suppose that $u \in C^{(2)}(\bar{\Omega})$, i.e., that the function $u(x)$ is continuous, together with its first and second derivatives, in the closed domain $\bar{\Omega}$.

We construct a differential expression M, which we call *formally adjoint to L*:

$$Mu = \frac{\partial^2 (A_{jk} u)}{\partial x_j \partial x_k} - \frac{\partial (A_k u)}{\partial x_k} + A_0 u. \tag{2}$$

We can conveniently transform L to the following form:

$$Lu = \frac{\partial}{\partial x_j}\left(A_{jk}\frac{\partial u}{\partial x_k}\right) + B_k\frac{\partial u}{\partial x_k} + Cu,$$

$$B_k = A_k - \frac{\partial A_{jk}}{\partial x_j}, \qquad C = A_0. \tag{3}$$

If L is written in this way, then M takes the form

$$Mu = \frac{\partial}{\partial x_j}\left(A_{jk}\frac{\partial u}{\partial x_k}\right) - \frac{\partial(B_k u)}{\partial x_k} + Cu. \tag{4}$$

Starting from this expression, we can easily verify that *formal adjointness is a reciprocal property*, i.e. that the expression, formally adjoint to M, is L. In fact,

$$Mu = \frac{\partial}{\partial x_j}\left(A_{jk}\frac{\partial u}{\partial x_k}\right) - B_k\frac{\partial u}{\partial x_k} + \left(C - \frac{\partial B_k}{\partial x_k}\right)u.$$

Let N be the differential expression adjoint to M. Then

$$Nu = \frac{\partial}{\partial x_j}\left(A_{jk}\frac{\partial u}{\partial x_k}\right) + \frac{\partial(B_k u)}{\partial x_k} + \left(C - \frac{\partial B_k}{\partial x_k}\right)u = Lu.$$

If $M \equiv L$, then the expression L is called *formally self-adjoint*.

Formulae (3) and (4) show that formally adjoint expressions differ only in their middle terms. Clearly, then, $M \equiv L$ if, and only if, $B_k \equiv 0$, $k = 1, 2, \ldots, m$. Therefore the differential expression L will be formally self-adjoint if and only if $B_k \equiv 0$, $k = 1, 2, \ldots, m$. It follows from this that a formally self-adjoint differential expression of the second-order can be reduced to the form

$$Lu = \frac{\partial}{\partial x_j}\left(A_{jk}\frac{\partial u}{\partial x_k}\right) + Cu, \qquad A_{jk} = A_{kj}. \tag{5}$$

The Laplace operator and the wave operator are formally self-adjoint; the operator of the heat-conduction equation is not formally self-adjoint.

§ 6. Green's Formulae

Let the differential expression L be defined by formula (5.3), its coefficients satisfying the conditions of § 5. Further, let the functions $u, v \in C^{(2)}(\bar{\Omega})$.

Construct the integral

$$\int_\Omega vLu\,dx = \int_\Omega v\frac{\partial}{\partial x_j}\left(A_{jk}\frac{\partial u}{\partial x_k}\right)dx + \int_\Omega v\left(B_k\frac{\partial u}{\partial x_k} + Cu\right)dx. \quad (1)$$

Applying the integration by parts formula (§ 1, Chap. 2) to the first term on the right-hand side, we obtain the so-called *Green's First Formula*:

$$\int_\Omega vLu\,dx = -\int_\Omega A_{jk}\frac{\partial v}{\partial x_j}\frac{\partial u}{\partial x_k}\,dx$$
$$+ \int_\Omega v\left(B_k\frac{\partial u}{\partial x_k} + Cu\right)dx + \int_\Gamma vA_{jk}\frac{\partial u}{\partial x_k}\cos(v,x_j)\,d\Gamma. \quad (2)$$

Here v is the normal to the surface Γ, outward with respect to the domain.

We write down Green's First Formula for the formally adjoint differential expression M, and interchange the roles of u and v:

$$\int_\Omega uMv\,dx = -\int_\Omega A_{jk}\frac{\partial u}{\partial x_j}\frac{\partial v}{\partial x_k}\,dx +$$
$$+ \int_\Omega u\left\{-B_k\frac{\partial v}{\partial x_k} + \left(C - \frac{\partial B_k}{\partial x_k}\right)v\right\}dx + \int_\Gamma uA_{jk}\frac{\partial v}{\partial x_k}\cos(v,x_j)\,d\Gamma. \quad (3)$$

Now subtract formula (3) from formula (2). It can be shown that all the volume integrals on the right-hand side vanish. Since $A_{jk} = A_{kj}$, the first integrals on the right-hand sides of (2) and (3) are identical. Furthermore, integrating by parts, we obtain

$$-\int_\Omega uB_k\frac{\partial v}{\partial x_k}\,dx = \int_\Omega v\frac{\partial(B_k u)}{\partial x_k}\,dx - \int_\Gamma B_k uv\cos(v,x_k)\,d\Gamma$$
$$= \int_\Omega \left\{vB_k\frac{\partial u}{\partial x_k} + uv\frac{\partial B_k}{\partial x_k}\right\}dx - \int_\Gamma B_k uv\cos(v,x_k)\,d\Gamma.$$

Hence it is clear that the volume integrals on the right in (2) and (3) are equal.

The subtraction thus gives *Green's Second Formula*:

$$\int_\Omega (vLu - uMv)\,dx = \int_\Gamma \left[A_{jk}\left(v\frac{\partial u}{\partial x_j} - u\frac{\partial v}{\partial x_j}\right) + B_k uv\right]\cos(v,x_k)\,d\Gamma. \quad (4)$$

Green's formulae can be somewhat simplified for formally self-adjoint differential expressions. In this case $B_k \equiv 0$, and we obtain the following, simpler results: Green's First Formula is

$$\int_\Omega vLu\,dx = -\int_\Omega A_{jk}\frac{\partial v}{\partial x_j}\frac{\partial u}{\partial x_k}dx + \int_\Omega Cuv\,dx + \int_\Gamma vA_{jk}\frac{\partial u}{\partial x_k}\cos(v, x_j)d\Gamma; \quad (5)$$

Green's Second Formula is

$$\int_\Omega (vLu - uLv)\,dx = \int_\Gamma A_{jk}\left(v\frac{\partial u}{\partial x_k} - u\frac{\partial v}{\partial x_k}\right)\cos(v, x_j)d\Gamma. \quad (6)$$

We write down Green's formulae for the three most important expressions (usually called operators) of mathematical physics: Laplace, heat-conduction and wave.

1. The Laplace operator

$$\Delta = \sum_{k=1}^{m}\frac{\partial^2}{\partial x_k^2}$$

is formally self-adjoint; its coefficients have the values $A_{jk} = \delta_{jk}$, $C = 0$. Substituting these values into formula (5), we obtain *Green's First Formula for the Laplace operator*:

$$\int_\Omega v\Delta u\,dx = -\int_\Omega \frac{\partial u}{\partial x_k}\frac{\partial v}{\partial x_k}dx + \int_\Gamma v\frac{\partial u}{\partial v}d\Gamma. \quad (7)$$

We note two special cases of formula (7).

When $u = v$, we obtain

$$\int_\Omega u\Delta u\,dx = -\int_\Omega \sum_{k=1}^{m}\left(\frac{\partial u}{\partial x_k}\right)^2 dx + \int_\Gamma u\frac{\partial u}{\partial v}d\Gamma. \quad (8)$$

As we have remarked earlier (§ 3, Chap. 4), the integral

$$\int_\Omega \sum_{k=1}^{m}\left(\frac{\partial u}{\partial x_k}\right)^2 dx$$

is called the *Dirichlet integral*.

If in (7) we put $v \equiv 1$, we obtain a result which will be important later:

$$\int_\Gamma \frac{\partial u}{\partial v}d\Gamma = \int_\Omega \Delta u\,dx. \quad (9)$$

Green's Second Formula for the Laplace operator has the form

$$\int_\Omega (v\Delta u - u\Delta v)\,dx = \int_\Gamma \left(v\frac{\partial u}{\partial v} - u\frac{\partial v}{\partial v}\right)d\Gamma. \quad (10)$$

2. The heat-conduction operator

$$L = \frac{\partial}{\partial x_m} - \sum_{k=1}^{m-1} \frac{\partial^2}{\partial x_k^2}$$

is not formally self-adjoint. For this operator

$$A_{mm} = 0; \quad A_{kk} = -1, \quad 1 \le k \le m-1; \quad A_{jk} = 0, \quad j \ne k;$$
$$B_m = 1; \quad B_k = 0, \quad 1 \le k \le m-1; \quad C = 0.$$

The operator M, formally adjoint to the heat-conduction operator, has the form

$$M = -\frac{\partial}{\partial x_m} - \sum_{k=1}^{m-1} \frac{\partial^2}{\partial x_k^2}.$$

From formulae (2) and (4) we find

$$\int_\Omega vLu\,dx = \int_\Omega \left(\sum_{k=1}^{m-1} \frac{\partial u}{\partial x_k}\frac{\partial v}{\partial x_k} + v\frac{\partial u}{\partial x_m}\right)dx - \int_\Gamma v\sum_{k=1}^{m-1}\frac{\partial u}{\partial x_k}\cos(v,x_k)\,d\Gamma; \quad (11)$$

$$\int_\Omega (vLu - uMv)\,dx =$$
$$= \int_\Gamma \left[\sum_{k=1}^{m-1}\left(u\frac{\partial v}{\partial x_k} - v\frac{\partial u}{\partial x_k}\right)\cos(v,x_k) + uv\cos(v,x_m)\right]d\Gamma. \quad (12)$$

3. The wave operator is often denoted by the symbol □

$$\Box = \frac{\partial^2}{\partial x_m^2} - \sum_{k=1}^{m-1}\frac{\partial^2}{\partial x_k^2}.$$

This operator is formally self-adjoint; the values of its coefficients are

$$A_{mm} = 1; \quad A_{kk} = -1, \quad 1 \le k \le m-1;$$
$$A_{jk} = 0, \quad j \ne k; \quad C = 0.$$

Formulae (5) and (6) for the wave operator take the following form:

$$\int_\Omega v\Box u\,dx = \int_\Omega \left[\sum_{k=1}^{m-1}\frac{\partial u}{\partial x_k}\frac{\partial v}{\partial x_k} - \frac{\partial u}{\partial x_m}\frac{\partial v}{\partial x_m}\right]dx$$
$$+ \int_\Gamma v\left[\frac{\partial u}{\partial x_m}\cos(v,x_m) - \sum_{k=1}^{m-1}\frac{\partial u}{\partial x_k}\cos(v,x_k)\right]d\Gamma, \quad (13)$$

$$\int_\Omega (v\Box u - u\Box v)\,dx = \int_\Gamma \left[\left(v\frac{\partial u}{\partial x_m} - u\frac{\partial v}{\partial x_m}\right)\cos(v,x_m)\right.$$
$$\left. - \sum_{k=1}^{m-1}\left(v\frac{\partial u}{\partial x_k} - u\frac{\partial v}{\partial x_k}\right)\cos(v,x_k)\right]d\Gamma. \quad (14)$$

PART V

Equations of Elliptic Type

Chapter 11. Laplace's Equation and Harmonic Functions

§ 1. Basic Concepts

We begin with the simplest and most important elliptic equation, namely, *Laplace's equation*. This has the form

$$-\Delta u = f(x), \tag{1}$$

where $f(x)$ is a prescribed function. If $f(x) \not\equiv 0$, equation (1) is called the *inhomogeneous Laplace equation*. When $f(x) \equiv 0$, we have the *homogeneous Laplace equation*

$$\Delta u = 0. \tag{2}$$

The inhomogeneous Laplace equation is often called *Poisson's equation*.

More explicit ways of writing Laplace's equation – inhomogeneous and homogeneous – are

$$-\sum_{k=1}^{m} \frac{\partial^2 u}{\partial x_k^2} = f(x)$$

and

$$\sum_{k=1}^{m} \frac{\partial^2 u}{\partial x_k^2} = 0,$$

respectively.

Consider some closed surface Γ, not necessarily connected, and let Γ be the boundary of a domain Ω, either finite (Fig. 12) or infinite (Fig. 13). In both cases we assume that the surface itself is finite. We shall investigate the behaviour of solutions of the homogeneous Laplace equation in such domains *.

---

* Although considerable interest also attaches to the study of solutions of elliptic equations in domains bounded by infinite surfaces, we shall not be concerned with this question in this book, except that in one place only (Chap. 18) we shall look at the case of a half-space.

The function $u(x)$ is said to be *harmonic in a finite domain* Ω, if it is twice continuously differentiable and satisfies the homogeneous Laplace equation in this domain.

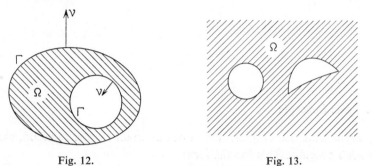

Fig. 12.   Fig. 13.

We shall say that the function $u(x)$ is *harmonic in an infinite domain* Ω, if, at every point of the domain which is at a finite distance from the origin, $u(x)$ is twice continuously differentiable, and satisfies the homogeneous Laplace equation, while, for sufficiently large $|x|$, it is subject to the inequality

$$|u(x)| \leq \frac{C}{|x|^{m-2}}, \tag{3}$$

where m is the dimension of the space, and C is some constant. In the case of a two-dimensional domain ($m = 2$), condition (3) implies that a harmonic function in an infinite domain is bounded at infinity.

We emphasize that the definition of a harmonic function applies only to the case of an *open domain* (i.e. of an open, connected set); if we speak of functions being harmonic in a closed domain, we mean by this that the function is harmonic in a larger open domain.

We also note that the definition of a harmonic function does not impose any restrictions on its behaviour at the boundary of the domain.

Example 1. If Ω is an infinite domain, then the function $u(x) \equiv 1$ is harmonic only for $m = 2$. If $m > 2$, this function is not harmonic in the infinite domain, but it is harmonic in any finite domain, for any m.

Example 2. In two-dimensions the function

$$u(x, y) = \operatorname{Re}\left(\frac{1}{z}\right) = \frac{x}{x^2 + y^2},$$

where $z = x+iy$, is harmonic in any region which does not contain the coordinate origin.

Example 3. The function $\operatorname{Re} \sqrt{z}$, $z = x+iy$, is harmonic in the circle $|z| < R$ (where R is any positive number), which is cut along some radius.

Example 4. The function of two variables $u = x^2+y^2$ is not harmonic in any domain, since it does not satisfy the homogeneous Laplace equation

$$\Delta(x^2+y^2) = 4 \neq 0.$$

Example 5. The function $u = x^2-y^2$ is harmonic in any finite domain.

In the plane, the homogeneous Laplace equation is invariant under a conformal transformation (cf. §3, Chap. 13). In the case of arbitrary m this is not so, but nevertheless there exists a transformation which maps a harmonic function into another harmonic function. This is *Kelvin's transformation*, which maps the point $x(x_1, x_2, \ldots, x_m)$ into a point $x'(x'_1, x'_2, \ldots, x'_m)$, the image of x with respect to a sphere of given radius R and centre at the origin of coordinates; it transforms the given function $u(x)$ into the function

$$w(x') = \frac{R^{m-2}}{|x'|^{m-2}} u(x). \tag{4}$$

We recall that points x and x' are called image points with respect to the sphere mentioned above, if they lie on the same ray, drawn from the origin, and if $|x| \cdot |x'| = R^2$. The cartesian coordinates of image points are related by the expression

$$x_k = x'_k \frac{R^2}{|x'|^2}. \tag{5}$$

A straight-forward, though quite tedious, calculation leads us to the relation

$$\Delta_{x'} w = \sum_{k=1}^{m} \frac{\partial^2 w}{\partial x_k'^2} = \frac{|x|^{m+2}}{R^{m+2}} \sum_{k=1}^{m} \frac{\partial^2 u}{\partial x_k^2} = \frac{|x|^{m+2}}{R^{m+2}} \Delta_x u,$$

so that if $\Delta_x u = 0$, then $\Delta_{x'} w = 0$.

§ 2. Singular Solution of Laplace's Equation

Let $x(x_1, x_2, \ldots, x_m)$ and $\xi(\xi_1, \xi_2, \ldots, \xi_m)$ be two points of an m-dimensional Euclidean space E_m. Write

$$r = |x-\xi| = \sqrt{\sum_{k=1}^{m} (x_k-\xi_k)^2} \tag{1}$$

and consider the function
$$v(x, \xi) = \frac{1}{r^{m-2}} \qquad (2)$$

with the condition that $m > 2$. We shall suppose the point ξ to be fixed, so that $v(x, \xi)$ can be regarded as a function of the point x.

The function $v(x, \xi)$ is discontinuous when $x = \xi$. We show that in any domain which does not contain the point ξ, $v(x, \xi)$ is harmonic. First of all, the function $v(x, \xi)$, together with its derivatives of arbitrary order, is continuous in such a domain. Next, at infinity

$$v(x, \xi) = O\left(\frac{1}{|x|^{m-2}}\right). \qquad (3)$$

In fact, $r = |x-\xi| \geqq |x|-|\xi|$. We are interested in the behaviour of the function v for sufficiently large $|x|$, so we may assume that $|x| > 2|\xi|$. Then $|\xi| < \frac{1}{2}|x|, r > \frac{1}{2}|x|$, and

$$v(x, \xi) < \frac{2^{m-2}}{|x|^{m-2}}.$$

The relation (3) is essential if the domain under consideration is infinite.

Finally, the function (2) satisfies the homogeneous Laplace equation. In this connection, we have $\partial r/\partial x_k = (x_k-\xi_k)/r$, from which

$$\frac{\partial v}{\partial x_k} = -\frac{m-2}{r^{m-1}}\frac{\partial r}{\partial x_k} = -\frac{(m-2)(x_k-\xi_k)}{r^m}.$$

Moreover,

$$\frac{\partial^2 v}{\partial x_k^2} = -\frac{m-2}{r^m} + \frac{m(m-2)(x_k-\xi_k)^2}{r^{m+2}} = \frac{m-2}{r^m}\left(\frac{m(x_k-\xi_k)^2}{r^2} - 1\right).$$

Adding, we find

$$\Delta v = \sum_{k=1}^{m} \frac{\partial^2 v}{\partial x_k^2} = \frac{m-2}{r^m}\left[\frac{m}{r^2}\sum_{k=1}^{m}(x_k-\xi_k)^2 - m\right] = 0.$$

The function $v(x, \xi)$ is called a *singular solution of Laplace's equation*.

As we shall see later, it is important in the application of singular solutions that they should tend to infinity at a definite rate as $x \to \xi$. When $m = 2$, the function (2) is identically equal to one; in this case the function cannot serve as a singular solution.

In the case $m = 2$, the singular solution of Laplace's equation is the function

$$v(x, \xi) = \ln \frac{1}{r}. \tag{4}$$

This is harmonic in any finite domain not containing the point ξ.

A singular solution of Laplace's equation is a symmetric function of x and ξ. Therefore, for fixed x the function $v(x, \xi)$ represents a harmonic function of ξ in any domain not containing the point x; in the two-dimensional case this domain must be finite.

Remark. The reader familiar with the concept of distributions (generalized functions) may observe that a singular solution of Laplace's equation satisfies the equation

$$-\Delta v = c\delta(x-\xi),$$

where δ is the Dirac function and c is some suitably chosen constant.

§ 3. Integral Representation of Functions of Class $C^{(2)}$

Consider a finite domain Ω in a space of m dimensions ($m > 2$), bounded by a piecewise-smooth surface Γ; and consider a function $u(\xi)$, where ξ is a variable point in this domain. Assume that $u \in C^{(2)}(\overline{\Omega})$. We prescribe in the domain Ω an arbitrary point x, and circumscribe this point with a sphere Σ_ε, centre x and radius ε; we denote the surface of the sphere Σ_ε by S_ε. Its radius ε is taken small enough so that the sphere Σ_ε is contained wholly inside the domain Ω (Fig. 14).

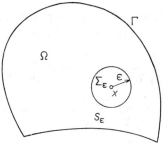

Fig. 14.

In the domain $\Omega^{(\varepsilon)} = \Omega \backslash \overline{\Sigma}_\varepsilon$, both the function $u(\xi)$ and the singular solution of Laplace's equation

$$v(x,\xi) = \frac{1}{r^{m-2}}$$

belong to the class $C^{(2)}(\overline{\Omega}^{(\varepsilon)})$, and we can apply to these functions Green's formula (6.10), Chap. 10:

$$\int_{\Omega\setminus\Sigma_\varepsilon}(v\Delta u - u\Delta v)\,d\xi = \int_\Gamma \left(v\frac{\partial u}{\partial \nu} - u\frac{\partial v}{\partial \nu}\right) d_\xi \Gamma + \int_{S_\varepsilon}\left(v\frac{\partial u}{\partial \nu} - u\frac{\partial v}{\partial \nu}\right) d_\xi S_\varepsilon. \quad (1)$$

The subscript ξ attached to the differential symbol means that integration is performed with respect to the variable point ξ.

We can somewhat simplify formula (1) if we observe that $\Delta v = 0$ in $\Omega\setminus\overline{\Sigma}_\varepsilon$. Moreover, on the sphere S_ε, $r = \varepsilon$, and the normal ν, being outwards with respect to the domain $\Omega\setminus\overline{\Sigma}_\varepsilon$, is in a direction opposite to the radius. Therefore, on S_ε,

$$v = \frac{1}{\varepsilon^{m-2}},$$

$$\frac{\partial v}{\partial \nu} = -\frac{\partial}{\partial r}\left(\frac{1}{r^{m-2}}\right)\bigg|_{r=\varepsilon} = \frac{m-2}{\varepsilon^{m-1}}.$$

We now use the well-known formula $dS_r = r^{m-1}\,dS_1$, where S_r is a sphere of radius r. For $r = \varepsilon$ we obtain

$$dS_\varepsilon = \varepsilon^{m-1}\,dS_1.$$

Finally, we introduce the notation

$$\theta = \frac{\xi - x}{\varepsilon}.$$

If $\xi \in S_\varepsilon$, then $|\theta| = 1$, and, consequently, $\theta \in S_1$, with $\xi = x + \varepsilon\theta$. It is now possible to convert the expression (1) into a more general form:

$$\int_{\Omega\setminus\Sigma_\varepsilon}\frac{1}{r^{m-2}}\Delta u(\xi)\,d\xi = \int_\Gamma\left(\frac{1}{r^{m-2}}\frac{\partial u(\xi)}{\partial \nu} - u(\xi)\frac{\partial}{\partial \nu}\frac{1}{r^{m-2}}\right)d_\xi\Gamma$$

$$+ \int_{S_1}\left[\varepsilon\frac{\partial u(x+\varepsilon\theta)}{\partial \nu} - (m-2)u(x+\varepsilon\theta)\right]dS_1. \quad (2)$$

In formula (2), take $\varepsilon \to 0$. The left-hand side in the limit gives the improper integral

$$\int_\Omega \frac{1}{r^{m-2}}\Delta u(\xi)\,d\xi.$$

On the right-hand side, the first integral does not depend on ε, while the second divides into two parts:

$$\int_{S_1} \left[\varepsilon \frac{\partial u(x+\varepsilon\theta)}{\partial v} - (m-2)u(x+\varepsilon\theta)\right] dS_1 =$$

$$= \varepsilon \int_{S_1} \frac{\partial u(x+\varepsilon\theta)}{\partial v} dS_1 - (m-2) \int_{S_1} u(x+\varepsilon\theta) dS_1. \quad (3)$$

The first derivatives of the function $u(\xi)$ are continuous in the closed domain $\bar{\Omega}$, and therefore bounded. Let

$$\left|\frac{\partial u}{\partial \xi_k}\right| \leq M = \text{const}.$$

Then

$$\left|\frac{\partial u}{\partial v}\right| = \left|\frac{\partial u}{\partial \xi_k} \cos(v, \xi_k)\right| \leq mM.$$

Hence

$$\left|\varepsilon \int_{S_1} \frac{\partial u(x+\varepsilon\theta)}{\partial v} dS_1\right| \leq \varepsilon m M |S_1| \underset{\varepsilon \to 0}{\to} 0.$$

In the second term on the right-hand side of formula (3), we can proceed to the limit under the integral sign, because the function u is continuous; the limit of this term, consequently, is equal to

$$-(m-2)\int_{S_1} u(x) dS_1 = -(m-2)|S_1|u(x).$$

The result of taking the limit in formula (2) is the relation

$$\int_\Omega \frac{1}{r^{m-2}} \Delta u(\xi) d\xi = \int_\Gamma \left(\frac{1}{r^{m-2}} \frac{\partial u(\xi)}{\partial v} - u(\xi) \frac{\partial}{\partial v} \frac{1}{r^{m-2}}\right) d_\xi \Gamma - (m-2)|S_1|u(x).$$

Hence

$$u(x) = \frac{1}{(m-2)|S_1|} \int_\Gamma \left(\frac{1}{r^{m-2}} \frac{\partial u(\xi)}{\partial v} - u(\xi) \frac{\partial}{\partial v} \frac{1}{r^{m-2}}\right) d_\xi \Gamma$$

$$- \frac{1}{(m-2)|S_1|} \int_\Omega \frac{1}{r^{m-2}} \Delta u(\xi) d\xi. \quad (4)$$

Formula (4) is called the *integral representation of a function of the class* $C^{(2)}$.

When $m = 3$, $|S_1| = 4\pi$, and the integral representation (4) takes the form

$$u(x) = \frac{1}{4\pi}\int_\Gamma \left(\frac{1}{r}\frac{\partial u(\xi)}{\partial v} - u(\xi)\frac{\partial}{\partial v}\frac{1}{r}\right)d_\xi \Gamma - \frac{1}{4\pi}\int_\Omega \frac{1}{r}\Delta u(\xi)\,d\xi. \quad (5)$$

When $m = 2$, formula (4) ceases to have any meaning. But if we begin with the corresponding singular solution $v(x, \xi) = \ln(1/r)$, and repeat the preceding derivation, we obtain the formula

$$u(x) = \frac{1}{2\pi}\int_\Gamma \left(\ln\frac{1}{r}\frac{\partial u(\xi)}{\partial v} - u(\xi)\frac{\partial}{\partial v}\ln\frac{1}{r}\right)d_\xi \Gamma - \frac{1}{2\pi}\int_\Omega \ln\frac{1}{r}\Delta u(\xi)\,d\xi, \quad (6)$$

which gives an integral representation of functions of the class $C^{(2)}$ in the case of two independent variables.

From the integral representation (4) there ensue a series of important results; the derivation of these will be the concern of the subsequent sections of this chapter.

§ 4. Integral Representation of a Harmonic Function

Let $u(x)$ be a harmonic function in the finite domain Ω having a piecewise-smooth boundary Γ; from the definition of a harmonic function, we have $u \in C^{(2)}(\Omega)$. If we suppose further that $u \in C^{(2)}(\overline{\Omega})$, then we can express the function $u(x)$ by the integral representation (3.4). In this case $\Delta u = 0$, the volume integral vanishes, and we obtain the formula

$$u(x) = \frac{1}{(m-2)|S_1|}\int_\Gamma \left(\frac{1}{r^{m-2}}\frac{\partial u(\xi)}{\partial v} - u(\xi)\frac{\partial}{\partial v}\frac{1}{r^{m-2}}\right)d_\xi \Gamma. \quad (1)$$

This is called the *integral representation of a harmonic function*.

In the case $m = 2$, we use (3.5), with $\Delta u = 0$, and obtain the corresponding result

$$u(x) = \frac{1}{2\pi}\int_\Gamma \left(\ln\frac{1}{r}\frac{\partial u(\xi)}{\partial v} - u(\xi)\frac{\partial}{\partial v}\ln\frac{1}{r}\right)d_\xi \Gamma. \quad (2)$$

THEOREM 11.4.1. *A function which is harmonic in some domain has derivatives of all orders in this domain.*

Let the function $u(x)$ be harmonic in a domain Ω, which may be finite or infinite. From the domain Ω we take an interior sub-domain Ω_1, chosen such that its boundary Γ_1 is piecewise-smooth. Obviously, $u \in C^{(2)}(\overline{\Omega}_1)$;

thus we can apply the integral representation (1) to the domain Ω_1:

$$u(x) = \frac{1}{(m-2)|S_1|} \int_{\Gamma_1} \left(\frac{1}{r^{m-2}} \frac{\partial u(\xi)}{\partial \nu} - u(\xi) \frac{\partial}{\partial \nu} \frac{1}{r^{m-2}} \right) d_\xi \Gamma, \qquad (3)$$

$$x \in \Omega_1.$$

Now construct a sub-domain Ω_2 which is interior with respect to Ω_1 (Fig. 15), and suppose that $x \in \bar{\Omega}_2$. Then the integrand function in (3) is continuous with respect to the totality of variables x and ξ, and, also, has

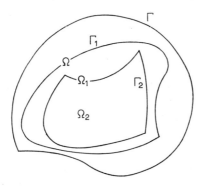

Fig. 15.

continuous derivatives of all orders with respect to the cartesian coordinates $x_1, x_2, \ldots, x_m$ of the point x. From the well-known theorem on differentiating an integral with respect to a parameter, we conclude that the function $u(x)$ has derivatives of all orders with respect to $x_1, x_2, \ldots, x_m$, and that these derivatives can be obtained by differentiating under the integral sign in equation (1).

To complete the proof, it remains to note the following: it it possible to choose the sub-domains Ω_1 and Ω_2 such that the arbitrary point $x \in \Omega_1$, referred to earlier, is located in Ω_2.

§ 5. The Concept of a Potential

The integral representation (3.4) enables us to introduce three integral operators of a special form.

Let Γ be a bounded, piecewise-smooth surface. In the integrals of formula (3.4), we substitute for the functions $\Delta u(\xi)$, $\partial u(\xi)/\partial \nu$ and $u(\xi)$ respectively the arbitrary functions $\rho(\xi)$, $\mu(\xi)$ and $\sigma(\xi)$. We then obtain three integrals which depend on x as a parameter:

$$\int_\Gamma \frac{1}{r^{m-2}} \mu(\xi) \, d_\xi \Gamma, \quad \int_\Gamma \frac{\partial}{\partial \nu} \frac{1}{r^{m-2}} \sigma(\xi) \, d_\xi \Gamma, \quad \int_\Omega \frac{1}{r^{m-2}} \rho(\xi) \, d\xi.$$

These are called respectively *single-layer potential*, *double-layer potential* and *Newtonian potential*. The functions $\mu(\xi)$, $\sigma(\xi)$ and $\rho(\xi)$ are known as the *densities* of the potentials.

We investigate the simplest properties of single-layer and double-layer potentials. In contrast with formula (3.4), where it was necessary to insist that the point x be situated inside Γ, we shall assume here that x can be located both outside and inside Γ. The case $x \in \Gamma$ requires special treatment, which is postponed until Chapter 18.

THEOREM 11.5.1. *If the densities are integrable on Γ, then the single- and double-layer potentials are harmonic in any domain – finite or infinite – which does not have a common point with the surface Γ.*

At any point $x \bar\in \Gamma$, the single- and double-layer potentials have derivatives of all orders – this can be verified by a direct repetition of the arguments used to prove Theorem 11.4.1. If D is the domain referred to in the statement of the present theorem, then both potentials have derivatives of all orders and, in particular, second derivatives, in D.

Furthermore, the single- and double-layer potentials satisfy the homogeneous Laplace equation. We know that if $x \bar\in \Gamma$, then we may differentiate under the integral sign. Putting

$$v(x) = \int_\Gamma \frac{1}{r^{m-2}} \mu(\xi) \, d_\xi \Gamma,$$

$$w(x) = \int_\Gamma \frac{\partial}{\partial \nu} \frac{1}{r^{m-2}} \sigma(\xi) \, d_\xi \Gamma,$$

we have

$$\Delta_x v(x) = \int_\Gamma \Delta_x \left(\frac{1}{r^{m-2}} \right) \mu(\xi) \, d_\xi \Gamma = 0,$$

where the subscript x attached to the symbol Δ means that differentiation is carried out with respect to the coordinates of the point x. Furthermore,

$$\Delta_x w(x) = \int_\Gamma \Delta_x \left(\frac{\partial}{\partial \nu} \frac{1}{r^{m-2}} \right) \sigma(\xi) \, d_\xi \Gamma = \int_\Gamma \sigma(\xi) \Delta_x \left(\frac{\partial}{\partial \xi_k} \frac{1}{r^{m-2}} \cos(\nu, x_k) \right) d_\xi \Gamma.$$

Since v is the normal drawn at the point ξ, $\cos(v, x_k)$ does not depend on x, and can be taken outside the differential sign Δ_x:

$$\Delta_x w(x) = \int_\Gamma \sigma(\xi) \cos(v, x_k) \Delta_x \left(\frac{\partial}{\partial \xi_k} \frac{1}{r^{m-2}} \right) d_\xi \Gamma$$

$$= \int_\Gamma \sigma(\xi) \cos(v, x_k) \frac{\partial}{\partial \xi_k} \Delta_x \left(\frac{1}{r^{m-2}} \right) d_\xi \Gamma = 0.$$

If D is a finite domain, the proof of the theorem is now complete. If, however, the domain D is infinite, then it still remains to show that $v(x)$ and $w(x)$ take the form (1.3) at infinity.

We locate the origin of coordinates inside Γ, and denote by H the greatest distance between points of the surface Γ.

Repeating the arguments used in § 2, we find that, when $|x| > 2H$, we have $r > \frac{1}{2}|x|$, and, consequently,

$$|v(x)| \leq \frac{2^{m-2}}{|x|^{m-2}} \int_\Gamma |\mu(\xi)| d_\xi \Gamma;$$

this last integral is bounded, because the function $\mu(\xi)$ is integrable on Γ. For the function $v(x)$ the relation (1.3) is thus established, with the value of the constant given by

$$C = 2^{m-2} \int_\Gamma |\mu(\xi)| d_\xi \Gamma.$$

Now consider the double-layer potential $w(x)$. We have

$$|w(x)| \leq \int_\Gamma |\sigma(\xi)| \left| \frac{\partial}{\partial v} \frac{1}{r^{m-2}} \right| d_\xi \Gamma$$

$$\leq (m-2) \int_\Gamma |\sigma(\xi)| \left| \frac{\xi_k - x_k}{r^m} \right| \cdot |\cos(v, x_k)| d_\xi \Gamma,$$

and since $|\xi_k - x_k| \leq r$, and $|\cos(v, x_k)| \leq 1$, therefore

$$|w(x)| \leq m(m-2) \int_\Gamma |\sigma(\xi)| \frac{d_\xi \Gamma}{r^{m-1}}.$$

If $|x| > 2H$, then $r > \frac{1}{2}|x|$, and, finally,

$$|w(x)| \leq \frac{2^{m-1} m(m-2)}{|x|^{m-1}} \int_\Gamma |\sigma(\xi)| d_\xi \Gamma.$$

The function $\sigma(\xi)$ is integrable on Γ, and the integral on the right is finite.

Thus, in the case of the double-layer potential, an even stronger inequality than (1.3) holds; the double-layer potential behaves at infinity like $|x|^{-(m-1)}$.

The theorem is now completely established.

The properties of single- and double-layer potentials will be fully examined in Chapter 18.

If the surface Γ divides the space into two domains – interior and exterior – then each of the single-layer and the double-layer potentials defines two harmonic functions: one harmonic in the interior domain, the other in the exterior.

§ 6. Properties of the Newtonian Potential

Let Ω be a finite domain in m-dimensional Euclidean space, bounded by a piecewise-smooth surface Γ. Consider the Newtonian potential

$$\phi(x) = \int_\Omega \frac{\rho(\xi)}{r^{m-2}} \, d\xi, \tag{1}$$

where x can denote an arbitrary point of the space E_m.

THEOREM 11.6.1. *If the density ρ is measurable and bounded in Ω, then the Newtonian potential* (1) *is continuous and continuously differentiable in the whole space E_m.*

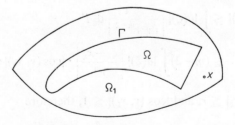

Fig. 16.

By the conditions of the theorem, the function $\rho(\xi)$ is bounded. Let $|\rho(\xi)| \leq C$. We extend the definition of this function by putting $\rho(\xi) \equiv 0$, $\xi \in \bar{\Omega}$. The function extended in this way, also satisfies the conditions of the theorem: it is bounded and measurable, with $|\rho(\xi)| \leq C$ as before. Let

x be an arbitrary given point of the space E_m. We construct any finite domain Ω_1 which contains within it both the point x and the domain Ω (Fig. 16); if $x \in \Omega$, then we can take $\Omega_1 = \Omega$. The function $\rho(\xi) \equiv 0$ in $\Omega_1 \backslash \Omega$, and so the potential (1) can be written in the form

$$\phi(x) = \int_{\Omega_1} \frac{\rho(\xi)}{r^{m-2}} \, d\xi. \tag{2}$$

The continuity of the function $\phi(x)$ at the point x follows from the result stated at the end of § 4, Chapter 7.

We now show that the Newtonian potential (1) has continuous first derivatives with respect to $x_1, x_2, \ldots, x_m$ at the point x. With this in view, we formally differentiate integral (2) with respect to x_k under the integral sign. This leads to the integral

$$-(m-2) \int_{\Omega_1} \rho(\xi) \frac{x_k - \xi_k}{r^m} \, d\xi, \tag{3}$$

which converges uniformly, since

$$\frac{|x_k - \xi_k|}{r^m} \leq \frac{1}{r^{m-1}}.$$

We rewrite integral (3) in the form

$$\int_{\Omega_1} \frac{A(x,\xi)}{r^{m-\frac{1}{2}}} \rho(\xi) \, d\xi; \quad A(x,\xi) = -\frac{(m-2)(x_k - \xi_k)}{\sqrt{r}}.$$

The function $A(x,\xi)$ is continuous, and $\rho(\xi)$ is bounded in $\bar{\Omega}_1$. From a result obtained earlier, in § 4, Chapter 7, it follows that the integral (3) is a continuous function of x in $\bar{\Omega}_1$. By the theorem on differentiation of integrals depending on a parameter, the derivative $\partial \phi / \partial x_k$ exists at the point $x \in \Omega_1$, and is equal to the integral (3):

$$\frac{\partial \phi}{\partial x_k} = \int_{\Omega_1} \rho(\xi) \frac{\partial}{\partial x_k} \frac{1}{r^{m-2}} \, d\xi.$$

The integral over $\Omega_1 \backslash \Omega$ is equal to zero. Therefore the latter expression becomes

$$\frac{\partial \phi}{\partial x_k} = \int_{\Omega} \rho(\xi) \frac{\partial}{\partial x_k} \frac{1}{r^{m-2}} \, d\xi. \tag{4}$$

Thus, the first derivatives of the Newtonian potential can be obtained by differentiating under the integral sign.

THEOREM 11.6.2. *If the density ρ is measurable and bounded in Ω, then in every domain complementary to Ω the Newtonian potential* (1) *is harmonic*.

If the boundary Γ of the domain Ω is not connected, then there exist several domains $\Omega_1, \Omega_2, \ldots, \Omega_n$, complementary to Ω (Fig. 17). Let Ω_j be one of these. Take an arbitrary subdomain Ω_j', which is interior with respect to Ω_j, and let $x \in \Omega_j'$. Then in the integral (1), the distance r is bounded from below by a positive number δ, equal to the shortest distance between points of the boundaries of the domains Ω_j and Ω_j' (Fig. 17). The integrand

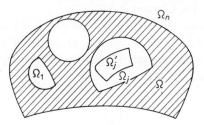

Fig. 17.

function $1/r^{m-2}$, and also its derivatives of any order with respect to $x_1, x_2, \ldots, x_m$, are continuous on the set of points $x \in \overline{\Omega}_j'$, and $\xi \in \overline{\Omega}$. Hence it follows that the function $\phi(x)$ has continuous derivatives of all orders in Ω_j', that these derivatives can be obtained by differentiating under the integral sign, and that in this domain

$$\Delta_x \phi(x) = \int_\Omega \rho(\xi) \Delta_x \frac{1}{r^{m-2}} \, d\xi = 0.$$

Since Ω_j' is an arbitrary subdomain, these conclusions are valid for the whole of the domain Ω_j. If this domain is finite, then the harmonic nature of the function (1) has been proven; if, however, the domain Ω_j is infinite, then we still have to establish the result (1.3) at infinity. This is done just as in Theorem 11.5.1; if H is the diameter of the boundary Γ, and if the origin of coordinates lies in Ω, then for $|x| > 2H$ we have $r > \frac{1}{2}|x|$ and $|\phi(x)| \leq 2^{m-2} C |\Omega|/|x|^{m-2}$.

We shall say that a function $f(x)$, defined on a set G, satisfies the *Lipschitz condition* with exponent α (denoted by $f \in \text{Lip}_\alpha(G)$), if for any two points $x, x' \in G$, we have the inequality

$$|f(x') - f(x)| \leq A|x' - x|^\alpha, \tag{5}$$

in which A and α are positive constants. It is easy to see that if $\alpha > 1$, then $f(x) \equiv \text{const.}$, so that we usually take $0 < \alpha \leq 1$.

The Lipschitz condition with $\alpha < 1$ is often called the Hölder condition.

THEOREM 11.6.3. *If the density $\rho \in \text{Lip}_\alpha(\bar{\Omega})$, then the Newtonian potential* (1) *has continuous second derivatives in the domain Ω, and satisfies in this domain the inhomogeneous Laplace equation*

$$-\Delta\phi = (m-2)|S_1|\rho(x), \quad m > 2, \qquad (6)$$
$$-\Delta\phi = 2\pi\rho(x), \quad m = 2.$$

Proof: We demonstrate this for $m > 2$; a similar procedure will apply for $m = 2$.

Let $x \in \Omega$. Surround the point x with a sphere of sufficiently small radius ε (as shown in Fig. 14, page 221), and write formula (4) in the form

$$\frac{\partial \phi}{\partial x} = \lim_{\varepsilon \to 0} \int_{\Omega \setminus \Sigma_\varepsilon} \rho(\xi) \frac{\partial}{\partial x_k} \frac{1}{r^{m-2}} d\xi. \qquad (7)$$

We take ε small enough so that the sphere Σ_ε and its boundary lie in Ω. We show that as $\varepsilon \to 0$ the expression

$$\frac{\partial}{\partial x_j} \int_{\Omega \setminus \Sigma_\varepsilon} \rho(\xi) \frac{\partial}{\partial x_k} \frac{1}{r^{m-2}} d\xi \qquad (8)$$

tends uniformly to a limit in any closed subdomain of the domain Ω.

Let $\bar{\Omega}'$ be this subdomain, and let $x \in \bar{\Omega}'$. Obviously we have to choose ε smaller than the shortest distance between the boundaries of the domains Ω and Ω'. We now carry out the differentiation in the expression (8). For the sake of brevity the integrand function will be denoted by $w(x, \xi)$, and we put $r' = |\xi - x'|$, where x' is the point having coordinates $x_1, \ldots, x_{j-1}, x_j + h, x_{j+1}, \ldots, x_m$. Then we have

$$\frac{\partial}{\partial x_j} \int_{\Omega \setminus \Sigma_\varepsilon} w(x, \xi) d\xi =$$
$$= \lim_{h \to 0} \frac{1}{h} \left[\int_{\Omega \setminus (r' < \varepsilon)} w(x', \xi) d\xi - \int_{\Omega \setminus (r < \varepsilon)} w(x, \xi) d\xi \right]$$
$$= \lim_{h \to 0} \frac{1}{h} \left[\int_{\Omega \setminus (r' < \varepsilon)} w(x', \xi) d\xi - \int_{\Omega \setminus (r' < \varepsilon)} w(x, \xi) d\xi \right]$$
$$+ \lim_{h \to 0} \frac{1}{h} \left[\int_{\Omega \setminus (r' < \varepsilon)} w(x, \xi) d\xi - \int_{\Omega \setminus (r < \varepsilon)} w(x, \xi) d\xi \right]. \qquad (9)$$

It is easy to see that the first limit on the right-hand side is equal to

$$\int_{\Omega\setminus(r<\varepsilon)} \frac{\partial w(x,\xi)}{\partial x_j}\,d\xi = \int_{\Omega\setminus\Sigma_\varepsilon} \rho(\xi)\frac{\partial^2}{\partial x_j\,\partial x_k}\frac{1}{r^{m-2}}\,d\xi,$$

and here the passage to the limit in $\bar{\Omega}'$ is uniform.

We now examine the second limit. The expression under the integral sign can be written in the form

$$\frac{1}{h}\int_{D_1} w(x,\xi)\,d\xi - \frac{1}{h}\int_{D_2} w(x,\xi)\,d\xi,$$

where D_1 and D_2 are the crescents shown in Fig. 18. Consider for example, the expression

$$\frac{1}{h}\int_{D_2} w(x,\xi)\,d\xi.$$

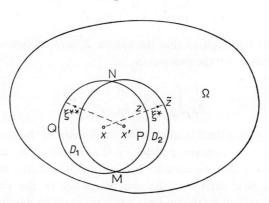

Fig. 18.

In D_2 we introduce spherical coordinates with centre at the point x. One of the coordinates is r; its smallest value in D_2 is equal to ε. On the portion MNP of the sphere $r = \varepsilon$ (Fig. 18), take any point z; its cartesian coordinates are

$$z_k = x_k + \varepsilon \cos(r, x_k).$$

We find the value of the quantity $r = \tilde{r}$ at the point $\tilde{z}$, lying on the intersection of the sphere $r' = \varepsilon$ and the ray

$$\xi = x + \lambda(z-x),$$

and passing through the points x and z (Fig. 18). The corresponding

quantity $\lambda = \tilde\lambda$ is obtained from the system of equations
$$\sum_{l\ne j}(\xi_l-x_l)^2+(\xi_j-x_j-h)^2 = \varepsilon^2,$$
$$\xi_l = x_l+\lambda(z_l-x_l), \qquad l = 1, 2, \ldots, m.$$
Taking account of the fact that $|z-x| = \varepsilon$, we find the equation for λ:
$$\lambda^2\varepsilon^2 - 2\lambda h\varepsilon \cos(r, x_j)+h^2-\varepsilon^2 = 0.$$
To the point $\tilde z$ there corresponds a value $\tilde\lambda > 1$, and so
$$\tilde\lambda = \frac{h}{\varepsilon}\cos(r, x_j)+\sqrt{\left\{\frac{h^2}{\varepsilon^2}\cos^2(r, x_j)+1-\frac{h^2}{\varepsilon^2}\right\}}$$
$$= 1+\frac{h}{\varepsilon}\cos(r, x_j)+O(h^2).$$
Now
$$\tilde r = |\tilde z-x| = \tilde\lambda|z-x| = \tilde\lambda\varepsilon = \varepsilon+h\cos(r, x_j)+O(h^2).$$

Integration over D_2 reduces to integration over the portion MNP of the sphere $r = \varepsilon$, and with respect to r between the limits ε and $\tilde\lambda\varepsilon$:
$$\frac{1}{h}\int_{D_2} w(x,\xi)\,d\xi = \frac{1}{h}\int_{MNP}\left\{\int_\varepsilon^{\tilde\lambda\varepsilon} w(x,\xi)\,dr\right\}dS_\varepsilon,$$
or, if we apply the Mean Value Theorem to the inner integral,
$$\frac{1}{h}\int_{D_2} w(x,\xi)\,d\xi = \int_{MNP} w(x,\xi^*)[\cos(r, x_j)+O(h)]\,dS_\varepsilon,$$
where ξ^* is some point of the interval $(z, \tilde z)$.

Similarly we find
$$\frac{1}{h}\int_{D_1} w(x,\xi)\,d\xi = \int_{MNQ} w(x,\xi^{**})[\cos(r', x_j)+O(h)]\,dS_\varepsilon.$$

If $h \to 0$, then in the limit MNP and MNQ give two hemispheres, which make together the sphere $r = \varepsilon$. As a result we obtain the desired differentiation formula:
$$\frac{\partial}{\partial x_j}\int_{\Omega\setminus\Sigma_\varepsilon} w(x,\xi)\,d\xi = \int_{\Omega\setminus\Sigma_\varepsilon}\frac{\partial w(x,\xi)}{\partial x_j}\,d\xi - \int_{r=\varepsilon} w(x,\xi)\cos(r, x_j)\,dS_\varepsilon, \qquad (10)$$
where in the second integral we have replaced z by ξ as the variable of integration.

Substituting now for $w(x, \xi)$, we obtain

$$\frac{\partial}{\partial x_j} \int_{\Omega \setminus \Sigma_\varepsilon} \rho(\xi) \frac{\partial}{\partial x_k} \frac{1}{r^{m-2}} \, d\xi =$$

$$= \int_{\Omega \setminus \Sigma_\varepsilon} \rho(\xi) \frac{\partial^2}{\partial x_j \partial x_k} \frac{1}{r^{m-2}} \, d\xi - \int_{r=\varepsilon} \rho(\xi) \frac{\partial}{\partial x_k} \frac{1}{r^{m-2}} \cos(r, x_j) \, dS_\varepsilon.$$

The second integral can be somewhat simplified. Put $\xi = x + \varepsilon\theta$, then $|\theta| = 1$, and the integration will be performed over the sphere S_1, with $dS_\varepsilon = \varepsilon^{m-1} \, dS_1$. Moreover,

$$\frac{\partial}{\partial x_k} \frac{1}{r^{m-2}} \bigg|_{r=\varepsilon} = -\frac{m-2}{\varepsilon^{m-1}} \frac{x_k - \xi_k}{r} \bigg|_{r=\varepsilon} = \frac{m-2}{\varepsilon^k} \cos(r, x_k),$$

and, finally,

$$\frac{\partial}{\partial x_j} \int_{\Omega \setminus \Sigma_\varepsilon} \rho(\xi) \frac{\partial}{\partial x_k} \frac{1}{r^{m-2}} \, d\xi = \int_{\Omega \setminus \Sigma_\varepsilon} \rho(\xi) \frac{\partial^2}{\partial x_j \partial x_k} \frac{1}{r^{m-2}} \, d\xi$$

$$- (m-2) \int_{S_1} \rho(x + \varepsilon\theta) \cos(r, x_j) \cos(r, x_k) \, dS_1. \quad (11)$$

We can now prove that the expression (8) or, equivalently, the right-hand side of equation (11), converges uniformly in $\bar{\Omega}'$ to some limit when $\varepsilon \to 0$. This is obviously true for the second term in (11), which tends uniformly to the value

$$-(m-2)\rho(x) \int_{S_1} \cos(r, x_j) \cos(r, x_k) \, dS_1. \quad (12)$$

Notice that the integral in (12) can be easily evaluated. We have

$$\cos(r, x_j)|_{S_1} = \frac{\xi_j - x_j}{r} \bigg|_{S_1} = \xi_j - x_j,$$

and by Gauss's Theorem

$$\int_{S_1} (\xi_j - x_j) \cos(r, x_k) \, dS_1 = \int_{S_1} (\xi_j - x_j) \cos(v, x_k) \, dS_1 = \int_{r<1} \frac{\partial(\xi_j - x_j)}{\partial \xi_k} \, d\xi,$$

where v is the outward normal to S_1. If $j \neq k$, then this last integral is equal to zero. When $j = k$ we have

$$\int_{r<1} d\xi = \int_{S_1} \left\{ \int_0^1 r^{m-1} \, dr \right\} dS_1 = \frac{|S_1|}{m}.$$

The expression (12) can ultimately be represented in the form

$$-\frac{(m-2)|S_1|}{m} S(x)\delta_{jk}. \tag{12'}$$

We now turn to the first term of (11). Let δ be the shortest distance between the boundaries of the domains Ω and Ω'. We have

$$\int_{\Omega\setminus\Sigma_\varepsilon} \rho(\xi) \frac{\partial^2}{\partial x_j \partial x_k} \frac{1}{r^{m-2}} d\xi = \int_{\Omega\setminus(r<\delta)} \rho(\xi) \frac{\partial^2}{\partial x_j \partial x_k} \frac{1}{r^{m-2}} d\xi +$$

$$+ \int_{\varepsilon<r<\delta} [\rho(\xi)-\rho(x)] \frac{\partial^2}{\partial x_j \partial x_k} \frac{1}{r^{m-2}} d\xi + \rho(x) \int_{\varepsilon<r<\delta} \frac{\partial^2}{\partial x_j \partial x_k} \frac{1}{r^{m-2}} d\xi.$$

It is easy to see that the last integral is equal to zero:

$$\int_{\varepsilon<r<\delta} \frac{\partial^2}{\partial x_j \partial x_k} \frac{1}{r^{m-2}} d\xi = -(m-2) \int_{\varepsilon<r<\delta} [\delta_{jk} - m \cos(r, x_j) \cos(r, x_k)] d\xi$$

$$= -(m-2) \int_{S_1} [\delta_{jk} - m \cos(r, x_j) \cos(r, x_k)] \left\{ \int_\varepsilon^\delta \frac{dr}{r} \right\} dS_1$$

$$= -(m-2) \ln \frac{\delta}{\varepsilon} \int_{S_1} [\delta_{jk} - m \cos(r, x_j) \cos(r, x_k)] dS_1,$$

which is equal to zero because of (12) and (12'). Thus,

$$\int_{\Omega\setminus\Sigma_\varepsilon} \rho(\xi) \frac{\partial^2}{\partial x_j \partial x_k} \frac{1}{r^{m-2}} d\xi = \int_{\Omega\setminus(r<\delta)} \rho(\xi) \frac{\partial^2}{\partial x_j \partial x_k} \frac{1}{r^{m-2}} d\xi$$

$$+ \int_{\varepsilon<r<\delta} [\rho(\xi)-\rho(x)] \frac{\partial^2}{\partial x_j \partial x_k} \frac{1}{r^{m-2}} d\xi. \tag{13}$$

Now we can easily show again that, as $\varepsilon \to 0$, the expression (13) tends uniformly to a limit. The first term on the right does not depend on ε, while the integrand function in the second term can be estimated in the following way:

$$\left| [\rho(\xi)-\rho(x)] \frac{\partial^2}{\partial x_j \partial x_k} \frac{1}{r^{m-2}} \right| =$$

$$= \frac{m-2}{r^m} |\delta_{jk} - m \cos(r, x_j) \cos(r, x_k)| \cdot |\rho(\xi)-\rho(x)| \leq \frac{(m-2)(m+1)A}{r^{m-\alpha}}.$$

The function on the right-hand side in this last inequality is integrable in Ω, and its integral converges uniformly with respect to x, since the power of r in the denominator is $m-\alpha < m$. But then the second integral in (13) also converges uniformly.

From the well-known theorem concerning differentiation of an integral with respect to a parameter, it now follows that the Newtonian potential $\phi(x)$ has all possible second derivatives, continuous in Ω; they can be expressed by the formula

$$\frac{\partial^2 \phi}{\partial x_j \partial x_k} = \lim_{\varepsilon \to 0} \int_{\Omega \setminus \Sigma_\varepsilon} \rho(\xi) \frac{\partial^2}{\partial x_j \partial x_k} \frac{1}{r^{m-2}} \, d\xi - \frac{(m-2)|S_1|}{m} \rho(x). \qquad (14)$$

Putting $j = k$ in this equation, and summing, we find

$$\Delta \phi = \lim_{\varepsilon \to 0} \int_{\Omega \setminus \Sigma_\varepsilon} \rho(\xi) \Delta_x \frac{1}{r^{m-2}} \, d\xi - (m-2)(S_1)\rho(x).$$

But $\Delta_x r^{2-m} = 0$, and consequently,

$$\Delta \phi = -(m-2)|S_1|\rho(x).$$

This completes the proof of the theorem.

Remark 1. It is possible to prove the following theorem: if $\rho \in \text{Lip}_\alpha(\overline{\Omega})$, $0 < \alpha < 1$, then $\partial^2 \phi / \partial x_j \partial x_k \in \text{Lip}_\alpha(\overline{\Omega}')$, where ϕ is the Newtonian potential with density ρ, and Ω' is any interior sub-domain of the domain Ω.

Remark 2. The following theorem is also true. Let the density $\rho \in L_p(\Omega)$, where p is a constant such that $1 < p < \infty$. Then the second derivatives of the Newtonian potential (1) exist as generalized derivatives; they are integrable in Ω to degree p, and can be evaluated from formula (14). The limit occurring in this formula exists almost everywhere in Ω. The potential (1) satisfies the inhomogeneous Laplace equation (6) almost everywhere.

Equation (6) enables us to construct a particular solution to the inhomogeneous Laplace equation, and thereby to reduce the latter to a homogeneous equation.

Let the inhomogeneous Laplace equation be given in the form

$$-\Delta u = f(x). \qquad (16)$$

Its general solution consists of the sum of some particular solution and the general solution of the homogeneous Laplace equation. If we assume, for example, that $f \in \text{Lip}_\alpha(\overline{\Omega})$, then the particular solution of equation can be represented by the formula

$$u_0(x) = \frac{1}{(m-2)|S_1|} \int_\Omega \frac{f(\xi)}{r^{m-2}} d\xi.$$

If we replace the unknown function in our equation through the formula $u = u_0 + w$, then we obtain the homogeneous Laplace equation

$$\Delta w = 0.$$

§ 7. Mean Value Theorem

THEOREM 11.7.1 (DIRECT MEAN VALUE THEOREM). *Let the function $u(x)$ be harmonic in some sphere, and continuous in it up to the boundary. Then the value of this function at the centre of the sphere is equal to the arithmetic mean of its values on the surface bounding the given sphere.*

Before proving this theorem, we mention one formula which will play an important role in what follows. In formula (6.9), Chapter 10, let $u(x)$ be a function of class $C^{(2)}(\bar\Omega)$, harmonic in Ω. Then $\Delta u = 0$, the right-hand side of this formula is equal to zero, and we obtain

$$\int_\Gamma \frac{\partial u}{\partial \nu} d\Gamma = 0. \tag{1}$$

We now proceed to the proof of the theorem. Denote by Σ_R the sphere mentioned in the statement of the theorem, and let x_0 and R be its centre and radius respectively. Denote by S_R the bounding surface of the spherical region Σ_R. Also, let $\Sigma_{R'}$ be a sphere of radius $R' < R$, concentric with Σ_R, and let $S_{R'}$ be the boundary of $\Sigma_{R'}$. Obviously $u \in C^{(2)}(\bar\Sigma_{R'})$. By formula (4.1),

$$u(x) = \frac{1}{(m-2)|S_1|} \int_{S_{R'}} \left(\frac{1}{r^{m-2}} \frac{\partial u(\xi)}{\partial \nu} - u(\xi) \frac{\partial}{\partial \nu} \frac{1}{r^{m-2}}\right) dS_{R'}; \quad x \in \Sigma_{R'}. \tag{2}$$

Put $x = x_0$ in formula (2). Then $r = R'$. Moreover, the normal ν is outward with respect to the sphere, and, consequently, is directed along the radius. Therefore,

$$\left.\frac{\partial}{\partial \nu} \frac{1}{r^{m-2}}\right|_{r=R'} = \left.\frac{\partial}{\partial r} \frac{1}{r^{m-2}}\right|_{r=R'} = -\frac{m-2}{R'^{m-1}}.$$

Formula (2) takes the form

$$u(x_0) = \frac{1}{(m-2)|S_1|R'^{m-2}} \int_{S_{R'}} \frac{\partial u}{\partial \nu} dS_{R'} + \frac{1}{|S_1|R'^{m-1}} \int_{S_{R'}} u \, dS_{R'}.$$

Because of formula (1), the first integral here vanishes, and we have

$$u(x_0) = \frac{1}{|S_1|R'^{m-1}} \int_{S_{R'}} u \, dS_{R'}.$$

Now let $R' \to R$. The function u is continuous in the sphere $\bar{\Sigma}_R$, and we may proceed to the limit under the integral sign. Finally,

$$u(x_0) = \frac{1}{|S_1|R^{m-1}} \int_{S_R} u \, dS_R. \tag{3}$$

The right-hand side of equation (3) is what is called the arithmetic mean of the values of the function u over the sphere S_R – it is the integral of this function over the sphere S_R divided by the surface area of this sphere. The direct Mean Value Theorem is thus proven.

THEOREM 11.7.2 (INVERSE MEAN VALUE THEOREM). *Let Ω be a finite domain of the space E_m, and let $u \in C(\Omega)$. If the function $u(x)$ satisfies equation (3) for any spherical region which, together with its boundary, lies entirely in the domain Ω, then this function is harmonic in Ω.*

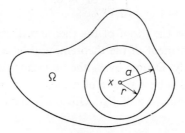

Fig. 19.

Take an arbitrary point $x \in \Omega$, and circumscribe it with a sphere Σ_a, of fixed radius a, which, together with its boundary, lies in Ω (Fig. 19). If $r \leq a$ and S_r is a sphere of radius r and with centre at x, then by the conditions of the theorem we have that

$$u(x) = \frac{1}{r^{m-1}|S_1|} \int_{S_r} u(\xi) \, dS_r. \tag{4}$$

Let $\omega_a(|\xi - x|) = \omega_a(r)$ be an averaging kernel, with averaging radius $h = a$. Multiply both sides of equation (4) by $r^{m-1} \omega_a(r) dr$, and integrate with respect to r between the limits 0 and a. The left-hand side gives the expression $cu(x)$, where $c = \int_0^a r^{m-1}\omega_a(r)\,dr$. By property 3 of the averaging kernel (§ 1, Chapter 1), $c = |S_1|^{-1}$, and we obtain the equation

$$u(x) = \int_0^a \int_{S_r} u(\xi)\omega_a(r)\,dr\,dS_r = \int_{\Sigma_a} u(\xi)\omega_a(r)\,d\xi. \tag{5}$$

Outside the sphere Σ_a, the averaging kernel $\omega_a(r) = 0$, so that formula (5) can be written in the form

$$u(x) = \int_\Omega u(\xi)\omega_a(r)\,d\xi. \tag{6}$$

This new form has the advantage that the domain of integration Ω does not depend on the choice of the point x.

Formula (5), and also formula (6), are valid for any point $x \in \Omega \backslash \Omega_a$, where Ω_a is a boundary strip of width a.

The averaging kernel has continuous derivatives of all orders and the function $u(\xi)$ is continuous, so that the integrand function in (6) has derivatives of all orders with respect to $x_1, x_2, \ldots, x_m$, which are continuous on the totality of points x and ξ. Thus the integral (6), i.e. the function $u(x)$, has continuous derivatives of all orders in the domain $\Omega \backslash \Omega_a$. This can be expressed by writing $u \in C^{(\infty)}(\Omega \backslash \Omega_a)$. But the number a can be taken as small as we please; therefore $u \in C^{(\infty)}(\Omega)$.

We next show that $\Delta u = 0$. In the sphere Σ_a, described above, we write the integral representation (3.4):

$$u(x) = \frac{1}{(m-2)|S_1|}\int_{S_a}\left(\frac{1}{r^{m-2}}\frac{\partial u(\xi)}{\partial \nu} - u(\xi)\frac{\partial}{\partial \nu}\frac{1}{r^{m-2}}\right)dS_a$$

$$- \frac{1}{(m-2)|S_1|}\int_{\Sigma_a}\frac{1}{r^{m-2}}\Delta u\,d\xi \tag{7}$$

where we recall that x is the centre of the sphere Σ_a.

Consider the second term on the right-hand side. As before

$$\frac{\partial}{\partial \nu}\frac{1}{r^{m-2}} = \frac{\partial}{\partial r}\frac{1}{r^{m-2}}\bigg|_{r=a} = -\frac{m-2}{a^{m-1}};$$

the second surface integral is equal to

$$\frac{1}{a^{m-1}|S_1|}\int_{S_a} u(\xi)\,dS_a,$$

and by formula (4) this is equal to $u(x)$. Formula (7) now gives

$$\frac{1}{a^{m-2}}\int_{S_a}\frac{\partial u}{\partial v}dS_a - \int_{\Sigma_a}\frac{1}{r^{m-2}}\Delta u\, d\xi = 0.$$

By equation (6.9), Chapter 10,

$$\int_{S_a}\frac{\partial u}{\partial v}dS_a = \int_{\Sigma_a}\Delta u\, d\xi.$$

Collecting together the terms in the volume integral, we obtain

$$\int_{\Sigma_a}\left(\frac{1}{a^{m-2}} - \frac{1}{r^{m-2}}\right)\Delta u(\xi)\, d\xi = 0. \tag{8}$$

Since $r = |\xi - x| \leq a$ in the sphere Σ_a, we have

$$\frac{1}{a^{m-2}} - \frac{1}{r^{m-2}} \leq 0.$$

Furthermore, the function Δu is continuous. We apply the Mean Value Theorem to the integral (8):

$$\Delta u(x')\int_{\Sigma_a}\left(\frac{1}{a^{m-2}} - \frac{1}{r^{m-2}}\right)d\xi = 0;$$

when x' is some interior point of the sphere Σ_a.

The integral of the negative function is different from zero; consequently $\Delta u(x') = 0$. Let a tend to zero; then $x' \to x$, and, from the continuity of the second derivative, $\Delta u(x) = 0$.

§ 8. Maximum Principle

THEOREM 11.8.1. *If a function, harmonic in a finite domain, attains a minimum or a maximum at an interior point of this domain, then this function is a constant.*

Let the function $u(x)$, harmonic in the finite domain Ω, attain a maximum at the point $x_0 \in \Omega$. We shall show that $u(x) \equiv u(x_0)$. Construct a sphere $\Sigma_R(x_0)$, of radius R and with centre at the point x_0. Let S_R be the boundary of this sphere, and let Σ_R with its boundary lie within Ω. First of all, we shall show that $u(\xi) \equiv u(x_0)$ on the spherical surface S_R. We write down the formula which expresses the Mean Value Theorem:

$$u(x_0) = \frac{1}{|S_R|} \int_{S_R} u(\xi) \, dS_R. \tag{1}$$

Here the surface area of the sphere S_R is denoted by $|S_R|$, and

$$|S_R| = R^{m-1}|S_1|.$$

From the conditions of the theorem, $u(\xi) \leq u(x_0)$, $\xi \in S_R$. Suppose that at some point $\xi' \in S_R$ we have the strict inequality $u(\xi') < u(x_0)$. The function u is continuous on S_R; therefore on this surface the inequality $u(\xi) < u(x_0)$ will be satisfied in some neighbourhood S' of the point ξ'. Putting $S_R \backslash S' = S''$, we have

$$u(x_0) = \frac{1}{|S_R|} \left\{ \int_{S'} u(\xi) \, dS_R + \int_{S''} u(\xi) \, dS_R \right\}.$$

But

$$\int_{S'} u(\xi) \, dS_R < u(x_0) \int_{S'} dS_R,$$

$$\int_{S''} u(\xi) \, dS_R \leq u(x_0) \int_{S''} dS_R;$$

hence

$$u(x_0) < \frac{1}{|S_R|} u(x_0) \left\{ \int_{S'} dS_R + \int_{S''} dS_R \right\} = \frac{1}{|S_R|} u(x_0) \int_{S_R} dS_R = u(x_0),$$

which is impossible.

Thus $u(\xi) \equiv u(x_0)$, $\xi \in S_R$. Replacing R by an arbitrarily small quantity, we see that this last identity holds through the whole sphere $\Sigma_R(x_0)$:

$$u(\xi) \equiv u(x_0), \quad \xi \in \Sigma_R(x_0). \tag{2}$$

We next show that the identity (2) holds throughout the whole domain Ω. Take an arbitrary point $x \in \Omega$, and join the points x_0 and x by an open polygon, lying entirely in Ω (Fig. 20). Let δ be the shortest distance from a point of this polygon to a point of the boundary Γ of the domain Ω, and let $\delta' = \frac{1}{2}\delta$. Then any sphere of radius δ' with centre on the polygon lies, together with its boundary, entirely within the domain Ω.

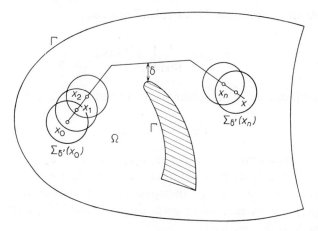

Fig. 20.

Let $\Sigma_{\delta'}(y)$ denote a sphere of radius δ', with centre at y. Inside the sphere $\Sigma_{\delta'}(x_0)$ we take a point x_1, on the polygon, such that $|x_1 - x_0| > \tfrac{1}{2}\delta'$, and construct a sphere $\Sigma_{\delta'}(x_1)$. Inside this new sphere, we take a point x_2, again on the polygon, such that $|x_2 - x_1| > \tfrac{1}{2}\delta'$, with the point x_1 lying between x_0 and x_2. Continuing this procedure, we easily see that the whole polygon can be enclosed by a finite number of such spheres. We call these spheres $\Sigma_{\delta'}(x_j), j = 0, 1, 2, \ldots, n$, and, from the construction, $x_j \in \Sigma_{\delta'}(x_{j-1})$, $j = 1, 2, \ldots, n$. As the identity (2) shows, the function u has a constant value, equal to its maximum, in the sphere $\Sigma_{\delta'}(x_0)$. But then this function also assumes its maximum value at the point x_1: $u(x_1) = u(x_0)$. Again using (2), we now have $u(\xi) = u(x_1) = u(x_0), \xi \in \Sigma_{\delta'}(x_1)$. Continuing this argument, we eventually arrive at the identity

$$u(\xi) = u(x_0), \qquad \forall \xi \in \Sigma_{\delta'}(x_n).$$

But $x \in \Sigma_{\delta'}(x_n)$, and therefore $u(x) = u(x_0)$. This is what we set out to prove.

The case of a minimum reduces to the case of a maximum by substituting $-u$ for u.

COROLLARY 11.8.1. *A harmonic function which is not identically constant cannot attain either a maximum or a minimum in a finite domain.*

COROLLARY 11.8.2. *If $u(x)$ is a harmonic function in a finite domain Ω, and if, in addition, $u \in C(\bar{\Omega})$, then $u(x)$ attains both its maximum and its minimum values on the boundary of the domain.*

Corollary 11.8.2 is called the *maximum principle for harmonic functions*.

§ 9. Convergence of Sequences of Harmonic Functions

THEOREM 11.9.1. (HARNACK'S THEOREM). *Let $\{u_n(x)\}$ be a sequence of harmonic functions in a finite domain Ω. Further, let the functions $u_n(x)$ be continuous in the closed domain $\bar{\Omega} = \Omega \cup \Gamma$, where Γ is the boundary of Ω. If the sequence $\{u_n(x)\}$ converges uniformly on Γ, then*
1) *the sequence $\{u_n(x)\}$ converges uniformly in the closed domain $\bar{\Omega}$;*
2) *the limit function is harmonic in Ω;*
3) *in any closed sub-domain Ω' of the domain Ω, the derivatives of any order of the function $u_n(x)$ converge uniformly to the corresponding derivatives of the limit function.*

Proof: The sequence $\{u_n(x)\}$ converges uniformly on the surface Γ. This means that for any $\varepsilon > 0$ we can find a number N such that, for all $n \geq N$, and for any positive integer p, the inequality

$$|u_{n+p}(x) - u_n(x)| < \varepsilon, \qquad \forall x \in \Gamma \tag{1}$$

holds. The left-hand side of (1) is the absolute value of the difference of two harmonic functions, and, consequently, is itself harmonic. Moreover, it is continuous in $\bar{\Omega}$. But such a function attains both its maximum and its minimum on the boundary of the domain.

It follows from inequality (1) that

$$-\varepsilon < \min_{\xi \in \Gamma} [u_{n+p}(\xi) - u_n(\xi)] \leq \max_{\xi \in \Gamma} [u_{n+p}(\xi) - u_n(\xi)] < \varepsilon,$$

and from the Maximum Principle we then have the inequality

$$\min_{\xi \in \Gamma} [u_{n+p}(\xi) - u_n(\xi)] \leq u_{n+p}(x) - u_n(x) \leq \max_{\xi \in \Gamma} [u_{n+p}(\xi) - u_n(\xi)]$$

for any $x \in \Omega$. But then

$$-\varepsilon < u_{n+p}(x) - u_n(x) < \varepsilon, \qquad \forall x \in \bar{\Omega},$$

or

$$|u_{n+p}(x) - u_n(x)| < \varepsilon, \qquad \forall x \in \bar{\Omega}.$$

This last inequality implies that the sequence $\{u_n(x)\}$ converges uniformly in the closed domain $\bar{\Omega}$. Consequently there exists a limit function

$$u(x) = \lim_{n \to \infty} u_n(x), \tag{2}$$

defined and continuous in $\bar{\Omega}$.

Let $x \in \Omega$. Construct a sphere $\Sigma_R(x)$ with centre at the point x, and with radius R small enough so that Σ_R, together with its boundary S_R, lies wholly inside Ω. From the direct Mean Value Theorem,

$$u_n(x) = \frac{1}{|S_R|}\int_{S_R} u_n(\xi)\,dS_R. \tag{3}$$

Let $n \to \infty$. Since the sequence $\{u_n(x)\}$ converges uniformly in $\bar{\Omega}$, we can take the limit under the integral sign:

$$u(x) = \frac{1}{|S_R|}\int_{S_R} u(\xi)\,d\xi. \tag{4}$$

Because of the inverse Mean Value Theorem, the function $u(x)$ is harmonic in Ω.

It remains to prove the uniform convergence of the derivatives. Consider an arbitrary interior sub-domain Ω' of the domain Ω, and let $2a$ denote the shortest distance between the boundaries of Ω and Ω'. We make use of formula (7.6); assuming that $x \in \bar{\Omega}'$, we apply it to the function $u_n(x)$:

$$u_n(x) = \int_\Omega u_n(\xi)\omega_a(r)\,d\xi. \tag{5}$$

Differentiating this expression with respect to the coordinates of the point x, we obtain

$$\frac{\partial^k u_n}{\partial x_1^{k_1}\partial x_2^{k_2}\ldots\partial x_m^{k_m}} = \int_\Omega u_n(\xi)\frac{\partial^k \omega_a(r)}{\partial x_1^{k_1}\partial x_2^{k_2}\ldots\partial x_m^{k_m}}\,d\xi. \tag{6}$$

Here k is any positive integer, and $k_1+k_2+\ldots+k_m = k$. Similarly,

$$\frac{\partial^k u}{\partial x_1^{k_1}\partial x_2^{k_2}\ldots\partial x_m^{k_m}} = \int_\Omega u(\xi)\frac{\partial^k \omega_a(r)}{\partial x_1^{k_1}\partial x_2^{k_2}\ldots\partial x_m^{k_m}}\,d\xi. \tag{7}$$

The sequence $\{u_n(\xi)\}$ converges uniformly in $\bar{\Omega}$ to the function $u(\xi)$, and the function

$$\frac{\partial^k \omega_a(r)}{\partial x_1^{k_1}\partial x_2^{k_2}\ldots\partial x_m^{k_m}}$$

is continuous for any x and ξ. In this case the integral in (6) tends uniformly to the integral (7), and, consequently,

$$\frac{\partial^k u_n(x)}{\partial x_1^{k_1}\partial x_2^{k_2}\ldots\partial x_m^{k_m}} \xrightarrow[n\to\infty]{} \frac{\partial^k u(x)}{\partial x_1^{k_1}\partial x_2^{k_2}\ldots\partial x_m^{k_m}}$$

uniformly in $\bar{\Omega}'$. The theorem is proven.

THEOREM 11.9.2. (THEOREM ON CONVERGENCE IN THE MEAN). *Let Ω be a finite domain. Let $\{u_n(x)\}$ be a sequence of harmonic functions in Ω, convergent in the metric of $L_p(\Omega)$, where $1 < p < \infty$. Then:* 1) *the limit function is harmonic in Ω;* 2) *in any interior sub-domain, both the given sequence, and the sequences obtained by differentiating its terms, are uniformly convergent.*

By the conditions of the theorem, there exists a limit function

$$u(x) = \lim_{n \to \infty} u_n(x)$$

in the sense of convergence in the mean with index p. This means that $u \in L_p(\Omega)$, and

$$\int_\Omega |u(\xi) - u_n(\xi)|^p \, d\xi \underset{n \to \infty}{\to} 0.$$

We use formula (5), and obtain

$$\left. \begin{array}{l} u_n(x) = \int_\Omega u_n(\xi) \omega_a(r) \, d\xi, \\ u_k(x) = \int_\Omega u_k(\xi) \omega_a(r) \, d\xi, \end{array} \right\} \quad \forall x \in \overline{\Omega}'.$$

Subtracting, and applying Hölder's inequality, we find

$$|u_n(x) - u_k(x)| \le \left\{ \int_\Omega |u_n(\xi) - u_k(\xi)|^p \, d\xi \right\}^{1/p} \left\{ \int_\Omega \omega_a^{p'}(r) \, d\xi \right\}^{1/p'}, \qquad (8)$$

$$\frac{1}{p} + \frac{1}{p'} = 1.$$

For a fixed averaging radius a, the function $\omega_a(r)$ is bounded. Moreover, by the conditions of the theorem the first integral on the right-hand side in (8) tends to zero. Then from inequality (8) it follows that the sequence $\{u_n(x)\}$ converges uniformly in $\overline{\Omega}'$. Hence we have that $u(x)$ is a continuous function in $\overline{\Omega}'$; since Ω' is an arbitrary interior sub-domain, $u(x)$ is continuous in the open domain Ω.

Let $n \to \infty$ in equation (3). This gives equation (4), and as before, it can be inferred from this that $u(x)$ is a harmonic function in Ω. The statement concerning convergence of the derivatives of the sequence $\{u_n(x)\}$ can be proved in the same way as for Theorem 11.9.1.

From the theorems of the present section there ensue two corollaries:

COROLLARY 11.9.1. *Let Ω be a finite domain. In the space $C(\bar{\Omega})$, harmonic functions generate a subspace. Convergence in this subspace implies uniform convergence of derivatives of any order in any interior closed sub-domain.*

COROLLARY 11.9.2. *Let Ω be a finite domain, and let p be a constant in the range $1 < p < \infty$. In the space $L_p(\Omega)$ harmonic functions generate a subspace. Convergence in this subspace implies uniform convergence both of the functions themselves, and of their derivatives of any order, in any interior closed sub-domain.*

§ 10. Extension to Equations with Variable Coefficients

The results of the present chapter can to a considerable degree be extended to elliptic equations with variable coefficients. In this section we shall not investigate the general problem, but shall mention – in many cases without proof – some basic facts in relation to the self adjoint elliptic equation

$$-\frac{\partial}{\partial x_j}\left(A_{jk}(x)\frac{\partial u}{\partial x_k}\right) + C(x)u = 0. \tag{1}$$

A more general exposition of the questions raised here is given in books by C. Miranda [10] and by the present author [11], cited in the bibliography at the end of the book.

Let Ω be a finite domain in m-dimensional Euclidean space, bounded by a piecewise-smooth surface Γ. We shall suppose that $A_{jk} \in C^{(3)}(\bar{\Omega})$, $C \in C^{(1)}(\bar{\Omega})$ (in some cases these conditions could be relaxed). We take equation (1) to be elliptic in $\bar{\Omega}$ – this means that all the eigenvalues of the matrix $A(x) = \|A_{jk}(x)\|_{j,k=1}^{j,k=m}$ are different from zero, and have the same sign; we shall take them to be positive. Let $\lambda_1(x)$ be the smallest eigenvalue of the matrix $A(x)$; this will be a root of the equation $\text{Det}\,(A(x) - \lambda I) = 0$, which has all its coefficients continuous functions of x in $\bar{\Omega}$, with the highest coefficient equal to $(-1)^m$. But then the roots of this equation are continuous in $\bar{\Omega}$; the function $\lambda_1(x)$, continuous and positive in $\bar{\Omega}$, has a positive lower bound:

$$\lambda_1(x) \geq \mu_0 = \text{const.} > 0, \quad \forall x \in \bar{\Omega}. \tag{2}$$

A well-known result in the theory of quadratic forms is that for any real numbers $t_1, t_2, \ldots, t_m$, the inequality

$$A_{jk}(x)t_j t_k \geq \lambda_1(x) \sum_{k=1}^{m} t_k^2$$

holds. Then from inequality (2) there follows the important relation

$$A_{jk}(x)t_j t_k \geq \mu_0 \sum_{k=1}^{m} t_k^2, \quad \forall x \in \bar{\Omega}. \tag{3}$$

We shall consider only those solutions of equation (1) which belong to the class $C^{(2)}(\Omega)$.

1. We shall demonstrate the Maximum Principle for equation (1) in the following weak form. For the stronger form of this principle see, for example, ref. [10].

THEOREM 11.10.1. *If $C(x) > 0$, $\forall x \in \Omega$, then inside the domain a solution of equation* (1) *cannot have either a negative minimum or a positive maximum. If $C(x) < 0$, $\forall x \in \Omega$, then inside the domain this solution cannot have either a positive minimum or a negative maximum.*

Proof. We shall establish the result relating to negative minimum when $C(x) > 0$ – the other cases can be treated similarly. Let $C(x) > 0$, $x \in \Omega$, and let the solution of (1) have a negative minimum at the point $x_0 \in \Omega$. Then at this point x_0,

$$u < 0, \quad \frac{\partial u}{\partial x_k} = 0, \quad \frac{\partial^2 u}{\partial x_k^2} \geq 0; \quad k = 1, 2, \ldots, m. \tag{4}$$

The relations (4) hold in any coordinate system. Take coordinate axes such that at the point x_0 equation (1) is in canonical form. Then

$$A_{kk}(x_0) > 0; \quad A_{jk}(x_0) = 0, \quad j \neq k. \tag{5}$$

Putting $x = x_0$ in equation (1), we obtain

$$-\sum_{k=1}^{m} A_{kk}(x_0)\left(\frac{\partial^2 u}{\partial x_k^2}\right)_{x=x_0} + C(x_0)u(x_0) = 0.$$

But this last equation is impossible: because of relations (4) and (5), the first term on the left is non-positive, and the second strictly negative, so that the left-hand side is negative.

Remark. The Maximum Principle holds also for a more general – non self-adjoint – equation of elliptic type, namely

$$-\frac{\partial}{\partial x_j}\left(A_{jk}\frac{\partial u}{\partial x_k}\right) + B_k\frac{\partial u}{\partial x_k} + Cu = 0.$$

The terms containing the coefficients B_k have no bearing on the arguments used above, because $\partial u/\partial x_k = 0$ at a maximum or minimum point.

2. Singular Solution. Let

$$a(x) = A^{-1}(x) = \|a_{jk}(x)\|_{j,\,k=1}^{j,\,k=m}.$$

Consider the function

$$\psi(x,\xi) = \frac{1}{(m-2)|S_1|\sqrt{D(\xi)}}[a_{jk}(\xi)(x_j-\xi_j)(x_k-\xi_k)]^{-\frac{1}{2}(m-2)}, \quad m > 2, \quad (6)$$

where $D(\xi) = \mathrm{Det}\, A(\xi)$. For fixed ξ, the function $\psi(x,\xi)$ satisfies the equation

$$-\frac{\partial}{\partial x_j}\left(A_{jk}(\xi)\frac{\partial \psi}{\partial x_k}\right) = 0.$$

The function (6) is called the *parametrix* of equation (1). When $m = 2$ the parametrix is defined by the formula

$$\psi(x,\xi) = \frac{1}{2\pi\sqrt{D(\xi)}} \ln\,[a_{jk}(\xi)(x_j-\xi_j)(x_k-\xi_k)]^{-\frac{1}{2}}. \quad (6')$$

The function $v(x,\xi)$ is called a *singular solution* of equation (1) if it has the following properties:

1) it can be represented in the form

$$v(x,\xi) = \psi(x,\xi) + \psi_1(x,\xi), \quad (7)$$

where for $x \to \xi$

$$\psi_1(x,\xi) = O(r^{\kappa+2-m}), \quad \frac{\partial \psi_1}{\partial x_k} = O(r^{\kappa+1-m}), \quad (8)$$

$$\frac{\partial^2 \psi_1}{\partial x_j \partial x_k} = O(r^{\kappa-m}); \quad \kappa = \mathrm{const.} > 0;$$

2) for fixed ξ, $\xi \neq x$, it satisfies equation (1).

A proof that the singular solution exists can be found in the above-mentioned book by Miranda [10].

There exist singular solutions of equation (1) which are symmetric with respect to x and ξ (cf. Miranda [10]). When speaking of a singular solution we shall henceforth assume that it is symmetric.

3. Integral Representation. Let Ω' be an interior sub-domain of the domain Ω, and let the boundary Γ' of Ω' be piecewise-smooth. Let the function $u \in C^{(2)}(\overline{\Omega}')$. Using the singular solution of equation (1), we can construct an integral representation similar to (3.4).

Let $x \in \Omega'$. Around the point x describe a domain σ_ε, defined by the inequality
$$a_{jk}(x)(x_j - \xi_j)(x_k - \xi_k) < \varepsilon^2;$$
the boundary of this domain is the ellipsoid
$$a_{jk}(x)(x_j - \xi_j)(x_k - \xi_k) = \varepsilon^2$$
which we denote by Γ_ε. Let $v(x, \xi)$ be a symmetric singular solution of equation (1). By Green's Formula (6.6), Chapter 10, we obtain

$$\int_{\Omega' \setminus \sigma_\varepsilon} vLu \, d\xi = \int_{\Gamma'} A_{jk}(\xi) \left(v \frac{\partial u}{\partial \xi_k} - u \frac{\partial v}{\partial \xi_k} \right) \cos(v, x_j) \, d_\xi \Gamma'$$
$$+ \int_{\Gamma_\varepsilon} A_{jk}(\xi) \left(v \frac{\partial u}{\partial \xi_k} - u \frac{\partial v}{\partial \xi_k} \right) \cos(v, x_j) \, d_\xi \Gamma_\varepsilon, \quad (9)$$

where L denotes the differential expression on the left-hand side of (1), and where we have taken into account that $Lv = 0$.

Let $\varepsilon \to 0$. Then

$$\int_{\Omega' \setminus \sigma_\varepsilon} vLu \, d\xi \to \int_{\Omega'} vLu \, d\xi,$$

$$\int_{\Gamma_\varepsilon} vA_{jk} \frac{\partial u}{\partial \xi_k} \cos(v, x_j) \, d_\xi \Gamma_\varepsilon \to 0,$$

$$\int_{\Gamma_\varepsilon} uA_{jk} \frac{\partial \psi_1}{\partial \xi_k} \cos(v, x_j) \, d_\xi \Gamma_\varepsilon \to 0.$$

It remains to find the limit of the integral

$$\int_{\Gamma_\varepsilon} u(\xi) A_{jk}(\xi) \frac{\partial \psi}{\partial \xi_k} \cos(v, x_j) \, d_\xi \Gamma_\varepsilon. \quad (10)$$

The singular solution is symmetric with respect to x and ξ, therefore we can write $v(x, \xi) = \psi(\xi, x) + \psi_1(\xi, x)$, and, in integral (10), use for ψ the expression *

$$\psi = \psi(\xi, x) = \frac{1}{(m-2)|S_1|\sqrt{D(x)}} [a_{jk}(x)(\xi_j - x_j)(\xi_k - x_k)]^{-\frac{1}{2}(m-2)}.$$

We now have

$$\frac{\partial \psi}{\partial \xi_k} = -\frac{1}{|S_1|\sqrt{D(x)}} \frac{a_{pk}(x)(\xi_p - x_p)}{[a_{jl}(x)(\xi_j - x_j)(\xi_l - x_l)]^{\frac{1}{2}m}},$$

or, if we utilize the equation of the ellipsoid Γ_ε,

$$\frac{\partial \psi}{\partial \xi_k} = \frac{1}{|S_1|\varepsilon^m \sqrt{D(x)}} a_{pk}(x)(x_p - \xi_p),$$

and the integral (10) takes the form

$$\frac{1}{|S_1|\varepsilon^m \sqrt{D(x)}} \int_{\Gamma_\varepsilon} u(\xi) a_{pk}(x) A_{jk}(x)(x_p - \xi_p) \cos(v, x_j) \, d_\xi \Gamma_\varepsilon. \tag{11}$$

The integral (11) can be simplified. First of all, the matrices A and a are mutually inverse and symmetric, therefore,

$$a_{pk}(x) A_{jk}(x) = a_{pk}(x) A_{kj}(x) = \delta_{pj},$$

and the integral (11) is transformed into

$$\frac{1}{|S_1|\varepsilon^m \sqrt{D(x)}} \int_{\Gamma_\varepsilon} u(\xi)(x_j - \xi_j) \cos(v, x_j) \, d_\xi \Gamma_\varepsilon.$$

If we substitute $u(x)$ for $u(\xi)$, then this last integral would be changed by a quantity which tends to zero as $\varepsilon \to 0$. Thus, the limit of the integral (10) coincides with the limit of the quantity

$$\frac{1}{|S_1|\varepsilon^m} \frac{u(x)}{\sqrt{D(x)}} \int_{\Gamma_\varepsilon} (x_j - \xi_j) \cos(v, x_j) \, d_\xi \Gamma_\varepsilon. \tag{12}$$

With the aid of Gauss's Theorem, we convert this last expression into a volume integral:

$$\int_{\Gamma_\varepsilon} (x_j - \xi_j) \cos(v, x_j) \, d_\xi \Gamma_\varepsilon = \int_{\sigma_\varepsilon} \frac{\partial (x_j - \xi_j)}{\partial \xi_j} \, d\xi = -m \int_{\sigma_\varepsilon} d\xi = -m|\sigma_\varepsilon|.$$

We now evaluate the quantity $|\sigma_\varepsilon|$.

* As usual, we understand that summation over j and k, from 1 to m, is performed inside the square brackets.

The volume of the ellipsoid is equal to the product of the volume of a unit sphere with all the semi-axes of the ellipsoid. If $\lambda_1(x), \lambda_2(x), \ldots, \lambda_m(x)$ are eigenvalues of the matrix $A(x)$, then the semi-axes of the ellipsoid Γ_ε are $\varepsilon\sqrt{\lambda_k(x)}$, $k = 1, 2, \ldots, m$. Now

$$|\sigma_\varepsilon| = \frac{|S_1|\varepsilon^m}{m}\sqrt{\prod_{k=1}^{m}\lambda_k(x)} = \frac{|S_1|\varepsilon^m}{m}\sqrt{D(x)};$$

the quantity (12) is independent of ε and is equal to $-u(x)$, and after taking the limit in formula (9) we obtain the following integral representation:

$$u(x) = \int_{\Gamma'} A_{jk}(\xi)\left(v\frac{\partial u}{\partial \xi_k} - u\frac{\partial v}{\partial \xi_k}\right)\cos(v, x_j)\,d_\xi\Gamma' - \int_{\Omega'} vLu\,d\xi. \quad (13)$$

If $u(x)$ satisfies equation (1), then $Lu = 0$, and we find an integral representation for the solution of equation (1):

$$u(x) = \int_{\Gamma'} A_{jk}(\xi)\left(v\frac{\partial u}{\partial \xi_k} - u\frac{\partial v}{\partial \xi_k}\right)\cos(v, x_j)\,d_\xi\Gamma'. \quad (14)$$

4. Mean Value Theorem. Let $u(x) \in C^{(2)}(\Omega)$ be a solution of equation (1). As the surface Γ' in equation (14), we take the equipotential surface of the functions $v(x, \xi)$, i.e. the surface

$$v(x, \xi) = v_0 = \text{const.}$$

We suppose that the constant v_0 is sufficiently large, and that the point x is fixed. Then

$$u(x) = v_0\int_{v=v_0} A_{jk}(\xi)\frac{\partial u}{\partial \xi_k}\cos(v, x_j)\,d_\xi\Gamma' - \int_{v=v_0} u(\xi)A_{jk}(\xi)\frac{\partial v}{\partial \xi_k}\cos(v, x_j)\,d_\xi\Gamma'.$$

By Gauss's Theorem, *

$$\int_{v=v_0} A_{jk}\frac{\partial u}{\partial \xi_k}\cos(v, x_j)\,d_\xi\Gamma' = \int_{v>v_0}\frac{\partial}{\partial x_j}\left(A_{jk}\frac{\partial u}{\partial \xi_k}\right)d\xi = \int_{v>v_0} C(\xi)u(\xi)\,d\xi.$$

For brevity we use the notation

$$-A_{jk}\frac{\partial v}{\partial \xi_k}\cos(v, x_j) = N(v),$$

* Obviously, the domain $v > v_0$ lies inside the surface $v = v_0$.

and obtain the formula

$$u(x) = \int_{v=v_0} u(\xi)N(v)\,\mathrm{d}_\xi \Gamma' + v_0 \int_{v>v_0} C(\xi)u(\xi)\,\mathrm{d}\xi, \tag{15}$$

which is analogous to the result for the Mean Value Theorem.

Formula (15) holds for solutions of equation (1). The converse statement is also true: if the function $u \in L_p(\Omega)$, $1 < p < \infty$, and satisfies relation (15), then $u \in C^{(2)}(\Omega)$ and this function satisfies equation (1). This statement will be verified in the course of proving the theorem of the next paragraph.

The right-hand side of equation (15) contains a surface integral. It is not difficult to construct a formula of the same type containing only a volume integral. Let $\Phi(\rho)$ be an infinitely differentiable function of a real variable ρ, and let this function be different from zero only on the interval $0 < \rho < \rho_1$, where ρ_1 is a sufficiently small positive constant. We shall suppose that if the distance $r = |x-\xi|$ is sufficiently small then $v(x, \xi) > 0$. We put

$$\rho(x, \xi) = [v(x, \xi)]^{-1/(m-2)}, \qquad \rho_0 = v_0^{-1/(m-2)}.$$

Obviously, $\rho(x, \xi) = 0(r)$, and $\mathrm{d}\xi = \mathrm{d}\rho\,\mathrm{d}_\xi \Gamma'/(\partial\rho/\partial v)$, where Γ' is the surface $v = \mathrm{const.}$, and v is the outward normal. We multiply both sides of equation (15) by $\{\Phi(\rho_0)/(\partial\rho/\partial v)_{\rho=\rho_0}\}\mathrm{d}\rho_0$ and integrate with respect to ρ_0 between the limits 0 and ∞. We thus arrive at a formula of the form

$$u(x) = \int_{\rho<\rho_0} K(x, \xi)u(\xi)\,\mathrm{d}\xi, \tag{16}$$

where $K(x, \xi)$ is a continuous function of the points x and ξ, having continuous first and second derivatives with respect to the coordinates of the point x.

5. Subspaces of Solutions.

THEOREM 11.10.2. *The set of solutions of equation* (1) *which belong to the intersection* $C^{(2)}(\Omega) \cap L_p(\Omega)$, *where* $1 < p < \infty$, *generates a subspace in* $L_p(\Omega)$. *Convergence in this subspace implies uniform convergence both of the functions themselves and of their derivatives of first and second order in any closed interior sub-domain.*

Let $\{u_n(x)\}$ be a sequence of solutions of equation (1), which belong to the intersection $C^{(2)}(\Omega) \cap L_p(\Omega)$, and let this sequence converge in the metric of $L_p(\Omega)$ to some function $u(x)$. Take an interior sub-domain Ω'

and choose a number $\rho_2 > 0$ small enough so that the domain $\rho(x, \xi) < \rho_2$ lies inside Ω whenever $x \in \bar{\Omega}'$. If $\rho_1 \leq \rho_2$, then for any point $x \in \bar{\Omega}'$ and for any of the functions $u_n(x)$, formula (16) holds:

$$u_n(x) = \int_{\rho < \rho_1} K(x, \xi) u_n(\xi) \, d\xi.$$

Let $n \to \infty$ in this equation. Then we find that the limit function $u(x)$ also satisfies relation (16). Since the kernel has continuous second derivatives, therefore $u \in C^{(2)}(\bar{\Omega}')$. It also follows from (16) that, in $\bar{\Omega}'$, the functions $u_n(x)$, and their derivatives of first and second order, tend uniformly to the function $u(x)$ and its respective derivatives.

Being solutions of equation (1), the functions $u_n(x)$ satisfy equation (15). Letting $n \to \infty$, we find that the same equation is satisfied by the limit function $u(x)$. Now let Γ' in the representation (13) be the surface $v = v_0$. Then from (13) and (15) there follows a relation for the limit function:

$$\int_{v > v_0} (v - v_0) Lu \, d\xi = 0.$$

Since $v - v_0 > 0$, we can apply the Mean Value Theorem to this integral; we obtain

$$(Lu)_{x=x'} \int_{v > v_0} (v - v_0) \, dx = 0,$$

where x' is some point of the domain $v(x, \xi) > v_0$. Hence $(Lu)_{x=x'} = 0$. Letting $v_0 \to \infty$, we obtain $(Lu)(x) = 0$, which is the required result.

In precisely the same way we can prove a theorem, analogous to Theorem 11.10.2, in which $L_p(\Omega)$ is replaced by $C(\bar{\Omega})$.

Chapter 12.
Dirichlet and Neumann Problems

§ 1. Formulation

We shall consider domains of two types: *finite* and *infinite*. In both cases we shall suppose that the boundaries of the domains are finite, and, as usual, that these boundaries consist of a finite number of piecewise-smooth surfaces (Figs. 12 and 13). In subsequent chapters we shall sometimes consider so-called *semi-infinite* domains, whose boundaries are infinite, but these will be specifically delineated in each case. The simplest example of a semi-infinite domain is a half-space.

A boundary-value problem for elliptic equations is said to be *interior* if the required function is to be determined in a finite domain, and *exterior* if the function is to be determined in an infinite domain.

The most important boundary-value problems for second-order elliptic equations are the *Dirichlet problem* (the first boundary-value problem) and the *Neumann problem* (the second boundary-value problem).

Consider an elliptic equation of the general form

$$-A_{jk}\frac{\partial^2 u}{\partial x_j \partial x_k} + A_k \frac{\partial u}{\partial x_k} + A_0 u = F(x). \qquad (1)$$

The *interior Dirichlet problem* for this equation can be formulated in the following way.

Let Ω be a finite domain with a piecewise-smooth boundary Γ, and let $\phi(x)$ be a function prescribed and continuous on the boundary Γ. It is required to find a solution of equation (1) which belongs to class $C^{(2)}(\Omega) \cap \cap C(\bar{\Omega})$, and which is equal to the given function $\phi(x)$ on the boundary:

$$u(x) = \phi(x), \qquad x \in \Gamma. \qquad (2)$$

The *interior Neumann problem* for the same equation (1) is formulated as follows.

To find a solution to equation (1) having the properties: $u \in C^{(2)}(\Omega) \cap$

$\cap\, C(\bar{\Omega})$, and satisfies the equation

$$\lim_{x'\to x} A_{jk}(x') \frac{\partial u(x')}{\partial x'_j} \cos(v, x_k) = \psi(x) \tag{3}$$

on the set of points $x \in \Gamma$, at which there exists a normal v to the surface Γ. Here x' is a point lying inside Ω on the normal v, x'_k are the cartesian coordinates of this point, and $\psi(x)$ is a function prescribed on the above-mentioned set of points of the surface Γ.

We shall later write the boundary condition for the Neumann problem in the more concise form

$$A_{jk}(x) \frac{\partial u}{\partial x_j} \cos(v, x_k)\big|_\Gamma = \psi(x). \tag{3'}$$

This expression (3') can be understood in a literal sense if $u \in C^{(1)}(\bar{\Omega})$.

If $A_{jk} = \delta_{jk}$, then the highest terms of equation (1) generate the Laplace operator; the equation then takes the form

$$-\Delta u + A_k \frac{\partial u}{\partial x_k} + A_0 u = F(x). \tag{4}$$

In this case the boundary condition (3') takes the especially simple form:

$$\frac{\partial u(x)}{\partial v}\bigg|_\Gamma = \psi(x). \tag{5}$$

Remark. The above formulations of the Dirichlet and Neumann problems are not completely general. Thus, for example, we could consider the case when the function $\phi(x)$ in the boundary condition (2) of the Dirichlet problem is discontinuous on Γ. In that case we could not require that $u \in C(\bar{\Omega})$ – this condition would have to be replaced by some other; the boundary condition (2) can be satisfied only at points where the function $\phi(x)$ is continuous. We could also dispense with the requirement of a piecewise-smooth boundary.

The *exterior problem* differs from the corresponding interior problem only in that an additional restriction is imposed on the unknown function, namely,

$$u(x) = O\left(\frac{1}{|x|^{m-2}}\right), \quad x \to \infty. \tag{6}$$

§ 2. Uniqueness Theorems for Laplace's Equation

THEOREM 12.2.1. *Both the interior and the exterior Dirichlet problem for Laplace's equation have not more than one solution.*

Suppose that the Dirichlet problem has two solutions: $u_1(x)$ and $u_2(x)$. Then the following two systems of equations hold:

$$-\Delta u_1 = F(x), \quad u_1|_\Gamma = \phi(x); \tag{1}$$

$$-\Delta u_2 = F(x), \quad u_2|_\Gamma = \phi(x). \tag{2}$$

Introduce the notation $u_1(x) - u_2(x) = v(x)$. Subtracting equation (1) from (2), we obtain

$$\Delta v = 0, \quad v|_\Gamma = 0. \tag{3}$$

Consider the interior Dirichlet problem. The question reduces to finding a function $v(x)$, harmonic in Ω and continuous in $\overline{\Omega}$. By the Maximum Principle, its greatest and least values are attained on the boundary, but then they are equal to each other, and equal to zero. Hence it follows that $v \equiv 0$, and $u_1(x) = u_2(x)$.

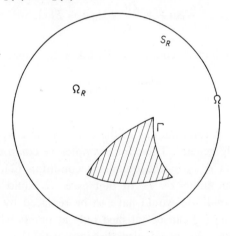

Fig. 21.

Now turn to the exterior Dirichlet problem. We shall consider only the case $m > 2$. Because of the condition (1.6), the difference $v = u_1 - u_2$ is harmonic in Ω. We enclose the boundary Γ within a circumscribed sphere, having radius R and bounding surface S_R, and consider the function $v(x)$ in the annular domain Ω_R included between Γ and S_R (Fig. 21). We know that $v|_\Gamma = 0$; in addition, at a sufficiently large distance from the origin, which is located at the centre of the sphere S_R, we have

$$|v(x)| \leq \frac{C}{|x|^{m-2}}, \quad C = \text{const.}$$

Consequently, on the surface of the sphere S_R, we have

$$|v(x)| \leq \frac{C}{R^{m-2}}$$

so long as the radius R is sufficiently large. Let $\varepsilon > 0$ be a given arbitrary number, and choose R large enough that $CR^{2-m} < \varepsilon$. In the annular domain Ω_R, the greatest and least values of the function $v(x)$ are attained either on Γ or on S_R; it follows that these values cannot exceed ε in magnitude.

Let x be an arbitrary point of the domain Ω. For sufficiently large R this point is situated in the domain Ω_R, and therefore $|v(x)| < \varepsilon$. But ε is an arbitrary positive number, and so $u_1(x) \equiv u_2(x)$.

Remark 1. If $m = 2$, the uniqueness theorem for the exterior Dirichlet problem remains valid. The proof can be established by making use of the invariance of a two-dimensional harmonic function with respect to a conformal transformation (cf. below, Chapter 13). An infinite domain can be transformed into a finite domain by means of a bilinear mapping. Thus, a function which is harmonic in an infinite domain transforms into a function which is harmonic in a finite domain, and, as before, the new function is equal to zero on the boundary of its domain. In this way we can reduce the problem to the question of uniqueness for an interior Dirichlet problem, and this has already been settled.

Remark 2. With regard to the conditions for uniqueness of the solution to the Dirichlet problem for the more general second-order elliptic equation (1.1), see the book by Miranda [10]. If the matrix of highest coefficients is positive definite in the closed domain $\overline{\Omega}$, and if $A_0(x) > 0$, then the uniqueness of the solution to the Dirichlet problem follows from the Maximum Principle (§ 10, Chapter 11).

We now turn to the Neumann problem. We shall say that the function $u(x)$, defined in the domain Ω, has on the boundary of this domain a *regular normal derivative* if the limit

$$\lim_{x' \to x} \frac{\partial u(x')}{\partial \nu}, \quad \forall x \in \Gamma,$$

exists and is continuous on Γ, where the approach to the limit is uniform with respect to x; as before, x' denotes a point of the domain lying on the normal ν which passes through the point x.

If the function $u \in C^{(1)}(\bar{\Omega})$ and the boundary Γ is smooth, then $u(x)$ obviously has a regular normal derivative on Γ.

Consider a finite domain Ω with boundary Γ. We pose the interior Neumann problem for this domain:

$$-\Delta u = F(x), \quad x \in \Omega, \quad u \in C^{(2)}(\Omega) \cap C(\bar{\Omega}),$$

$$\lim_{x' \to x} \frac{\partial u(x')}{\partial \nu} = \psi(x), \quad x \in \Gamma. \tag{4}$$

We shall here impose the requirement that *the unknown function must have a regular normal derivative*. Moreover, we shall assume that the boundary Γ is a *regular surface*. This means the following: 1) at every point x of the surface there exists a defined normal; 2) if at any point $x \in \Gamma$ we construct a local coordinate system, in which the x_m-axis is directed along the normal, and the axes $x_1, x_2, \ldots, x_{m-1}$ lie in the tangent plane, then near the point x we can prescribe the surface Γ by an explicit equation $x_m = f(x_1, x_2, \ldots, x_{m-1})$; 3) for sufficiently small $x_1, x_2, \ldots, x_{m-1}, f \in C^{(2)}$.

It follows from our statement of the Neumann problem that $F \in C(\Omega)$ and $\psi \in C(\Gamma)$. Moreover, we assume that $F(x)$ is integrable in Ω.

We show that in general the Neumann problem in the form (4) is insoluble, and derive a necessary condition that it has a solution.

At each point x of the surface Γ we construct a normal pointing into the domain, and on the normal we cut off an interval of fixed length h, one end of which coincides with the point x. The set of the other end generates a surface Γ_h, which is said to be *parallel* to Γ. It is well-known from differential geometry that if h is sufficiently small, then the surface Γ_h is smooth, and the normal to one parallel surface is a normal to the other. Denote by $\Omega^{(h)}$ the domain contained inside Γ_h; obviously $\Omega^{(h)} = \Omega \backslash \Omega_h$, where Ω_h is the boundary strip of width h for the domain Ω.

If u is the solution of problem (4), then obviously $u \in C^{(2)}(\bar{\Omega}^{(h)})$, and we can apply Green's Formula (6.9), Chapter 10, to the functions u and $v \equiv 1$. This formula then gives

$$-\int_{\Omega^{(h)}} F(x) dx = \int_{\Gamma_h} \frac{\partial u}{\partial \nu} d\Gamma_h.$$

The function u has a regular normal derivative, therefore $(\partial u/\partial \nu)|_{\Gamma_h}$ tends uniformly to $(\partial u/\partial \nu)|_\Gamma = \psi(x)$, as $h \to 0$. Taking the limit $h \to 0$ in the last formula, we obtain

$$\int_\Omega F(x)\,dx + \int_\Gamma \psi(x)\,d\Gamma = 0. \tag{5}$$

This is the relation which must necessarily be satisfied by our problem. Thus we see that in the general case the Neumann problem for Laplace's equation has no solution. A solution can exist only if condition (5) is fulfilled.

If either the boundary conditions, or the differential equation, are homogeneous, the above necessary condition becomes

$$\int_\Omega \Delta u\,dx = \int_\Omega F(x)\,dx = 0, \tag{6}$$

or

$$\int_\Gamma \frac{\partial u}{\partial \nu}\,d\Gamma = \int_\Gamma \psi(x)\,d\Gamma = 0, \tag{7}$$

respectively.

It is obvious that the solution to the interior Neumann problem for Laplace's equation (if it exists) is not unique; if the function $u(x)$ satisfies the problem (4), then it is easy to see that the function $u(x)+C$, where C is an arbitrary constant, satisfies the same problem. The uniqueness theorem in this case is the statement that the expression $u(x)+C$ exhausts all the solutions to this problem.

THEOREM 12.2.2. *Two solutions of the interior Neumann problem for Laplace's equation can differ only by a constant term.*

In the proof of this theorem we shall assume that the surface Γ is regular.

Let the problem (4) have two solutions. Their difference $v(x)$ satisfies the relation

$$\Delta v = 0, \qquad \left.\frac{\partial v}{\partial \nu}\right|_\Gamma = 0. \tag{8}$$

Apply Green's Formula (6.8), Chapter 10, to the function v and the domain $\Omega^{(h)}$ (cf. above):

$$\int_{\Omega^{(h)}} \sum_{k=1}^m \left(\frac{\partial v}{\partial x_k}\right)^2 dx = \int_{\Gamma_h} v\,\frac{\partial v}{\partial \nu}\,d\Gamma_h. \tag{9}$$

The function v is continuous, and, consequently, bounded in $\overline{\Omega}$. At the same time the quantity $(\partial v/\partial \nu)|_{\Gamma_h}$ tends uniformly to zero. Taking the limit as $h \to 0$ in equation (9), we obtain

$$\int_\Omega \sum_{k=1}^m \left(\frac{\partial v}{\partial x_k}\right)^2 dx = 0.$$

Hence $\partial v/\partial x_k \equiv 0$, $k = 1, 2, \ldots, m$, and, consequently, $v = $ const.

Remark 1. If it is assumed that the required solution has continuous first derivatives in $\bar{\Omega}$, then it is easy to prove Theorem 12.2.2, subject only to the requirement that the surface Γ is piecewise-smooth.

Remark 2. Theorem 12.2.2 can be proved under quite broad assumptions. Cf. V. I. Smirnov [17], Vol. IV, par. 205, where the case $m = 3$ is discussed.

The uniqueness theorem for the exterior Neumann problem will be proved later, in § 7.

§ 3. Solution of the Dirichlet Problem for a Sphere

Here and later in this chapter (with the exception of § 7) we shall consider only the homogeneous Laplace equation – the inhomogeneous equation can be reduced to the homogeneous by the method discussed in § 6, Chapter 11; we recall that this method is based on the construction of a particular solution of Laplace's equation in the form of a volume potential.

Suppose, then, that we are given a sphere Σ_R, of radius R and centre at the origin of coordinates. We pose the problem of finding a function $u \in C(\bar{\Sigma}_R)$, which is harmonic in the sphere and satisfies the boundary condition

$$u|_{S_R} = \phi(x), \qquad (1)$$

where S_R is the boundary of the sphere, and $\phi(x)$ is a function prescribed and continuous on the surface S_R.

The problem will be solved by the following method. We assume that a solution exists which satisfies some stricter conditions, and then construct a formula which determines the solution to the given problem. After doing this, we show that the formula obtained in this way actually does solve the problem.

Let the problem stated above have a solution $u(x)$ which belongs to the class $C^{(2)}(\bar{\Sigma}_R)$.

We write down the integral representation of this solution (formula

§ 3] DIRICHLET PROBLEM FOR A SPHERE

(4.1), Chapter 11)

$$u(x) = \frac{1}{(m-2)|S_1|} \int_{S_R} \left(\frac{1}{r^{m-2}} \frac{\partial u}{\partial v} - u \frac{\partial}{\partial v} \frac{1}{r^{m-2}} \right) d_\xi S_R. \tag{2}$$

Let x be a point inside the sphere, and x' the image point of x with respect to the sphere S_R (Fig. 22). This means that the points x and x' lie on the one ray passing through the centre of the sphere, and that

$$|x| \cdot |x'| = R^2. \tag{3}$$

Put

$$r = |x - \xi|, \qquad r' = |x' - \xi|.$$

Note that $r' \neq 0$ when the point ξ varies inside the sphere or along its surface. Introduce the function

$$v(\xi) = \frac{1}{r'^{m-2}}; \tag{4}$$

it is harmonic in any domain which does not contain the point x'. In particular, the function (4) is harmonic in the sphere Σ_R.

We apply Green's Formula (6.10), Chapter 10, to the pair of functions u and v. Both functions are harmonic, so that the volume integral vanishes, and we obtain

$$\int_{S_R} \left(\frac{1}{r'^{m-2}} \frac{\partial u}{\partial v} - u \frac{\partial}{\partial v} \frac{1}{r'^{m-2}} \right) d_\xi S_R = 0. \tag{5}$$

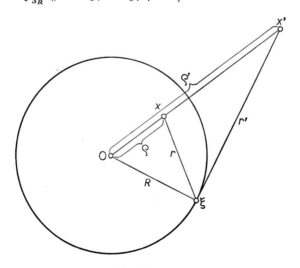

Fig. 22.

In what follows it is important to note that the first terms in the integrals (2) and (5) differ only by a factor which is independent of ξ. This can be proved from the simple fact that the triangles $Ox\xi$ and $Ox'\xi$ are similar (Fig. 22). For these triangles the angle at the point O is common, and, because of relation (3), the sides which include this angle are proportional. From the similarity of the triangles it follows that

$$\frac{r}{r'} = \frac{|x|}{R}.$$

Hence

$$\frac{1}{r} = \frac{R}{|x|r'}$$

and, consequently,

$$\frac{1}{r^{m-2}} = \left(\frac{R}{|x|}\right)^{m-2} \frac{1}{r'^{m-2}},$$

so that $1/r^{m-2}$ and $1/r'^{m-2}$ differ by a factor $(R/|x|)^{m-2}$, independent of ξ.

We shall use the notation $|x| = \rho$, $|x'| = \rho'$.

Multiplying formula (5) by

$$\frac{1}{(m-2)|S_1|}\left(\frac{R}{\rho}\right)^{m-2}$$

and subtracting from formula (3):

$$u(x) = \frac{1}{(m-2)|S_1|}\int_{S_R} u(\xi)\left[\left(\frac{R}{\rho}\right)^{m-2}\frac{\partial}{\partial \nu}\frac{1}{r'^{m-2}} - \frac{\partial}{\partial \nu}\frac{1}{r^{m-2}}\right] d_\xi S_R.$$

Noting that because of the boundary condition for the Dirichlet problem we have

$$u(\xi)|_{S_R} = \phi(\xi),$$

we obtain a formula for the solution (under the assumption that it exists and belongs to the class $C^{(2)}(\bar{\Sigma}_R)$):

$$u(x) = \frac{1}{(m-2)|S_1|}\int_{S_R} \phi(\xi)\left[\left(\frac{R}{\rho}\right)^{m-2}\frac{\partial}{\partial \nu}\frac{1}{r'^{m-2}} - \frac{\partial}{\partial \nu}\frac{1}{r^{m-2}}\right] d_\xi S_R. \qquad (6)$$

Formula (6) can be simplified. First of all, the outward normal and the radius coincide for a sphere, so that

$$\cos(\nu, x_k) = \frac{\xi_k}{R}$$

and
$$\frac{\partial}{\partial v} = \frac{\xi_k}{R}\frac{\partial}{\partial \xi_k}.$$

We also note the formula
$$\frac{\partial r}{\partial \xi_k} = \frac{\xi_k - x_k}{r}, \quad \frac{\partial r'}{\partial \xi_k} = \frac{\xi_k - x'_k}{r'};$$

where x_k and x'_k are coordinates of the points x and x' respectively.

The second term under the integral sign in (6) is easily evaluated:
$$\frac{\partial}{\partial v}\frac{1}{r^{m-2}} = -(m-2)\frac{\xi_k}{R}\frac{1}{r^{m-1}}\frac{\partial r}{\partial \xi_k}$$
$$= -\frac{m-2}{r^m R}\xi_k(\xi_k - x_k) = -\frac{m-2}{r^m R}(R^2 - \xi_k x_k). \tag{7}$$

Similarly,
$$\frac{\partial}{\partial v}\frac{1}{r'^{m-2}} = -\frac{m-2}{r'^m R}(R^2 - \xi_k x'_k).$$

We multiply this expression by $(R/\rho)^{m-2}$; using the relation $R/\rho r' = 1/r$, which was obtained earlier, we find
$$\left(\frac{R}{\rho}\right)^{m-2}\frac{\partial}{\partial v}\frac{1}{r'^{m-2}} = -\frac{m-2}{r^m R}\left(\rho^2 - \xi_k x'_k \frac{\rho^2}{R^2}\right).$$

The points x and x' lie on the same ray, passing through the origin; therefore
$$x_k = x'_k \frac{|x|}{|x'|} = x'_k \frac{\rho^2}{\rho \rho'} = x'_k \frac{\rho^2}{R^2}$$

and consequently,
$$\left(\frac{R}{\rho}\right)^{m-2}\frac{\partial}{\partial v}\frac{1}{r'^{m-2}} = -\frac{m-2}{r^m R}(\rho^2 - \xi_k x_k). \tag{8}$$

Substituting expressions (7) and (8) into the integral (6), we finally obtain
$$u(x) = \frac{1}{|S_1|}\int_{S_R} \phi(\xi)\frac{R^2 - \rho^2}{Rr^m}\, d_\xi S_R. \tag{9}$$

Formula (9) is called *Poisson's formula*, and the expression
$$\frac{R^2 - \rho^2}{Rr^m}, \quad \rho \leq R,$$

is called *Poisson's kernel*.

The foregoing arguments imply that Poisson's formula is always valid for harmonic functions of the class $C^{(2)}(\bar{\Sigma}_R)$.

We note some properties of the Poisson kernel.

1. Poisson's kernel is non-negative. When $\rho = R$ it is equal to zero everywhere except at the point $x = \xi$ in the neighbourhood of which it is unbounded.

2. If the point x varies inside the sphere, then the Poisson kernel is a harmonic function of x.

We shall prove this. If the point x lies inside the sphere, then $r \neq 0$ and Poisson's kernel has continuous derivatives of all orders. It remains to show that it satisfies the homogeneous Laplace equation. By Leibnitz's theorem,

$$\frac{\partial^2}{\partial x_k^2} \frac{R^2-\rho^2}{r^m} = \frac{1}{r^m} \frac{\partial^2(R^2-\rho^2)}{\partial x_k^2}$$
$$+ 2 \frac{\partial(R^2-\rho^2)}{\partial x_k} \frac{\partial}{\partial x_k}\left(\frac{1}{r^m}\right) + (R^2-\rho^2) \frac{\partial^2}{\partial x_k^2}\left(\frac{1}{r^m}\right).$$

Noting that

$$\frac{\partial \rho}{\partial x_k} = \frac{x_k}{\rho}, \qquad \frac{\partial r}{\partial x_k} = \frac{x_k - \xi_k}{r},$$

and summing with respect to k, we obtain

$$\Delta \frac{R^2-\rho^2}{r^m} = \frac{2m}{r^m} \left[-1 + \frac{1}{r^2}(R^2+\rho^2 - 2x_k\xi_k) \right],$$

which is equal to zero, since

$$r^2 = (\xi - x, \xi - x) = R^2 + \rho^2 - 2(\xi, x) = R^2 + \rho^2 - 2x_k\xi_k.$$

3. The formula

$$\frac{1}{|S_1|} \int_{S_R} \frac{R^2-\rho^2}{Rr^m} d_\xi S_R \equiv 1, \qquad \rho < R \tag{10}$$

holds.

We can certainly find a function harmonic in the sphere and taking the value 1 on the boundary. By virtue of the uniqueness theorem, the solution of this Dirichlet problem will be equal to 1 everywhere. Obviously, $1 \in C^{(2)}(\bar{\Sigma}_R)$, and obeys Poisson's formula, which here coincides with equation (10).

We shall now prove that *if the function $\phi(x)$ is continuous on the sphere*

S_R, then Poisson's formula gives a harmonic function in Σ_R, which at any point x_0 of the sphere S_R has the limiting value $\phi(x_0)$.

Let $u(x)$ be a function of the point x, defined inside the sphere Σ_R by Poisson's formula (9). This function is obviously continuous, and has derivatives of all orders, inside the sphere. It is easy to see that it is harmonic:

$$\Delta u = \frac{1}{|S_1|} \int_{S_R} \phi(\xi) \Delta_x \frac{R^2-\rho^2}{Rr^m} d_\xi S_R = 0.$$

Let the point x tend, from inside the sphere S_R, to a point x_0 on this sphere. We multiply equation (10) by $\phi(x_0)$, and then subtract this from formula (9):

$$u(x)-\phi(x_0) = \frac{1}{|S_1|} \int_{S_R} [\phi(\xi)-\phi(x_0)] \frac{R^2-\rho^2}{Rr^m} d_\xi S_R. \qquad (11)$$

The function $\phi(x)$ is continuous on the sphere S_R; we choose on S_R a spherical neighbourhood σ of the point x_0, small enough so that

$$|\phi(\xi)-\phi(x_0)| < \tfrac{1}{2}\varepsilon, \qquad \forall \xi \in \sigma,$$

where ε is an arbitrarily chosen positive number. We observe that in $S_R \setminus \sigma$

$$|\xi-x_0| \geq \delta,$$

where δ is the radius of the neighbourhood σ.

We evaluate the difference $u(x)-\phi(x_0)$, for which the integral (11) divides into two: over σ and over $S_R \setminus \sigma$

$$u(x)-\phi(x_0) = \frac{1}{|S_R|} \int_\sigma \frac{R^2-\rho^2}{Rr^m} [\phi(\xi)-\phi(x_0)] d_\xi S_R$$
$$+ \frac{1}{|S_1|} \int_{S_R \setminus \sigma} \frac{R^2-\rho^2}{Rr^m} [\phi(\xi)-\phi(x_0)] d_\xi S_R.$$

For the first integral we have

$$\frac{1}{|S_1|} \left| \int_\sigma \frac{R^2-\rho^2}{Rr^m} [\phi(\xi)-\phi(x_0)] d_\xi S_R \right| < \frac{\varepsilon}{2|S_1|} \int_\sigma \frac{R^2-\rho^2}{Rr^m} d_\xi S_R$$
$$< \frac{\varepsilon}{2|S_1|} \int_{S_R} \frac{R^2-\rho^2}{Rr^m} d_\xi S_R = \tfrac{1}{2}\varepsilon.$$

We have obtained a result for the first integral which is independent of the position of the point x. The second integral can be made small owing to the closeness of the points x and x_0. We take these points sufficiently close together so that the inequality $|x-x_0| < \frac{1}{2}\delta$ is satisfied. Then

$$r = |\xi-x| = |(\xi-x_0)+(x_0-x)| \geq |\xi-x_0|-|x_0-x| \geq \frac{1}{2}\delta,$$

from which

$$\frac{1}{r} < \frac{2}{\delta}.$$

Now

$$\frac{R^2-\rho^2}{Rr^m} = \frac{(R+\rho)(R-\rho)}{Rr^m} < \frac{2^{m+1}(R-\rho)}{\delta^m}.$$

The function ϕ is continuous on a compact set, and therefore bounded. Let $|\phi(\xi)| \leq M = \text{const.}$; then $|\phi(\xi)-\phi(x_0)| \leq 2M$. We now have

$$|u(x)-\phi(x_0)| < \tfrac{1}{2}\varepsilon + \frac{2^{m+2}M(R-\rho)}{\delta^m} \frac{1}{|S_1|}\int_{S_R\setminus\sigma} dS_R$$
$$< \tfrac{1}{2}\varepsilon + \frac{2^{m+2}MR^{m-1}(R-\rho)}{\delta^m}.$$

Let $h > 0$ be a number sufficiently small that

$$\frac{2^{m+2}MR^{m-1}h}{\delta^m} < \tfrac{1}{2}\varepsilon.$$

Then, if $|x_0-x| < h$, it follows that $R-\rho = |x_0|-|x| \leq |x_0-x| < h$, and $|u(x)-\phi(x_0)| < \varepsilon$. From this we have

$$\lim_{x \to x_0} u(x) = \phi(x_0), \qquad \forall x_0 \in S_R. \tag{12}$$

The function $u(x)$ is defined in the open sphere of Poisson's formula (9), and we extend its definition on to the surface S_R by putting $u(x) = \phi(x)$, $x \in S_R$. The function extended in this way is harmonic inside the sphere, continuous in the closed sphere (because of formula (12)), and satisfies the boundary condition (4). Thus we have a solution to the Dirichlet problem for the sphere.

Formula (2), and, in fact, the whole of the proof, require that $m > 2$. But Poisson's formula is equally valid when $m = 2$. In the latter case it is possible to derive a formula starting from the integral representation (4.2). An alternative derivation of Poisson's formula for the case $m = 2$ will be given in § 1, Chapter 13.

§ 4. Liouville's Theorem

THEOREM 12.4.1 (LIOUVILLE'S THEOREM). *A function which is harmonic in any finite domain, and which is bounded above or below, is a constant.*

If the function $u(x)$ is harmonic and $u(x) \leq M$, where $M = $ const., then $-u(x)$ is also harmonic, and $-u(x) \geq -M$. Consequently, it is sufficient to consider the case when the harmonic function is bounded below: $u(x) \geq \mu = $ const. We can assume that $\mu > 0$; if not then we can simply add to $u(x)$ a sufficiently large positive constant.

We fix an arbitrary point x. Construct a sphere with centre at the origin, and with radius R sufficiently large that the point x falls inside the sphere. The given function $u(x)$, being harmonic in any finite domain, is also harmonic in the sphere, and therefore obeys Poisson's formula in it:

$$u(x) = \frac{1}{|S_1|} \int_{S_R} \frac{R^2 - \rho^2}{Rr^m} u(\xi) \, d_\xi S_R,$$

where S_R is the boundary of the sphere.

It is easy to see that $R - \rho \leq r \leq R + \rho$, and since the function $u(x)$ is positive we obtain the following bounds on it:

$$\frac{R-\rho}{R(R+\rho)^{m-1}} \frac{1}{|S_1|} \int_{S_R} u(\xi) \, dS_R \leq u(x)$$

$$\leq \frac{R+\rho}{R(R-\rho)^{m-1}} \frac{1}{|S_1|} \int_{S_R} u(\xi) \, dS_R. \quad (1)$$

By the Mean Value Theorem,

$$u(0) = \frac{1}{|S_1| R^{m-1}} \int_{S_R} u(\xi) \, dS_R,$$

and inequality (1) takes the form

$$\frac{(R-\rho) R^{m-2}}{(R+\rho)^{m-1}} u(0) \leq u(x) \leq \frac{(R+\rho) R^{m-2}}{(R-\rho)^{m-1}} u(0).$$

Letting R tend to infinity, we arrive at the inequality

$$u(0) \leq u(x) \leq u(0).$$

Hence $u(x) = u(0)$, and the theorem is proven.

§ 5. Dirichlet Problem for the Exterior of a Sphere

Let Ω be the exterior of a sphere of radius R having the boundary S_R. It is required to find a harmonic function in Ω which satisfies the boundary condition

$$u|_{S_R} = \phi(x). \tag{1}$$

We shall prove that the solution to this problem is given by *Poisson's formula*

$$u(x) = \frac{1}{|S_1|} \int_{S_R} \frac{\rho^2 - R^2}{Rr^m} \phi(\xi) \, d_\xi S_R, \qquad \rho > R, \tag{2}$$

where, as in § 3, $r = |\xi - x|$ and $\rho = |x|$.

As in § 3, it can be shown that the function $u(x)$, defined by formula (2), has continuous derivatives of all orders and satisfies Laplace's equation outside the sphere S_R. We examine the behaviour of this function at infinity. Obviously, $r \geq \rho - R$. Hence

$$|u(x)| \leq c \frac{\rho + R}{(\rho - R)^{m-1}}; \qquad c = \frac{1}{|S_1|R} \int_{S_R} |\phi(\xi)| \, dS_R.$$

We are interested in large values of ρ, and we shall therefore suppose that $\rho > 2R$. Then $R < \tfrac{1}{2}\rho$, and $\rho - R > \tfrac{1}{2}\rho$. Now

$$|u(x)| < \frac{2^m c}{\rho^{m-2}},$$

and so the function $u(x)$ is harmonic outside the sphere.

It remains to verify the limit statement

$$\lim_{x \to x_0} u(x) = \phi(x_0), \qquad \forall x_0 \in S_R. \tag{3}$$

To do this we evaluate the integral (2), taking $\phi(\xi) \equiv 1$.

Introduce a point x', the image of x with respect to the sphere S_R. We have (putting $\rho = |x|, \rho' = |x'|, r' = |\xi - x'|$)

$$\rho^2 = \frac{R^4}{\rho'^2}, \qquad \frac{1}{r} = \frac{1}{r'} \cdot \frac{R}{\rho},$$

and the Poisson kernel can be represented in the form

$$\frac{\rho^2 - R^2}{Rr^m} = \frac{R^{m-2}}{\rho^{m-2}} \cdot \frac{R^2 - \rho'^2}{Rr'^m}.$$

The point x' lies inside the sphere S_R, and by formula (3.10),

$$\frac{1}{|S_1|}\int_{S_R} \frac{\rho^2-R^2}{Rr^m} dS_R = \frac{R^{m-2}}{\rho^{m-2}} \frac{1}{|S_1|}\int_{S_R} \frac{R^2-\rho'^2}{Rr'^m} dS_R = \frac{R^{m-2}}{\rho^{m-2}}. \qquad (4)$$

Multiply equation (4) by $\phi(x_0)$, and subtract from Poisson's formula (2)

$$u(x) - \frac{R^{m-2}}{\rho^{m-2}}\phi(x_0) = \frac{1}{|S_1|}\int_{S_R} \frac{\rho^2-R^2}{Rr^m} [\phi(\xi)-\phi(x_0)] dS_R.$$

Repeating exactly the arguments of § 3, we obtain

$$\lim_{x \to x_0} \left[u(x) - \frac{R^{m-2}}{\rho^{m-2}}\phi(x_0) \right] = 0.$$

Hence

$$|u(x)-\phi(x_0)| \leq \left| u(x) - \frac{R^{m-2}}{\rho^{m-2}}\phi(x_0) \right|$$
$$+ |\phi(x_0)| \left(1 - \frac{R^{m-2}}{\rho^{m-2}}\right) \underset{x \to x_0}{\to} 0,$$

and the relation (3) is proved.

§ 6. Derivatives of Harmonic Functions at Infinity

THEOREM 12.6.1. *Let $u(x)$ be a harmonic function in the infinite domain Ω having a finite boundary Γ, and let $D^k u$ be any of the k-th derivatives of the function u. Then for sufficiently large $|x|$ the inequality*

$$|D^k u| \leq \frac{C_k}{|x|^{m-2+k}}, \qquad (1)$$

holds, where C_k is independent of x.

Proof. We shall prove the theorem for $k = 1$; the general case can be treated analogously.

Since the boundary Γ is finite, it is possible to construct a sphere S_R of radius R sufficiently large that Γ lies wholly inside the sphere. For a function $u(x)$ exterior to the sphere S_R, Poisson's formula holds:

$$u(x) = \frac{1}{|S_1|}\int_{S_R} \frac{\rho^2-R^2}{Rr^m} u(\xi) d_\xi S_R, \qquad \rho = |x|.$$

Let us find any one of the first derivatives, for example, the derivative with respect to x_1:

$$\frac{\partial u}{\partial x_1} = \frac{1}{|S_1|} \int_{S_R} u(\xi) \frac{\partial}{\partial x_1} \frac{\rho^2 - R^2}{Rr^m} d_\xi S_R. \qquad (2)$$

We evaluate the derivative occurring in the integrand:

$$\frac{\partial}{\partial x_1} \frac{\rho^2 - R^2}{Rr^m} = 2\rho \frac{\partial \rho}{\partial x_1} \cdot \frac{1}{Rr^m} - \frac{m(\rho^2 - R^2)}{Rr^{m+1}} \cdot \frac{\partial r}{\partial x_1}$$

$$= \frac{2x_1}{Rr^m} - \frac{m(\rho^2 - R^2)}{Rr^{m+1}} \cdot \frac{x_1 - \xi_1}{r}.$$

Let ρ be sufficiently large, for example, let $\rho > 2R$. Then $r > \rho - R > \tfrac{1}{2}\rho$, and we obtain the following bound on the kernel of integral (2):

$$\left| \frac{\partial}{\partial x_1} \frac{\rho^2 - R^2}{Rr^m} \right| \leq \frac{2^{m+1}}{R\rho^{m-1}} + \frac{2^{m+1}m}{R\rho^{m-1}} = \frac{C_1}{\rho^{m-1}}.$$

Thus the theorem is proved.

Note that when $m = 2$ it is possible to obtain a bound on the derivatives in which the order of the denominator is one greater than in formula (1).

§ 7. Uniqueness Theorem for the Exterior Neumann Problem

THEOREM 12.7.1. *In the case $m > 2$ the exterior Neumann problem for Laplace's equation has not more than one solution.*

Proof. We prove the theorem with the assumption that the boundary of the domain under consideration is regular.

Let the infinite domain Ω have a regular boundary Γ, and let the Neumann problem have two solutions in this domain; their difference satisfies the relations

$$\Delta v = 0, \qquad x \in \Omega; \qquad v(x) = O(|x|^{-m+2}), \qquad x \to \infty; \qquad (1)$$

$$\left. \frac{\partial v}{\partial \nu} \right|_\Gamma = 0. \qquad (2)$$

Construct a surface Γ_h parallel to Γ and lying inside Ω (Fig. 23). Let S_R be a sphere having its centre at the origin, and radius R sufficiently large that the whole of the surface Γ_h is located inside S_R. Denote by $\Omega_R^{(h)}$ the domain bounded by the surfaces Γ_h and S_R, and by Ω_R the domain bounded by the surfaces Γ and S_R.

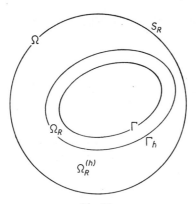

Fig. 23.

The domain $\Omega_R^{(h)}$ is finite, so that $v \in C^{(2)}(\bar{\Omega}_R^{(h)})$; consequently it is possible to apply Green's formula (6.8), Chapter 10:

$$\int_{\Omega^{(h)}_R} v\Delta v\,dx = -\int_{\Omega^{(h)}_R} \sum_{k=1}^{m}\left(\frac{\partial v}{\partial x_k}\right)^2 dx + \int_{\Gamma_h} v\frac{\partial v}{\partial \nu}\,d\Gamma_h + \int_{S_R} v\frac{\partial v}{\partial \nu}\,dS_R.$$

Taking the limit as $h \to 0$, and making use of equations (1) and (2), we arrive at the equation

$$\int_{\Omega_R} \sum_{k=1}^{m}\left(\frac{\partial v}{\partial x_k}\right)^2 dx = \int_{S_R} v\frac{\partial v}{\partial \nu}\,dS_R. \qquad (3)$$

Let $R \to \infty$. The expressions (1) show that v is harmonic in Ω, and for sufficiently large R,

$$|v(x)|_{|x|=R} \leq \frac{C}{R^{m-2}}, \qquad C = \text{const.}$$

Furthermore, by the inequality (6.1),

$$\left|\frac{\partial v}{\partial \nu}\right|_{|x|=R} \leq \frac{C_1}{R^{m-1}}, \qquad C_1 = \text{const.}$$

and for the right-hand side of formula (3) we obtain the bound

$$\left|\int_{S_R} v\frac{\partial v}{\partial \nu}\,dS_R\right| \leq \frac{CC_1|S_R|}{R^{2m-3}} = \frac{CC_1|S_1|}{R^{m-2}}.$$

By our assumption, $m > 2$; therefore

$$\int_{\Omega} \sum_{k=1}^{m}\left(\frac{\partial v}{\partial x_k}\right)^2 dx = \lim_{R\to\infty}\int_{S_R} v\frac{\partial v}{\partial \nu}\,dS_R \leq \lim_{R\to\infty}\frac{CC_1|S_1|}{R^{m-2}} = 0;$$

hence it follows in an obvious way that

$$\frac{\partial v}{\partial x_k} \equiv 0, \quad v \equiv \text{const.}$$

Recalling that $v(x)$ tends to zero at infinity, we obtain $v(x) \equiv 0$. The theorem is proved.

Let $m = 2$. In § 6 we observed that in this special case the first derivatives decrease more quickly than in the general case. Making use of this fact, we easily obtain the equation

$$v \equiv \text{const.}$$

At the same time the condition $|v(x)| \leq C/|x|^{m-2}$ only gives boundedness at infinity. Hence it follows that for $m = 2$ the uniqueness of the exterior Neumann problem only holds to within a constant term.

Chapter 13. Elementary Solutions of the Dirichlet and Neumann Problems

§ 1. Dirichlet and Neumann Problems for a Circle

In this section and the next we shall consider the homogeneous Laplace equation in a plane. In contrast with our practice in other parts of this book, we shall here denote the cartesian coordinates of a point by x and y, or by ξ and η; the points themselves will be denoted by the symbols (x, y) or (ξ, η) respectively. We shall also make use of the notation $z = x+iy$, $\zeta = \xi+i\eta$, where $i = \sqrt{-1}$.

The connection between harmonic and analytic functions is well-known: if $f(z) = u(x, y)+iv(x, y)$ is a function holomorphic in some domain, then its real part $u(x, y)$ and imaginary part $v(x, y)$ are both harmonic in this domain. On the other hand, if the real function is harmonic in a *simply-connected* domain, then it is possible to find a function $v(x, y)$ harmonic in the same domain (called the *conjugate* of the function $u(x, y)$) such that the sum $u(x, y)+iv(x, y)$ is a holomorphic function of z in the domain. If the domain is multiply-connected, then in general the above-mentioned sum will be multiple-valued.

If n is a positive integer, then the function z^n is holomorphic in any finite domain; if n is a negative integer, then the function is holomorphic in any domain not containing the origin. Hence it follows that the polynomials

$$\text{Re}\,(z^n), \qquad \text{Im}\,(z^n), \qquad n \geq 0 \tag{1}$$

are harmonic in any finite domain, and the rational functions

$$\text{Re}\,(z^{-n}), \qquad \text{Im}\,(z^{-n}), \qquad n \geq 1 \tag{2}$$

are harmonic in any domain not containing the origin.

Introduce polar coordinates ρ and θ, with pole at the origin. Then $z = \rho e^{i\theta}$, and the functions (1) and (2) take the respective forms

$$\rho^n \cos n\theta, \qquad \rho^n \sin n\theta, \qquad n \geq 0 \tag{1'}$$

and
$$\frac{\cos n\theta}{\rho^n}, \quad -\frac{\sin n\theta}{\rho^n}, \quad n \geq 1. \tag{2'}$$

1. We pose the Dirichlet problem for a circle. It is required to find a function $u(x, y)$ which is harmonic in the circle $|z| < R$, and which is equal to a given continuous function $\phi(\theta)$ on the circumference of this circle:

$$u|_{\rho=R} = \phi(\theta). \tag{3}$$

Suppose that the function $\phi(\theta)$ can be expanded in a Fourier series which converges for all θ. Let

$$\phi(\theta) = a_0 + \sum_{n=1}^{\infty} (a_n \cos n\theta + b_n \sin n\theta). \tag{4}$$

It is easy to write down a formal solution to our problem:

$$u(x, y) = a_0 + \sum_{n=1}^{\infty} \frac{\rho^n}{R^n} (a_n \cos n\theta + b_n \sin n\theta). \tag{5}$$

The series (5) converges in the circle $|z| < R$, and its sum in this circle is harmonic (prove this!). If in this series we suppose that the limit $\rho \to R$ can be taken term-by-term, then

$$u(x, y)|_{\rho=R} = a_0 + \sum_{n=1}^{\infty} (a_n \cos n\theta + b_n \sin n\theta) = \phi(\theta)$$

and the expression (5) does in fact solve the problem.

It is well-known from Abel's theorem that a limiting process of this kind is permissible for those values of θ for which the series (4) converges. This series converges to $\phi(\theta)$ for all θ if, for example, $\phi(\theta)$ is periodic with period 2π, is absolutely continuous, and has a derivative $\phi'(\theta) \in L_2(0, 2\pi)$. Hence for the family of functions having these properties the series (5) is in fact a solution of the Dirichlet problem.

We sum the series (5). From the well-known formula for Fourier coefficients:

$$a_0 = \frac{1}{2\pi} \int_0^{2\pi} \phi(\omega) \, d\omega, \quad a_n = \frac{1}{\pi} \int_0^{2\pi} \phi(\omega) \cos n\omega \, d\omega,$$

$$b_n = \frac{1}{\pi} \int_0^{2\pi} \phi(\omega) \sin n\omega \, d\omega.$$

Substituting these into the series (5), we obtain

$$u(x, y) = \frac{1}{2\pi} \int_0^{2\pi} \phi(\omega) \left[1 + 2 \sum_{n=1}^{\infty} \frac{\rho^n}{R^n} \cos n(\omega - \theta) \right] d\omega.$$

Here we have interchanged the order of summation and integration; it is not difficult to show that this is permissible so long as $\rho < R$.

We have $\rho e^{i\theta} = z$, and we also put $Re^{i\omega} = \zeta$. Then

$$\sum_{n=1}^{\infty} \frac{\rho^n}{R^n} \cos n(\omega - \theta) = \operatorname{Re} \sum_{n=1}^{\infty} \frac{z^n}{\zeta^n} = \operatorname{Re} \frac{z}{\zeta - z}$$

and

$$u(x, y) = \frac{1}{2\pi} \int_0^{2\pi} \phi(\omega) \operatorname{Re} \frac{\zeta + z}{\zeta - z} d\omega = \operatorname{Re} \frac{1}{2\pi i} \int_{|\zeta|=R} \phi(\omega) \frac{\zeta + z}{\zeta - z} \frac{d\zeta}{\zeta}.$$

This result is the well-known *Schwartz's formula*. Now

$$\operatorname{Re} \frac{\zeta + z}{\zeta - z} = \operatorname{Re} \frac{(\zeta + z)(\bar{\zeta} - \bar{z})}{|\zeta - z|^2} = \frac{R^2 - \rho^2}{r^2}; \quad r = |\zeta - z|,$$

and so finally

$$u(x, y) = \frac{1}{2\pi} \int_0^{2\pi} \frac{R^2 - \rho^2}{r^2} \phi(\omega) d\omega. \tag{6}$$

This is Poisson's formula for a circle. It is easy to see that $r^2 = R^2 + \rho^2 - 2R\rho \cos(\omega - \theta)$, and we arrive at a more usual representation of Poisson's formula:

$$u(x, y) = \frac{1}{2\pi} \int_0^{2\pi} \frac{R^2 - \rho^2}{R^2 + \rho^2 - 2R\rho \cos(\omega - \theta)} \phi(\omega) d\omega.$$

2. The solution to the Dirichlet problem for the region exterior to the circle $|z| > R$ and with the same boundary conditions (3) is given by the series

$$u(x, y) = a_0 + \sum_{n=1}^{\infty} \frac{R^n}{\rho^n} (a_n \cos n\theta + b_n \sin n\theta). \tag{5'}$$

This series can be summed, and the result is Poisson's formula for the exterior of a circle:

$$u(x, y) = \frac{1}{2\pi} \int_0^{2\pi} \frac{\rho^2 - R^2}{r^2} \phi(\omega) d\omega$$

$$= \frac{1}{2\pi} \int_0^{2\pi} \frac{\rho^2 - R^2}{R^2 + \rho^2 - 2R\rho \cos(\omega - \theta)} \phi(\omega) d\omega. \tag{6'}$$

3. The Neumann problem for the circle $|z| < R$, with boundary conditions

$$\left.\frac{\partial u}{\partial v}\right|_{\rho=R} = \psi(\theta) \tag{7}$$

is solved in the following way. Expand $\psi(\theta)$ in a Fourier series:

$$\psi(\theta) = \alpha_0 + \sum_{n=1}^{\infty} (\alpha_n \cos n\theta + \beta_n \sin n\theta).$$

If the solution $u(x, y)$ exists, then because of formula (2.7), Chapter 12, we have

$$\alpha_0 = \frac{1}{2\pi}\int_0^{2\pi} \psi(\theta)\,d\theta = \frac{1}{2\pi R}\int_{\rho=R} \psi(\theta)\,d\Gamma = \frac{1}{2\pi R}\int_{\rho=R} \frac{\partial u}{\partial v}\,d\Gamma = 0; \tag{8}$$

here $d\Gamma$ denotes an element of length of the circumference $|z| = R$: $d\Gamma = R\,d\theta$. Now

$$\psi(\theta) = \sum_{n=1}^{\infty} (\alpha_n \cos n\theta + \beta_n \sin n\theta). \tag{9}$$

Using the fact that for the circle $\rho = |z| = R$

$$\frac{\partial}{\partial v} = \left.\frac{\partial}{\partial \rho}\right|_{\rho=R},$$

we can write down a formal solution to the Neumann problem in the following form:

$$u(x, y) = C + \sum_{n=1}^{\infty} \frac{\rho^n}{nR^{n-1}}(\alpha_n \cos n\theta + \beta_n \sin n\theta); \tag{10}$$

where C is an arbitrary constant. The series (10) will converge and will permit term-by-term differentiation in the closed circle $|z| \leq R$ if, for example, the function $\psi(\theta)$ satisfies the same conditions as were earlier imposed on the function $\phi(\theta)$; in addition the function $\psi(\theta)$ must satisfy equation (8). Under these conditions, the series (10) gives a solution to the Neumann problem for the circle $|z| < R$.

4. If $\psi(\theta)$ satisfies all the conditions just enumerated, then the Neumann problem for the exterior of the circle $|z| > R$, with the same boundary condition (7), is solved by the formula

$$u(x, y) = C - \sum_{n=1}^{\infty} \frac{R^{n+1}}{n\rho^n}(\alpha_n \cos n\theta + \beta_n \sin n\theta), \tag{11}$$

where, as before, C is an arbitrary constant.

5. It is not difficult to sum the series (10). From the formulae for Fourier coefficients, we have

$$\alpha_n = \frac{1}{\pi}\int_0^{2\pi} \psi(\omega)\cos n\omega\, d\omega, \qquad \beta_n = \frac{1}{\pi}\int_0^{2\pi} \psi(\omega)\sin n\omega\, d\omega.$$

Substituting this into (10), and changing the order of integration and summation (the validity of which is easily established), we obtain

$$u(x,y) = C + \frac{1}{\pi}\int_0^{2\pi} \psi(\omega) \sum_{n=1}^{\infty} \frac{\rho^n}{nR^{n-1}} \cos n(\omega-\theta)\, d\omega$$

$$= C + \frac{R}{\pi}\int_0^{2\pi} \psi(\omega) \sum_{n=1}^{\infty} \frac{1}{n}\operatorname{Re}\left\{\frac{\rho^n}{R^n} e^{in(\theta-\omega)}\right\} d\omega$$

$$= C + \frac{R}{\pi}\int_0^{2\pi} \psi(\omega) \operatorname{Re} \sum_{n=1}^{\infty} \frac{z^n}{n\zeta^n}\, d\omega; \qquad \zeta = Re^{i\omega}.$$

Recalling that

$$\sum_{n=1}^{\infty} \frac{t^n}{n} = \ln\frac{1}{1-t}, \qquad |t| < 1,$$

and that $\operatorname{Re}\ln\tau = \ln|\tau|$, we find

$$u(x,y) = C + \frac{R}{\pi}\int_0^{2\pi} \ln\frac{R}{r}\psi(\omega)\, d\omega = C + \frac{R}{\pi}\int_0^{2\pi} \ln\frac{1}{\rho}\psi(\omega)\, d\omega. \qquad (12)$$

Equation (12) is known as *Dini's formula*.

In a similar manner we can also sum series (11).

§ 2. Dirichlet Problem for a Circular Annulus

Let the annulus be defined by the inequality $R_0 < \rho < R_1$; $\rho = |z|$. Denote by $\phi_0(\theta)$ and $\phi_1(\theta)$ the values of the boundary functions on the circumferences $\rho = R_0$ and $\rho = R_1$ respectively. Then the boundary conditions for the required harmonic function $u(x,y)$ can be written in the form

$$u|_{\rho=R_0} = \phi_0(\theta), \qquad u|_{\rho=R_1} = \phi_1(\theta). \qquad (1)$$

This harmonic function $u(x,y)$ is the real part of some analytic function $f(z)$:

$$u(x,y) = \operatorname{Re} f(z).$$

The annulus is a doubly-connected domain, and the function $f(z)$ is in general not holomorphic. It can be shown that

$$f(z) = C \ln z + \sum_{n=-\infty}^{+\infty} C_n z^n,$$

where C is a real constant. Putting

$$C_n = A_n - iB_n,$$

we see that we can look for the function $u(x, y)$ in the form

$$u(x, y) = C \ln \rho + A_0 + \sum_{n=1}^{\infty} \{(A_n \rho^n + A_{-n} \rho^{-n}) \cos n\theta$$
$$+ (B_n \rho^n - B_{-n} \rho^{-n}) \sin n\theta\}. \tag{2}$$

The functions $\phi_0(\theta)$ and $\phi_1(\theta)$ can be expanded in Fourier series. Let

$$\phi_0(\theta) = a_0 + \sum_{n=1}^{\infty} (a_n \cos n\theta + b_n \sin n\theta),$$
$$\phi_1(\theta) = \alpha_0 + \sum_{n=1}^{\infty} (\alpha_n \cos n\theta + \beta_n \sin n\theta). \tag{3}$$

Substitute the expansions (2) and (3) into the boundary condition (1), and equate coefficients of terms $\cos n\theta$ and $\sin n\theta$. This leads us to a sequence of second-order algebraic systems:

$$C \ln R_0 + A_0 = a_0, \qquad C \ln R_1 + A_0 = \alpha_0;$$
$$A_n R_0^n + A_{-n} R_0^{-n} = a_n, \qquad A_n R_1^n + A_{-n} R_1^{-n} = \alpha_n;$$
$$B_n R_0^n - B_{-n} R_0^{-n} = b_n, \qquad B_n R_1^n - B_{-n} R_1^{-n} = \beta_n.$$

The determinants of these systems are different from zero, and so all the coefficients of expansion (2) can be evaluated. Thus a formal solution to the problem can be constructed. The substantiation of this is left to the reader.

The same method can be used to solve the Neumann problem for the circular annulus.

§ 3. Application of Conformal Mapping

Let the holomorphic function $z = z(\zeta) = x(\xi, \eta) + iy(\xi, \eta)$ conformally map a domain D of the ζ-plane into a domain Ω of the z-plane. Further, let $u(x, y)$ be a harmonic function in Ω. Then the function $\tilde{u}(\xi, \eta) = u(x(\xi, \eta), y(\xi, \eta))$ is harmonic in D.

§3] APPLICATION OF CONFORMAL MAPPING

To prove this, we evaluate the quantity

$$\Delta_\zeta \tilde{u} = \frac{\partial^2 \tilde{u}}{\partial \xi^2} + \frac{\partial^2 \tilde{u}}{\partial \eta^2}.$$

We have

$$\frac{\partial \tilde{u}}{\partial \xi} = \frac{\partial u}{\partial x}\frac{\partial x}{\partial \xi} + \frac{\partial u}{\partial y}\frac{\partial y}{\partial \xi}$$

and

$$\frac{\partial^2 \tilde{u}}{\partial \xi^2} = \frac{\partial^2 u}{\partial x^2}\left(\frac{\partial x}{\partial \xi}\right)^2 + 2\frac{\partial^2 u}{\partial x \partial y}\frac{\partial x}{\partial \xi}\frac{\partial y}{\partial \xi} + \frac{\partial^2 u}{\partial y^2}\left(\frac{\partial y}{\partial \xi}\right)^2 + \frac{\partial u}{\partial x}\frac{\partial^2 x}{\partial \xi^2} + \frac{\partial u}{\partial y}\frac{\partial^2 y}{\partial \xi^2}.$$

Similarly

$$\frac{\partial^2 \tilde{u}}{\partial \eta^2} = \frac{\partial^2 u}{\partial x^2}\left(\frac{\partial x}{\partial \eta}\right)^2 + 2\frac{\partial^2 u}{\partial x \partial y}\frac{\partial x}{\partial \eta}\frac{\partial y}{\partial \eta} + \frac{\partial^2 u}{\partial y^2}\left(\frac{\partial y}{\partial \eta}\right)^2 + \frac{\partial u}{\partial x}\frac{\partial^2 x}{\partial \eta^2} + \frac{\partial u}{\partial y}\frac{\partial^2 y}{\partial \eta^2}.$$

We add these last equations, noting in the process that $x(\xi, \eta)$ and $y(\xi, \eta)$ are respectively the real and imaginary parts of the holomorphic function $z(\zeta)$; therefore x and y are harmonic functions of ξ and η, related by the Cauchy-Riemann equations. That is, they satisfy

$$\frac{\partial x}{\partial \xi} = \frac{\partial y}{\partial \eta}, \qquad \frac{\partial x}{\partial \eta} = -\frac{\partial y}{\partial \xi},$$

$$\Delta_\zeta x = \Delta_\zeta y = 0.$$

Now

$$\Delta_\zeta \tilde{u} = \left[\left(\frac{\partial x}{\partial \xi}\right)^2 + \left(\frac{\partial y}{\partial \xi}\right)^2\right]\Delta_z u = |z'(\zeta)|^2 \Delta_z u;$$

where

$$\Delta_z u = \frac{\partial^2 u}{\partial x^2} + \frac{\partial^2 u}{\partial y^2}.$$

If $\Delta_z u = 0$, then obviously $\Delta_\zeta \tilde{u} = 0$, and our assertion is proved.

In short, we can say that a conformal mapping transforms one harmonic function into another. It can be shown that a conformal mapping also transforms a Dirichlet problem into another Dirichlet problem, and a Neumann problem into a Neumann problem.

Suppose that a Dirichlet problem is posed in the domain Ω;

$$\Delta u = 0, \qquad u|_\Gamma = \phi(s). \tag{1}$$

Here Γ is the contour of the domain Ω, and s a parameter defining the position of points on Γ.

Let the function $z(\zeta)$, which conformally maps the domain D into the domain Ω, be continuous in the closed domain * $\bar{D} = D \cup \Gamma_1$, where Γ_1 is the contour of the domain D. As we have seen, a conformal transformation does not change Laplace's equation, so that the transformed function $\tilde{u}$ satisfies the equation

$$\Delta_\zeta \tilde{u} = \frac{\partial^2 \tilde{u}}{\partial \xi^2} + \frac{\partial^2 \tilde{u}}{\partial \eta^2} = 0. \tag{2}$$

It is well-known that if the mapping is conformal, then the correspondence between the contours is one-to-one: the same parameter s can be used to define the position of a point on Γ_1, and, consequently, boundary condition (1) does not change under the conformal transformation. Thus we have

$$u|_{\Gamma_1} = \phi(s). \tag{3}$$

The equations (2) and (3) taken together show that the transformed function $\tilde{u}$ is a solution of the Dirichlet problem in the domain D.

We turn to the Neumann problem. It is well-known, for example, that the derivative $z'(\zeta)$ remains continuous on regular ** segments of the contour. As before, the transformed function satisfies equation (2); we now clarify what boundary condition it satisfies.

Let the function $u(x, y)$ satisfy the condition of the Neumann problem

$$\left.\frac{\partial u}{\partial v}\right|_\Gamma = \psi(s). \tag{4}$$

Denote by v_1 the normal to the contour Γ_1 at the point corresponding to a value s of the parameter. Under conformal mapping the contour transforms into a contour, and, consequently, tangential directions become tangential directions; since angles are preserved in this process, the direction of the normal v becomes the direction of the normal v_1. Let us elucidate this last assertion. Let z and ζ be points of the contours Γ and Γ_1, corresponding to each other under a conformal transformation. If in the domain Ω we draw a normal v, which intersects the contour Γ at the point z, then under conformal transformation this normal becomes a curve which is tangential at the point ζ to the normal v_1 (Fig. 24). On the normal v we take a point $z_1 = x_1 + iy_1$. Let the point which corresponds to this in the domain D be

---

* For this it is necessary and sufficient that the domain D be bounded by a Jordan curve.
** The definition of a regular boundary is given on page 258.

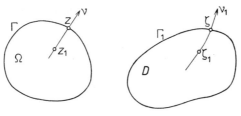

Fig. 24.

$\zeta_1 = \xi_1 + i\eta_1$. We write $h = |z-z_1|, k = |\zeta-\zeta_1|$. We have

$$\frac{\partial u}{\partial v} = \lim_{z_1 \to z} \frac{u(x,y)-u(x_1,y_1)}{h} = \lim_{\zeta_1 \to \zeta} \frac{\tilde{u}(\xi,\eta)-\tilde{u}(\xi_1,\eta_1)}{k}\frac{k}{h} = \frac{\partial \tilde{u}}{\partial v_1}\lim_{\zeta_1 \to \zeta}\frac{k}{h}.$$

Since the mapping is conformal,

$$\lim_{\zeta_1 \to \zeta} \frac{k}{h} = \frac{1}{|z'(\zeta)|},$$

and finally

$$\frac{\partial \tilde{u}}{\partial v_1} = \frac{\partial u}{\partial v}|z'(\zeta)|. \tag{5}$$

If the function $\tilde{u}$ satisfies the boundary condition (4) then the transformed function $\tilde{u}$ satisfies the boundary condition

$$\left.\frac{\partial \tilde{u}}{\partial v_1}\right|_{\Gamma_1} = \psi(s)|z'(\zeta)|, \tag{6}$$

i.e. the boundary condition for the Neumann problem again.

The solutions to the Dirichlet and Neumann problems for the circle and circular annulus are known – they have been constructed in the preceding sections. It follows from the foregoing discussion that if we know the conformal mapping of a given domain into a circle or a circular annulus, then we can write down the solutions to the Dirichlet and Neumann problems for this domain. We enumerate some of these domains.

1. A half-plane is transformed into a circle by a bilinear transformation.

2. The exterior of an ellipse is mapped on to a circle by a transformation of the form

$$z = a\zeta + \frac{b}{\zeta}, \tag{7}$$

where a, b are constants.

3. A polygon is mapped on to a circle with the aid of a Schwartz-Christoffel transformation.

4. The annular domain between two circumferences is mapped on to a circular annulus by means of a bilinear transformation.

5. The annular region between two confocal ellipses can be mapped on to a circular annulus with the aid of transformation (7).

A number of other such examples can be written down.

§ 4. Spherical Harmonics and their Properties

A full account of the theory and applications of spherical harmonics can be found in books by V. I. Smirnov [17] Vol. III, E. V. Hobson [3], and N. Ya. Vilenkin [1]. Here we only state the definition and (without proof) the simplest properties of spherical harmonics.

Take an m-dimensional Euclidean space of points having coordinates $x_1, x_2, \ldots, x_m$. Consider in this space homogeneous polynomials of degree n, satisfying Laplace's equation; here n is an arbitrary non-negative integer. These polynomials are obviously harmonic in any finite domain. It is not difficult to construct harmonic, homogeneous polynomials; we need only take a homogeneous polynomial of the required degree n with arbitrary coefficients, form its Laplace operator, and set the result equal to zero. This gives a number of relations between the coefficients; polynomials whose coefficients satisfy these relations will be harmonic.

As an example, consider the case of three independent variables x_1, x_2, x_3. Obviously any polynomial of degree zero or one is harmonic. A homogeneous polynomial of second degree has the general form

$$\sum_{j,k=1}^{3} a_{jk} x_j x_k, \qquad a_{jk} = a_{kj}.$$

Its Laplace operator is equal to $2(a_{11}+a_{22}+a_{33})$. Setting this equal to zero, we obtain

$$a_{33} = -(a_{11}+a_{22}).$$

This gives the general form of a second-degree harmonic polynomial with three independent variables:

$$a_{11}(x_1^2-x_3^2)+a_{22}(x_2^2-x_3^2)+2a_{11}x_1x_2+2a_{13}x_1x_3+2a_{23}x_2x_3.$$

Among other things this last formula shows that of the class of polynomials under discussion five are linearly independent. These are, for example, the polynomials

$$x_1^2-x_3^2, \quad x_1^2-x_2^2, \quad x_1 x_2, \quad x_1 x_3, \quad x_2 x_3.$$

There exist $2n+1$ linearly-independent, homogeneous, harmonic polynomials of degree n having three independent variables. In the general case of m independent variables, the number of linearly-independent, homogeneous, harmonic polynomials of degree n is equal to

$$(2n+m-2)\frac{(m+n-3)!}{(m-2)!n!}. \tag{1}$$

The quantity (1) will henceforth be denoted $k_{n,m}$.

In the case $m = 2, n > 0$, there exist only two linearly-independent, homogeneous, harmonic polynomials of degree n, and they are the polynomials $\mathrm{Re}(z^n)$ and $\mathrm{Im}(z^n)$, where $z = x_1 + ix_2$.

We transform from cartesian coordinates $x_1, x_2, \ldots, x_m$ to spherical coordinates $\rho, \vartheta_1, \vartheta_2, \ldots, \vartheta_{m-2}, \vartheta_{m-1}$, by means of the formulae

$$\begin{aligned}
x_1 &= \rho \cos \vartheta_1, \\
x_2 &= \rho \sin \vartheta_1 \cos \vartheta_2, \\
&\cdots \cdots \cdots \\
x_{m-1} &= \rho \sin \vartheta_1 \sin \vartheta_2 \cdots \sin \vartheta_{m-2} \cos \vartheta_{m-1}, \\
x_m &= \rho \sin \vartheta_1 \sin \vartheta_2 \cdots \sin \vartheta_{m-2} \sin \vartheta_{m-1}.
\end{aligned} \tag{2}$$

The ranges of variation of spherical coordinates are

$$0 \leq \rho < \infty; \quad 0 \leq \vartheta_k \leq \pi, \quad k \leq m-2; \quad 0 \leq \vartheta_{m-1} \leq 2\pi.$$

If $\rho = 1$, we obtain a point on the unit sphere; this point is completely defined by its angular coordinates $\vartheta_1, \vartheta_2, \ldots, \vartheta_{m-1}$. Conversely, prescribing the values of the angular coordinates completely defines a point on the unit sphere.

Let $P_{n,m}(x)$ be a homogeneous, harmonic, n-th degree polynomial of the variables $x_1, x_2, \ldots, x_m$. If we replace these coordinates by spherical coordinates, with the aid of formula (2), then the polynomial, being homogeneous of degree n, takes the form

$$P_{n,m}(x) = \rho^n Y_{n,m}(\theta). \tag{3}$$

Here θ denotes the point on the unit sphere with angular coordinates $\vartheta_1, \vartheta_2, \ldots, \vartheta_{m-1}$.

The function $Y_{n,m}(\theta)$ is called an *m-dimensional spherical harmonic of degree n*. In what follows the dimensionality m of the space will remain fixed, and we shall omit the word "*m*-dimensional" from the description of the spherical harmonic.

We now enumerate the most important properties of spherical harmonics. Some of these properties are obvious; the others require proof which, however, we shall not give here.

1. Spherical harmonics are polynomials of sines and cosines of the angular coordinates.

2. $Y_{0,m}(\theta) = \text{const.}$

3. Spherical harmonics of different orders are orthogonal over the unit sphere:

$$\int_{S_1} Y_{n,m}(\theta) Y_{n',m}(\theta) \, dS_1 = 0, \qquad n \neq n'. \tag{4}$$

4. If $n \neq 0$, there exist $k_{n,m}$ linearly-independent spherical harmonics of order n. We shall denote these functions by $Y_{n,m}^{(k)}(\theta)$, $k = 1, 2, \ldots, k_{n,m}$. To ensure uniqueness of the notation, we shall take $k_{0,m} = 1$, and shall write $Y_{0,m}^{(1)}(\theta)$ in place of $Y_{0,m}(\theta)$.

5. For any given n, it is possible to subject the functions $Y_{n,m}^{(k)}(\theta)$ to an orthogonalization process. If we suppose that the orthogonalization has been carried out, then the system of functions

$$Y_{n,m}^{(k)}(\theta); \; n = 0, 1, 2, \ldots, k = 1, 2, \ldots, k_{n,m}$$

is orthonormal over the unit sphere S_1:

$$\int_{S_1} Y_{n,m}^{(k)}(\theta) Y_{n',m}^{(k')}(\theta) \, dS_1 = \begin{cases} 0, & n \neq n' \text{ or } k \neq k', \\ 1, & n = n' \text{ and } k = k'. \end{cases} \tag{5}$$

6. The system of spherical harmonics (5) is complete in $L_2(S_1)$. Hence it follows that any function, defined almost everywhere and square-integrable on the unit sphere S_1, can be expanded in a series of spherical harmonics, and on the sphere S_1 this series will converge in the mean to the given function. If $f(\theta)$ is a given function, then its series expansion in spherical harmonics has the form

$$f(\theta) = \sum_{n=0}^{\infty} \sum_{k=1}^{k_{n,m}} a_n^{(k)} Y_{n,m}^{(k)}(\theta), \tag{6}$$

where

$$a_n^{(k)} = \int_{S_1} f(\theta) Y_{n,m}^{(k)}(\theta) \, dS_1. \tag{7}$$

7. If $m = 2$, then there exist two orthogonal spherical harmonics of order $n > 0$. These functions can be taken to be $\cos n\vartheta$ and $\sin n\vartheta$, for example, where ϑ is the polar angle in the plane.

8. The functions

$$\rho^n Y_{n,m}^{(k)}(\theta), \quad \frac{Y_{n,m}^{(k)}(\theta)}{\rho^{m+n-2}}, \quad n \geq 0 \tag{8}$$

are harmonic, the first in any finite domain, and the second in any domain not containing the origin.

§ 5. Dirichlet and Neumann Problems: Solutions in Spherical Harmonics

In this section we shall construct formal solutions of our two basic problems in the form of series of spherical harmonics. The word "formal" means here that we shall not investigate the convergence of these series, their term-by-term differentiability, the process of limit-taking, etc. If the series we construct do converge, and if all the steps taken are valid, then the formal solution of the problem is in fact its solution.

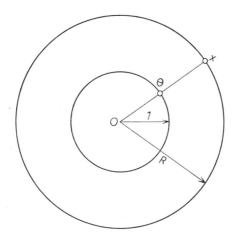

Fig. 25.

1. Dirichlet problem for the sphere. Suppose we are required to find a function $u(x)$, harmonic in the sphere $\rho = |x| < R$, and equal to a given function $\phi(x)$ on the surface of the sphere $\rho = R$. On $\rho = R$ the quantity ρ is constant, and the position of a point x is defined simply by prescribing its angular coordinates, or, in other words, by prescribing the point θ on the unit sphere which lies on the same radial line as the point x (Fig. 25). This enables us to regard $\phi(x)$ as a function of the point θ; we shall therefore write $\phi(\theta)$ in place of $\phi(x)$, and express the boundary condition for the Dirichlet problem in the form

$$u|_{\rho=R} = \phi(\theta). \tag{1}$$

Let the function $\phi(\theta)$ be expanded in a series of spherical harmonics:

$$\phi(\theta) = \sum_{n=0}^{\infty} \sum_{k=1}^{k_{n,m}} a_n^{(k)} Y_{n,m}^{(k)}(\theta). \tag{2}$$

The solution to the Dirichlet problem for the sphere $\rho < R$ is then given by the series

$$u(x) = \sum_{n=0}^{\infty} \frac{\rho^n}{R^n} \sum_{k=1}^{k_{n,m}} a_n^{(k)} Y_{n,m}^{(k)}(\theta). \tag{3}$$

2. Dirichlet problem for the exterior of a sphere. It is required to find a function $u(x)$, harmonic in the domain $\rho = |x| > R$, and satisfying boundary condition (1). If the function $\phi(\theta)$ is expanded in the series (2), then the required function is represented by the series

$$u(x) = \sum_{n=0}^{\infty} \frac{R^{n+m-2}}{\rho^{n+m-2}} \sum_{k=1}^{k_{n,m}} a_n^{(k)} Y_{n,m}^{(k)}(\theta). \tag{4}$$

3. Neumann problem for the sphere. Required to find a function $u(x)$, harmonic in the sphere $\rho < R$, and satisfying the boundary condition

$$\left.\frac{\partial u}{\partial \nu}\right|_{\rho=R} = \psi(\theta); \tag{5}$$

here ν is the outward normal to the sphere $\rho = R$; it coincides in direction with the radius passing through the point x. Thus $\partial/\partial \nu = \partial/\partial \rho|_{\rho=R}$, and conditions (5) can be written in the form

$$\left.\frac{\partial u}{\partial \rho}\right|_{\rho=R} = \psi(\theta). \tag{6}$$

From formula (2.7), Chapter 12, concerning the solubility of the Neumann problem, it is necessary that

$$\int_{\rho=R} \psi(\theta)\, dS_R = 0.$$

But $dS_R = R^{m-1} dS_1$, where S_1 is the unit sphere and the last condition is equivalent to

$$\int_{S_1} \psi(\theta) \, dS_1 = 0. \tag{7}$$

We expand the function $\psi(\theta)$ in a series of spherical harmonics:

$$\psi(\theta) = \sum_{n=1}^{\infty} \sum_{k=1}^{k_{n,m}} b_n^{(k)} Y_{n,m}^{(k)}(\theta). \tag{8}$$

The term corresponding to $n=0$ vanishes by virtue of formula (7), (4.7) and property 2 of the spherical harmonics (§ 4).

The solution of the Neumann problem is given by the series

$$u(x) = C + \sum_{n=1}^{\infty} \frac{\rho^n}{nR^{n-1}} \sum_{k=1}^{k_{n,m}} b_n^{(k)} Y_{n,m}^{(k)}(\theta), \tag{9}$$

where C is an arbitrary constant.

4. Neumann problem for the exterior of a sphere. We seek a function $u(x)$, harmonic in the domain $\rho > R$, and satisfying boundary condition (5). For an infinite domain the condition (2.7), Chapter 12, is not necessary, and the expansion of $\psi(\theta)$ in a series of spherical harmonics can also contain the zero-order term. Let

$$\psi(\theta) = \sum_{n=0}^{\infty} \sum_{k=1}^{k_{n,m}} b_n^{(k)} Y_{n,m}^{(k)}(\theta). \tag{10}$$

Then the solution to the Neumann problem is given by the formula

$$u(x) = -\sum_{n=0}^{\infty} \frac{R^{n+m-1}}{(n+m-2)\rho^{n+m-2}} \sum_{k=1}^{k_{n,m}} b_n^{(k)} Y_{n,m}^{(k)}(\theta). \tag{11}$$

5. Dirichlet and Neumann problems for a spherical layer. We are required to find a function harmonic in the spherical layer $R_0 < \rho < R_1$ and satisfying the Dirichlet boundary conditions

$$u|_{\rho=R_0} = \phi_0(\theta), \qquad u|_{\rho=R_1} = \phi_1(\theta). \tag{12}$$

The functions $\phi_0(\theta)$ and $\phi_1(\theta)$ are expanded in series of spherical harmonics

$$\phi_0(\theta) = \sum_{n=0}^{\infty} \sum_{k=1}^{k_{n,m}} a_{0n}^{(k)} Y_{n,m}^{(k)}(\theta),$$

$$\phi_1(\theta) = \sum_{n=0}^{\infty} \sum_{k=1}^{k_{n,m}} a_{1n}^{(k)} Y_{n,m}^{(k)}(\theta). \tag{13}$$

The solution is sought in the form of a series

$$u(x) = \sum_{n=0}^{\infty} \sum_{k=1}^{k_{n,m}} \left(A_n^{(k)} \rho^n + \frac{B_n^{(k)}}{\rho^{n+m-2}} \right) Y_{n,m}^{(k)}(\theta). \tag{14}$$

Putting $\rho = R_0$ and $\rho = R_1$ in formula (14), and equating the resultant series with the series (13), we obtain a system of equations for the unknown coefficients $A_n^{(k)}$ and $B_n^{(k)}$:

$$\begin{aligned} A_n^{(k)} R_0^n + B_n^{(k)} R_0^{2-m-n} &= a_{0n}^{(k)}, \\ A_n^{(k)} R_1^n + B_n^{(k)} R_1^{2-m-n} &= a_{1n}^{(k)}. \end{aligned} \tag{15}$$

The determinant of system (15) is equal to

$$R_0^n R_1^n (R_0^{2-m-2n} - R_1^{2-m-2n}).$$

This is different from zero, and so the coefficients $A_n^{(k)}$, $B_n^{(k)}$ can be determined, which leads to the solution of the Dirichlet problem.

In the case of the Neumann problem, the boundary conditions have the form

$$\left. \frac{\partial u}{\partial v} \right|_{\rho=R_0} = \psi_0(\theta), \quad \left. \frac{\partial u}{\partial v} \right|_{\rho=R_1} = \psi_1(\theta). \tag{16}$$

The relation (2.7), Chapter 12, implies in this case that the functions $\psi_0(\theta)$ and $\psi_1(\theta)$ must satisfy the equation

$$R_0^{m-1} \int_{S_1} \psi_0(\theta) \, dS_1 + R_1^{m-1} \int_{S_1} \psi_1(\theta) \, dS_1 = 0 \tag{17}$$

— without this the Neumann problem does not have a solution.

We expand the functions $\psi_0(\theta)$ and $\psi_1(\theta)$ in series of spherical harmonics:

$$\psi_0(\theta) = \sum_{n=0}^{\infty} \sum_{k=1}^{k_{n,m}} b_{0n}^{(k)} Y_{n,m}^{(k)}(\theta), \quad \psi_1(\theta) = \sum_{n=0}^{\infty} \sum_{k=1}^{k_{n,m}} b_{1n}^{(k)} Y_{n,m}^{(k)}(\theta). \tag{18}$$

Equation (17), together with formula (4.7), leads to the following relationship between the free terms of the series (18):

$$R_0^{m-1} b_{00}^{(1)} + R_1^{m-1} b_{10}^{(1)} = 0. \tag{19}$$

The solution is sought in the form of the series (14). Taking into account the fact that $\partial/\partial v = -\partial/\partial \rho|_{\rho=R_0}$ on the inner sphere, and $\partial/\partial v = \partial/\partial \rho|_{\rho=R_1}$ on the outer sphere, we differentiate the series (14) with respect to the normal, and equate the resulting series to the series (18). We then obtain the following system of equations:

$$-nR_0^{n-1}A_n^{(k)} + \frac{n+m-2}{R_0^{n+m-1}} B_n^{(k)} = b_{0n}^{(k)},$$

$$nR_1^{n-1}A_n^{(k)} - \frac{n+m-2}{R_1^{n+m-1}} B_n^{(k)} = b_{1n}^{(k)}.$$
(20)

The determinant of system (20), equal to

$$\frac{n(n+m-2)}{(R_0 R_1)^{n+m-1}} (R_0^{2n+m-2} - R_1^{2n+m-2}),$$

is non-zero for $n \neq 0$. Hence it follows that the coefficients $A_n^{(k)}$ and $B_n^{(k)}$ can be uniquely determined for $n > 0$. When $n = 0$ the system (20) gives two values for the same quantity $B_0^{(1)}$:

$$B_0^{(1)} = R_0^{m-1} \frac{b_{00}^{(1)}}{m-2}, \qquad B_0^{(1)} = -R_1^{m-1} \frac{b_{10}^{(1)}}{m-2}.$$

Because of the relationship (19), these values coincide. As might have been expected, the coefficient $A_0^{(1)}$ remains arbitrary.

Exercises

1. Solve the Neumann problem for a circular annulus.
2. Solve the mixed problem

$$u\bigg|_{\rho=R_0} = \phi(\theta), \qquad \frac{\partial u}{\partial \nu}\bigg|_{\rho=R_1} = \psi(\theta),$$

for the circular annulus $R_0 < |z| < R_1$.

Chapter 14. Variational Method for the Dirichlet Problem. Other Positive Definite Problems

In this chapter we study a method of solving the Dirichlet problem which is different from the elementary methods of the preceding chapters. This new method is based on the solution to the problem of minimizing a quadratic functional, which was discussed in Chapter 5. The method enables us to solve the Dirichlet problem for a formally self-adjoint, non-degenerate elliptic equation under quite general conditions, as well as a series of other problems in the theory of partial differential equations.

As we have just remarked, the variational method gives quite general results; at the same time it imposes some essential limitations on the problem. The Dirichlet problem has a sufficiently complete solution only in a finite domain. In the case of inhomogeneous boundary conditions, the function given on the boundary has to satisfy some quite stringent restrictions (cf. below § 5).

The variational method will also be used in subsequent chapters: in Chapter 15 we use this method to analyze the spectrum of the Dirichlet problem and of the Neumann problem in Chapter 16. In Chapter 17 we deal with the somewhat more complex problem of formally non-self-adjoint elliptic equations.

§ 1. Friedrichs's Inequality

Let Ω be a finite domain in m-dimensional Euclidean space; for simplicity we suppose that its boundary is piecewise smooth. Let the function $u \in C^{(1)}(\Omega)$ satisfy the boundary condition

$$u|_\Gamma = 0. \tag{1}$$

Then it can be shown that the function u satisfies the so-called *Friedrichs's inequality*

$$\int_\Omega u^2 \, dx \leq \kappa \int_\Omega \sum_{k=1}^m \left(\frac{\partial u}{\partial x_k}\right)^2 dx, \tag{2}$$

where κ is a constant independent of u and is completely determined by the domain Ω.

Take a coordinate system such that the domain Ω is located in that part of the space where all the coordinates are positive. Place this domain inside some parallelepiped Π, defined by the inequalities

$$0 \leq x_k \leq a_k; \quad k = 1, 2, \ldots, m$$

(fig. 26).

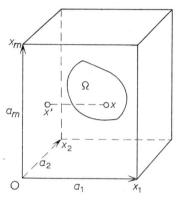

Fig. 26.

We extend the definition of the function $u(x)$ by putting it identically equal to zero in the domain $\Pi \setminus \Omega$. The function remains continuous after doing this, since $u|_\Gamma = 0$. The derivative of this function can be discontinuous in passing through the boundary. For functions of this type, the Newton-Leibnitz formula holds.

In the parallelepiped Π, take a point $x(x_1, x_2, \ldots, x_m)$ and project it on to the coordinate surface orthogonal to the Ox_1-axis. Denote this projection by x'; then x' can be regarded as a point of an $(m-1)$-dimensional Euclidean space, having coordinates $x_2, \ldots, x_m$. We shall use the notation $x = (x_1, x')$ and analogously for other points.

By the Newton-Leibnitz formula

$$u(x) - u(0, x') = \int_0^{x_1} \frac{\partial u(\xi, x')}{\partial \xi} \, d\xi.$$

But the point $(0, x')$ lies outside the domain Ω, therefore $u(0, x') = 0$ and $u(x) = \int_0^{x_1} \{\partial u(\xi, x')/\partial \xi\} d\xi$.

By the Schwartz-Bunyakovskii inequality

$$u^2(x) \leq \int_0^{x_1} d\xi \int_0^{x_1} \left(\frac{\partial u(\xi, x')}{\partial \xi}\right)^2 d\xi \leq a_1 \int_0^{a_1} \left(\frac{\partial u(\xi, x')}{\partial \xi}\right)^2 d\xi.$$

We integrate this inequality over the parallelepiped Π:

$$\int_\Pi u^2(x)\,dx \leq a_1^2 \int_0^{a_1} d\xi \int_0^{a_2} dx_2 \ldots \int_0^{a_m} \left(\frac{\partial u(\xi, x')}{\partial \xi}\right)^2 dx_m$$

$$= a_1^2 \int_0^{a_1} dx_1 \int_0^{a_2} dx_2 \ldots \int_0^{a_m} \left(\frac{\partial u(x_1, x')}{\partial x_1}\right)^2 dx_m = a_1^2 \int_\Pi \left(\frac{\partial u(x)}{\partial x_1}\right)^2 dx.$$

On both left- and right-hand sides, the integrals over $\Pi\backslash\Omega$ are equal to zero, and so are discarded; in addition, we add to the integrand function on the right-hand side the non-negative sum

$$\sum_{k=2}^m \left(\frac{\partial u(x)}{\partial x_k}\right)^2.$$

This leads to the inequality

$$\int_\Omega u^2(x)\,dx \leq a_1^2 \int_\Omega \sum_{k=1}^m \left(\frac{\partial u}{\partial x_k}\right)^2 dx, \qquad (3)$$

which coincides with inequality (2), if in the latter we put $\kappa = a_1^2$. It should be understood that κ can be taken equal to any of the numbers a_k^2 (and in particular, the smallest of them). The question of what is the smallest possible value of κ will be settled in the next chapter.

§ 2. Operator of the Dirichlet Problem

Let

$$Lu = -\frac{\partial}{\partial x_j}\left(A_{jk}(x)\frac{\partial u}{\partial x_k}\right) + C(x)u \qquad (1)$$

be a differential expression whose coefficients are defined in some finite domain Ω of m-dimensional Euclidean space E_m. The boundary Γ of the domain Ω is taken to be piecewise-smooth. Also, suppose that $A_{jk} \in C^{(1)}(\bar{\Omega})$, and $C \in C(\bar{\Omega})$.

We shall assume that the differential expression (1) is elliptic in the closed domain $\bar{\Omega}$. If this is the case, then all the eigenvalues $\lambda_1(x), \lambda_2(x), \ldots, \lambda_m(x)$ of the matrix $A_{jk}(x)$ of highest coefficients have the same sign. By changing, if necessary, the sign of the expression L, we can always assume that $\lambda_k(x) > 0$, $x \in \bar{\Omega}$.

The equation

$$\begin{vmatrix} A_{11}-\lambda & A_{12} & \cdots & A_{1m} \\ A_{21} & A_{22}-\lambda & \cdots & A_{2m} \\ \cdot & \cdot & \cdot & \cdot \\ A_{m1} & A_{m2} & \cdots & A_{mm}-\lambda \end{vmatrix} = 0$$

has as its highest coefficient $(-1)^m$, constant and non-zero; the other coefficients in this equation are continuous in $\bar{\Omega}$. Hence it follows that the roots $\lambda_k(x)$ of this equation are continuous functions of x in $\bar{\Omega}$. Being positive on a compact set $\bar{\Omega}$, they are bounded below in this domain by some positive constant, which we denote by μ_0:

$$\lambda_k(x) \geq \mu_0, \quad \forall x \in \bar{\Omega}; \quad \mu_0 = \text{const.} > 0. \tag{2}$$

An elliptic differential equation satisfying the inequality (2) is called *non-degenerate* in Ω.

Let $t_1, t_2, \ldots, t_m$ be arbitrary real numbers. If $\lambda_1(x)$ is the smallest eigenvalue of the matrix $\|A_{jk}\|_{j,k=1}^{j,k=m}$, then we know that

$$A_{jk}(x) t_j t_k \geq \lambda_1(x) \sum_{k=1}^{m} t_k^2.$$

Utilizing inequality (2), we obtain

$$A_{jk}(x) t_j t_k \geq \mu_0 \sum_{k=1}^{m} t_k^2. \tag{3}$$

Inequality (3) also characterizes a non-degenerate elliptic expression. This inequality will play an important role in what follows.

We also require of the differential expression (1) that

$$C(x) \geq 0, \quad \forall x \in \bar{\Omega}. \tag{4}$$

Consider now the Dirichlet problem with a homogeneous boundary condition:

$$Lu = -\frac{\partial}{\partial x_j} \left(A_{jk} \frac{\partial u}{\partial x_k} \right) + Cu = f(x), \tag{5}$$

$$u|_\Gamma = 0. \tag{6}$$

We shall assume that $f \in L_2(\Omega)$, and shall look for a solution which also belongs to the space $L_2(\Omega)$. As do all boundary-value problems, the problem

(5)–(6) generates an operator, which we denote by $\mathfrak{A}$. This operator acts according to the formula

$$\mathfrak{A}u = Lu = -\frac{\partial}{\partial x_j}\left(A_{jk}\frac{\partial u}{\partial x_j}\right) + Cu;$$

for its domain of definition $D(\mathfrak{A})$, we can take a set of those functions from $C^{(2)}(\overline{\Omega})$ which satisfy the boundary condition (6). Clearly $\mathfrak{A}$ can be regarded as an operator acting in the space $L_2(\Omega)$.

We now show that *the operator $\mathfrak{A}$ is positive definite* in $L_2(\Omega)$. In accordance with the definition (§ 2, Chapter 5), it is sufficient to establish three facts: 1) the set $D(\mathfrak{A})$ is dense in $L_2(\Omega)$; 2) the operator $\mathfrak{A}$ is symmetric:

$$(\mathfrak{A}u, v) = (u, \mathfrak{A}v); \qquad u, v \in D(\mathfrak{A}); \tag{7}$$

3) the operator $\mathfrak{A}$ satisfies the positive-definiteness inequality

$$(\mathfrak{A}u, u) \geqq \gamma^2 \|u\|^2; \qquad \gamma^2 = \text{const.} > 0. \tag{8}$$

The set $D(\mathfrak{A})$ is certainly dense in $L_2(\Omega)$ – this follows immediately from Corollary 1.3.1, since $D(\mathfrak{A})$ obviously contains the set of all functions finite in Ω.

We next prove the symmetry of the operator $\mathfrak{A}$. Let $u, v \in D(\mathfrak{A})$. This means that $u, v \in C^{(2)}(\overline{\Omega})$, and

$$u|_\Gamma = v|_\Gamma = 0. \tag{9}$$

Now construct the difference

$$(\mathfrak{A}u, v) - (u, \mathfrak{A}v) = (Lu, v) - (u, Lv) = \int_\Omega (vLu - uLv)\,dx.$$

To this last integral we apply Green's Second Formula ((6.6), Chapter 10), and obtain

$$(\mathfrak{A}u, v) - (u, \mathfrak{A}v) = -\int_\Gamma A_{jk}\left(v\frac{\partial u}{\partial x_k} - u\frac{\partial v}{\partial x_k}\right)\cos(v, x_j)\,d\Gamma.$$

Because of equations (9), the integral on the right is equal to zero. Hence equation (7) holds, and consequently the symmetry of the operator $\mathfrak{A}$ is proved.

It remains to establish inequality (8). We have

$$(\mathfrak{A}u, u) = (Lu, u) = -\int_\Omega u\frac{\partial}{\partial x_j}A_{jk}\frac{\partial u}{\partial x_k}\,dx + \int_\Omega Cu^2\,dx.$$

We apply Green's First Formula ((6.5), Chapter 10) to the first integral here. By virtue of boundary condition (6), the surface integral vanishes, and we obtain

$$(\mathfrak{A}u, u) = \int_\Omega \left[A_{jk} \frac{\partial u}{\partial x_j} \frac{\partial u}{\partial x_k} + Cu^2 \right] dx. \tag{10}$$

We find a lower bound for the integral (10). First of all, we discard the non-negative term Cu^2. Next, we utilize inequality (3), with $t_k = \partial u/\partial x_k$:

$$A_{jk} \frac{\partial u}{\partial x_j} \frac{\partial u}{\partial x_k} \geq \mu_0 \sum_{k=1}^m \left(\frac{\partial u}{\partial x_k} \right)^2.$$

Now

$$(\mathfrak{A}u, u) \geq \mu_0 \int_\Omega \sum_{k=1}^m \left(\frac{\partial u}{\partial x_k} \right)^2 dx. \tag{11}$$

For functions $u \in D(\mathfrak{A})$ it is obvious that the Friedrichs inequality (1.2) holds, and so finally

$$(\mathfrak{A}u, u) \geq \frac{\mu_0}{\kappa} \int_\Omega u^2 \, dx = \frac{\mu_0}{\kappa} \|u\|^2. \tag{12}$$

Inequality (8) is now established (with the constant $\gamma^2 = \mu_0/\kappa$); hence we have proved that $\mathfrak{A}$ is a positive definite operator.

Remark. The operator $\mathfrak{A}$ is also positive definite when $C(x) = 0$. This enables some relaxation of condition (4). Denote by $\mathfrak{A}_0$ the operator which $\mathfrak{A}$ becomes when $C(x) \equiv 0$, and let $\gamma_0^2 > 0$ be the lower bound of the operator $\mathfrak{A}_0$. Then

$$(\mathfrak{A}_0 u, u) \geq \gamma_0^2 \|u\|^2. \tag{13}$$

Obviously $\mathfrak{A}u = \mathfrak{A}_0 u + C(x)u$. Hence

$$(\mathfrak{A}u, u) = (\mathfrak{A}_0 u, u) + (Cu, u) \geq \gamma_0^2 \|u\|^2 + (Cu, u). \tag{14}$$

Suppose that $C(x)$ satisfies the inequality

$$C(x) \geq \varepsilon - \gamma_0^2, \quad \forall x \in \bar{\Omega}, \tag{15}$$

where ε is a positive constant. Then

$$(Cu, u) = \int_\Omega C(x) u^2(x) \, dx \geq (\varepsilon - \gamma_0^2) \int_\Omega u^2(x) \, dx = (\varepsilon - \gamma_0^2) \|u\|^2.$$

Substituting this into (14), we find that

$$\mathfrak{A}(u, u) \geq \varepsilon ||u||^2 \qquad (16)$$

and the operator $\mathfrak{A}$ is positive definite. Thus condition (4) can be replaced by the weaker condition (15).

§ 3. Energy Space for the Dirichlet Problem

The operator $\mathfrak{A}$ of the Dirichlet problem, like every positive definite operator, has associated with it an energy space $H_\mathfrak{A}$. We shall examine this space more closely.

THEOREM 14.3.1. *The energy space $H_\mathfrak{A}$ consists of those, and only those, functions, which 1) are square-integrable and have square-integrable generalized first derivatives in Ω, 2) satisfy boundary conditions (2.6) in the following sense: if $u \in H_\mathfrak{A}$, then there exists a sequence of functions $u_n \in D(\mathfrak{A})$, such that*

$$\int_\Omega (u_n - u)^2 \, dx \underset{n \to \infty}{\to} 0, \qquad \int_\Omega \sum_{k=1}^m \left(\frac{\partial u_n}{\partial x_k} - \frac{\partial u}{\partial x_k}\right)^2 dx \underset{n \to \infty}{\to} 0. \qquad (1)$$

In the space $H_\mathfrak{A}$, the energy product and energy norm are defined by the formulae

$$[u, v]_\mathfrak{A} = \int_\Omega \left(A_{jk} \frac{\partial u}{\partial x_j} \frac{\partial v}{\partial x_k} + Cuv\right) dx, \qquad (2)$$

$$|||u|||_\mathfrak{A}^2 = \int_\Omega \left(A_{jk} \frac{\partial u}{\partial x_j} \frac{\partial u}{\partial x_k} + Cu^2\right) dx. \qquad (3)$$

We first show that if the function $u \in H_\mathfrak{A}$, then it has the properties enumerated in the theorem.

The function $u \in L_2(\Omega)$ – this follows immediately from Theorem 5.3.1, according to which all the elements of the energy space $H_\mathfrak{A}$ belong to the original space $L_2(\Omega)$.

As with every space obtained by completion, the energy space $H_\mathfrak{A}$ consists of old elements – in this case the element of the domain $D(\mathfrak{A})$ – and ideal elements. If u is an ideal element, then there exists a sequence $\{u_n\}$ of old elements such that

$$||u_n - u|| \underset{n \to \infty}{\to} 0, \qquad |||u_n - u|||_\mathfrak{A} \underset{n \to \infty}{\to} 0. \qquad (4)$$

Suppose for the time being that u and v are old elements. Utilizing Green's First Formula ((6.5), Chapter 10) and equation (2.9), we obtain

$$[u, v]_\mathfrak{A} = (\mathfrak{A}u, v) = \int_\Omega \left(A_{jk} \frac{\partial u}{\partial x_k} \frac{\partial v}{\partial x_j} + Cuv \right) dx.$$

But $A_{jk} = A_{kj}$, and the last integral is the same as the integral in formula (2), which proves the latter for old elements.

Putting $v = u$ in formula (2), we see that formula (3) also holds for old elements.

Now let u be an ideal element of the space $H_\mathfrak{A}$. We construct a sequence of elements $u_n \in D(\mathfrak{A})$, $n = 1, 2, \ldots$, such that formulae (4) hold. The second of these formulae gives immediately that

$$|||u_n - u_s|||^2 = (\mathfrak{A}(u_n - u_s), u_n - u_s)$$

$$= \int_\Omega \left[A_{jk} \left(\frac{\partial u_n}{\partial x_j} - \frac{\partial u_s}{\partial x_j} \right) \left(\frac{\partial u_n}{\partial x_k} - \frac{\partial u_s}{\partial x_k} \right) + C(u_n - u_s)^2 \right] dx \underset{n, s \to \infty}{\to} 0.$$

It now follows from inequality (2.11) that

$$\int_\Omega \left(\frac{\partial u_n}{\partial x_k} - \frac{\partial u_s}{\partial x_k} \right)^2 dx = \left\| \frac{\partial u_n}{\partial x_k} - \frac{\partial u_s}{\partial x_k} \right\|^2 \underset{n, s \to \infty}{\to} 0, \quad k = 1, 2, \ldots, m,$$

i.e. the sequences of derivatives $\partial u_n/\partial x_k$, $n = 1, 2, \ldots$, converge in themselves in the metric of $L_2(\Omega)$.

Hence it follows that there exist limits

$$v_k = \lim_{n \to \infty} \frac{\partial u_n}{\partial x_k}, \quad k = 1, 2, \ldots, m; \quad v_k \in L_2(\Omega).$$

By virtue of the first of relations (4), Theorem 2.3.1 gives

$$v_k = \frac{\partial u}{\partial x_k}. \tag{5}$$

It remains to show that formulae (2) and (3) are also valid for ideal elements. Let u be an ideal element, and let the sequence $u_n \in D(\mathfrak{A})$ satisfy the relations (4). Then

$$|||u|||_\mathfrak{A}^2 = \lim_{n \to \infty} |||u_n|||_\mathfrak{A}^2 = \lim_{n \to \infty} \int_\Omega \left[A_{jk} \frac{\partial u_n}{\partial x_j} \frac{\partial u_n}{\partial x_k} + Cu_n^2 \right] dx. \tag{6}$$

We show that the latter limit is equal to

$$\int_\Omega \left[A_{jk} \frac{\partial u}{\partial x_j} \frac{\partial u}{\partial x_k} + Cu^2 \right] dx.$$

To do this, we estimate the quantity

$$J_n = \left| \int_\Omega \left[A_{jk} \left(\frac{\partial u_n}{\partial x_j} \frac{\partial u_n}{\partial x_k} - \frac{\partial u}{\partial x_j} \frac{\partial u}{\partial x_k} \right) + C(u_n^2 - u^2) \right] dx \right|.$$

The functions $A_{jk}(x)$ and $C(x)$ are continuous in a closed domain, and hence bounded. Let $|A_{jk}(x)| \leq M, |C(x)| \leq M; M = \text{const}$. Then

$$J_n \leq M \int_\Omega \sum_{j,k=1}^m \left| \frac{\partial u_n}{\partial x_j} \frac{\partial u_n}{\partial x_k} - \frac{\partial u}{\partial x_j} \frac{\partial u}{\partial x_k} \right| dx + M \int_\Omega |u_n^2 - u^2| dx. \quad (7)$$

The second integral is estimated as follows:

$$\int_\Omega |u_n^2 - u^2| dx = \int_\Omega |u_n + u| \cdot |u_n - u| dx$$

$$\leq \left\{ \int_\Omega (u_n + u)^2 dx \right\}^{\frac{1}{2}} \cdot \left\{ \int_\Omega (u_n - u)^2 dx \right\}^{\frac{1}{2}} = \|u_n + u\| \cdot \|u_n - u\|.$$

The second quantity tends to zero, and the first tends to a limit (equal to $2\|u\|$) and so is bounded; consequently the second term in (7) tends to zero.

The first term can be estimated in a similar way:

$$\int_\Omega \sum_{j,k=1}^m \left| \frac{\partial u_n}{\partial x_j} \frac{\partial u_n}{\partial x_k} - \frac{\partial u}{\partial x_j} \frac{\partial u}{\partial x_k} \right| dx$$

$$= \int_\Omega \sum_{j,k=1}^m \left| \frac{\partial u_n}{\partial x_j} \left(\frac{\partial u_n}{\partial x_k} - \frac{\partial u}{\partial x_k} \right) + \frac{\partial u}{\partial x_k} \left(\frac{\partial u_n}{\partial x_j} - \frac{\partial u}{\partial x_j} \right) \right| dx$$

$$\leq \sum_{j,k=1}^m \left\{ \left\| \frac{\partial u_n}{\partial x_j} \right\| \cdot \left\| \frac{\partial u_n}{\partial x_k} - \frac{\partial u}{\partial x_k} \right\| + \left\| \frac{\partial u}{\partial x_k} \right\| \cdot \left\| \frac{\partial u_n}{\partial x_j} - \frac{\partial u}{\partial x_j} \right\| \right\}.$$

The first quantities on the right-hand side are bounded and the second tend to zero, so the whole expression tends to zero. Thus, finally,

$$\|\|u\|\|_{\mathfrak{A}}^2 = \int_\Omega \left[A_{jk} \frac{\partial u}{\partial x_j} \frac{\partial u}{\partial x_k} + Cu^2 \right] dx,$$

and formula (3) is true for ideal elements of the energy space.

If now u and v are two such elements, then

$$[u, v]_\mathfrak{A} = \tfrac{1}{4}\{|||u+v|||_\mathfrak{A}^2 - |||u-v|||_\mathfrak{A}^2\}.$$

Substituting for the norms on the right-hand side from formula (3), and carrying out some elementary simplifications, we arrive at formula (2), which is thus proven for ideal elements as well.

We now prove the converse: if the function $u \in L_2(\Omega)$ has generalized derivatives $\partial u/\partial x_k \in L_2(\Omega)$, and if there exists a sequence $\{u_n\}$, $u_n \in D(\mathfrak{A})$, satisfying relations (1), then $u \in H_\mathfrak{A}$.

The sequence $\{u_n\}$ converges in itself in the energy metric. In fact,

$$|||u_n - u_s|||_\mathfrak{A}^2 = \int_\Omega \left[A_{jk} \left(\frac{\partial u_n}{\partial x_j} - \frac{\partial u_s}{\partial x_j} \right) \left(\frac{\partial u_n}{\partial x_k} - \frac{\partial u_s}{\partial x_k} \right) + C(u_n - u_s)^2 \right] dx.$$

The coefficients A_{jk} and C are bounded by the constant M. The eigenvalues of the matrix of highest coefficients, being continuous functions of the coefficients A_{jk}, are also bounded; let $N = \text{const.}$ be their upper bound. Then

$$A_{jk} t_j t_k \leq N \sum_{k=1}^m t_k^2$$

and, consequently,

$$|||u_n - u_s|||_\mathfrak{A}^2 \leq N \int_\Omega \sum_{k=1}^m \left(\frac{\partial u_n}{\partial x_k} - \frac{\partial u_s}{\partial x_k} \right)^2 dx + M \int_\Omega (u_n - u_s)^2 dx;$$

because of the relations (1) this tends to zero as $n, s \to \infty$.

The energy space is complete, therefore there exists in it an element w, such that $|||w - u_n|||_\mathfrak{A} \to 0$. The first of relations (1) shows that $w = u$. Thus, finally, $u \in H_\mathfrak{A}$.

§ 4. Generalized Solution of the Dirichlet Problem

The problem (2.5)–(2.6) can be written as a single operational equation

$$\mathfrak{A} u = f; \tag{*}$$

in so far as we regard $\mathfrak{A}$ as an operator in $L_2(\Omega)$, it is natural to take $f \in L_2(\Omega)$. It is clear that, in general, equation (*) does not have a solution: if $u \in D(\mathfrak{A})$, then the function f must be continuous in $\bar{\Omega}$. But the operator $\mathfrak{A}$ is positive definite in $L_2(\Omega)$. It was shown in § 5, Chapter 5, that for any $f \in L_2(\Omega)$ the problem under discussion has one and only one generalized solution $u_0 \in H_{\mathfrak{A}}$. By Theorem 14.3.1, the function u_0 is square-integrable, has square-integrable generalized first derivatives, and is equal to zero on the boundary in the sense of formulae (3.1).

Equation (2.5) is of second order, and it would be interesting to clarify whether second derivatives of the generalized solution of the Dirichlet problem exist. For the Laplace operator a partial answer to this question will be given below, in § 6. A more complete answer is to be found in the books [8] and [11].

The generalized solution is the solution to the problem of minimizing the functional

$$F(u) = |||u|||_{\mathfrak{A}}^2 - 2(u,f) = \int_\Omega \left[A_{jk} \frac{\partial u}{\partial x_j} \frac{\partial u}{\partial x_k} + Cu^2 - 2fu \right] dx \qquad (1)$$

subject to boundary conditions (2.6). This solution can be represented in the form of a series (cf. § 5, Chapter 5)

$$u_0(x) = \sum_{n=1}^\infty (f, \omega_n) \omega_n(x), \qquad (2)$$

where $\{\omega_n(x)\}$ is a complete, orthonormal sequence in the space $H_{\mathfrak{A}}$.

Example. Consider the Dirichlet problem for Laplace's equation

$$-\Delta u = f(x), \qquad (3)$$

$$u|_\Gamma = 0, \qquad (4)$$

in the parallelepiped Ω, defined by the inequalities

$$0 \leq x_k \leq a_k; \quad k = 1, 2, \ldots, m; \qquad (5)$$

here Γ denotes the surface of the parallelepiped (5). In the present case $A_{jk} = \delta_{jk}$, $C = 0$; formulae (3.2) and (3.3) can be simplified and take the following form:

$$[u, v]_{\mathfrak{A}} = \int_\Omega \frac{\partial u}{\partial x_k} \frac{\partial v}{\partial x_k} dx,$$

$$|||u|||_{\mathfrak{A}}^2 = \int_\Omega \sum_{k=1}^m \left(\frac{\partial u}{\partial x_k} \right)^2 dx. \qquad (6)$$

The system of functions

$$\frac{2^{\frac{1}{2}m}}{\pi \sqrt{|\Omega|}} \left\{ \sum_{k=1}^m \frac{n_k^2}{a_k^2} \right\}^{-\frac{1}{2}} \prod_{k=1}^m \sin \frac{n_k \pi x_k}{a_k}, \quad n_k = 1, 2, 3, \ldots, \qquad (7)$$

is orthonormal and complete in $H_\mathfrak{A}$ (prove!); here $|\Omega| = \prod_{k=1}^{m} a_k$ is the volume of the parallelepiped (5).

From formula (2) we find

$$u_0(x) = \frac{2^m}{\pi^2 |\Omega|} \sum_{n_1, n_2, \ldots, n_m = 1}^{\infty} \left(\sum_{k=1}^{m} \frac{n_k^2}{a_k^2} \right)^{-1} a_{n_1 n_2 \ldots n_m} \prod_{k=1}^{m} \sin \frac{n_k \pi x_k}{a_k};$$

$$a_{n_1 n_2 \ldots n_m} = \int_\Omega f(x) \prod_{k=1}^{m} \sin \frac{n_k \pi x_k}{a_k} \, dx. \tag{8}$$

§ 5. Dirichlet Problem for the Homogeneous Equation

Let the domain Ω and coefficients $A_{jk}(x)$, $C(x)$ satisfy the conditions of § 2, and let the function $\phi(x)$ be prescribed on the surface Γ – the boundary of Ω.

Consider the Dirichlet problem in Ω for the homogeneous elliptic equation

$$\frac{\partial}{\partial x_j} \left(A_{jk} \frac{\partial v}{\partial x_k} \right) - Cv = 0 \tag{1}$$

with inhomogeneous boundary conditions

$$v|_\Gamma = \phi(x). \tag{2}$$

In the classical formulation of the Dirichlet problem it is assumed that the given function $\phi \in C(\Gamma)$, and the required solution $u \in C(\bar{\Omega})$ and has continuous second derivatives in Ω. We shall solve the Dirichlet problem in a modified formulation, which we now proceed to describe.

For simplicity we retain the condition that $\phi \in C(\Gamma)$, but make the following assumption. There exists a function $\psi(x)$ which satisfies three conditions: 1) $\psi \in C(\bar{\Omega})$; 2) the function ψ has generalized derivatives $\partial \psi / \partial x_k \in L_2(\Omega)$, $k = 1, 2, \ldots, m$; 3) $\psi|_\Gamma = \phi(x)$. *

We now pose the problem of minimizing the homogeneous quadratic functional

$$\Phi(v) = \int_\Omega \left[A_{jk} \frac{\partial v}{\partial x_j} \frac{\partial v}{\partial x_k} + Cv^2 \right] dx \tag{3}$$

---

* In this connection, it would be sufficient to require that $\psi \in W_2^{(1)}(\Omega)$ (c.f. § 5, Chap. 1), and that the equality $\psi|_\Gamma = \phi(x)$ be satisfied in some generalized sense. Then it would not be necessary to insist that $\phi \in C(\Gamma)$.

If no function $\psi(x)$ having the stated properties exists, then the set $D(\Phi)$ is empty and the variational problem (2)–(3) loses its meaning. The necessary and sufficient conditions that have to be imposed on the function $\phi(x)$ in order that the set $D(\Phi)$ be not empty are given in reference [16].

on the set $D(\Phi)$ of functions which are square-integrable in Ω together with their generalized first derivatives, and which satisfy boundary condition (2) in the following sense:

$$(v-\psi) \in H_\mathfrak{A}, \tag{4}$$

where $H_\mathfrak{A}$ is the energy space of the Dirichlet problem (2.5)–(2.6). Put $v(x) - \psi(x) = u(x)$. Then $u \in H_\mathfrak{A}$, and

$$\Phi(v) = \Phi(u) + 2\Phi(u, \psi) + \Phi(\psi),$$

where $\Phi(u, \psi)$ is the bilinear functional

$$\Phi(u, \psi) = \int_\Omega \left[A_{jk} \frac{\partial u}{\partial x_j} \frac{\partial \psi}{\partial x_k} + Cu\psi \right] dx. \tag{5}$$

Obviously $\Phi(u, \psi)$ is a linear functional of u.

Since $\Phi(\psi)$ is constant, it is clear that the variational problem (2)–(3) is equivalent to the following problem: find a function in the space $H_\mathfrak{A}$ which achieves for the functional

$$\Phi(u) + 2\Phi(u, \psi) \tag{6}$$

its smallest value.

We show that the latter problem has a solution, and, moreover, a unique one. From conditions (2.3) and (2.4) it follows that the homogeneous quadratic functional Φ is non-negative, and so obeys Cauchy's inequality

$$|\Phi(u, \psi)| \leq \sqrt{\Phi(u)} \sqrt{\Phi(\psi)}. \tag{7}$$

If $u \in H_\mathfrak{A}$, then $\sqrt{\Phi(u)} = |||u|||_\mathfrak{A}$ (formula (3.3)), and, consequently,

$$|\Phi(u, \psi)| \leq \sqrt{\Phi(\psi)}\, |||u|||_\mathfrak{A}. \tag{8}$$

The function ψ is fixed; therefore $\sqrt{\Phi(\psi)}$ is a constant quantity, and inequality (8) shows that the functional $\Phi(u, \psi)$ is bounded in $H_\mathfrak{A}$. It now follows from the results of § 9, Chapter 5, that the problem of minimizing the functional (6) has one and only one solution in the space $H_\mathfrak{A}$.

Denote this solution by $u_0(x)$, and let $v_0(x) = u_0(x) + \psi(x)$. Obviously the function $v_0(x)$ solves the variational problem (2)–(3).

THEOREM 14.5.1. *If v_0 is the solution of the variational problem* (2)–(3), *and $v_0 \in C^{(2)}(\Omega)$, then this function satisfies the differential equation* (1).

As formula (4) shows, the domain of definition $D(\Phi)$ of the functional Φ is a linear manifold (cf. § 2, Chapter 3) of functions of the form

§ 5] THE HOMOGENEOUS EQUATION 303

$v = \psi + u$, $u \in H_{\mathfrak{A}}$. Putting $v - v_0 = \eta$, we can also write $v = v_0 + \eta$, $\eta \in H_{\mathfrak{A}}$. Since v_0 achieves a minimum of the functional Φ, so at the point v_0 the variation of this functional is equal to zero (cf. § 4, Chapter 3)

$$\delta\Phi(v_0, \eta) = 0. \tag{9}$$

The functional Φ is homogeneous and quadratic. Hence it is easily inferred that

$$\delta\Phi(v, \eta) = 2\Phi(v, \eta) = 2 \int_\Omega \left[A_{jk} \frac{\partial v}{\partial x_j} \frac{\partial \eta}{\partial x_k} + Cv\eta \right] dx.$$

Thus, equation (9) is equivalent to the following:

$$\int_\Omega \left[A_{jk} \frac{\partial v_0}{\partial x_j} \frac{\partial \eta}{\partial x_k} + Cv_0 \eta \right] dx = 0, \quad \forall \eta \in H_{\mathfrak{A}}. \tag{10}$$

Given a number $\delta > 0$, we construct a boundary strip Ω_δ of width δ, and take for η a function which is finite in $\Omega \backslash \Omega_\delta$ and equal to zero in Ω_δ; this function is therefore finite in Ω also.

In equation (10) we discard the integral over Ω_δ, it being equal to zero. The result is

$$\int_{\Omega \backslash \Omega_\delta} \left[A_{jk} \frac{\partial v_0}{\partial x_k} \frac{\partial \eta}{\partial x_j} + Cv_0 \eta \right] dx = 0.$$

The first term is integrated by parts, so as to leave η undifferentiated. Because $\eta = 0$ on the boundary of the domain $\Omega \backslash \Omega_\delta$, the surface integral here vanishes, and we obtain

$$\int_{\Omega \backslash \Omega_\delta} \left[-\frac{\partial}{\partial x_j}\left(A_{jk} \frac{\partial v_0}{\partial x_k} \right) + Cv_0 \right] \eta \, dx = 0. \tag{11}$$

The set of finite functions is dense in $L_2(\Omega \backslash \Omega_\delta)$. On the other hand, since the function $v_0 \in C^{(2)}(\Omega)$, the expression in square brackets in integral (11) is an element of the space $L_2(\Omega \backslash \Omega_\delta)$. Being orthogonal to a dense set of finite functions, this expression is identically equal to zero; this implies that in the domain $\Omega \backslash \Omega_\delta$ the function v_0 satisfies the equation

$$\frac{\partial}{\partial x_j}\left(A_{jk} \frac{\partial v_0}{\partial x_k} \right) - Cv_0 = 0. \tag{12}$$

But δ can be chosen as small as we please. Hence it follows that equation (12) is satisfied almost everywhere in Ω. The theorem is thus proved.

Theorem 14.5.1 suggests the following definition.

The function $v_0(x)$ which solves the variational problem (2)–(3) is called the *generalized solution of the Dirichlet problem* (1)–(2). The word "generalized" will frequently be omitted, but understood.

§ 6. On the Existence of Second Derivatives in the Solution of the Dirichlet Problem

THEOREM 14.6.1. *The generalized solution, in a finite domain Ω, of the Dirichlet problem for the homogeneous Laplace equation with inhomogeneous boundary conditions is a harmonic function in Ω.*

For Laplace's equation $A_{jk} = \delta_{jk}$, $C = 0$, and equation (5.10) takes the form

$$\int_\Omega \frac{\partial v_0}{\partial \xi_k} \frac{\partial \eta}{\partial \xi_k} \, d\xi = 0, \tag{1}$$

with x replaced by ξ.

Take an arbitrary point $x \in \Omega$, and put $\eta = \omega_h(r)$ in equation (1), where $r = |\xi - x|$ and ω_h is an averaging kernel (§ 1, Chapter 1); the averaging radius h is taken smaller than the distance from x to Γ – the boundary of the domain Ω – so that $\omega_h(r)|_\Gamma = 0$. The function $\omega_h(r)$ depends only on the difference $\xi - x$, and therefore

$$\frac{\partial \omega_h(r)}{\partial \xi_k} = -\frac{\partial \omega_h(r)}{\partial x_k},$$

and equation (1) can be expressed in the following form:

$$\frac{\partial}{\partial x_k} \int_\Omega \frac{\partial v_0}{\partial \xi_k} \omega_h(r) \, d\xi = 0.$$

By Theorem 2.2.1,

$$\int_\Omega \frac{\partial v_0}{\partial \xi_k} \omega_h(r) \, d\xi = \frac{\partial v_{0h}(x)}{\partial x_k}$$

and, consequently,

$$\Delta v_{0h} = 0. \tag{2}$$

If $h \to 0$, then $v_{0h} \to v_0$ in the metric of $L_2(\Omega)$ (Theorem 1.3.3). By Theorem 11.9.2, the function $v_0(x)$ is harmonic in Ω.

THEOREM 14.6.2. *If* $f \in \operatorname{Lip}_\alpha(\bar{\Omega})$, $0 < \alpha \leq 1$, *and* $u_0(x)$ *is a generalized solution to the Dirichlet problem for Laplace's equation*

$$-\Delta u = f(x), \qquad u|_\Gamma = 0, \qquad (3)$$

then $u_0(x) \in C^{(2)}(\Omega)$.

Introduce the Newtonian potential

$$\psi(x) = \frac{1}{(m-2)|S_1|} \int_\Omega f(\xi) \frac{1}{r^{m-2}} \, d\xi.$$

From the results of § 6, Chapter 11, it follows that

$$\psi \in C^{(1)}(\bar{\Omega}) \cap C^{(2)}(\Omega)$$

and that $-\Delta \psi = f(x)$.

The function $u_0(x)$ solves the problem of minimizing the functional

$$F(u) = \int_\Omega \left[\sum_{k=1}^m \left(\frac{\partial u}{\partial \xi_k} \right)^2 - 2fu \right] d\xi, \qquad u \in H_\mathfrak{A};$$

therefore the variation of the functional F at the point u_0 is zero:

$$\delta F(u_0, \eta) = 2 \int_\Omega \left[\frac{\partial u_0}{\partial \xi_k} \frac{\partial \eta}{\partial \xi_k} - f\eta \right] d\xi = 0, \qquad \forall \eta \in H_\mathfrak{A}.$$

We substitutei $u_0 = v_0 + \psi$

$$\int_\Omega \left[\frac{\partial v_0}{\partial \xi_k} \frac{\partial \eta}{\partial \xi_k} + \frac{\partial \psi}{\partial \xi_k} \frac{\partial \eta}{\partial \xi_k} - f\eta \right] d\xi = 0. \qquad (4)$$

In equation (4) put $\eta = \omega_h(r)$, where ω_h is the averaging kernel, $r = |\xi - x|$, x a point of the domain Ω, and the averaging radius h is smaller than the distance from the point x to Γ – the boundary of the domain Ω.

Integrate by parts the second term in (4):

$$\int_\Omega \frac{\partial \psi}{\partial \xi_k} \frac{\partial \eta}{\partial \xi_k} d\xi = -\int_\Omega \eta \Delta \psi \, d\xi + \int_\Gamma \eta \frac{\partial \psi}{\partial \nu} d\Gamma = \int_\Omega f\eta \, d\xi,$$

since the integral over Γ obviously vanishes. Equation (4) takes the form

$$\int_\Omega \frac{\partial v_0}{\partial \xi_k} \frac{\partial \omega_h(r)}{\partial \xi_k} d\xi = 0.$$

The same transformation which was used in the preceding theorem gives that $\Delta v_{0h} = 0$, and, consequently, the function v_0 is harmonic in Ω. Moreover, $v_0 \in C^{(2)}(\Omega)$. But then $u_0 = v_0 + \psi \in C^{(2)}(\Omega)$ also. The theorem is proved.

Remark. There exist stronger theorems than 14.6.2 concerning the differential properties of generalized solutions of the Dirichlet problem. We state some of these.

  1. If the function $f(x)$ in equation (3) satisfies in $\bar{\Omega}$ the Lipschitz condition with exponent α, $0 < \alpha < 1$, then the second derivatives of the generalized solution $u_0(x)$ satisfy the same condition in any interior, closed sub-domain.
  2. If $f \in L_p(\Omega)$, $1 < p < \infty$, then the function u_0 has all possible second generalized derivatives $\partial^2 u_0 / \partial x_j \partial x_k \in L_p(\Omega')$, where Ω' is any interior sub-domain of Ω.

Both these theorems follow from the properties of Newtonian potential, stated in the Remarks to § 6, Chapter 11 (page 236). For further information see reference [12].

§ 7. Elliptic Equations of Higher Order and Systems of Equations

The variational method enables us to solve problems considerably more general and more complex than the Dirichlet problem for a second-order elliptic equation. As an example, we consider the first boundary-value problem (with homogeneous boundary conditions) for elliptic equations of order higher than the second.

In the most general case it is possible to write a formally self-adjoint equation of order $2s$ in the space E_m in the following form:

$$\sum_{k=0}^{s}(-1)^k \sum \frac{\partial^k}{\partial x_{i_1} \partial x_{i_2} \ldots \partial x_{i_k}} \left(A_{j_1 j_2 \ldots j_k}^{i_1 i_2 \ldots i_k}(x) \frac{\partial^k u}{\partial x_{j_1} \partial x_{j_2} \ldots \partial x_{j_k}} \right) = f(x). \quad (1)$$

In the inner sum, addition is performed over all possible values of the indices $i_1, i_2, \ldots, i_k$ and $j_1, j_2, \ldots, j_k$, each of which ranges over the values $1, 2, \ldots, m$ independently of the others. The coefficients $A_{j_1 j_2 \ldots j_k}^{i_1 i_2 \ldots i_k}$ remain unchanged under any permutations of either the subscripts or the superscripts among themselves, or under an exchange where the subscripts become superscripts, and vice versa.

As in the case of a second-order equation, the fact that equation (1) is of elliptic type is due to the nature of its highest coefficients, corresponding to the value of the index $k = s$. Equation (1) is said to be *non-degenerate elliptic* in a domain $\Omega \subset E_m$ if the following condition is satisfied. Let $t_{i_1 i_2 \ldots i_s}$ be real variables which are invariant with respect to permutations among the subscripts $i_1, i_2, \ldots, i_s$; there exists a constant $\mu_0 > 0$ such that for any $x \in \bar{\Omega}$ and for any values of the variables $t_{i_1 i_2 \ldots i_s}$ the inequality

$$\sum A_{j_1 j_2 \ldots j_s}^{i_1 i_2 \ldots i_s}(x) t_{i_1 i_2 \ldots i_s} t_{j_1 j_2 \ldots j_s} \geq \mu_0 \sum t_{i_1 i_2 \ldots i_s}^2 \tag{2}$$

holds. We shall assume henceforth that this condition is fulfilled.

The *first boundary-value problem* (with homogeneous boundary conditions) for equation (1) is the problem of integrating this equation in the given domain Ω subject to the boundary conditions:

$$u|_\Gamma = 0, \quad \frac{\partial u}{\partial x_{i_1}}\bigg|_\Gamma = 0, \quad \frac{\partial^2 u}{\partial x_{i_1} \partial x_{i_2}}\bigg|_\Gamma = 0, \ldots,$$

$$\frac{\partial^{s-1} u}{\partial x_{i_1} \partial x_{i_2} \ldots \partial x_{i_{s-1}}}\bigg|_\Gamma = 0, \tag{3}$$

where Γ denotes the boundary of the domain Ω and the subscripts $i_1, i_2, \ldots, i_{s-1}$ take all the values $1, 2, \ldots, m$ independently of one another. We shall suppose that the domain Ω is finite and that the boundary Γ is piecewise-smooth. We assume that $A_{j_1 j_2 \ldots j_k}^{i_1 i_2 \ldots i_k} \in C^{(k)}(\bar{\Omega})$, $k = 0, 1, \ldots, s$.

The problem (1), (3) generates a unique operator, which we shall denote by $\mathfrak{A}_s$. Its domain of definition $D(\mathfrak{A}_s)$ we take to be the set of functions belonging to the class $C^{(2s)}(\bar{\Omega})$ and satisfying condition (3); this operator acts according to the formula

$$\mathfrak{A}_s u = \sum_{k=0}^s (-1)^k \sum \frac{\partial^k}{\partial x_{i_1} \partial x_{i_2} \ldots \partial x_{i_k}} \left(A_{j_1 j_2 \ldots j_k}^{i_1 i_2 \ldots i_k}(x) \frac{\partial^k u}{\partial x_{j_1} \partial x_{j_2} \ldots \partial x_{j_k}} \right).$$

We shall prove that the operator $\mathfrak{A}_s$ is positive definite in the space $L_2(\Omega)$ if the lowest terms of equation (1) satisfy the inequality

$$\sum A_{j_1 j_2 \ldots j_k}^{i_1 i_2 \ldots i_k}(x) t_{i_1 i_2 \ldots i_k} t_{j_1 j_2 \ldots j_k} \geq 0, \quad k = 0, 1, \ldots, s-1, \tag{4}$$

where $t_{i_1 i_2 \ldots i_k}$ are arbitrary real numbers, invariant under a permutation of the subscripts $i_1, i_2, \ldots, i_k$. We form the scalar product

$$(\mathfrak{A}_s u, u) = \int_\Omega u \sum_{k=0}^s (-1)^k \sum \frac{\partial^k}{\partial x_{i_1} \partial x_{i_2} \ldots \partial x_{i_k}} \left(A_{j_1 j_2 \ldots j_k}^{i_1 i_2 \ldots i_k} \frac{\partial^k u}{\partial x_{j_1} \partial x_{j_2} \ldots \partial x_{j_k}} \right) dx.$$

Integrating by parts and using the fact that the surface integrals vanish (because of the boundary conditions (3)), we obtain

$$(\mathfrak{A}_s u, u) = \int_\Omega \sum_{k=0}^s \sum A_{j_1 j_2 \ldots j_s}^{i_1 i_2 \ldots i_s} \frac{\partial^k u}{\partial x_{i_1} \partial x_{i_2} \ldots \partial x_{i_k}} \frac{\partial^k u}{\partial x_{j_1} \partial x_{j_2} \ldots \partial x_{j_k}} dx. \tag{5}$$

On the right-hand side we discard the non-negative sums corresponding to values $k = 0, 1, \ldots, s-1$ of the index, and estimate the remaining sum with the aid of inequality (2). This gives the relation

$$(\mathfrak{A}_s u, u) \geq \mu_0 \int_\Omega \sum \left(\frac{\partial^s u}{\partial x_{i_1} \partial x_{i_2} \ldots \partial x_{i_s}} \right)^2 dx. \tag{6}$$

As the conditions (3) clearly show, the function $u(x)$ itself, and all its derivatives which occur in (3), obey the Friedrichs inequality. This yields the following chain of inequalities:

$$\|u\|^2 = \int_\Omega u^2 \, dx \leq \kappa \int_\Omega \sum \left(\frac{\partial u}{\partial x_{i_1}} \right)^2 dx \leq \kappa^2 \int_\Omega \sum \left(\frac{\partial^2 u}{\partial x_{i_1} \partial x_{i_2}} \right)^2 dx$$

$$\ldots \leq \kappa^s \int_\Omega \sum \left(\frac{\partial^s u}{\partial x_{i_1} \partial x_{i_2} \ldots \partial x_{i_s}} \right)^2 dx. \tag{7}$$

Comparing the relations (6) and (7), we obtain the inequality

$$(\mathfrak{A}_s u, u) \geq \gamma^2 \|u\|^2, \qquad \gamma = \sqrt{\frac{\mu_0}{\kappa^s}}, \tag{8}$$

which shows that the operator $\mathfrak{A}_s$ is positive definite. Hence it follows that the problem (1), (3) has one, and only one, generalized solution; this can be obtained by solving the problem of minimizing the functional

$$\int_\Omega \left[\sum_{k=0}^s \sum A_{j_1 j_2 \ldots j_k}^{i_1 i_2 \ldots i_k} \frac{\partial^k u}{\partial x_{i_1} \partial x_{i_2} \ldots \partial x_{i_k}} \frac{\partial^k u}{\partial x_{j_1} \partial x_{j_2} \ldots \partial x_{j_k}} - 2fu \right] dx \tag{9}$$

in the corresponding energy space.

The expression (1) can also be used to represent a system of some number N of equations with N unknown functions, if we understand by $u(x)$ and $f(x)$ vector-functions with N components, and by $A_{j_1 j_2 \ldots j_k}^{i_1 i_2 \ldots i_k}(x)$ square matrices of order N. We shall assume that these matrices are invariant with respect to any permutation of either the superscripts or subscripts amongst themselves, while if the superscripts are replaced by the subscripts, and vice versa, a matrix becomes its transpose. It follows that condition (2) for the system can be written as

$$(\sum A_{j_1 j_2 \ldots j_s}^{i_1 i_2 \ldots i_s}(x) t_{i_1 i_2 \ldots i_s}, t_{j_1 j_2 \ldots j_s}) \geq \mu_0 \sum \|t_{i_1 i_2 \ldots i_s}\|^2. \tag{10}$$

Here μ_0 is a positive constant, $t_{i_1 i_2 \ldots i_s}$ is an arbitrary N-component vector invariant under permutations of the subscripts, and the symbols

(,) and $\| \ \|$ denote respectively scalar product and norm of vectors in N-dimensional Euclidean space. Condition (4) is amended in a similar way.

Systems of the form (1) which satisfy condition (10) belong to the class of so-called "strongly elliptic" systems *.

All the results of the present section can be extended without difficulty to systems of the form (1) which satisfy (10) and a suitably amended version of condition (4).

§ 8. Dirichlet Problem for an Infinite Domain

In the elliptic equation

$$-\frac{\partial}{\partial x_j}\left(A_{jk}\frac{\partial u}{\partial x_k}\right) + Cu = f(x)$$

let the matrix of the coefficients $A_{jk}(x)$ be positive, and the coefficient $C(x)$ satisfy the inequality

$$C(x) \geq C_0 = \text{const.} > 0.$$

Then it can easily be verified that the operator of the corresponding Dirichlet problem continues to be positive definite when the domain is infinite, and that this problem (with homogeneous boundary conditions) has a generalized solution.

The interesting case is when $C(x) \equiv 0$.

For simplicity we restrict ourselves to Laplace's equation with homogeneous boundary conditions. More complicated cases involving infinite domains are treated in reference [13].

Let Ω be an infinite domain with a *finite* piecewise-smooth boundary Γ. In this domain we pose the Dirichlet problem

$$-\Delta u = f(x), \tag{1}$$

$$u|_\Gamma = 0. \tag{2}$$

Denote by $\mathfrak{B}$ the operator generated by this problem. For its domain of definition $D(\mathfrak{B})$ it is convenient to take the set of functions of class $C^{(2)}(\bar{\Omega})$ which are equal to zero on the boundary and in some neighbourhood (separately defined for each function) of the point at infinity; the operator $\mathfrak{B}$ acts according to the formula $\mathfrak{B}u = -\Delta u$. We shall regard $\mathfrak{B}$ as an oper-

---

* For a more extensive treatment of strongly-elliptic systems cf. the article by M. I. Vishik [2].

ator in $L_2(\Omega)$. We now prove that this operator is positive, but not positive definite. Construct the scalar product

$$(\mathfrak{B}u, v) = -\int_\Omega v \Delta u \, dx; \quad u, v \in D(\mathfrak{B}). \tag{3}$$

The function $u(x)$ is different from zero only in some finite domain, over which the integration in (3) is then actually performed. Therefore we can apply Green's formula (formula (6.7), Chapter 10) to the integral in (3). Using the fact that $u = 0$ on the boundary of the domain in question, we obtain

$$(\mathfrak{B}u, v) = \int_\Omega \frac{\partial u}{\partial x_k} \frac{\partial v}{\partial x_k} \, dx,$$

and so the operator $\mathfrak{B}$ is symmetric. When $v = u$ we have

$$(\mathfrak{B}u, u) = \int_\Omega \sum_{k=1}^m \left(\frac{\partial u}{\partial x_k}\right)^2 dx \geq 0. \tag{4}$$

If $(\mathfrak{B}u, u) = 0$, then, obviously,

$$\frac{\partial u}{\partial x_k} \equiv 0, \quad k = 1, 2, \ldots, m,$$

and

$$u(x) \equiv \text{const.}$$

But $u|_\Gamma = 0$, so that $u(x) \equiv 0$. This proves that the operator is positive.

In the infinite domain Ω with a finite boundary we can construct a tube of side a, arbitrarily large. We choose the coordinate system such that the cube is defined by the inequalities

$$0 \leq x_k \leq a, \quad k = 1, 2, \ldots, m.$$

Consider the function

$$u_a(x) = \begin{cases} \prod_{k=1}^m \sin^3 \frac{\pi x_k}{a} & \text{inside the cube,} \\ 0 & \text{outside the cube.} \end{cases} \tag{5}$$

Obviously $u_a(x) \in D(\mathfrak{B})$. A simple calculation shows that

$$\frac{(\mathfrak{B}u_a, u_a)}{\|u_a\|^2} = \frac{c}{a^2}, \quad c = \text{const.}$$

Hence
$$\lim_{a\to\infty}\frac{(\mathfrak{B}u_a, u_a)}{\|u_a\|^2} = 0$$

and consequently

$$\inf_{u\in D(\mathfrak{B})}\frac{(\mathfrak{B}u, u)}{\|u\|^2} = 0. \tag{6}$$

Equation (6) implies that $\mathfrak{B}$ is not a positive definite operator.

In accordance with the discussion of § 10, Chapter 5, we can associate with the operator $\mathfrak{B}$ an energy space $H_\mathfrak{B}$. Since the operator $\mathfrak{B}$ is merely positive, not all elements of the space $H_\mathfrak{B}$ belong to the original space $L_2(\Omega)$. It can be shown (this is left to the reader) that the space $H_\mathfrak{B}$ consists of precisely those functions which 1) are defined almost everywhere in Ω, 2) have generalized first derivatives whose squares are integrable in Ω, and 3) satisfy boundary condition (2) in the following sense: if $u \in H_\mathfrak{B}$, then there exists a sequence of functions $\{u_n(x)\}$, $u_n \in C^{(2)}(\overline{\Omega})$ such that $u_n(x) = 0$ on the surface Γ and for sufficiently large $|x|$, and satisfy the relation

$$\int_\Omega \sum_{k=1}^m \left(\frac{\partial u_n}{\partial x_k} - \frac{\partial u}{\partial x_k}\right)^2 dx \underset{n\to\infty}{\to} 0.$$

It was shown in § 10, Chapter 5, that the problem (1)–(2) has a solution if and only if the functional (u, f) is bounded in $H_\mathfrak{B}$, where $u \in H_\mathfrak{B}$. It can be shown in turn that for this to be true it is necessary and sufficient that there exists a vector $F(x)$ such that $f(x) = \operatorname{div} F(x)$ and $\|F(x)\| \in L_2(\Omega)$. Here the divergence is understood in a generalized sense (similar to generalized derivative), and the symbol $\|F(x)\|$ denotes the norm of the vector $F(x)$ in m-dimensional Euclidean space.

This can be stated more simply, but only as a sufficient condition: the generalized solution of the problem (1)–(2) exists if the space has dimension $m > 2$, and if the integral

$$\int_\Omega |x|^2 f^2(x)\,dx$$

converges.

Exercises

1. Show that the series (4.8) can be differentiated twice term-by-term, and that this leads to series which converge in the metric of L_2. Hence deduce that when $f \in L_2(\Omega)$ the generalized solution of problem (4.3)–(4.4) has generalized second derivatives, square-integrable in Ω.

2. Show that the energy space $H_{\mathfrak{A}_s}$ of the operator $\mathfrak{A}_s$ (§ 7) consists of precisely those functions which satisfy the following conditions:

1) both the functions themselves and all their generalized derivatives of order $\leqq s$ belong to the class $L_2(\Omega)$;

2) these functions satisfy boundary conditions (7.2) in the following sense: for any function $u \in H_{\mathfrak{A}_s}$ there exists a sequence of functions $u_n(x) \in C^{(s)}(\bar{\Omega})$, satisfying conditions (7.2) and the limiting relations

$$\left\| \frac{\partial^k (u_n - u)}{\partial x_{i_1} \partial x_{i_2} \ldots \partial x_{i_k}} \right\|_{n \to \infty} \to 0.$$

Here the norm is understood in the sense of the metric of $L_2(\Omega)$, $k = 0, 1, \ldots, s$, and the subscripts $i_1, i_2, \ldots, i_k$ take the values $1, 2, \ldots, m$, independently of one another.

3) The simplest elliptic equation of order $2s$ is the so-called *polyharmonic* equation

$$(-1)^s \Delta^s u = f(x).$$

A singular solution of the homogeneous polyharmonic equation

$$\Delta^s u = 0$$

is the function

$$v(x, \xi) = \begin{cases} cr^{2s-m} \ln \dfrac{1}{r}, & m \text{ even, } 2s > m, \\ cr^{2s-m} & \text{otherwise.} \end{cases}$$

Here c is a constant, and $r = |x - \xi|$.

Show that if $f \in C^{(1)}(\bar{\Omega})$, and

$$\psi(x) = \int_\Omega v(x, \xi) f(\xi) \, d\xi,$$

then $\psi \in C^{(2s-1)}(\bar{\Omega}) \cap C^{(2s)}(\Omega)$, and, for suitably chosen c, $(-1)^s \Delta^s \psi(x) = f(x)$. Find this value of c.

4. Prove that if $u_0(x)$ is a solution of the equation $(-1)^s \Delta^s u = f(x)$ with boundary conditions (7.3), and if $f \in C^{(1)}(\Omega)$, then $u_0 \in C^{(2s)}(\Omega)$.

Chapter 15.
Spectrum of the Dirichlet Problem

1. Integral Representation of Functions which Vanish on the Boundary of a Finite Domain

Let Ω be a finite domain of m-dimensional Euclidean space, bounded by a piecewise-smooth surface Γ. Let the function $u \in C^{(1)}(\bar{\Omega})$, and

$$u|_\Gamma = 0. \tag{1}$$

Denote by ξ a variable point of the domain Ω. In Ω we take some point x and surround it with a sphere Σ_ε (Fig. 14, page 221), whose radius ε is small enough so that the sphere lies wholly inside the domain. In the domain $\Omega \setminus \Sigma_\varepsilon$, the function $u(\xi)$ and the function $v(\xi) = 1/r^{m-2}$, $r = |x-\xi|$, are continuously differentiable right up to the boundary. Hence we can apply Green's First Formula for the Laplace operator ((6.7), Chapter 10) to these functions:

$$\int_{\Omega \setminus \Sigma_\varepsilon} u \Delta v \, d\xi = -\int_{\Omega \setminus \Sigma_\varepsilon} \frac{\partial u}{\partial \xi_k} \frac{\partial v}{\partial \xi_k} d\xi + \int_\Gamma u \frac{\partial v}{\partial v} d\Gamma + \int_{S_\varepsilon} u \frac{\partial v}{\partial v} dS_\varepsilon, \tag{2}$$

where S_ε is the boundary of the sphere Σ_ε.

The function $v(\xi)$ is harmonic in $\Omega \setminus \Sigma_\varepsilon$, so that the integral on the left vanishes. Because of condition (1) the integral over Γ also vanishes. Formula (2) is thus simplified and takes the following form:

$$\int_{\Omega \setminus \Sigma_\varepsilon} \frac{\partial u}{\partial \xi_k} \frac{\partial v}{\partial \xi_k} d\xi = \int_{S_\varepsilon} u \frac{\partial v}{\partial v} dS_\varepsilon. \tag{3}$$

In the limit as $\varepsilon \to 0$ the left-hand side of equation (3) becomes the improper integral

$$\int_\Omega \frac{\partial u}{\partial \xi_k} \frac{\partial v}{\partial \xi_k} d\xi = \int_\Omega \frac{\partial u}{\partial \xi_k} \frac{\partial}{\partial \xi_k} \frac{1}{r^{m-2}} d\xi.$$

The limit of the right-hand side has been evaluated previously, in § 3, Chapter 11, in the process of deriving an integral representation for a function of class $C^{(2)}$; the limit is $(m-2)|S_1|u(x)$. Equating these two limits, we arrive at the required integral representation:

$$u(x) = \frac{1}{(m-2)|S_1|} \int_\Omega \frac{\partial u}{\partial \xi_k} \frac{\partial}{\partial \xi_k} \frac{1}{r^{m-2}} d\xi. \tag{4}$$

Formula (4) is valid for $m > 2$; when $m = 2$ it is replaced by the expression

$$u(x) = \frac{1}{2\pi} \int_\Omega \frac{\partial u}{\partial \xi_k} \frac{\partial}{\partial \xi_k} \ln \frac{1}{r} d\xi. \tag{5}$$

The integral representation (4) (also (5)) has been derived with the assumption that the function $u \in C^{(1)}(\bar{\Omega})$, and that it equals zero on Γ. For our subsequent work it is important that we should be able to extend this representation to functions belonging to the energy space $H_\mathfrak{A}$ of the Dirichlet problem (§ 3, Chapter 14).

We have

$$\frac{\partial}{\partial \xi_k} \frac{1}{r^{m-2}} = -\frac{m-2}{r^{m-1}} \frac{\partial r}{\partial \xi_k} = -\frac{m-2}{r^{m-1}} \frac{\xi_k - x_k}{r}.$$

Hence

$$\left| \frac{\partial}{\partial \xi_k} \frac{1}{r^{m-2}} \right| \leq \frac{m-2}{r^{m-1}}.$$

This last estimate shows that the integral on the right-hand side of equation (4) is an operator with a weak singularity (§ 3, Chapter 7) on $\partial u/\partial \xi_k$; consequently it is bounded in $L_2(\Omega)$ (Theorem 7.3.1).

From Theorem 14.3.1, if $u \in H_\mathfrak{A}$ there exists a sequence $\{u_n\}$ such that $u_n \in C^{(1)}(\bar{\Omega})$, $u_n|_\Gamma = 0$, and

$$\|u_n - u\|_{L_2} \underset{n \to \infty}{\to} 0; \quad \left\| \frac{\partial u_n}{\partial \xi_k} - \frac{\partial u}{\partial \xi_k} \right\|_{L_2} \underset{n \to \infty}{\to} 0, \quad k = 1, 2, \ldots, m.$$

This proves the representation (4) for the functions u_n:

$$u_n(x) = \frac{1}{(m-2)|S_1|} \int_\Omega \frac{\partial u_n}{\partial \xi_k} \frac{\partial}{\partial \xi_k} \frac{1}{r^{m-2}} d\xi.$$

As $n \to \infty$ the left-hand side of this equation tends to $u(x)$ in the metric of $L_2(\Omega)$. At the same time $\partial u_n/\partial \xi_k \to \partial u/\partial \xi_k$ in this metric. The integral operator on the right is bounded, and so the limit can be taken under the integral sign. This again brings us to formula (4), in this case proven for functions from the space $H_{\mathfrak{A}}$.

§ 2. Spectrum of the Dirichlet Problem for a Finite Domain

THEOREM 15.2.1. *For a finite domain with a piecewise-smooth boundary, the operator of the Dirichlet problem for a non-degenerate, self-adjoint, elliptic equation has a discrete spectrum.*

Let M be a bounded set in the space $H_{\mathfrak{A}}$:

$$|||u|||_{\mathfrak{A}} \leq c = \text{const.}, \qquad \forall u \in M.$$

From formula (3.3), Chapter 14, we have

$$\int_\Omega \left[A_{jk} \frac{\partial u}{\partial x_j} \frac{\partial u}{\partial x_k} + Cu^2 \right] dx \leq c^2.$$

Using now the relation (2.3), Chapter 14, we obtain the inequality

$$\int_\Omega \sum_{k=1}^m \left(\frac{\partial u}{\partial x_k} \right)^2 dx \leq \frac{c^2}{\mu_0}.$$

All the more

$$\left\| \frac{\partial u}{\partial x_k} \right\| = \left\{ \int_\Omega \left(\frac{\partial u}{\partial x_k} \right)^2 dx \right\}^{\frac{1}{2}} \leq \frac{c}{\sqrt{\mu_0}}.$$

Thus, the derivatives of the function $u \in M$ generate a bounded set in $L_2(\Omega)$. The integral operator (1.4), acting on $\partial u/\partial \xi_k$, is completely continuous in $L_2(\Omega)$ (Theorem 7.3.2); it transforms the above-mentioned set of derivatives into a compact set in $L_2(\Omega)$. But this latter set coincides with M, because the operator (1.4) recovers any function of M from its first derivative. Hence it follows that *the set M is compact in the space $L_2(\Omega)$.*

Thus any set bounded in $H_{\mathfrak{A}}$ is compact in $L_2(\Omega)$. By Theorem 6.6.1, operator $\mathfrak{A}$ has a discrete spectrum.

From this theorem, together with Theorem 6.6.1, we arrive at the following conclusion:

There exists an enumerable set $\{\lambda_n\}$ of values of the parameter λ for which the problem

$$\Delta u + \lambda u = 0, \quad x \in \Omega, \quad u|_\Gamma = 0 \qquad (1)$$

has a non-trivial solution. The quantities λ_n are eigenvalues of the operator of the Dirichlet problem (more briefly, eigenvalues of the Dirichlet problem), and the corresponding non-trivial solutions of problem (1) are eigenfunctions associated with the eigenvalues λ_n. To each eigenvalue there corresponds only a finite number of linearly-independent eigenfunctions. If several eigenfunctions are associated with one eigenvalue, we shall repeat that eigenvalue the appropriate number of times. All the eigenvalues $\lambda_n > 0$, and

$$\lambda_n \underset{n \to \infty}{\to} \infty. \qquad (2)$$

The system $\{u_n\}$ of eigenfunctions can be assumed to be orthonormal in $L_2(\Omega)$

$$(u_j, u_k) = \delta_{jk}; \quad j, k = 1, 2, \ldots \qquad (3)$$

It is also orthogonal, but not normalized, in $H_\mathfrak{A}$; we have

$$[u_j, u_k]_\mathfrak{A} = 0, \quad j \neq k; \quad [u_k, u_k]_\mathfrak{A} = |||u_k|||_\mathfrak{A}^2 = \lambda_k. \qquad (4)$$

The system of eigenfunctions $\{u_n\}$ is complete in both the spaces $L_2(\Omega)$ and $H_\mathfrak{A}$.

§ 3. Elementary Cases

To construct the actual spectrum of a positive definite operator on the basis of Theorem 6.6.1 is extremely difficult. This is obvious because to do so it is necessary to select a convergent subsequence from a minimizing sequence. Therefore we are interested in examining special cases in which the spectrum of the Dirichlet problem can be constructed explicitly. We therefore now consider examples in which elementary expressions for the eigenvalues and eigenfunctions of the Dirichlet problem for Laplace's equation can be obtained.

1. Parallelepiped in m-dimensional space. If n is a positive integer, then the function

$$u_n(t) = \sin \frac{n\pi t}{a} \qquad (1)$$

of the real variable t, satisfies the differential equation

$$u_n'' + \frac{n^2\pi^2}{a^2} u_n = 0$$

and boundary conditions $u_n(0) = u_n(a) = 0$. Hence it follows that the function

$$u_{n_1 n_2 \ldots n_m}(x) = c \prod_{k=1}^m \sin \frac{n_k \pi x_k}{a_k}, \qquad c = \text{const.} \tag{2}$$

satisfies the differential equation

$$\Delta u + \lambda u = 0, \qquad \lambda = \pi^2 \sum_{k=1}^m \frac{n_k^2}{a_k^2}$$

and is zero on the surface of the parallelepiped

$$0 < x_k < a_k, \qquad k = 1, 2, \ldots, m. \tag{3}$$

Thus, the functions (2) are eigenfunctions of the Dirichlet problem for Laplace's equation in the parallelepiped (3); the corresponding eigenvalues are

$$\lambda_{n_1 n_2 \ldots n_m} = \pi^2 \sum_{k=1}^m \frac{n_k^2}{a_k^2}. \tag{4}$$

The system (2) is orthogonal in the metric of $L_2(\Omega)$, where Ω is here the parellelepiped (3); the system can also be normalized if we put

$$c = 2^{\frac{1}{2}m} \prod_{k=1}^m a_k^{-\frac{1}{2}}. \tag{5}$$

The system (2) is complete in $L_2(\Omega)$ – this is easily proved, starting from the fact that the system of functions (1) is complete in $L_2(0, a)$, and using the ideas of § 2, Chap. 7. Hence it follows that formula (2) exhausts the system of eigenfunctions of the Dirichlet problem for Laplace's equation in a parallelepiped.

2. Circle in the plane. Take polar coordinates ρ, θ, and consider the equation

$$\Delta u + \lambda u = 0,$$

which, when written out in full, becomes *

$$\frac{\partial^2 u}{\partial \rho^2} + \frac{1}{\rho}\frac{\partial u}{\partial \rho} + \frac{1}{\rho^2}\frac{\partial^2 u}{\partial \theta^2} + \lambda u = 0, \qquad (6)$$

with boundary conditions

$$u|_{\rho=1} = 0. \qquad (7)$$

We look for a non-trivial solution of the problem (6)–(7) by the so-called *method of separation of variables*; that is, we look for a solution having the form

$$u = f(\rho)\,\phi(\theta). \qquad (8)$$

Equation (7) becomes a boundary condition for the function $f(\rho)$, namely

$$f(1) = 0. \qquad (9)$$

Substituting from (8) into equation (6), we easily bring the latter to the form

$$\frac{\rho^2}{f}\left(f'' + \frac{1}{\rho}f' + \lambda f\right) = -\frac{\phi''}{\phi}. \qquad (10)$$

The left-hand side in equation (10) is independent of θ, while the right-hand side is independent of ρ; being equal to each other, the two sides of

---

* If x and y are cartesian coordinates in the plane, then $\rho = \sqrt{(x^2+y^2)}$, $\theta = \tan^{-1}(y/x)$. Hence

$$\frac{\partial u}{\partial x} = \cos\theta\,\frac{\partial u}{\partial \rho} - \frac{1}{\rho}\sin\theta\,\frac{\partial u}{\partial \theta}, \qquad \frac{\partial u}{\partial y} = \sin\theta\,\frac{\partial u}{\partial \rho} + \frac{1}{\rho}\cos\theta\,\frac{\partial u}{\partial \rho}$$

and, moreover,

$$\frac{\partial^2 u}{\partial x^2} = \cos^2\theta\,\frac{\partial^2 u}{\partial \rho^2} - \frac{2}{\rho}\cos\theta\sin\theta\,\frac{\partial^2 u}{\partial \rho\partial\theta}$$

$$+ \frac{1}{\rho^2}\sin^2\theta\,\frac{\partial^2 u}{\partial \theta^2} + \frac{\sin^2\theta}{\rho}\frac{\partial u}{\partial \rho} - \frac{\sin 2\theta}{\rho^2}\frac{\partial u}{\partial \theta},$$

$$\frac{\partial^2 u}{\partial y^2} = \sin^2\theta\,\frac{\partial^2 u}{\partial \rho^2} + \frac{2}{\rho}\cos\theta\sin\theta\,\frac{\partial^2 u}{\partial \rho\partial\theta}$$

$$+ \frac{1}{\rho^2}\cos^2\theta\,\frac{\partial^2 u}{\partial \theta^2} + \frac{\cos^2\theta}{\rho}\frac{\partial u}{\partial \rho} + \frac{\sin 2\theta}{\rho^2}\frac{\partial u}{\partial \theta}.$$

Hence

$$\Delta u = \frac{\partial^2 u}{\partial x^2} + \frac{\partial^2 u}{\partial y^2} = \frac{\partial^2 u}{\partial \rho^2} + \frac{1}{\rho}\frac{\partial u}{\partial \rho} + \frac{1}{\rho^2}\frac{\partial^2 u}{\partial \theta^2}.$$

(10) must be independent of both ρ and θ, and, consequently, they must be equal to some constant, which we denote by n^2. Equation (10) now separates into two ordinary differential equations

$$\phi''(\theta)+n^2\phi(\theta) = 0 \tag{11}$$

and

$$f''(\rho)+\frac{1}{\rho}f'(\rho)+\left(\lambda-\frac{n^2}{\rho^2}\right)f(\rho) = 0. \tag{12}$$

The general solution of (11) has the form

$$\phi(\theta) = C_1 \cos n\theta + C_2 \sin n\theta. \tag{13}$$

The function (8) must be a single-valued function of any point in the plane, and therefore n must be an integer. It is sufficient to consider only non-negative n – a change in the sign of n would only alter the constant C_2.

The general solution of (12) is expressed through Bessel functions *

$$f(\rho) = C_3 J_n(\sqrt{\lambda}\rho)+C_4 Y_n(\sqrt{\lambda}\rho).$$

By definition, the eigenfunction of the Dirichlet problem belongs to the energy space $H_\mathfrak{A}$ (§ 3, Chapter 14), and, consequently, the first derivatives of this function are square-integrable in the circle $\rho < 1$. The Bessel function of the second kind, $Y_n(\sqrt{\lambda}\rho)$, does not have this property, so that we must have $C_4 = 0$. The value of the constant C_3 is not required in what follows (except that it is necessary to have $C_3 \neq 0$), and so we put $C_3 = 1$. Now

$$f(\rho) = J_n(\sqrt{\lambda}\rho).$$

From condition (9) it now follows that

$$\lambda = j_{n,k}^2, \tag{14}$$

where $j_{n,k}$ is the kth positive zero of the Bessel function of the first kind J_n. Finally,

$$f(\rho) = J_n(j_{n,k}\rho). \tag{15}$$

Formulae (13)–(15) lead to the conclusion that the quantities

$$j_{n,k}^2; \quad n = 0, 1, 2, \ldots; \quad k = 1, 2, 3, \ldots, \tag{16}$$

---

* For the definition and properties of Bessel functions, see G. N. Watson, A Treatise on the Theory of Bessel Functions.

are eigenvalues of the Dirichlet problem for the Laplace operator in the unit circle; to each of the eigenvalues $j_{0,k}^2$ there corresponds one eigenfunction

$$J_0(j_{0,k}\rho), \qquad (17_1)$$

to each eigenvalue $j_{n,k}^2$, $n > 0$, there correspond two eigenfunctions

$$J_n(j_{n,k}\rho)\cos n\theta, \qquad J_n(j_{n,k}\rho)\sin n\theta. \qquad (17_2)$$

If Ω denotes the circle $\rho < 1$, then the functions (17) are orthogonal (but not normalized) in $L_2(\Omega)$ and generate a complete system – both these facts follow from well-known properties of Bessel functions, and from our results on completeness of the system of products (Theorem 7.2.2). Hence it follows that formulae (16) and (17) exhaust the totality of eigenvalues and eigenfunctions of the Dirichlet problem for Laplace's equation in a circle.

Remark. The eigenfunctions of the Dirichlet problem for the Laplace operator in the case of a sphere in any number of dimensions are expressible through Bessel functions and spherical harmonics.

§ 4. Estimate of the Magnitude of the Eigenvalues

1. Laplace's equation in an m-dimensional cube. In the m-dimensional cube Q, defined by the inequality

$$0 \leq x_k \leq a, \qquad k = 1, 2, \ldots, m,$$

the eigenvalues of the Dirichlet problem for Laplace's equation are (formula (3.4))

$$\lambda_{n_1 n_2 \ldots n_m} = \frac{\pi^2}{a^2} \sum_{k=1}^m n_k^2. \qquad (1)$$

We write down the quantities (1) in increasing (strictly, non-decreasing) order of magnitude, and denote them by λ_n, where $\lambda_1 \leq \lambda_2 \leq \ldots$. Our problem consists in finding an estimate for the order of magnitude of the quantity λ_n as $n \to \infty$.

Instead of the quantities λ_n it is somewhat more convenient to estimate the quantities

$$\lambda'_n = \frac{a^2}{\pi^2}\lambda_n = \sum_{k=1}^m n_k^2. \qquad (2)$$

Consider a sphere Σ_N, of radius N and with its centre at the origin, in an m-dimensional Euclidean space; we take the quantity N to be sufficiently large, and moreover, to be such that the value of one of the quantities (2) is $\lambda'_n = N^2$. Let v_1 and v_2 denote respectively the least and greatest values of n for which $\lambda'_n = N^2$. The value of v_2 indicates how many of the quantities (2) lie in the closed sphere $\bar{\Sigma}_N$. Similarly, formula (2) shows that v_2 is equal to the number of points having positive-integer coordinates and contained in the sphere $\bar{\Sigma}_N$. This number is equal to the volume of the cubic lattice T_m with unit edge which can be drawn in the first octant of the sphere Σ_N

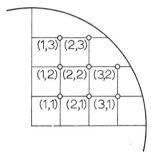

Fig. 27.

(Fig. 27 illustrates the two-dimensional case). The volume of the lattice T_m is less than the volume of the octant itself:

$$v_2 < \frac{|S_1|}{2^m m} N^m. \qquad (3)$$

Now construct a sphere Σ_{N-2}, of radius $N-2$, with centre at the origin. It is not difficult to see that the first octant of this sphere is contained within the lattice T_m. It is sufficient to show that for any boundary vertex $(n_1, n_2, \ldots, n_m)$ of the lattice T_m,

$$\sum_{k=1}^{m} n_k^2 > (N-2)^2. \qquad (4)$$

A boundary vertex is characterised by the fact that it satisfies simultaneously

$$n_1^2 + n_2^2 + \ldots + n_m^2 = \sum_{k=1}^{m} n_k^2 \leqq N^2$$
$$\sum_{k=1}^{m} (n_k+1)^2 = \sum_{k=1}^{m} n_k^2 + 2\sum_{k=1}^{m} n_k + m > N^2. \qquad (5)$$

From the second inequality (5) it follows that

$$\sum_{k=1}^{m} n_k^2 > N^2 - 2\sum_{k=1}^{m} n_k - m \geq N^2 - 2Nm - m = (N-m)^2 - m^2 - m,$$

which is greater than $(N-m-1)^2$ if N is sufficiently large.

Now we have

$$v_2 > \frac{|S_1|}{2^m m}(N-m-1)^m, \tag{6}$$

and inequalities (3) and (6) show that, when N is sufficiently large, we obtain the estimate

$$v_2 = \frac{|S_1|}{2^m m} N^m + o(N^m) = \frac{\pi^{\frac{1}{2}m} N^m}{2^{m-1} m \Gamma(\frac{1}{2}m)} + o(N^m). \tag{7}$$

We now estimate the value of v_1. Obviously $v_1 = v_2 - \sigma$, where σ denotes the number of points having positive-integer coordinates and lying on the spherical surface

$$\sum_{k=1}^{m} n_k^2 = N^2,$$

which bounds the sphere Σ_N. Through these points we draw straight lines parallel to the m-axis. The first $n-1$ coordinates on these lines are positive integers; the intersection of the lines with the surface $n_m = 0$ gives a lattice of points, which enables us to construct a cubic lattice T_{m-1}, similar to T_m, but of one dimension less. It is clear from this that the number σ is equal to the volume of the $((m-1)$-dimensional) lattice T_{m-1}, and hence we obtain a formula analogous to (7):

$$\sigma = \frac{\pi^{\frac{1}{2}(m-1)} N^{m-1}}{2^{m-2}(m-1)\Gamma(\frac{1}{2}(m-1))} + o(N^{m-1}). \tag{8}$$

From (7) and (8), and the equation $v_1 = v_2 - \sigma$, it follows that

$$v_1 = \frac{\pi^{\frac{1}{2}m} N^m}{2^{m-1} m \Gamma(\frac{1}{2}m)} + o(N^m). \tag{9}$$

We recall that we have denoted by n any number for which $\lambda'_n = N^2$, and that $v_1 \leq n \leq v_2$. Equations (7) and (9) give

$$n = \frac{\pi^{\frac{1}{2}m} N^m}{2^{m-1} m \Gamma(\frac{1}{2}m)} + o(N^m).$$

Hence
$$\lambda_n = c_m n^{2/m} + o(n^{2/m}), \tag{10}$$
where c_m is a constant, which is easy to evaluate.

2. **General case.** Consider the operator $\mathfrak{A}$ of the Dirichlet problem (§ 2, Chapter 14); let the coefficients A_{jk} and C satisfy conditions (2.3) and (2.4) of Chapter 14, so that inequality (2.11), Chapter 14, holds. For what follows it is important that reverse inequality should hold.

The coefficients $A_{jk}(x)$ are continuous, and, consequently, bounded in the closed domain $\bar{\Omega}$. It follows from this that the largest eigenvalue of the matrix of the coefficients $A_{jk}(x)$ is also bounded. Let M_0 be its upper bound. Then

$$A_{jk}(x) t_j t_k \leq M_0 \sum_{k=1}^{m} t_k^2. \tag{11}$$

The coefficient $C(x)$ is also continuous and bounded in $\bar{\Omega}$. Let $C(x) \leq M_1$. Thus, if $u \in D(\mathfrak{A})$, then

$$(\mathfrak{A}u, u) \leq M_0 \int_\Omega \sum_{k=1}^{m} \left(\frac{\partial u}{\partial x_k}\right)^2 dx + M_1 \int_\Omega u^2 \, dx.$$

Applying Friedrichs's inequality to the second integral, we eventually obtain

$$(\mathfrak{A}u, u) \leq \mu_1 \int_\Omega \sum_{k=1}^{m} \left(\frac{\partial u}{\partial x_k}\right)^2 dx; \qquad \mu_1 = M_0 + \kappa M_1. \tag{12}$$

Construct two cubes Q_1 and Q_2, the former containing the domain Ω and the latter contained within this domain. Denote by $\mathfrak{A}_1$ and $\mathfrak{A}_2$ respectively the operators of the Dirichlet problem for Laplace's equation in the cubes Q_1 and Q_2.

Denote also by λ_n, $\lambda_n^{(1)}$, $\lambda_n^{(2)}$ the eigenvalues of the operators $\mathfrak{A}$, $\mathfrak{A}_1$, $\mathfrak{A}_2$ written out in non-decreasing order of magnitude. These operators are related by the inequalities (prove!)

$$\mu_0 \mathfrak{A}_1 \leq \mathfrak{A} \leq \mu_1 \mathfrak{A}_2.$$

Here μ_0 is the constant of inequality (2.11), Chapter 14. By virtue of the Minimax Principle,

$$\mu_0 \lambda_n^{(1)} \leq \lambda_n \leq \mu_1 \lambda_n^{(2)}. \tag{13}$$

But the quantities $\lambda_n^{(1)}$ and $\lambda_n^{(2)}$ satisfy relations of the form (10), and so we obtain for the λ_n – the eigenvalues of the operator $\mathfrak{A}$ – the two-sided

estimate
$$c_1 n^{2/m} \leq \lambda_n \leq c_2 n^{2/m}; \qquad c_1, c_2 = \text{const.} > 0. \tag{14}$$

Remark. A more precise statement can be made: the limit
$$\lim \frac{n}{\lambda_n^{\frac{1}{2}m}}$$
exists, and this limit is equal to the quantity
$$c'_m \int_\Omega \frac{dx}{\sqrt{\text{Det } A(x)}}, \qquad c'_m = \text{const.},$$
where $A(x)$ is the matrix of the coefficients $A_{jk}(x)$, and c'_m depends only on the dimensionality m of the space. Cf. the article by T. Carleman [5] and the book by V. I. Smirnov [17], Vol. IV.

Chapter 16. The Neumann Problem

§ 1. The Case of Positive $C(x)$

Consider the formally self-adjoint equation of elliptic type

$$-\frac{\partial}{\partial x_j}\left(A_{jk}(x)\frac{\partial u}{\partial x_k}\right) + C(x)u = f(x), \tag{1}$$

whose solution is sought in a finite domain $\Omega \subset E_m$ having a piecewise-smooth boundary Γ.

We shall assume that

$$A_{jk} \in C^{(1)}(\bar{\Omega}), \qquad C \in C(\bar{\Omega})$$

and that equation (1) is non-degenerate in $\bar{\Omega}$. Then for any real positive numbers $t_1, t_2, \ldots, t_m$ we have the inequality

$$A_{jk}(x)t_j t_k \geq \mu_0 \sum_{k=1}^{m} t_k^2, \tag{2}$$

where μ_0 is a positive constant.

We now pose for equation (1) the homogeneous boundary condition of the Neumann problem

$$A_{jk}\frac{\partial u}{\partial x_k}\cos(v, x_j)\Big|_\Gamma = 0, \tag{3}$$

where v is the outward normal to Γ.

Denote by $\mathfrak{N}$ the operator generated by the Neumann problem, and take for the domain of definition $D(\mathfrak{N})$ a set of functions from $C^{(2)}(\bar{\Omega})$, satisfying condition (3). On this set the operator $\mathfrak{N}$ acts according to the formula

$$\mathfrak{N}u = -\frac{\partial}{\partial x_j}\left(A_{jk}\frac{\partial u}{\partial x_k}\right) + Cu.$$

We shall consider $\mathfrak{N}$ as an operator in $L_2(\Omega)$, and show that it is symmetric.

Its domain of definition is dense in $L_2(\Omega)$, because it obviously contains a set of finite functions which is dense in $L_2(\Omega)$.

We next construct the scalar product

$$(\mathfrak{N}u, v) = -\int_\Omega v \frac{\partial}{\partial x_j}\left(A_{jk}\frac{\partial u}{\partial x_k}\right)dx + \int_\Omega Cuv\,dx; \qquad u, v \in D(\mathfrak{N}).$$

Integrating by parts the first term on the right, we obtain

$$(\mathfrak{N}u, v) = -\int_\Gamma vA_{jk}\frac{\partial u}{\partial x_k}\cos(v, x_j)\,d\Gamma$$
$$+ \int_\Omega A_{jk}\frac{\partial u}{\partial x_k}\frac{\partial v}{\partial x_j}\,dx + \int_\Omega Cuv\,dx.$$

Because of the boundary condition (3) the first integral on the right-hand side vanishes, and the two remaining integrals are obviously symmetric. We thus have

$$(\mathfrak{N}u, v) = \int_\Omega \left[A_{jk}\frac{\partial v}{\partial x_j}\frac{\partial u}{\partial x_k} + Cuv\right]dx, \tag{4}$$

and, on putting $v = u$

$$(\mathfrak{N}u, u) = \int_\Omega \left[A_{jk}\frac{\partial u}{\partial x_j}\frac{\partial u}{\partial x_k} + Cu^2\right]dx. \tag{5}$$

It is clear that

$$\int_\Omega A_{jk}\frac{\partial u}{\partial x_j}\frac{\partial u}{\partial x_k}\,dx \geqq \mu_0 \int_\Omega \sum_{k=1}^m \left(\frac{\partial u}{\partial x_k}\right)^2 dx \geqq 0,$$

and we easily obtain the following sufficient condition for positive definiteness of the operator $\mathfrak{N}$:

$$C(x) \geqq C_0 = \text{const.} > 0; \tag{6}$$

in this case

$$(\mathfrak{N}u, u) \geqq C_0 \int_\Omega u^2\,dx = C_0\|u\|^2,$$

and this is just the positive-definiteness inequality, with the constant $\gamma = \sqrt{C_0}$. Naturally when $C(x)$ has the required form the theory developed earlier is applicable and the Neumann problem also has one and only one generalized solution.

§ 2. The Case $C(x) \equiv 0$

This case, to which all the remaining sections of the present chapter are devoted, is more interesting and somewhat more difficult. If $C(x) \equiv 0$, then

$$\mathfrak{N}u = -\frac{\partial}{\partial x_j}\left(A_{jk}\frac{\partial u}{\partial x_k}\right) \tag{1}$$

and

$$(\mathfrak{N}u, u) = \int_\Omega A_{jk} \frac{\partial u}{\partial x_j} \frac{\partial u}{\partial x_k} dx; \tag{2}$$

the boundary conditions (1.3) remain unchanged.

In the case under consideration, the operator $\mathfrak{N}$ is not only not positive definite, but not even positive. To see this, it is sufficient to look at the function $u_0(x) \equiv 1$. This function has all derivatives and satisfies the boundary condition (1.3); hence $u_0 \in D(\mathfrak{N})$. Also, it is obvious that $\|u_0\| > 0$. At the same time $(\mathfrak{N}u_0, u_0) = 0$, which would be impossible if $\mathfrak{N}$ were a positive operator.

The Neumann problem

$$\mathfrak{N}u = f \tag{3}$$

or, writing it out in full,

$$\begin{gathered}-\frac{\partial}{\partial x_j}\left(A_{jk}(x)\frac{\partial u}{\partial x_k}\right) = f(x),\\ A_{jk}(x)\frac{\partial u}{\partial x_k}\cos(v, x_j)|_\Gamma = 0,\end{gathered} \tag{3'}$$

has no solution unless the function $f(x)$ is subject to some special condition, which we now elucidate.

Suppose that problem (3) has a solution $u \in D(\mathfrak{N})$.

Integrate both sides of the differential equation (3') over the domain Ω. Integrating the left-hand side by parts, and utilizing the boundary condition (3'), we obtain the required condition

$$\int_\Omega f(x) dx = (f, 1) = 0. \tag{4}$$

Thus it is expedient to consider equation (3), not for arbitrary $f \in L_2(\Omega)$, but only for those f which belong to the subspace orthogonal to unity. We shall denote this subspace by $\tilde{L}_2(\Omega)$. The condition (4) just obtained also permits the following formulation: when $C(x) \equiv 0$, the operator $\mathfrak{N}$ transforms any function of $D(\mathfrak{N})$ into a function in $\tilde{L}_2(\Omega)$.

On the other hand *, if problem (3) has a solution, then this solution is not unique: if the function $u_0(x)$ solves problem (3), then the same problem is also solved by the function $u_0(x)+c$, where c is an arbitrary constant. Other solutions do not exist. In fact, if $u_0(x)$ and $u_1(x)$ are two solutions of problem (3), then the difference $v(x) = u_0(x)-u_1(x)$ satisfies the homogeneous equation $\mathfrak{N}v = 0$. From formula (1.5) – with $C(x) \equiv 0$ – we have

$$\int_\Omega A_{jk} \frac{\partial v}{\partial x_j} \frac{\partial v}{\partial x_k} \, dx = 0.$$

The matrix of the coefficients A_{jk} is positive definite, so that necessarily $\partial v/\partial x_k \equiv 0$, $k = 1, 2, \ldots, m$, and $v(x) \equiv \text{const.}$, which we set out to prove.

The solution to the Neumann problem can be made unique if we require that it should belong to the subspace $\tilde{L}_2(\Omega)$ introduced above – this requirement defines the constant c in a unique way.

This last condition can be formulated as follows: we contract the operator $\mathfrak{N}$ by replacing its domain of definition $D(\mathfrak{N})$ with the more restricted domain $D(\mathfrak{N}) \cap \tilde{L}_2(\Omega)$. We denote this contracted operator by $\mathfrak{N}_0$. Its domain of definition is then $D(\mathfrak{N}_0) = D(\mathfrak{N}) \cap \tilde{L}_2(\Omega)$, and its domain of values belongs to $\tilde{L}_2(\Omega)$ as before.

Thus, both the domain of definition and the domain of values of the operator $\mathfrak{N}_0$ belong to the subspace $\tilde{L}_2(\Omega)$. But then $\tilde{L}_2(\Omega)$ can be regarded as the space in which the operator $\mathfrak{N}_0$ acts. It is not difficult to see that the operator $\mathfrak{N}_0$ is positive in this space. In fact, equation (1.5) obviously remains valid for this operator:

$$(\mathfrak{N}_0 u, u) = \int_\Omega A_{jk} \frac{\partial u}{\partial x_j} \frac{\partial u}{\partial x_k} \, dx. \tag{5}$$

Hence $(\mathfrak{N}_0 u, u) \geq 0$. If $(\mathfrak{N}_0 u, u) = 0$, then $\partial u/\partial x_k \equiv 0$, $k = 1, 2, \ldots, m$ and $u(x) = c = \text{const.}$ But $u \in \tilde{L}_2(\Omega)$, so that

$$0 = (c, 1) = c \int_\Omega d\Omega = c|\Omega|.$$

Hence $c = 0$ and $u(x) \equiv 0$, which was to be proven.

Our next task is to prove that the operator $\mathfrak{N}_0$ is positive definite. This is the object of the following three sections.

---

* An argument analogous to the following was used in Chapter 12 for Laplace's equation.

§ 3. Integral Representation of S. L. Sobolev

Consider a finite domain $\Omega \subset E_m$ which has the following properties: any point in this domain or on its boundary can be made the vertex of a spherical sector, having a fixed radius R, angle 2α, and lying wholly (except perhaps for the vertex) inside the domain Ω (Fig. 28). It is convenient to speak of such domains as *satisfying the cone condition*.

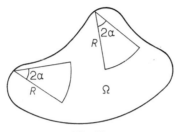

Fig. 28.

Introduce a function $\psi(t)$ of a real variable t, infinitely differentiable on the segment $0 \leq t \leq 1$, for which

$$\psi(t) = \begin{cases} 1, & 0 \leq t \leq \tfrac{1}{3}, \\ 0, & \tfrac{2}{3} \leq t \leq 1. \end{cases}$$

This function can be constructed, for example, by averaging the function

$$\psi_0(t) = \begin{cases} 1, & t < \tfrac{1}{2}, \\ 0, & t > \tfrac{1}{2} \end{cases}$$

with an averaging radius $h < \tfrac{1}{6}$.

Let $u \in C^{(1)}(\Omega)$. Take a point x in Ω, and construct a spherical sector with vertex x, lying in Ω. Introduce spherical coordinates* $r = |x-\xi|$, $\vartheta_1, \vartheta_2, \ldots, \vartheta_{m-1}$ with centre at the point x; these coordinates are chosen such that the conical surface of the spherical sector has equation $\vartheta_1 = \alpha$. Put

$$v(\xi) = v(r, \theta) = v(r, \vartheta_1, \vartheta_2, \ldots, \vartheta_{m-1}) = u(\xi)\psi\left(\frac{r}{R}\right). \tag{1}$$

Obviously

$$v(0, \theta) = u(x), \qquad v(R, \theta) = 0, \tag{2}$$

---

* We assume here that $m > 2$; the subsequent argument can easily be modified for the case $m = 2$.

$$\frac{\partial v}{\partial r} = u(\xi)\frac{\partial \psi(r/R)}{\partial r} + \psi\left(\frac{r}{R}\right)\frac{\partial u(\xi)}{\partial r}. \tag{3}$$

From formula (2) it follows that

$$u(x) = -\int_0^R \frac{\partial v}{\partial r}\,dr. \tag{4}$$

Integrate equation (4) over that portion S_1' of the unit sphere S_1 in which $0 \leq \vartheta_1 \leq \alpha$ (i.e. over that portion cut out by a lateral surface of the spherical sector). Then the left-hand side will be multiplied by the positive constant $b = |S_1'|$; dividing by b, we obtain the formula

$$u(x) = -\frac{1}{b}\int_{V_x} \frac{\partial v}{\partial r}\,dr\,dS_1 = -\frac{1}{b}\int_{V_x} \frac{\partial v}{\partial r}\frac{d\xi}{r^{m-1}}; \tag{5}$$

here V_x is a spherical sector with vertex x.

We substitute for $\partial v/\partial r$ from formula (3), and take into account that

$$\frac{\partial u}{\partial r} = \frac{\partial u}{\partial \xi_k}\cos(\xi_k, r),$$

where ξ_k are fixed cartesian coordinate axes. Also, we introduce the notation

$$B_0(x, \xi) = \begin{cases} -\dfrac{1}{b}\dfrac{\partial \psi(r/R)}{\partial r}, & \xi \in V_x, \\ 0, & \xi \bar{\in} V_x; \end{cases}$$

$$B_k(x, \xi) = \begin{cases} -\dfrac{1}{b}\psi\left(\dfrac{r}{R}\right)\cos(\xi_k, r), & \xi \in V_x, \\ 0, & \xi \bar{\in} V_x, \end{cases}$$

$$k = 1, 2, \ldots, m;$$

the functions $B_0(x, \xi)$ and $B_k(x, \xi)$ are obviously bounded. We can now write equation (5) in the form

$$u(x) = \int_\Omega \frac{B_0(x, \xi)}{r^{m-1}} u(\xi)\,d\xi + \int_\Omega \frac{B_k(x, \xi)}{r^{m-1}}\frac{\partial u}{\partial \xi_k}\,d\xi. \tag{6}$$

This is the integral representation due to S. L. Sobolev.

It is important for what follows that the integrals in (6) should be completely continuous operators in $L_2(\Omega)$ on u and $\partial u/\partial \xi_k$ respectively. Now when $r < \frac{1}{3}R$ the function $B_0(x, \xi) \equiv 0$, the kernel of the first integral in (6) is bounded, and this integral is a Fredholm operator on u. As regards the other integrals in (6), they are integral operators with a weak singularity on the derivatives $\partial u/\partial \xi_k$.

In the subsequent sections of this chapter it is assumed that the domain Ω is finite and satisfies the cone condition.

The representation (6) has been obtained for functions of the class $C^{(1)}(\Omega)$. It is not difficult to extend this representation to functions which are integrable in Ω and have generalized derivatives, integrable to some degree $p > 1$. We leave this to the reader, remarking only that in this connection use should be made of the theorem which states that the integral operator with a weak singularity is completely continuous in the space $L_p(\Omega)$ for any p, $1 < p < \infty$.

Remark. The representation (6) is in fact a special case of a more general representation obtained by S. L. Sobolev (cf. [18], [19]). Sobolev's integral representation is used as the basis for the derivation of the embedding theorems (§ 5, Chapter 2).

§ 4. Investigation of the Operator $\mathfrak{N}_0$

Put
$$\mathfrak{N}_1 = \mathfrak{N}_0 + I, \tag{1}$$

where I is the identity operator in the space $\tilde{L}_2(\Omega)$. Obviously $D(\mathfrak{N}_1) = D(\mathfrak{N}_0)$. Moreover,

$$(\mathfrak{N}_1 u, u) = (\mathfrak{N}u, u) + \|u\|^2 = \int_\Omega A_{jk} \frac{\partial u}{\partial x_j} \frac{\partial u}{\partial x_k} dx + \|u\|^2. \tag{2}$$

The integral is non-negative so that

$$(\mathfrak{N}_1 u, u) \geq \|u\|^2, \tag{3}$$

and the operator $\mathfrak{N}_1$ is positive definite.

Repeating the arguments of § 3, Chapter 14, we can easily show that functions which are contained in the energy space $H_{\mathfrak{N}_1}$ of the operator $\mathfrak{N}_1$ are square-integrable in the domain Ω and have generalized first derivatives, also square-integrable, in this domain. We can also show without difficulty that equation (2) gives the following expressions for the energy norm and energy product:

$$|||u|||_{\mathfrak{N}_1}^2 = \int_\Omega \left[A_{jk} \frac{\partial u}{\partial x_j} \frac{\partial u}{\partial x_k} + u^2 \right] dx; \tag{4}$$

$$[u, v]_{\mathfrak{N}_1} = \int_\Omega \left[A_{jk} \frac{\partial v}{\partial x_j} \frac{\partial u}{\partial x_k} + uv \right] dx. \tag{5}$$

We now prove that the spectrum of the operator $\mathfrak{N}_1$ is discrete.

Because of Theorem 6.6.1, it is sufficient to show that any set of bounded functions in the energy metric of the operator $\mathfrak{N}_1$ is compact in $\tilde{L}_2(\Omega)$, or, equivalently, in $L_2(\Omega)$.

Let the set $M \subset H_{\mathfrak{N}_1}$, and let

$$|||u|||_{\mathfrak{N}_1} \leq c = \text{const.}, \quad \forall u \in M.$$

This inequality implies that

$$\int_\Omega A_{jk} \frac{\partial u}{\partial x_j} \frac{\partial u}{\partial x_k} dx + ||u||^2 \leq c^2.$$

By inequality (1.2),

$$\mu_0 \int_\Omega \sum_{k=1}^m \left(\frac{\partial u}{\partial x_k} \right)^2 dx + ||u||^2 \leq c^2.$$

Hence

$$\left\| \frac{\partial u}{\partial x_k} \right\| \leq \frac{c}{\sqrt{\mu_0}}, \quad ||u|| \leq c.$$

Thus, the set M itself, and the sets of first derivatives of the functions of M, are bounded in the metric of $L_2(\Omega)$. We write down the integral representation (3.6) for the function $u \in M$. The completely continuous integral operators which occur in this representation transform the above-mentioned bounded sets into compact sets. This shows that the set M is compact in $L_2(\Omega)$.

Let

$$v_1 \leq v_2 \leq \ldots \leq v_n \leq \ldots$$

be the eigenvalues of the operator $\mathfrak{N}_1$, and $u_1, u_2, \ldots, u_n, \ldots$ the corresponding eigenfunctions.

From formula (3.5), Chapter 6 we have

$$v_1 = \frac{|||u_1|||_{\mathfrak{N}_1}^2}{||u_1||^2} \geq 1.$$

We now show that equality in this expression is impossible and, consequently,

§ 4]   INVESTIGATION OF THE OPERATOR $\mathfrak{N}_0$

$v_1 > 1$. Suppose the contrary: let $v_1 = 1$. Then it follows from formula (4) that

$$\int_\Omega A_{jk} \frac{\partial u_1}{\partial x_j} \frac{\partial u_1}{\partial x_k} dx = 0.$$

Hence

$$\frac{\partial u_1}{\partial x_k} \equiv 0, \quad k = 1, 2, \ldots, m; \quad u_1 \equiv c_1 = \text{const.}$$

But $u_1 \in \tilde{L}_2(\Omega)$, so that

$$0 = (u_1, 1) = (c_1, 1) = \int_\Omega c_1 \, dx = c_1 |\Omega|,$$

and $c_1 = 0$. Hence $u_1(x) \equiv 0$, which contradicts the definition of an eigenfunction. Our assertion is thus proved.

As a consequence we obtain the following important theorem.

THEOREM 16.4.1. *The operator $\mathfrak{N}_0$ is positive definite.*

Now

$$\inf_{u \in H_{\mathfrak{N}_1}} \frac{|||u|||_{\mathfrak{N}_1}^2}{||u||^2} = v_1.$$

Hence

$$|||u|||_{\mathfrak{N}_1}^2 \geq v_1 ||u||^2, \quad u \in H_{\mathfrak{N}_1}. \tag{6}$$

In particular, if $u \in D(\mathfrak{N}_1) = D(\mathfrak{N}_0)$, then

$$|||u|||_{\mathfrak{N}_1}^2 = (\mathfrak{N}_1 u, u) = (\mathfrak{N}_0 u, u) + ||u||^2 \tag{7}$$

and, consequently, $(\mathfrak{N}_0 u, u) \geq (v_1 - 1) ||u||^2$; since $v_1 > 1$, this last inequality implies that $\mathfrak{N}_0$ is a positive definite operator.

Consider now the energy space $H_{\mathfrak{N}_0}$ of the positive definite operator $\mathfrak{N}_0$. We show that the energy spaces $H_{\mathfrak{N}_1}$ and $H_{\mathfrak{N}_0}$ consist of precisely the same elements, with

$$|||u|||_{\mathfrak{N}_1}^2 = |||u|||_{\mathfrak{N}_0}^2 + ||u||^2. \tag{8}$$

Let $u \in H_{\mathfrak{N}_0}$. By Theorem 5.3.2, there exists a sequence $u_n \in D(\mathfrak{N}_0) = D(\mathfrak{N}_1)$, such that

$$|||u_n - u_m|||_{\mathfrak{N}_0} \underset{m, n \to \infty}{\to} 0, \quad ||u_n - u|| \underset{n \to \infty}{\to} 0.$$

Writing equation (7) for the difference $u_n - u_m$, we see that at the same time

the relations
$$|||u_n-u_m|||_{\mathfrak{N}_1} \underset{m,n\to\infty}{\to} 0, \qquad ||u_n-u|| \underset{n\to\infty}{\to} 0,$$
are also satisfied. They show that $u \in H_{\mathfrak{N}_1}$. The converse statement can be proved in exactly the same way: if $u \in H_{\mathfrak{N}_1}$, then $u \in H_{\mathfrak{N}_0}$ at the same time.

It follows from (7) that relation (8) holds for elements of the domain $D(\mathfrak{N}_0) = D(\mathfrak{N}_1)$; passage to the limit also establishes its validity for the ideal elements of the energy space.

Formula (2.5) shows that the equations
$$|||u|||_{\mathfrak{N}_0}^2 = \int_\Omega A_{jk} \frac{\partial u}{\partial x_j} \frac{\partial u}{\partial x_k} \, dx \tag{9}$$
and
$$[u, v]_{\mathfrak{N}_0} = \int_\Omega A_{jk} \frac{\partial v}{\partial x_j} \frac{\partial u}{\partial x_k} \, dx \tag{10}$$
hold for functions $u \in D(\mathfrak{N}_0)$. In the limit they are extended to any functions of $H_{\mathfrak{N}_0}$.

THEOREM 16.4.2. *The operator $\mathfrak{N}_0$ has a discrete spectrum; its eigenvalues and eigenfunctions are $v_k - 1$ and u_k, where v_k and u_k are the eigenvalues and eigenfunctions of the operator $\mathfrak{N}_1$.*

The quantity v_k and function $u_k(x)$ are the generalized eigenvalue and corresponding generalized eigenfunction of the operator $\mathfrak{N}_1$. Thus they are connected by the relation
$$[u_k, \eta]_{\mathfrak{N}_1} = v_k(u_k, \eta), \qquad \forall \eta \in H_{\mathfrak{N}_1}. \tag{11}$$

From the expression (8) for the norm there follows an analogous expression for the scalar product:
$$[u, v]_{\mathfrak{N}_1} = [u, v]_{\mathfrak{N}_0} + (u, v).$$

If in addition we recall that the spaces $H_{\mathfrak{N}_1}$ and $H_{\mathfrak{N}_0}$ consist of exactly the same elements, we can rewrite (11) in the form
$$[u_k, \eta]_{\mathfrak{N}_0} = (v_k - 1)(u_k, \eta), \qquad \forall \eta \in H_{\mathfrak{N}_1}. \tag{12}$$

Equation (12) shows that the operator $\mathfrak{N}_0$ has an enumerable set of positive eigenvalues $v_k - 1$, which increase without limit; to these eigenvalues there correspond generalized eigenfunctions $u_k(x)$, constituting a complete, orthonormal system in $\tilde{L}_2(\Omega)$. The theorem is proved.

§ 5. Generalized Solution of the Neumann Problem

Consider now the equation

$$\mathfrak{N}_0 u = f, \qquad f \in \tilde{L}_2(\Omega), \tag{1}$$

or, equivalently, the differential equation

$$-\frac{\partial}{\partial x_j}\left(A_{jk}\frac{\partial u}{\partial x_k}\right) = f(x) \tag{2}$$

with the boundary conditions

$$A_{jk}\frac{\partial u}{\partial x_k}\cos(v, x_j)\big|_\Gamma = 0, \tag{3}$$

where the coefficients $A_{jk} \in C^{(1)}(\bar{\Omega})$ satisfy the inequality

$$A_{jk}(x)t_j t_k \geq \mu_0 \sum_{k=1}^{m} t_k^2; \qquad x \in \bar{\Omega}, \qquad \mu_0 = \text{const.} > 0.$$

We recall that, according to our assumption, Ω is a finite domain obeying the cone condition, and that the condition $f \in \tilde{L}_2(\Omega)$ means that

$$\int_\Omega f(x)\,dx = 0, \qquad \int_\Omega f^2(x)\,dx < \infty. \tag{4}$$

The operator $\mathfrak{N}_0$ is positive definite in $\tilde{L}_2(\Omega)$, and therefore equation (1) has one and only one generalized solution $u_0(x)$ in $H_{\mathfrak{N}_0}$. In common with any element of the energy space $H_{\mathfrak{N}_0}$, $u_0 \in \tilde{L}_2(\Omega)$; i.e. $u_0(x)$ is square-integrable in Ω, and

$$\int_\Omega u_0(x)\,dx = 0. \tag{4_1}$$

Further, $u_0(x)$ has generalized first derivatives which are square-integrable in Ω. The function $u_0(x)$ achieves a minimum for the functional

$$|||u|||_{\mathfrak{N}_0}^2 - 2(u, f) = \int_\Omega \left[A_{jk}\frac{\partial u}{\partial x_j}\frac{\partial u}{\partial x_k} - 2uf\right]dx, \qquad u \in H_{\mathfrak{N}_0}. \tag{5}$$

Henceforth we limit our considerations to Laplace's equation

$$-\Delta u = f(x). \tag{6}$$

The boundary condition takes the simpler form

$$\frac{\partial u}{\partial v}\bigg|_\Gamma = 0. \tag{7}$$

Conditions (4) and (4_1) remain unchanged. As in the general case, there exists a general solution $u_0(x)$ to the problem (6)–(7); it achieves a minimum for the functional

$$\int_\Omega \left[\sum_{k=1}^m \left(\frac{\partial u}{\partial x_k} \right)^2 - 2uf \right] dx, \qquad u \in H_{\mathfrak{N}_0}, \tag{8}$$

and consequently the variation of the functional (8) is zero at the point u_0. This gives the equation

$$\int_\Omega \left(\frac{\partial u_0}{\partial \xi_k} \frac{\partial \eta}{\partial \xi_k} - f\eta \right) d\xi = 0, \qquad \forall \eta \in H_{\mathfrak{N}_0}. \tag{9}$$

THEOREM 16.5.1. *If* $f \in \mathrm{Lip}_\alpha(\overline{\Omega})$, $0 < \alpha \leq 1$, *then* $u_0 \in C^{(2)}(\Omega)$.

The proof is almost identical with that for Theorem 14.6.2: introduce the Newtonian potential

$$\psi(x) = -\frac{1}{(m-2)|S_1|} \int_\Omega f(\xi) \frac{1}{r^{m-2}} d\xi,$$

so that $-\Delta\psi = f(x)$, and $\psi \in C^{(1)}(\overline{\Omega}) \cap C^{(2)}(\Omega)$. Making the substitution $u_0 = v_0 + \psi$ in equation (9), we obtain the new equation

$$\int_\Omega \frac{\partial v_0}{\partial \xi} \frac{\partial \eta}{\partial \xi} d\xi + \int_\Gamma \eta \frac{\partial \psi}{\partial \nu} d\Gamma = 0, \qquad \forall \eta \in H_{\mathfrak{N}_0}. \tag{10}$$

Put $\eta(\xi) = \omega_h(r)$, with $r = |x - \xi|$, $x \in \Omega$ and h less than the distance from x to the boundary of the domain Ω. This can be done because: 1) $\omega_h(r)$ is infinitely differentiable; 2) close to the boundary $\omega_h(r) \equiv 0$, and, consequently, $\omega_h(r)$ satisfies boundary condition (7). Thus $\omega_h(r) \in D(\mathfrak{N}_0)$, and, moreover $\omega_h(r) \in H_{\mathfrak{N}_0}$. The surface integral vanishes, and we obtain

$$\int_\Omega \frac{\partial v_0}{\partial \xi_k} \frac{\partial \omega_h(r)}{\partial \xi_k} d\xi = 0.$$

It follows, just as in Theorem 14.6.2, that v_{0h}, and therefore v_0 also, are harmonic in Ω. But then

$$v_0 \in C^{(2)}(\Omega), \qquad u_0 = (v_0 + \psi) \in C^{(2)}(\Omega),$$

and the theorem is proved.

The remarks made after the proof of Theorem 14.6.2 are also valid in the present case.

Exercises

1. Show that any function $u \in \tilde{L}_2(\Omega)$ whose generalized first derivatives $\partial u/\partial x_k \in L_2(\Omega)$, $k = 1, 2, \ldots, m$, belongs to the energy space $H_{\mathfrak{N}_0}$ (§ 4). Hence derive *Poincaré's inequality*

$$\int_\Omega u^2 \, dx \leq c_1 \int_\Omega \sum_{k=1}^m \left(\frac{\partial u}{\partial x_k}\right)^2 dx + c_2 \left(\int_\Omega u \, dx\right)^2; \quad c_1, c_2 = \text{const.}$$

The domain Ω is assumed to be finite and to satisfy the cone condition.

2. Beginning with Sobolev's integral representation, prove that:

If the function $u(x)$ has generalized first derivatives $\partial u/\partial x_k \in L_2(\Omega)$, $k = 1, 2, \ldots, m$, then it is equivalent to a function which is defined almost everywhere and is square-integrable on any piecewise-smooth, $(m-1)$-dimensional surface $\Sigma \subset \Omega$.

3. Show that the generalized solution $u_0(x)$ of the Neumann problem

$$-\Delta u = f(x), \quad f \in \tilde{L}_2(\Omega) \cap C^{(1)}(\bar{\Omega}), \quad \left.\frac{\partial u}{\partial v}\right|_\Gamma = 0$$

satisfies the boundary condition in the following sense: let Ω_n be an interior subdomain of the domain Ω with a piecewise-smooth boundary Γ_n, and $|\Omega \backslash \Omega_n| \to 0$. Then

$$\int_{\Gamma_n} \eta \frac{\partial u}{\partial v} \, d\Gamma_n \underset{n \to \infty}{\to} 0, \quad \forall \eta \in H_{\mathfrak{N}_0}.$$

4. Let the coefficient $C(x)$ be continuous and non-negative in the closed domain $\bar{\Omega}$, where $C(x) > 0$ on some set of positive measure. Suppose also that the coefficients $A_{jk}(x)$ have the properties enumerated in § 1, and that the domain Ω is finite and satisfies the cone condition. Show that the operator $\mathfrak{N}$ (§ 1) is positive definite and its spectrum is discrete.

Chapter 17.
Non-Self-Adjoint Elliptic Equations

§ 1. Generalized Solution

Consider the equation

$$-\frac{\partial}{\partial x_j}\left(A_{jk}(x)\frac{\partial u}{\partial x_k}\right) + B_k(x)\frac{\partial u}{\partial x_k} + C(x)u = f(x) \qquad (1)$$

for which we pose the Dirichlet problem

$$u|_\Gamma = 0, \qquad (2)$$

where Γ is a piecewise-smooth boundary of a finite domain Ω in the space E_m of variables $x_1, x_2, \ldots, x_m$. We shall assume that

$$A_{jk} \in C^{(1)}(\bar{\Omega}); \qquad B_k, C \in C(\bar{\Omega}),$$

and also that our equation is non-degenerate elliptic in $\bar{\Omega}$, so that

$$A_{jk}(x)t_j t_k \geq \mu_0 \sum_{k=1}^{m} t_k^2, \qquad \mu_0 = \text{const.} > 0.$$

Consider now the problem with a formally self-adjoint elliptic expression on the left-hand side:

$$-\frac{\partial}{\partial x_j}\left(A_{jk}\frac{\partial u}{\partial x_k}\right) = f(x), \qquad u|_\Gamma = 0, \qquad (3)$$

and denote the operator of this problem by $\mathfrak{A}$. It was shown in Chapter 14 that the operator $\mathfrak{A}$ is positive definite, so that if $f \in L_2(\Omega)$, then there exists a generalized solution to problem (3) belonging to the energy space $H_\mathfrak{A}$. Thus, to every element $f \in L_2(\Omega)$ there corresponds an element $u_0 \in H_\mathfrak{A}$ – the generalized solution of problem (3). The correspondence generates an operator which we denote by G:

$$Gf = u_0. \qquad (4)$$

Obviously $D(G) = L_2(\Omega)$ and $R(G) \subset H_\mathfrak{A}$.

The operator G can be expressed in explicit form: if $\{\omega_n\}$ is a complete, orthonormal system in $H_\mathfrak{A}$, then (cf. formula (5.11), Chapter 5)

$$Gf = u_0 = \sum_{n=1}^{\infty} (f, \omega_n)\omega_n. \tag{5}$$

By definition the operator G acts from $L_2(\Omega)$ into $H_\mathfrak{A}$; it is clear that G can also be regarded as an operator acting from $L_2(\Omega)$ into $L_2(\Omega)$.

The generalized solution achieves a minimum for the functional

$$F(u) = [u, u]_\mathfrak{A} - 2(u, f)$$

and, consequently, its variation is equal to zero:

$$\delta F(u_0, \eta) = 2[u_0, \eta]_\mathfrak{A} - 2(f, \eta) = 0, \quad \forall \eta \in H_\mathfrak{A}.$$

Substituting Gf for u_0 here, we obtain the formula

$$[Gf, \eta]_\mathfrak{A} = (f, \eta) \tag{6}$$

which is valid for any elements $f \in L_2(\Omega)$ and $\eta \in H_\mathfrak{A}$. *

Let K denote an operator whose domain of definition coincides with $H_\mathfrak{A}$, and which acts according to the formula

$$Ku = B_k \frac{\partial u}{\partial x_k} + Cu. \tag{7}$$

We can now write the problem (1)–(2) in the following form:

$$-\frac{\partial}{\partial x_j}\left(A_{jk}\frac{\partial u}{\partial x_k}\right) = f(x) - Ku, \quad u|_\Gamma = 0. \tag{8}$$

Its solution (if it exists and belongs to $H_\mathfrak{A}$) can be written in the form

$$u = G(f - Ku) = Gf - GKu. \tag{9}$$

Introduce the notation

$$Gf = F, \quad GK = T.$$

Obviously $F \in H_\mathfrak{A}$, and the operator T is defined on the whole of the space $H_\mathfrak{A}$ and acts from $H_\mathfrak{A}$ into $H_\mathfrak{A}$. Equation (9) takes the form

$$u + Tu = F. \tag{10}$$

---

* Note that our statements concerning the operator G and its properties are true in the case of any positive definite operator; it is only necessary to replace the space $L_2(\Omega)$ by the space H in which the latter operator acts.

We thus introduce the following definition:

The function $u \in H_\mathfrak{A}$ satisfying equation (10) is called the generalized solution of problem (1)–(2).

§ 2. Fredholm's Theorems

LEMMA 17.2.1. *The operator G is completely continuous in $L_2(\Omega)$.*

The spectrum of the operator $\mathfrak{A}$ is discrete (cf. Chapter 15). Let λ_n and $u_n(x)$ be eigenvalues and eigenfunctions of this operator, where, as usual, $\lambda_1 \leq \lambda_2 \leq \ldots \leq \lambda_n \leq \ldots$, and the functions u_n are orthonormal in $L_2(\Omega)$. Then they are orthogonal, but not normalized, in $H_\mathfrak{A}$; i.e.

$$[u_j, u_k]_\mathfrak{A} = 0, \quad j \neq k; \quad |||u_k|||^2_\mathfrak{A} = \lambda_k.$$

We also recall that the system $\{u_n\}$ is complete in $H_\mathfrak{A}$. It is now clear that the system $\{u_n/\sqrt{\lambda_n}\}$ is orthonormal and complete in $H_\mathfrak{A}$.

In formula (1.5), we put

$$\omega_n = \frac{u_n}{\sqrt{\lambda_n}}.$$

This gives the following representation for the operator G:

$$Gf = \sum_{n=1}^\infty \frac{(f, u_n)}{\lambda_n} u_n. \tag{1}$$

Introduce the notation

$$G = G'_n + G''_n, \quad G'_n f = \sum_{k=1}^n \frac{(f, u_k)}{\lambda_k} u_k,$$

$$G''_n f = \sum_{k=n+1}^\infty \frac{(f, u_k)}{\lambda_k} u_k. \tag{2}$$

The first operator is finite-dimensional and, consequently, is completely continuous. We estimate the norm of the second operator. We have

$$\|G''_n f\|^2 = \sum_{k=n+1}^\infty \frac{(f, u_k)^2}{\lambda_k^2} \leq \frac{1}{\lambda_{n+1}^2} \sum_{k=n+1}^\infty (f, u_k)^2.$$

By Bessel's inequality,

$$\|G''_n f\|^2 \leq \frac{1}{\lambda_{n+1}^2} \|f\|^2.$$

Hence
$$\|G_n''\| \leq \frac{1}{\lambda_{n+1}} \underset{n \to \infty}{\to} 0$$
and, consequently,
$$\|G - G_n'\| \underset{n \to \infty}{\to} 0.$$

The operator G is now completely continuous in $L_2(\Omega)$, as the limit (in the sense of convergence in the norm) of completely continuous operators.

THEOREM 17.2.1. *The operator $T = GK$ is completely continuous in $H_{\mathfrak{A}}$.*

The operator G is completely continuous in $L_2(\Omega)$. Therefore if M is a bounded set in $L_2(\Omega)$, then it is possible to select from this set a sequence $\{v_n\}$, such that the sequence $\{Gv_n\}$ converges, and, consequently,
$$\|Gv_n - Gv_k\| \underset{n,k \to \infty}{\to} 0.$$

Consider now an arbitrary set $N \subset H_{\mathfrak{A}}$, bounded in the energy metric
$$\|\|u\|\|_{\mathfrak{A}} \leq a, \quad \forall u \in N, \quad a = \text{const.}$$

We show that if the element u ranges over the set N then Ku ranges over the set $K(N)$, bounded in $L_2(\Omega)$. Now,
$$a^2 \geq \|\|u\|\|_{\mathfrak{A}}^2 = \int_\Omega A_{jk} \frac{\partial u}{\partial x_j} \frac{\partial u}{\partial x_k} dx \geq \mu_0 \int_\Omega \sum_{k=1}^m \left(\frac{\partial u}{\partial x_k}\right)^2 dx = \mu_0 \sum_{k=1}^m \left\|\frac{\partial u}{\partial x_k}\right\|^2.$$

Hence
$$\left\|\frac{\partial u}{\partial x_k}\right\| \leq \frac{a}{\sqrt{\mu_0}}, \quad k = 1, 2, \ldots, m.$$

At the same time
$$\|u\| \leq \frac{1}{\gamma} \|\|u\|\| \leq \frac{a}{\gamma},$$
where γ is the lower bound of the operator $\mathfrak{A}$.

It was assumed that the coefficients $B_k(x)$ and $C(x)$ were continuous and, consequently, bounded in the closed domain $\bar{\Omega}$; let $|B_k(x)| \leq b$, $|C(x)| \leq b$, $b = \text{const.}$ Then
$$\|Ku\| \leq b \left(\sum_{k=1}^m \left\|\frac{\partial u}{\partial x_k}\right\| + \|u\|\right) \leq ab\left(\frac{m}{\sqrt{\mu_0}} + \frac{1}{\gamma}\right) = c = \text{const.},$$
which proves the result.

As we showed earlier, it is possible to select from the set $K(N)$, bounded in $L_2(\Omega)$, a sequence $\{Kv_n\}$, such that

$$\|GKv_n - GKv_k\| = \|Tv_n - Tv_k\| \underset{k,n\to\infty}{\to} 0. \tag{3}$$

Equation (3) implies that T is completely continuous as an operator from $H_\mathfrak{A}$ into $L_2(\Omega)$. It remains to show that T is completely continuous as an operator from $H_\mathfrak{A}$ into $H_\mathfrak{A}$. To do this it is sufficient to prove that

$$\|\|Tv_n - Tv_k\|\|_\mathfrak{A} \underset{n,k\to\infty}{\to} 0. \tag{4}$$

We estimate the square of this norm. Using formula (1.6), we obtain

$$\|\|Tv_n - Tv_k\|\|_\mathfrak{A}^2 = [T(v_n-v_k), T(v_n-v_k)]_\mathfrak{A} = [GK(v_n-v_k), T(v_n-v_k)]_\mathfrak{A}$$
$$= (Kv_n - Kv_k, Tv_n - Tv_k) \leq \|Kv_n - Kv_k\| \cdot \|Tv_n - Tv_k\|$$
$$\leq 2c\|Tv_n - Tv_k\| \underset{n,k\to\infty}{\to} 0.$$

The theorem is proved.

With the same boundary conditions (1.2), we consider an equation somewhat more general than (1.1):

$$-\frac{\partial}{\partial x_j}\left(A_{jk}\frac{\partial u}{\partial x_k}\right) + \lambda\left(B_k\frac{\partial u}{\partial x_k} + Cu\right) = f(x). \tag{5}$$

Obviously this problem reduces to the following:

$$u + \lambda Tu = F. \tag{6}$$

It follows from the general theory developed earlier (Part III), that there exists not more than an enumerable set of eigenvalues for this problem, which can have an accumulation point only at infinity. If λ is an eigenvalue then, in general, a solution does not exist. In this case it is necessary and sufficient for the existence of a solution that the function F should satisfy a finite number of orthogonality conditions. That is, if $w_1, w_2, \ldots, w_s$ are eigenfunctions of the equation $w + \bar{\lambda}T^*w = 0$, then it is necessary and sufficient for the solubility of equation (6) that

$$[F, w_j]_\mathfrak{A} = 0, \quad j = 1, 2, \ldots, s.$$

Using formula (1.6) and the relation $F = Gf$, we can write this condition in the form

$$(f, w_j) = 0, \quad j = 1, 2, \ldots, s. \tag{7}$$

If the orthogonality condition is fulfilled, then the generalized solution exists, but it is not unique. Let the orthogonality conditions be satisfied, and let u_0 be any particular solution of equation (6). Then $u = u_0 + \tilde{u}$, where $\tilde{u}$ is the general solution of the corresponding homogeneous equation

$$\tilde{u} + \lambda T\tilde{u} = 0 \qquad (8)$$

and has the form $\tilde{u} = \sum_{j=1}^{s} c_j \tilde{u}_j$. Here the c_j are arbitrary constants, and the $\tilde{u}_j$ are linearly-independent solutions of equation (8).

All these results for the Dirichlet problem are also valid for the Neumann problem if we put $\mathfrak{A} = \mathfrak{N}_1$, where $\mathfrak{N}_1$ is the operator discussed in § 4, Chapter 16.

Exercise

Show that the eigenvalues of the operator T are situated in the half-plane $\operatorname{Re} \lambda > -q$, where q is a sufficiently large constant.

Chapter 18. Method of Potentials for the Homogeneous Laplace Equation

In the preceding chapters we have been considering the variational method for solving the Dirichlet and Neumann problems. This method, despite its many advantages, has also some deficiencies. For example, it solves the homogeneous Dirichlet problem only if the function $\psi(x)$, discussed in § 5, Chapter 14, exists. A more serious limitation, however, is the fact that the variational method requires the operator for the problem to be positive definite (or, as in the previous chapter, to differ from a positive definite operator only in minor components). For this reason the variational method is really not suitable for the case of an infinite domain. If instead of a single elliptic equation we consider an elliptic system of general form, then even the Dirichlet problem for a finite domain ceases to be positive, and the variational method becomes inapplicable.

In this chapter we shall concern ourselves with the method of potentials. The value of this method lies in the fact that it frees us from many of the difficulties just enumerated. However, it is a method which leads to heavy calculations, so that its application in the case of a domain with a non-smooth boundary is extremely complicated. We shall therefore only consider Laplace's equation, which we can take to be homogeneous (cf. § 6, Chapter 11). The domain under consideration will be required to have a defined smoothness; this requirement will be formulated specifically in the subsequent sections. It will be assumed everywhere in this chapter, except in §§ 13 and 14, that the dimensionality of space is $m > 2$.

§ 1. Lyapunov Surfaces

A surface Γ in a space E_m is called a *Lyapunov surface* if it satisfies the following two *Lyapunov conditions*:
 1. There exists a defined normal at any point of the surface Γ.
 2. Let x and ξ be points of the surface Γ, $r = |x-\xi|$, n and ν be normals to Γ at the points x and ξ respectively, and ϑ be the angle between these

normals. There exist positive numbers a and α such that

$$\vartheta \leq ar^\alpha. \tag{1}$$

THEOREM 18.1.1. *Let Γ be a closed Lyapunov surface. There exists a constant $d > 0$ such that if an arbitrary point $x \in \Gamma$ is taken as the centre of a sphere of radius d, then a straight line parallel to the normal to the surface Γ at the point x can intersect Γ inside the sphere at only one point.*

The sphere referred to in this theorem will be called the *Lyapunov sphere*, and will be denoted by $S(x)$.

Choose d sufficiently small that

$$ad^\alpha < 1. \tag{2}$$

We shall show that the theorem is true for this value of d. Suppose the contrary: let the sphere of radius d with centre at some point $x \in \Gamma$ cut out from the surface Γ a portion $\Gamma'(x)$ such that a line n' parallel to the normal n, drawn at the point x, meets Γ at two points, ξ' and ξ (Fig. 29). We shall suppose that the direction of n is along the outwards normal. Let the line n' intersect Γ in such a way that it emerges from the domain bounded by the surface Γ at the point ξ', and enters the domain at the point ξ. Construct the tangent plane and the outward normal ν to Γ at the point ξ. The normals ν and $n' = n$ point in opposite senses with respect to the tangent plane; since ν is perpendicular to this plane, the angle $\vartheta = (\nu, n) > \frac{1}{2}\pi$. This is impossible, since $|x-\xi| < d$, and by inequality (2), $(\nu, n) < 1 < \frac{1}{2}\pi$.

It could be the case that at the point ξ the line n' is tangential to the surface Γ (Fig. 30). Then $(\nu, n) = \frac{1}{2}\pi$ which again contradicts inequality (2).

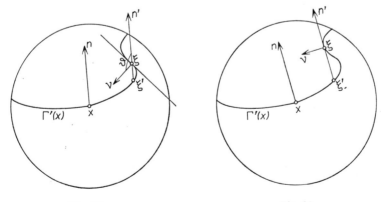

Fig. 29.   Fig. 30.

We shall suppose henceforth that the radius d of the Lyapunov sphere satisfies inequality (2).

Take an arbitrary point x on the surface Γ, and construct there a local system of coordinates: the point x is the origin of this system, the axis ξ_m is directed along the normal n to the surface Γ at the point x, and the axes $\xi_1, \xi_2, \ldots, \xi_{m-1}$ are situated in the tangent plane to Γ at the same point. From Theorem 18.1.1, we have the following consequences: that portion of the surface Γ which is included in the Lyapunov sphere $S(x)$ can be given by an explicit equation in the local coordinate system, namely,

$$\xi_m = f(\xi_1, \xi_2, \ldots, \xi_{m-1}), \quad f \in C^{(1)}. \tag{3}$$

Here, obviously,

$$f(0, 0, \ldots, 0) = 0; \quad f_{\xi_j}(0, 0, \ldots, 0) = 0, \quad j = 1, 2, \ldots, m-1. \tag{4}$$

The next problem is to estimate the order of smallness of the function f and its first derivatives inside the sphere $S(x)$. We shall denote by $\Gamma'(x)$ the portion of the surface Γ enclosed within the sphere $S(x)$.

Let $\xi \in \Gamma'(x)$ and let v be the normal to Γ at the point ξ. First of all we estimate the direction cosines of the normal v in the local coordinate system. We have

$$\cos(v, \xi_m) = \cos(v, n) = \cos \vartheta = 1 - \frac{\vartheta^2}{2!} + \frac{\vartheta^4}{4!} - \cdots$$

Because of inequalities (1) and (2), $\vartheta < 1$, and this series is alternating in sign, with monotonically decreasing terms; if we retain a finite number of terms of the series, then the remainder has the sign of the first neglected term. Hence

$$\cos(v, \xi_m) \geq 1 - \tfrac{1}{2}\vartheta^2. \tag{5}$$

Moreover, by inequality (1),

$$\cos(v, \xi_m) \geq 1 - \tfrac{1}{2}a^2 r^{2\alpha}. \tag{6}$$

From inequality (2) we have $a^2 r^{2\alpha} \leq a^2 d^{2\alpha} < 1$, and

$$\cos(v, \xi_m) \geq \tfrac{1}{2}. \tag{7}$$

Both estimates (6) and (7) will be required later.

If a surface has the equation

$$F(\xi_1, \xi_2, \ldots, \xi_m) = 0,$$

then the direction cosines of the normal to this surface are defined by the formula

$$\cos(v, \xi_k) = \pm \frac{\partial F}{\partial \xi_k} \bigg/ \sqrt{\sum_{j=1}^{m} \left(\frac{\partial F}{\partial \xi_j}\right)^2}.$$

The equation of the surface $\Gamma(x)$ has the form

$$\xi_m - f(\xi_1, \xi_2, \ldots, \xi_{m-1}) = 0,$$

and, moreover, $\cos(v, \xi_m) > 0$. Hence it follows that

$$\cos(v, \xi_k) = -\frac{\partial f}{\partial \xi_k} \bigg/ \sqrt{\left\{1 + \sum_{j=1}^{m-1} \left(\frac{\partial f}{\partial \xi_j}\right)^2\right\}}, \qquad k = 1, 2, \ldots, m-1, \qquad (8)$$

and

$$\cos(v, \xi_m) = 1 \bigg/ \sqrt{\left\{1 + \sum_{k=1}^{m-1} \left(\frac{\partial f}{\partial \xi_k}\right)^2\right\}}. \qquad (9)$$

From inequalities (6) and (2),

$$\sqrt{\left\{1 + \sum_{k=1}^{m-1} \left(\frac{\partial f}{\partial \xi_k}\right)^2\right\}} \leq \frac{1}{1 - \tfrac{1}{2} a^2 r^{2\alpha}} = 1 + \frac{\tfrac{1}{2} a^2 r^{2\alpha}}{1 - \tfrac{1}{2} a^2 r^{2\alpha}}$$

$$< 1 + \frac{\tfrac{1}{2} a^2 r^{2\alpha}}{1 - \tfrac{1}{2} a^2 d^{2\alpha}} < 1 + a^2 r^{2\alpha}.$$

Squaring this, we obtain

$$\sum_{k=1}^{m-1} \left(\frac{\partial f}{\partial \xi_k}\right)^2 < 2a^2 r^{2\alpha} + a^4 r^{4\alpha}.$$

But $r \leq d$; because of inequality (2) we have

$$\sum_{k=1}^{m-1} \left(\frac{\partial f}{\partial \xi_k}\right)^2 < 3a^2 r^{2\alpha}, \qquad (10)$$

and, consequently

$$\left|\frac{\partial f}{\partial \xi_k}\right| < \sqrt{3}\, a r^{\alpha}, \qquad k = 1, 2, \ldots, m-1. \qquad (11)$$

From formula (8) there now follows the estimate

$$|\cos(v, \xi_k)| \leq \left|\frac{\partial f}{\partial \xi_k}\right| < \sqrt{3}\, a r^{\alpha}, \qquad k = 1, 2, \ldots, m-1. \qquad (12)$$

We next estimate the quantity $|\xi_m|$ on the portion $\Gamma'(x)$ of the surface Γ. Write

$$\rho^2 = \sum_{k=1}^{m-1} \xi_k^2.$$

Obviously,
$$r^2 = \rho^2 + \xi_m^2. \tag{13}$$

In formula (11) ξ_k can denote any direction in the tangent plane to Γ, and in particular the direction ρ. Therefore

$$\left|\frac{\partial f}{\partial \rho}\right| < \sqrt{3}ar^\alpha \leq \sqrt{3}ad^\alpha \leq \sqrt{3}.$$

Hence
$$|f| \leq \int_0^\rho \left|\frac{\partial f}{\partial \rho}\right| d\rho < \sqrt{3}\rho,$$

and, consequently,
$$|\xi_m| < \sqrt{3}\rho. \tag{14}$$

This estimate can be substantially improved. From formulae (13) and (14) it follows that

$$r \leq 2\rho. \tag{15}$$

Hence
$$\left|\frac{\partial f}{\partial \rho}\right| < 2^\alpha \sqrt{3} a \rho^\alpha.$$

Now
$$|f| = |\xi_m| \leq a_1 \rho^{\alpha+1}, \qquad a_1 = \frac{2^\alpha \sqrt{3} a}{\alpha+1}. \tag{16'}$$

But $\rho \leq r$, and, finally,
$$|\xi_m| \leq a_1 r^{\alpha+1}. \tag{16''}$$

We shall henceforth denote by the letter r both the distance between the points x and ξ, and also the directed vector from x to ξ. We deduce an estimate for $\cos(v, r)$ under the assumption that $x \in \Gamma$ and $\xi \in \Gamma'(x)$. In the local coordinate system,

$$\cos(v, r) = \cos(r, x_k) \cos(v, x_k) = \frac{\xi_k - x_k}{r} \cos(v, x_k)$$

$$= \sum_{k=1}^{m-1} \frac{\xi_k}{r} \cos(v, x_k) + \frac{\xi_m}{r} \cos(v, x_m).$$

Using inequalities (12) and (16), as well as the fact that $|\xi_k|/r = |\cos(r, x_k)| \leq 1$ and $|\cos(v, x_m)| \leq 1$, we find

$$|\cos(v, r)| \leq cr^\alpha, \tag{17}$$

where

$$c = \sqrt{3}(m-1)a + a_1 = \text{const.}$$

§ 2. Solid Angle

Consider a piecewise-smooth surface Σ, not necessarily closed, with respect to which a positive direction of the normal is defined.

Denote by ξ an arbitrary point of the surface Σ, and by v the normal to Σ constructed at the point ξ. Let the point $x \in \Gamma$ be situated such that at any point $\xi \in \Sigma$ the radius vector r, drawn from the point x to the point ξ, forms an acute angle, or at most a right angle, with the normal v, so that

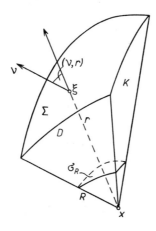

Fig. 31.

$\cos(r, v) \geq 0$. From the point x draw radius vectors to all the points of the surface Σ. These radius vectors fill the domain bounded by the surface Σ and the conical surface K, the latter being generated by those radius vectors which terminate at points on the edge of the surface Σ (Fig. 31). Note that if Σ is a closed surface then the point x must be located inside Σ (otherwise the angle (r, v) could be obtuse); in this case the above-mentioned domain coincides with the interior of Σ. Taking the point x as centre, draw a sphere of arbitrary radius R. Denote by σ_R the portion of this sphere enclosed in the cone referred to above. The ratio

$$\omega_x(\Sigma) = \frac{|\sigma_R|}{R^{m-1}} \tag{1}$$

is independent of R. It is called the *solid angle subtended by the surface Σ at the point x.*

The construction we have just described can also be used in the case when $\cos(r,v) \leq 0$ on the surface Σ. In this case the solid angle $\omega_x(\Sigma)$ subtended by the surface Σ at the point x is given by expression (1) with a minus sign added.

In the general case when $\cos(r,v)$ can change sign we shall assume that the surface Σ can be subdivided into portions $\Sigma_1, \Sigma_2, \ldots$, on each of which $\cos(r,v)$ retains the same sign. For a surface of this kind the solid angle is defined by the formula

$$\omega_x(\Sigma) = \sum \omega_x(\Sigma_k), \tag{2}$$

provided only that the series (2) converges absolutely (for example, if the portions Σ_k are finite in number).

We now show that in all these cases the solid angle $\omega_x(\Sigma)$ is defined by the formula

$$\omega_x(\Sigma) = -\frac{1}{m-2} \int_\Sigma \frac{\partial}{\partial v} \frac{1}{r^{m-2}} \, d_\xi \Sigma. \tag{3}$$

It is sufficient to consider the case when $\cos(r,v)$ is of one sign on the surface.

We first derive one auxiliary formula which turns out to be useful in what follows. We have

$$\frac{\partial}{\partial v} \frac{1}{r^{m-2}} = -\frac{m-2}{r^{m-1}} \frac{\partial r}{\partial \xi_k} \cos(v, \xi_k).$$

But

$$\frac{\partial r}{\partial \xi_k} = \frac{\xi_k - x_k}{r} = \cos(r, \xi_k).$$

Hence

$$\frac{\partial}{\partial v} \frac{1}{r^{m-2}} = -\frac{m-2}{r^{m-1}} \cos(r, \xi_k) \cos(v, \xi_k)$$

or, finally,

$$\frac{\partial}{\partial v} \frac{1}{r^{m-2}} = -\frac{m-2}{r^{m-1}} \cos(r, v). \tag{4}$$

Let $\cos(r, v) \geq 0$. Take the radius R sufficiently small so that the surfaces Σ and σ_R have no common points (Fig. 31).

Consider the domain D, bounded by the surfaces Σ, σ_R, and the portion of the cone K enclosed between them. In this domain, $1/r^{m-2}$ is a harmonic function of the point ξ, and therefore (formula (6.9), Chapter 10)

$$\int_\Sigma \frac{\partial}{\partial v^*} \frac{1}{r^{m-2}} d_\xi \Sigma + \int_{\sigma_R} \frac{\partial}{\partial v^*} \frac{1}{r^{m-2}} d_\xi \sigma_R + \int_K \frac{\partial}{\partial v^*} \frac{1}{r^{m-2}} d_\xi K = 0.$$

Here v^* denotes the normal to the surface, outwards with respect to the domain D; since $\cos(r, v) \geq 0$ on Σ, so $v^* = v$ on this surface.

On the surface K, $\cos(r, v^*) = 0$, since r is directed along a generator, while v^* is perpendicular to it. Because of formula (4) the integral over K vanishes, and

$$\int_\Sigma \frac{\partial}{\partial v} \frac{1}{r^{m-2}} d_\xi \Sigma = -\int_{\sigma_R} \frac{\partial}{\partial v^*} \frac{1}{r^{m-2}} d_\xi \sigma_R.$$

On σ_R, the direction of the normal v^* is opposite to that of the radius, and therefore

$$\frac{\partial}{\partial v^*} \frac{1}{r^{m-2}} = -\frac{\partial}{\partial r} \frac{1}{r^{m-2}} \bigg|_{r=R} = \frac{m-2}{R^{m-1}}.$$

Thus

$$\int_\Sigma \frac{\partial}{\partial v} \frac{1}{r^{m-2}} d_\xi \Sigma = -\frac{m-2}{R^{m-1}} \int_{\sigma_R} d_\xi \sigma_R = -(m-2) \frac{|\sigma_R|}{R^{m-1}}$$
$$= -(m-2)\omega_x(\Sigma),$$

which we were required to prove.

If $\cos(r, v) \leq 0$, then $v^* = -v$ on Σ; arguing as before, we obtain in this case

$$\int_\Sigma \frac{\partial}{\partial v} \frac{1}{r^{m-2}} d_\xi \Sigma = \frac{m-2}{R^{m-1}} \int_{\sigma_R} d_\xi \sigma_R = (m-2) \frac{|\sigma_R|}{R^{m-1}} = -(m-2)\omega_x(\Sigma).$$

If the surface Σ is subdivided into portions $\Sigma_1, \Sigma_2, \ldots$, of constant sign for $\cos(r, v)$, then, obviously

$$\sum |\omega_x(\Sigma_k)| = \int_\Sigma \left| \frac{\partial}{\partial v} \frac{1}{r^{m-2}} \right| d_\xi \Sigma. \tag{5}$$

THEOREM 18.2.1. *If Γ is a closed Lyapunov surface, then there exists a constant C such that*

$$\int_\Gamma \left| \frac{\partial}{\partial v} \frac{1}{r^{m-2}} \right| d_\xi \Gamma \leq C, \qquad \forall x \in E_m. \tag{6}$$

Let d be the radius of the Lyapunov sphere for the surface Γ. One possibility is that the distance from the point x to the surface Γ is not less than $\tfrac{1}{2}d$. In this case $r = |x-\xi| \geq \tfrac{1}{2}d$; from formula (4)

$$\left| \frac{\partial}{\partial v} \frac{1}{r^{m-2}} \right| = (m-2) \frac{|\cos(r, v)|}{r^{m-1}} \leq \frac{2^{m-1}(m-2)}{d^{m-1}}$$

and

$$\int_\Gamma \left| \frac{\partial}{\partial v} \frac{1}{r^{m-2}} \right| d_\xi \Gamma \leq \frac{2^{m-1}(m-2)}{d^{m-1}} |\Gamma|. \tag{7}$$

Now let the distance from the point x to the surface Γ be less than $\tfrac{1}{2}d$. There exists a point $x_0 \in \Gamma$ such that

$$|x-x_0| = \min_{\xi \in \Gamma} |x-\xi| < \tfrac{1}{2}d. \tag{8}$$

As we know, the point x lies on the normal n_0 to Γ, passing through the point x_0.

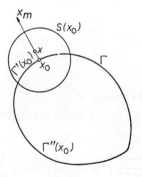

Fig. 32.

Construct the Lyapunov sphere $S(x_0)$. Denote by $\Gamma'(x_0)$ and $\Gamma''(x_0)$ respectively the portions of the surface Γ which lie inside and outside the sphere $S(x_0)$ (Fig. 32). If $\xi \in \Gamma''(x_0)$, then $|\xi - x_0| \geq d$ and

$$|\xi - x| \geq |\xi - x_0| - |x - x_0| > \tfrac{1}{2}d.$$

Hence

$$\int_{\Gamma''(x_0)} \left| \frac{\partial}{\partial v} \frac{1}{r^{m-2}} \right| d_\xi \Gamma \leq \frac{2^{m-1}(m-2)|\Gamma''(x_0)|}{d^{m-1}} < \frac{2^{m-1}(m-2)|\Gamma|}{d^{m-1}}. \quad (9)$$

It remains to consider the integral

$$\int_{\Gamma'(x_0)} \left| \frac{\partial}{\partial v} \frac{1}{r^{m-2}} \right| d_\xi \Gamma = (m-2) \int_{\Gamma'(x_0)} \frac{|\cos(r, v)|}{r^{m-1}} d_\xi \Gamma.$$

Construct a local coordinate system with origin at the point x_0; take the x_m axis along the normal to Γ at x_0. Put $|x-x_0| = \delta$, $\delta < \frac{1}{2}d$. The local coordinates of the point x are $(0, 0, \ldots, 0, \pm\delta)$. Denote by $G'(x_0)$ the projection of the surface $\Gamma'(x_0)$ on to the tangent plane at the point x_0. Then

$$\int_{\Gamma'(x_0)} \frac{|\cos(r, v)|}{r^{m-1}} d_\xi \Gamma = \int_{G'(x_0)} \frac{|\cos(r, v)| d\xi_1 d\xi_2 \ldots d\xi_{m-1}}{r^{m-1} \cos(v, \xi_m)}$$

$$\leq 2 \int_{G'(x_0)} \frac{|\cos(r, v)|}{r^{m-1}} d\xi_1 d\xi_2 \ldots d\xi_{m-1}, \quad (10)$$

where we have used here the estimate (1.7).

Put $\rho^2 = \xi_1^2 + \xi_2^2 + \ldots + \xi_{m-1}^2$; ρ is the distance from the point x_0 to the projection of the point ξ on to the tangent plane through x_0. From the formula

$$r^2 = \sum_{k=1}^{m} (\xi_k - x_k)^2 = \sum_{k=1}^{m-1} \xi_k^2 + (\xi_m - x_m)^2 = \rho^2 + (\xi_m \pm \delta)^2$$

it follows that $r \geq \rho$. On the other hand, the domain $G'(x_0)$ is defined by the inequality $r < d$. But $r \geq \rho$, and so $\rho < d$ also. This means that the domain $G'(x_0)$ lies in the $(m-1)$-dimensional sphere $\rho \leq d$, and from formula (10) we obtain

$$\int_{\Gamma'(x_0)} \frac{|\cos(r, v)|}{r^{m-1}} d_\xi \Gamma \leq 2 \int_{\rho < d} \frac{|\cos(r, v)|}{r^{m-1}} d\xi_1 d\xi_2 \ldots d\xi_{m-1}. \quad (11)$$

If $\delta = 0$ (i.e. $x = x_0 \in \Gamma$), then by inequality (1.17)

$$\frac{|\cos(r, v)|}{r^{m-1}} \leq \frac{c}{r^{m-1-\alpha}} \leq \frac{c}{\rho^{m-1-\alpha}}, \quad (12)$$

and, consequently,

$$\int_{\Gamma'(x_0)} \frac{|\cos(r,v)|}{r^{m-1}} d_\xi \Gamma \leq 2c \int_{\rho<d} \frac{d\xi_1 d\xi_2 \ldots d\xi_{m-1}}{\rho^{m-1-\alpha}} = \text{const.} \quad (13)$$

We have proved in passing that the integral (6) exists for any $x \in \Gamma$. Now let $\delta > 0$. We have

$$|\cos(r,v)| = |\cos(r,\xi_k)\cos(v,\xi_k)|$$

$$\leq \sqrt{3}(m-1)ar_0^\alpha + \frac{|\xi_m \pm \delta|}{r} \leq cr_0^\alpha + a_1 \frac{r_0^{\alpha+1}}{r} + \frac{\delta}{r}.$$

Here $r_0 = |\xi - x_0|$, and we have made use of the estimates (1.12), (1.16) and (1.17). We now have

$$2\int_{\rho<d} \frac{|\cos(r,v)|}{r^{m-1}} d\xi_1 d\xi_2 \ldots d\xi_{m-1} \leq 2c \int_{\rho<d} \frac{r_0^\alpha d\xi_1 d\xi_2 \ldots d\xi_{m-1}}{r^{m-1}} +$$

$$+ 2a_1 \int_{\rho<d} \frac{r_0^{\alpha+1} d\xi_1 d\xi_2 \ldots d\xi_{m-1}}{r^m} + 2\delta \int_{\rho<d} \frac{d\xi_1 d\xi_2 \ldots d\xi_{m-1}}{r^m}. \quad (14)$$

We estimate the values of r_0 and r in terms of ρ. The value of r_0 can be estimated by means of formula (1.15), since in the latter r is the distance from the point ξ to the origin of the local coordinate system. Thus, $r_0 \leq 2\rho$. In order to estimate r, we proceed as follows. We have

$$r^2 = \rho^2 + \xi_m^2 + \delta^2 \pm 2\xi_m \delta.$$

Furthermore, $|2\xi_m \delta| \leq \tfrac{1}{2}\delta^2 + 2\xi_m^2$. Hence

$$r^2 \geq \rho^2 + \tfrac{1}{2}\delta^2 - \xi_m^2.$$

By formula (1.16″),

$$|\xi_m| \leq a_1 \rho^{\alpha+1} \leq a_1 d^\alpha \rho.$$

The radius d can be taken as small as we please. Let it be such that

$$a_1 d^\alpha \leq \frac{1}{\sqrt{2}}.$$

Then

$$r^2 \geq \tfrac{1}{2}(\rho^2 + \delta^2).$$

It is now not difficult to estimate the integrals on the right-hand side of

§ 2] SOLID ANGLE 355

(14). The first two are estimated quite simply:

$$\int_{\rho<d} \frac{r_0^\alpha \, d\xi_1 \, d\xi_2 \ldots d\xi_{m-1}}{r^{m-1}} \leqq \int_{\rho<d} \frac{2^{m-1+\alpha} \rho^\alpha \, d\xi_1 \, d\xi_2 \ldots d\xi_{m-1}}{(\rho^2+\delta^2)^{\frac{1}{2}(m-1)}}$$

$$\leqq 2^{m-1+\alpha} \int_{\rho<d} \frac{d\xi_1 \, d\xi_2 \ldots d\xi_{m-1}}{\rho^{m-1-\alpha}} = \text{const.}$$

and, similarly,

$$\int_{\rho<d} \frac{r_0^{\alpha+1} \, d\xi_1 \, d\xi_2 \ldots d\xi_{m-1}}{r^m} \leqq 2^{m+1+\alpha} \int_{\rho<d} \frac{d\xi_1 \, d\xi_2 \ldots d\xi_{m-1}}{\rho^{m-1-\alpha}} = \text{const.}$$

We now proceed to consider the last integral. Using the estimate for r, we find

$$\delta \int_{\rho<d} \frac{d\xi_1 \, d\xi_2 \ldots d\xi_{m-1}}{r^m} \leqq 2^{\frac{1}{2}m} \delta \int_{\rho<d} \frac{d\xi_1 \, d\xi_2 \ldots d\xi_{m-1}}{(\rho^2+\delta^2)^{\frac{1}{2}m}}$$

$$< 2^{\frac{1}{2}m} \delta \int_{E_{m-1}} \frac{d\xi_1 \, d\xi_2 \ldots d\xi_{m-1}}{(\rho^2+\delta^2)^{\frac{1}{2}m}},$$

where E_{m-1} is the Euclidean space of $m-1$ dimensions. In this last integral we put

$$\xi_k = \delta \eta_k, \quad k = 1, 2, \ldots, m-1; \quad \sum_{k=1}^{m-1} \eta_k^2 = \rho_1^2.$$

Then

$$\delta \int_{E_{m-1}} \frac{d\xi_1 \, d\xi_2 \ldots d\xi_{m-1}}{(\rho^2+\delta^2)^{\frac{1}{2}m}} = \int_{E_{m-1}} \frac{d\eta_1 \, d\eta_2 \ldots d\eta_{m-1}}{(\rho_1^2+1)^{\frac{1}{2}m}}.$$

The latter converges, and is equal to some constant.

It is now clear that when $0 < \delta < \frac{1}{2}d$, the integral (14) does not exceed some constant value. But this statement is also true when $\delta = 0$ (formula (13)). There exists then a constant C' such that

$$\int_{\Gamma'(x_0)} \frac{|\cos(r, v)|}{r^m} \, d_\xi \Gamma \leqq C', \quad 0 \leqq \delta < \frac{1}{2}d.$$

Using inequality (9), we see that Theorem 18.2.1 is also true for $\delta < \frac{1}{2}d$; accordingly we can write

$$C = \frac{2^{m-1}(m-2)|\Gamma|}{d^{m-1}} + C'.$$

§ 3. The Double-Layer Potential and its Direct Value

In § 5, Chapter 11, we defined the double-layer potential as an integral of the form

$$W(x) = \int_\Gamma \sigma(\xi) \frac{\partial}{\partial v} \frac{1}{r^{m-2}} \, d_\xi \Gamma, \tag{1}$$

where v is the outward normal to the surface Γ at the point ξ. It was shown that the function $W(x)$ is harmonic both inside and outside Γ; but the potential (1) was not defined on the surface itself.

We shall suppose now that Γ is a closed Lyapunov surface, and that the density $\sigma(\xi)$ is continuous on this surface. Under these conditions the following theorem holds.

THEOREM 18.3.1. *The double-layer potential* (1) *has a fully defined value for any x lying on the surface Γ. This value varies continuously as x varies over the surface Γ.*

It is a simple matter to show that the integral (1) exists for $x \in \Gamma$. The density $\sigma(\xi)$ is continuous on the compact set Γ, and is therefore bounded. Let $|\sigma(\xi)| \leq M = \text{const}$. Then

$$\left| \sigma(\xi) \frac{\partial}{\partial v} \frac{1}{r^{m-2}} \right| \leq M \left| \frac{\partial}{\partial v} \frac{1}{r^{m-2}} \right|. \tag{2}$$

It has been shown in the preceding section that the integral (2.6) exists for $x \in \Gamma$, i.e., that the function $|(\partial/\partial v)(1/r^{m-2})|$ is integrable on Γ. But then the left-hand side of inequality (2) is also integrable on Γ, and, consequently, the integral (1) exists for $x \in \Gamma$.

We now show that integral (1) is continuous for $x \in \Gamma$. The estimate (1.17) implies that the potential (1) is an integral operator with a weak singularity on the function $\sigma(\xi)$; the kernel of this operator,

$$\frac{\partial}{\partial v} \frac{1}{r^{m-2}} = - \frac{(m-2) \cos(r, v)}{r^{m-1}}$$

can be represented in the form

$$\frac{\partial}{\partial v} \frac{1}{r^{m-2}} = \frac{(m-2) r^{-\frac{1}{2}\alpha} \cos(r, v)}{r^{m-1-\frac{1}{2}\alpha}}.$$

Denoting the numerator by $A(x, \xi)$, we see that when $x \neq \xi$, the function $A(x, \xi)$ is continuous on Γ. If $x \in \Gamma$ and $x \to \xi$, then because of inequality

(1.17), $A(x, \xi) \to 0$. Put $A(x, x) = 0$. Then $A(x, \xi)$ is continuous on Γ for any position of the points x and ξ. By Theorem 7.4.1, the integral operator (1) transforms a continuous function into a continuous function. But $\sigma(\xi)$ is continuous, by hypothesis, and therefore the double-layer potential also varies continuously as the point x moves over the surface Γ. The theorem is thus proved.

The value of the double-layer potential for $x \in \Gamma$ is called its *direct value*. Theorem 18.3.1 can obviously be formulated in the following way:

If Γ is a closed Lyapunov surface, and the density $\sigma(\xi)$ is continuous on Γ, the direct value of the double-layer potential is continuous on Γ.

§ 4. Gauss's Integral

This is the name given to the double-layer potential whose density is identically equal to one:

$$W_0(x) = \int_\Gamma \frac{\partial}{\partial v} \frac{1}{r^{m-2}} \, d_\xi \Gamma. \tag{1}$$

Here Γ is a closed surface, and v its outward normal at the point ξ. For this integral we have the following theorem.

THEOREM 18.4.1. *If Γ is a closed Lyapunov surface, then the value of Gauss's integral is defined by the formula*

$$W_0(x) = \begin{cases} -(m-2)|S_1| = -\dfrac{2(m-2)\pi^{\frac{1}{2}m}}{\Gamma(\frac{1}{2}m)} & \text{for } x \text{ inside } \Gamma, \\ 0 & \text{for } x \text{ outside } \Gamma, \\ -\frac{1}{2}(m-2)|S_1| = -\dfrac{(m-2)\pi^{\frac{1}{2}m}}{\Gamma(\frac{1}{2}m)} & \text{for } x \in \Gamma. \end{cases} \tag{2}$$

We note at once that the first two of equations (2) are valid for any closed, piecewise-smooth surface Γ – we shall prove this also.

Let Γ be a closed, piecewise-smooth surface, and let the point x lie inside Γ. Around this point we describe a sphere S_ε of radius ε, taking the latter sufficiently small that the sphere S_ε lies inside Γ (Fig. 33). In the domain bounded by Γ and S_ε, the function $1/r^{m-2}$ is harmonic, and therefore, by

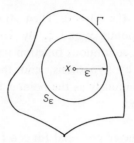

Fig. 33.

virtue of formula (6.9), Chapter 10,

$$\int_\Gamma \frac{\partial}{\partial \nu} \frac{1}{r^{m-2}} d\Gamma + \int_{S_\varepsilon} \frac{\partial}{\partial \nu} \frac{1}{r^{m-2}} dS_\varepsilon = 0. \tag{3}$$

The direction of the normal ν on S_ε is opposite to that of the radius. Hence

$$\int_{S_\varepsilon} \frac{\partial}{\partial \nu} \frac{1}{r^{m-2}} dS_\varepsilon = \int_{S_\varepsilon} \frac{m-2}{\varepsilon^{m-1}} dS_\varepsilon = (m-2)\frac{|S_\varepsilon|}{\varepsilon^{m-1}} = (m-2)|S_1|, \tag{4}$$

which proves the first of equations (2). It is even simpler to prove the second of (2): if the point x lies outside Γ then the function $1/r^{m-2}$ is harmonic inside Γ, and

$$\int_\Gamma \frac{\partial}{\partial \nu} \frac{1}{r^{m-2}} d\Gamma = 0.$$

We now turn to the third of equations (2). Let Γ be a closed Lyapunov surface, and let $x \in \Gamma$. We are interested in the direct value of the Gauss integral, whose existence follows from the theorem of the preceding section. It remains to calculate this value, which can be done in the following way.

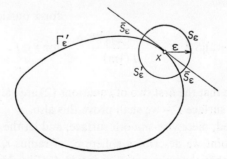

Fig. 34.

Take a number ε, $0 < \varepsilon < d$, and around the point $x \in \Gamma$ describe a sphere S_ε of radius ε. Denote by Γ'_ε the portion of the surface Γ which lies outside the sphere, and by S'_ε the portion of the sphere S_ε lying inside Γ (Fig. 34). Since Gauss's integral converges for $x \in \Gamma$, we have

$$W_0(x) = \lim_{\varepsilon \to 0} \int_{\Gamma'_\varepsilon} \frac{\partial}{\partial \nu} \frac{1}{r^{m-2}} d_\xi \Gamma. \tag{5}$$

The point x lies outside the domain bounded by the surfaces Γ'_ε and S'_ε; the function $1/r^{m-2}$ is harmonic in this domain, and, consequently,

$$\int_{\Gamma'_\varepsilon} \frac{\partial}{\partial \nu} \frac{1}{r^{m-2}} d_\xi \Gamma + \int_{S'_\varepsilon} \frac{\partial}{\partial \nu} \frac{1}{r^{m-2}} d_\xi S_\varepsilon = 0.$$

But then

$$W_0(x) = -\lim_{\varepsilon \to 0} \int_{S'_\varepsilon} \frac{\partial}{\partial \nu} \frac{1}{r^{m-2}} d_\xi S_\varepsilon. \tag{6}$$

In the integral (6) the direction of the normal ν is opposite to that of the radius; therefore

$$\int_{S'_\varepsilon} \frac{\partial}{\partial \nu} \frac{1}{r^{m-2}} d_\xi S_\varepsilon = \frac{m-2}{\varepsilon^{m-1}} \int_{S'_\varepsilon} d_\xi S_\varepsilon = (m-2) \frac{|S'_\varepsilon|}{\varepsilon^{m-1}}. \tag{7}$$

For ε sufficiently small the surface S'_ε is almost a hemisphere, with its base on the tangent plane (Fig. 34); the value of $|S'_\varepsilon|$ differs from the area of the hemisphere surface, $\pi^{\frac{1}{2}m}\varepsilon^{m-1}/\Gamma(\frac{1}{2}m)$, by the area (with appropriate sign attached) of the surface of the zonal region $\bar{S}_\varepsilon$ contained between Γ and the tangent plane. The height of this zonal region is equal to the maximum value of $|\xi_m|$ at the points of intersection of the surface Γ with the sphere S_ε. Since $\varepsilon < d$, these points lie inside the Lyapunov sphere, and are subject to the estimate (1.16''):

$$|\xi_m| \leq a_1 r^{\alpha+1} = a_1 \varepsilon^{\alpha+1}.$$

From this it is not difficult to see (the details are left to the reader) that the surface area of the zonal region is of order $O(\varepsilon^{m-1+\alpha})$. It is now clear that

$$\int_{S'_\varepsilon} \frac{\partial}{\partial \nu} \frac{1}{r^{m-2}} d_\xi S_\varepsilon = \frac{(m-2)\pi^{\frac{1}{2}m}}{\Gamma(\frac{1}{2}m)} + O(\varepsilon^\alpha),$$

and, consequently,

$$W_0(x) = -\frac{(m-2)\pi^{\frac{1}{2}m}}{\Gamma(\frac{1}{2}m)}, \quad x \in \Gamma.$$

The theorem is now completely proved.

§ 5. Limiting Values of the Double-Layer Potential

From the example of the Gauss integral it is clear that, in general, the double-layer potential has a discontinuity when the point x crosses the surface Γ. In addition, as we shall now see, limiting values of the double-layer potential exist under quite wide conditions when the point x tends to an arbitrary point $x_0 \in \Gamma$ either from inside or from outside Γ. We denote by $W_i(x_0)$ and $W_e(x_0)$ respectively the limiting values of the double-layer potential $W(x)$ at the point x_0 as $x \to x_0$ from inside and outside Γ. The direct value of this potential at the point x_0 is denoted by $\overline{W(x_0)}$.

THEOREM 18.5.1. *Let Γ be a closed Lyapunov surface, and let $\sigma(\xi)$ be the density, continuous on Γ. Then the following limiting relations hold for the double-layer potential* (3.1):

$$\left. \begin{array}{l} W_i(x_0) = -\tfrac{1}{2}(m-2)|S_1|\sigma(x_0) + \overline{W(x_0)}, \\ W_e(x_0) = \tfrac{1}{2}(m-2)|S_1|\sigma(x_0) + \overline{W(x_0)}, \end{array} \right\} x_0 \in \Gamma. \qquad (1)$$

We write (3.1) in the form

$$W(x) = W_1(x) + \sigma(x_0) W_0(x). \qquad (2)$$

Here W_0 is the Gauss integral, and

$$W_1(x) = \int_\Gamma [\sigma(\xi) - \sigma(x_0)] \frac{\partial}{\partial \nu} \frac{1}{r^{m-2}} \, d_\xi \Gamma; \qquad (3)$$

$W_1(x)$ is the double-layer potential whose density $\sigma(\xi) - \sigma(x_0)$ is equal to zero when $\xi = x_0$. We now show that this potential is continuous at the point x_0.

Taking the point x_0 as centre, we describe a sphere of some radius η; this divides the surface Γ into two portions: $\Gamma = \Gamma' \cup \Gamma''$, of which Γ' lies inside the sphere, and Γ'' outside it. The potential $W_1(x)$ is correspondingly divided into two parts:

$$W_1(x) = W_1'(x) + W_1''(x),$$

where

$$W_1'(x) = \int_{\Gamma'} [\sigma(\xi) - \sigma(x_0)] \frac{\partial}{\partial \nu} \frac{1}{r^{m-2}} \, d_\xi \Gamma,$$

$$W_1''(x) = \int_{\Gamma''} [\sigma(\xi) - \sigma(x_0)] \frac{\partial}{\partial \nu} \frac{1}{r^{m-2}} \, d_\xi \Gamma.$$

We write the obvious inequality

$$|W_1(x) - \overline{W_1(x_0)}| \leq |W_1'(x)| + |\overline{W_1'(x_0)}| + |W_1''(x) - \overline{W_1''(x_0)}|; \quad (4)$$

a line above the symbols here denotes the direct value of the corresponding potential.

We estimate the right-hand side of inequality (4). Choose η such that when $|\xi - x_0| < \eta$ the inequality

$$|\sigma(\xi) - \sigma(x_0)| < \frac{\varepsilon}{3C}$$

is satisfied; here ε is an arbitrary, given, positive number, and C is the constant in equation (2.6). This choice of η is possible because the density $\sigma(\xi)$ is continuous. Then for any $x \in E_m$ we have

$$|W_1'(x)| \leq \int_{\Gamma'} |\sigma(\xi) - \sigma(x_0)| \left| \frac{\partial}{\partial \nu} \frac{1}{r^{m-2}} \right| d\Gamma$$

$$< \frac{\varepsilon}{3C} \int_{\Gamma'} \left| \frac{\partial}{\partial \nu} \frac{1}{r^{m-2}} \right| d\Gamma \leq \frac{\varepsilon}{3C} \int_{\Gamma} \left| \frac{\partial}{\partial \nu} \frac{1}{r^{m-2}} \right| d\Gamma \leq \tfrac{1}{3}\varepsilon. \quad (5)$$

In particular,

$$|\overline{W_1'(x_0)}| < \tfrac{1}{3}\varepsilon. \quad (6)$$

We fix the radius η and assume that the point x is sufficiently close to x_0, namely, that $|x - x_0| \leq \tfrac{1}{2}\eta$. Then on the surface Γ''

$$r = |\xi - x| \geq |\xi - x_0| - |x - x_0| \geq \eta - \tfrac{1}{2}\eta = \tfrac{1}{2}\eta;$$

the integrand function in the integral $W_1''(x)$ is continuous, and so the integral itself is continuous. Then there exists a number $\delta > 0$ such that, for $|x - x_0| < \delta$, we necessarily have

$$|W_1''(x) - W_1''(x_0)| = |W_1''(x) - \overline{W_1''(x_0)}| < \tfrac{1}{3}\varepsilon. \quad (7)$$

From relations (4)–(7) it follows that

$$|W_1(x) - \overline{W_1(x_0)}| < \varepsilon, \quad \text{if} \quad |x - x_0| < \delta, \quad (8)$$

i.e. that the potential $W_1(x)$ is continuous at the point x_0. If this is so, then at this point the limiting values of the potential $W_1(x)$ and its direct value coincide:

$$W_{1i}(x_0) = W_{1e}(x_0) = \overline{W_1(x_0)}. \quad (9)$$

Formula (4.2) shows that the limiting values of Gauss's integral $W_0(x)$ exist and are respectively equal to

$$W_{0i}(x_0) = -(m-2)|S_1|, \qquad W_{0e}(x_0) = 0,$$

and the direct value is

$$\overline{W_0(x_0)} = -\tfrac{1}{2}(m-2)|S_1|.$$

It now follows from formulae (2) and (9) that the limiting values $W_i(x_0)$ and $W_e(x_0)$ exist, and

$$\begin{aligned}W_i(x_0) &= \overline{W_1(x_0)} + \sigma(x_0)W_{0i}(x_0) = \overline{W_1(x_0)} - (m-2)|S_1|\sigma(x_0),\\ W_e(x_0) &= \overline{W_1(x_0)} + \sigma(x_0)W_{0e}(x_0) = \overline{W_1(x_0)}.\end{aligned} \qquad (10)$$

Moreover,

$$\begin{aligned}\overline{W_1(x_0)} &= \int_\Gamma [\sigma(\xi) - \sigma(x_0)] \frac{\partial}{\partial \nu} \frac{1}{r^{m-2}} \, d_\xi \Gamma \\ &= \overline{W(x_0)} + \tfrac{1}{2}(m-2)|S_1|\sigma(x_0), \qquad r_0 = |\xi - x_0|,\end{aligned} \qquad (11)$$

and the formulae (1) follow immediately from relations (10) and (11). The theorem is proved.

From formulae (1) there ensues a simple expression relating the density of the double-layer potential with its limiting values:

$$W_i(x_0) - W_e(x_0) = -(m-2)|S_1|\sigma(x_0). \qquad (12)$$

Remark 1. If the density $\sigma(\xi)$ is not continuous on Γ, but only integrable, then it can be shown that the limiting values of the double-layer potential exist almost everywhere on Γ and are given by the same formulae (1).

Remark 2. The following theorem, due to Lyapunov, holds for the normal derivative of the double-layer potential. Let Γ be a closed Lyapunov surface, and let the density $\sigma(\xi)$ be continuous on Γ. If the normal derivative of the double-layer potential has a limit as $x \to x_0 \in \Gamma$ from inside (outside), then this derivative also has a limit from the outside (inside), and the two are equal.

THEOREM 18.5.2. *For a closed Lyapunov surface, on which the density is continuous, the double-layer potential tends uniformly to its limiting values both from inside and from outside the surface.*

We retain the notation used in the proof of Theorem 18.5.1. The density $\sigma(\xi)$ is continuous, and therefore also uniformly continuous on Γ. In this case the radius η can be chosen independently of the position of the point x_0 on the surface Γ. Furthermore, the function $W_1''(x)$ is actually a function of two points: the point $x \in E_m$ and the point $x_0 \in \Gamma$. If the radius η is fixed, then on the bounded, closed set defined by the relations

$$x_0 \in \Gamma, \quad |x - x_0| \leq \tfrac{1}{2}\eta,$$

the function under discussion is continuous, and, consequently, uniformly continuous. Therefore the quantity δ, which figures in inequality (8), can be chosen so that it depends only on ε. The same relation (8) then shows that $W_1(x) \xrightarrow[x \to x_0]{} W_1(x_0)$ uniformly with respect to $x_0 \in \Gamma$.

Finally, the potential $W_0(x)$ is constant both inside and outside Γ, so that if $x \to x_0$ either from inside or from outside Γ then $W_0(x)$ tends uniformly to its limiting value.

It is clear from formula (2) that the potential $W(x)$ has the same properties.

§ 6. Continuity of the Single-Layer Potential

Theorem 18.6.1. *If Γ is a closed Lyapunov surface, and if the density $\mu(\xi)$ is measurable and bounded, then the single-layer potential*

$$V(x) = \int_\Gamma \mu(\xi) \frac{1}{r^{m-2}} \, d_\xi \Gamma \tag{1}$$

is continuous in the whole of the space E_m.

Continuity of the potential (1) is obvious for $x \bar{\in} \Gamma$, and it is only necessary to consider the case $x \in \Gamma$.

We show first of all that for $x \in \Gamma$ the integral (1) converges, and, consequently, that the single-layer potential $V(x)$ is defined on the surface Γ. Construct the Lyapunov sphere $S(x)$, and let $\Gamma'(x)$ be the portion of Γ enclosed within the sphere $S(x)$. We have

$$V(x) = \int_{\Gamma'(x)} \mu(\xi) \frac{1}{r^{m-2}} \, d_\xi \Gamma + \int_{\Gamma \setminus \Gamma'(x)} \mu(\xi) \frac{1}{r^{m-2}} \, d_\xi \Gamma.$$

The integrand of the second integral is continuous here, and it is sufficient to consider the first integral. Introduce a local coordinate system

$(\xi_1, \xi_2, \ldots, \xi_m)$ with centre at the point x and with the axis ξ_m pointing along the normal to Γ at this point. Put

$$\rho^2 = \xi_1^2 + \xi_2^2 + \ldots + \xi_{m-1}^2.$$

Denote by $G'(x)$ the projection of the surface $\Gamma'(x)$ on to the plane $\xi_m = 0$ which is tangent to Γ at the point x. Then we have

$$\int_{\Gamma'(x)} \mu(\xi) \frac{1}{r^{m-2}} \, d\xi = \int_{G'(x)} \mu(\xi) \frac{d\xi_1 d\xi_2 \ldots d\xi_{m-1}}{r^{m-2} \cos(v, \xi_m)} ; \qquad (2)$$

if $|\mu(\xi)| \leq M = \text{const.}$, then the integrand does not exceed the value

$$\frac{2M}{\rho^{m-2}};$$

this estimate shows that integral (2) converges.

We now show that the potential (1) is continuous at any point $x \in \Gamma$. Let y be an arbitrary point of the space E_m. Around the point x describe a sphere of radius $\eta < d$; denote by Γ'_η and Γ''_η the portions of the surface Γ contained respectively inside and outside the sphere. The integral (1) splits into two parts, associated with Γ'_η and Γ''_η; denote these integrals by V' and V'' respectively. Then obviously

$$|V(y) - V(x)| \leq |V'(y)| + |V'(x)| + |V''(y) - V''(x)|. \qquad (3)$$

We estimate the first term in (3). Since $\eta < d$, so $\Gamma'_\eta \subset \Gamma'(x)$, and we can introduce on Γ'_η a local coordinate system with origin at x. Denote by y'

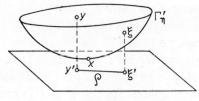

Fig. 35.

the projection of the point y on to the tangent plane at x (Fig. 35). Let the local coordinates of the point y' be $(y_1, y_2, \ldots, y_{m-1}, 0)$. Put

$$\rho^2 = \sum_{k=1}^{m-1} (\xi_k - y_k)^2,$$

where ρ is the length of the projection on to the tangent plane at x of the interval joining the point y and ξ. Clearly, $\rho \leq |\xi-y|$. Moreover,

$$|V'(y)| = \left|\int_{\Gamma'_n} \mu(\xi) \frac{d_\xi \Gamma}{|\xi-y|^{m-2}}\right|$$

$$\leq M \int_{G'_n} \frac{d\xi_1 \ldots d\xi_{m-1}}{\rho^{m-2} \cos(v, \xi_m)} \leq 2M \int_{G'_n} \frac{d\xi_1 \ldots d\xi_{m-1}}{\rho^{m-2}},$$

where G'_n is the projection of the surface Γ'_n on to the tangent plane at x. Take the point y close enough to x so that $|y-x| < \frac{1}{2}\eta$. Then, if $\xi \in \Gamma'_n$,

$$\rho \leq |\xi-y| \leq |\xi-x|+|x-y| < \tfrac{3}{2}\eta.$$

This means that the domain G'_n lies wholly in the $(m-1)$-dimensional sphere $\rho < \tfrac{3}{2}\eta$, and, consequently,

$$|V'(y)| \leq 2M \int_{\rho < \frac{3}{2}\eta} \frac{d\xi_1 \ldots d\xi_{m-1}}{\rho^{m-2}}. \tag{4}$$

In the $(m-1)$-dimensional space we introduce spherical coordinates with centre at the point y'. Then $d\xi_1 \ldots d\xi_{m-1} = \rho^{m-2} d\rho d\sigma_1$ where we denote by σ_1 the unit sphere in this space, and by $d\sigma_1$ an element of area of its surface. Formula (4) takes the form

$$|V'(y)| \leq 2M \int_{\sigma_1} d\sigma_1 \int_0^{\frac{3}{2}\eta} d\rho = 3M|\sigma_1|\eta.$$

Let ε be an arbitrary, given, positive number, and take $\eta = \tfrac{1}{9}\varepsilon/M|\sigma_1|$. Then, if $|y-x| < \tfrac{1}{18}\varepsilon/M|\sigma_1|$, we have $|V'(y)| < \tfrac{1}{3}\varepsilon$.

This last inequality obviously holds for $y = x$, so that

$$|V'(x)| < \tfrac{1}{3}\varepsilon.$$

Inequality (3) now gives

$$|V(y)-V(x)| < \tfrac{2}{3}\varepsilon + |V''(y)-V''(x)|.$$

Choose a number $\delta > 0$ small enough that $\delta < \tfrac{1}{18}\varepsilon/M|\sigma_1|$ and that $|V''(y)-V''(x)| < \tfrac{1}{3}\varepsilon$ for $|y-x| < \delta$. Then $|V(y)-V(x)| < \varepsilon$. The theorem is proved.

§ 7. Normal Derivative of the Single-Layer Potential

As before, we shall consider the single-layer potential (6.1) under the assumption that Γ is a closed Lyapunov surface.

Let x be an arbitrary point of the space E_m, and let n be the outward normal to the surface Γ passing through the point x. If $x \bar{\in} \Gamma$, then it is possible to calculate the derivative of the potential (6.1) along the direction of the normal n simply by differentiating under the integral sign

$$\frac{\partial V(x)}{\partial n} = \int_\Gamma \mu(\xi) \frac{\partial}{\partial n} \frac{1}{r^{m-2}} \, d_\xi \Gamma. \tag{1}$$

A calculation similar to that of § 2 leads to the formula

$$\frac{\partial}{\partial n} \frac{1}{r^{m-2}} = \frac{m-2}{r^{m-1}} \cos(r, n). \tag{2}$$

Hence

$$\frac{\partial V(x)}{\partial n} = (m-2) \int_\Gamma \mu(\xi) \frac{\cos(r, n)}{r^{m-1}} \, d_\xi \Gamma. \tag{3}$$

Let $x \in \Gamma$. If the density $\mu(\xi)$ is measurable and bounded, $|\mu(\xi)| \leq M = $ const., then the integral (3) converges. We now show this. We select a portion $\Gamma'(x)$ of the surface Γ, lying inside the Lyapunov sphere $S'(x)$. It is sufficient to prove the convergence of the integral

$$\int_{\Gamma'(x)} \mu(\xi) \frac{\cos(r, n)}{r^{m-1}} \, d_\xi \Gamma.$$

Introducing local coordinates with origin at the point x, we can reduce the latter integral to the form

$$\int_{G'(x)} \mu(\xi) \frac{\cos(r, n)}{r^{m-1}} \frac{d\xi_1 \ldots d\xi_{m-1}}{\cos(v, \xi_m)}, \tag{4}$$

where $G'(x)$ is the projection of $\Gamma'(x)$ on the tangent plane at the point x. The integrand in (4) is bounded by the quantity

$$\frac{2M}{\rho^{m-1}} |\cos(r, n)|, \qquad \rho^2 = \xi_1^2 + \xi_\eta^2 + \ldots + \xi_{m-1}^2.$$

Moreover,

$$\cos(r, n) = \cos(r, \xi_m) = \frac{\xi_m}{r};$$

from the inequalities (1.16) and (1.15) we have

$$|\cos(n, r)| \leq a_1 r^\alpha \leq 2^\alpha a_1 \rho^\alpha, \tag{5}$$

and so for the integrand in (4) we eventually obtain the estimate

$$\frac{2^\alpha M a_1}{\rho^{m-1-\alpha}},$$

which proves that the integral (3) converges.

As we shall see a little later, the value of the integral (3) when $x \in \Gamma$ cannot be regarded as the normal derivative of the potential (6.1). The value of integral (3) when $x \in \Gamma$ is called the *direct value of the normal derivative of the single-layer potential,* and is denoted by the symbol $\overline{\partial V(x)/\partial n}$.

We shall denote by $\partial V(x_0)/\partial n_i$ and $\partial V(x_0)/\partial n_e$ the limiting values (if they exist) of the normal derivative $\partial V(x)/\partial n$, as $x \to x_0 \in \Gamma$ from inside and from outside Γ respectively.

THEOREM 18.7.1. *If Γ is a closed Lyapunov surface, and if the density $\mu(\xi)$ is continuous on Γ, then the single-layer potential has on the surface Γ a regular normal derivative both from inside and from outside Γ. The limiting values of the normal derivative of the single-layer potential are given by the formulae*

$$\frac{\partial V(x_0)}{\partial n_i} = \tfrac{1}{2}(m-2)|S_1|\mu(x_0) + \overline{\frac{\partial V(x_0)}{\partial n}},$$

$$\frac{\partial V(x_0)}{\partial n_e} = -\tfrac{1}{2}(m-2)|S_1|\mu(x_0) + \overline{\frac{\partial V(x_0)}{\partial n}}. \tag{6}$$

Consider the double-layer potential with density μ

$$W(x) = \int_\Gamma \mu(\xi) \frac{\partial}{\partial \nu} \frac{1}{r^{m-2}} d_\xi \Gamma \tag{7}$$

and construct the sum

$$\frac{\partial V(x)}{\partial n} + W(x) = \int_\Gamma \mu(\xi) \left[\frac{\partial}{\partial n} \frac{1}{r^{m-2}} + \frac{\partial}{\partial \nu} \frac{1}{r^{m-2}} \right] d_\xi \Gamma.$$

We shall show that this sum varies continuously as the point x, moving along the normal to the surface Γ, crosses through it.

Let x_0 be a point on the surface Γ, n the normal to Γ at this point, and x an arbitrary point on the normal n, either inside or outside Γ. Around the point x_0 we describe a sphere of radius $\eta < d$ and denote by Γ'_η the portion of the surface which lies inside this sphere. The arguments of the preceding section show that it is sufficient to establish the following fact: the inequality

$$A = \left| \int_{\Gamma'_\eta} \mu(\xi) \left[\frac{\partial}{\partial n} \frac{1}{r^{m-2}} + \frac{\partial}{\partial \nu} \frac{1}{r^{m-2}} \right] d_\xi \Gamma \right| < \tfrac{1}{3}\varepsilon$$

holds for sufficiently small η. Here ε is an arbitrary, given, positive number. We have

$$\frac{\partial}{\partial n} \frac{1}{r^{m-2}} = \frac{m-2}{r^{m-1}} \frac{\xi_k - x_k}{r} \cos(n, \xi_k),$$

$$\frac{\partial}{\partial \nu} \frac{1}{r^{m-2}} = -\frac{m-2}{r^{m-1}} \frac{\xi_k - x_k}{r} \cos(\nu, \xi_k).$$

Introduce a local coordinate system with centre at the point x_0. In this system $x_k = 0$, $\cos(n, \xi_k) = 0$ for $1 \leq k \leq m-1$, and $\cos(n, \xi_m) = 1$. But then

$$\frac{\partial}{\partial n} \frac{1}{r^{m-2}} + \frac{\partial}{\partial \nu} \frac{1}{r^{m-2}} =$$

$$= \frac{m-2}{r^{m-1}} \frac{\xi_m - x_m}{r} [1 - \cos(\nu, \xi_m)] - \frac{m-2}{r^{m-1}} \sum_{k=1}^{m-1} \frac{\xi_k}{r} \cos(\nu, \xi_k).$$

Furthermore, by inequalities (1.12) and (1.6),

$$|\cos(\nu, \xi_k)| \leq \sqrt{3}\, a r_0^\alpha, \qquad k = 1, 2, \ldots, m-1,$$

$$\cos(\nu, \xi_m) \geq 1 - \tfrac{1}{2} a^2 r_0^{2\alpha},$$

where $r_0 = |\xi - x_0|$. It follows from the last inequality that

$$1 - \cos(\nu, \xi_m) \leq \tfrac{1}{2} a^2 r_0^{2\alpha} < \tfrac{1}{2} a r_0^\alpha.$$

It is now easy to see that

$$\left| \frac{\partial}{\partial n} \frac{1}{r^{m-2}} + \frac{\partial}{\partial \nu} \frac{1}{r^{m-2}} \right| \leq \frac{c_1 r_0^\alpha}{r^{m-1}}, \qquad c_1 = \text{const.} \tag{8}$$

Put $\rho^2 = \xi_1^2 + \xi_2^1 + \ldots + \xi_{m-1}^2$ and denote by G'_η the projection of the surface Γ'_η on the tangent plane at the point x_0. By formula (1.15), $r_0 \leq 2\rho$. On the other hand,

$$r^2 = \sum_{k=1}^{m} (\xi_k - x_k)^2 = \sum_{k=1}^{m-1} \xi_k^2 + (\xi_m - x_m)^2 \geq \rho^2.$$

Hence $r \geq \rho$. It now follows from formula (8) that

$$\left| \frac{\partial}{\partial n} \frac{1}{r^{m-2}} + \frac{\partial}{\partial \nu} \frac{1}{r^{m-2}} \right| \leq \frac{2^\alpha c_1}{\rho^{m-1-\alpha}}. \qquad (9)$$

Let $|\mu(\xi)| \leq M$. Then

$$A \leq 2^\alpha c_1 M \int_{\Gamma'_\eta} \frac{d_\xi \Gamma}{\rho^{m-1-\alpha}} = 2^\alpha c_1 M \int_{G'_\eta} \frac{d\xi_1 \, d\xi_2 \ldots d\xi_{m-1}}{\rho^{m-1-\alpha} \cos(\nu, \xi_m)}$$

$$\leq 2^{\alpha+1} c_1 M \int_{G'_\eta} \frac{d\xi_1 \, d\xi_2 \ldots d\xi_{m-1}}{\rho^{m-1-\alpha}}.$$

In the domain G'_η we have the inequality $\rho \leq r \leq \eta$, and therefore G'_η lies in the sphere $\rho \leq \eta$. Consequently,

$$A \leq 2^{\alpha+1} c_1 M \int_{\rho<\eta} \frac{d\xi_1 \, d\xi_2 \ldots d\xi_{m-1}}{\rho^{m-1-\alpha}} = 2^{\alpha+1} c_1 M \int_{\sigma_1} d\sigma_1 \int_0^\eta \rho^{\alpha-1} d\rho$$

$$= \frac{2^{\alpha+1} c_1 M |\sigma_1|}{\alpha} \eta^\alpha,$$

where σ_1 is a unit sphere in $(m-1)$-dimensional space. Obviously we only need take

$$\eta < \left[\frac{\alpha \varepsilon}{3 \cdot 2^{\alpha+1} c_1 M |\sigma_1|} \right]^{1/\alpha}$$

and we obtain

$$A < \tfrac{1}{3}\varepsilon. \qquad (10)$$

Repeating the arguments of §§ 3 and 6, and using the estimate (10), we see that the sum $\partial V(x)/\partial n + W(x)$ is continuous for the passage of the point x through the surface Γ. But then the limiting values and direct value coincide for this sum:

$$\frac{\partial V(x_0)}{\partial n_i} + W_i(x_0) = \frac{\partial V(x_0)}{\partial n_e} + W_e(x_0) = \overline{\frac{\partial V(x_0)}{\partial n}} + \overline{W(x_0)}.$$

From this the required equations ensue:

$$\frac{\partial V(x_0)}{\partial n_i} = \overline{W(x_0)} - W_i(x_0) + \overline{\frac{\partial V(x_0)}{\partial n}} = \tfrac{1}{2}(m-2)|S_1|\mu(x_0) + \overline{\frac{\partial V(x_0)}{\partial n}},$$

$$\frac{\partial V(x_0)}{\partial n_e} = \overline{W(x_0)} - W_e(x_0) + \overline{\frac{\partial V(x_0)}{\partial n}} = -\tfrac{1}{2}(m-2)|S_1|\mu(x_0) + \overline{\frac{\partial V(x_0)}{\partial n}}.$$

If we subtract equations (6), we obtain a formula connecting the density of the single-layer potential with the limiting values of its normal derivative

$$\frac{\partial V(x_0)}{\partial n_i} - \frac{\partial V(x_0)}{\partial n_e} = (m-2)|S_1|\mu(x_0). \tag{11}$$

It remains to prove that $\partial V(x)/\partial n_i$ and $\partial V(x)/\partial n_e$ are regular normal derivatives. We have

$$\frac{\partial V(x)}{\partial n} = \left[\frac{\partial V(x)}{\partial n} + W(x)\right] - W(x). \tag{12}$$

It has already been shown that the expression in square brackets is continuous, and so tends uniformly to its limiting value; in this regard it is immaterial whether the point x tends to the point $x_0 \in \Gamma$ from inside or from outside Γ. Moreover, if $x \to x_0$ either from inside or from outside Γ, then by Theorem 18.5.2 the double-layer potential $W(x)$ tends uniformly to its limiting values. Because of formula (12) the quantity $\partial V(x)/\partial n$ also has the same property. By definition (§ 2, Chapter 12), the potential (6.1) therefore has a regular normal derivative both from inside and from outside Γ.

§ 8. Reduction of the Dirichlet and Neumann Problems to Integral Equations

Consider a closed Lyapunov surface which is the boundary of two domains: the domain Ω lies inside it and the domain Ω' outside it. We pose simultaneously four boundary-value problems for the homogeneous Laplace equation: to find a function $u(x)$, which is harmonic in Ω or Ω', and which satisfies either the Dirichlet boundary conditions

$$u|_\Gamma = \phi(x), \tag{1}$$

or the Neumann boundary conditions

$$\left.\frac{\partial u}{\partial n}\right|_\Gamma = \psi(x). \tag{2}$$

The functions $\phi(x)$ and $\psi(x)$ will be assumed continuous on Γ.

We shall designate the interior and exterior Dirichlet problems by D_i and D_e respectively, and the interior and exterior Neumann problems by N_i and N_e respectively. We shall solve these problems by seeking a solution in the form of a potential. More precisely, we shall look for a solution of the

Dirichlet problem in the form of a double-layer potential

$$u(x) = \int_\Gamma \sigma(\xi) \frac{\partial}{\partial \nu} \frac{1}{r^{m-2}} \, d_\xi \Gamma, \tag{3}$$

and for a solution to the Neumann problem in the form of a single-layer potential

$$u(x) = \int_\Gamma \mu(\xi) \frac{1}{r^{m-2}} \, d_\xi \Gamma. \tag{4}$$

In this connection we shall require that the densities $\sigma(\xi)$ and $\mu(\xi)$ are continuous on Γ.

By representing the solutions in these forms, we automatically obtain harmonic functions in the respective domains, and it only remains to take care of the boundary conditions. We note, however, that we can anticipate certain difficulties in the case of the problem D_e; the solution can be of order $O(|x|^{2-m})$ at infinity, but the potential (3) decreases more rapidly – it is of order $O(|x|^{1-m})$, and, consequently, not every function harmonic in Ω' can be represented in the form (3).

Consider, for example, the interior Dirichlet problem D_i. The boundary condition (1) can be understood in the following sense: if $x \in \Omega$ and $x \to x_0 \in \Gamma$, then

$$\lim_{x \to x_0} u(x) = \phi(x_0). \tag{5}$$

But $u(x)$ is the potential of a double layer whose density, by hypothesis, is continuous. Therefore, by formula (5.1),

$$\lim_{x \to x_0} u(x) = -\tfrac{1}{2}(m-2)|S_1|\sigma(x_0) + \int_\Gamma \sigma(\xi) \frac{\partial}{\partial \nu} \frac{1}{|\xi - x_0|^{m-2}} \, d_\xi \Gamma.$$

Substituting this in formula (5), replacing x_0 by x, and dividing by $-\tfrac{1}{2}(m-2)|S_1|$, we obtain an integral equation for the unknown function $\sigma(x)$:

$$\sigma(x) - \frac{2}{(m-2)|S_1|} \int_\Gamma \sigma(\xi) \frac{\partial}{\partial \nu} \frac{1}{r^{m-2}} \, d_\xi \Gamma = -\frac{2}{(m-2)|S_1|} \phi(x), \quad x \in \Gamma.$$

Using formulae (5.1) and (7.6) for the limiting values, and also boundary conditions (1) and (2), we obtain integral equations for the three remaining problems. For convenience, we write all four integral equations together:

$$(\text{D}_i) \quad \sigma(x) - \frac{2}{(m-2)|S_1|} \int_\Gamma \sigma(\xi) \frac{\partial}{\partial \nu} \frac{1}{r^{m-2}} \, d_\xi \Gamma = -\frac{2}{(m-2)|S_1|} \phi(x), \qquad (6)$$

$$(\text{D}_e) \quad \sigma(x) + \frac{2}{(m-2)|S_1|} \int_\Gamma \sigma(\xi) \frac{\partial}{\partial \nu} \frac{1}{r^{m-2}} \, d_\xi \Gamma = \frac{2}{(m-2)|S_1|} \phi(x), \qquad (7)$$

$$(\text{N}_i) \quad \mu(x) + \frac{2}{(m-2)|S_1|} \int_\Gamma \mu(\xi) \frac{\partial}{\partial n} \frac{1}{r^{m-2}} \, d_\xi \Gamma = \frac{2}{(m-2)|S_1|} \psi(x), \qquad (8)$$

$$(\text{N}_e) \quad \mu(x) - \frac{2}{(m-2)|S_1|} \int_\Gamma \mu(\xi) \frac{\partial}{\partial n} \frac{1}{r^{m-2}} \, d_\xi \Gamma = -\frac{2}{(m-2)|S_1|} \psi(x). \qquad (9)$$

In equations (6)–(9), $x \in \Gamma$.

We note the following properties of equations (6)–(9).

1. The estimate (1.17), together with formulae (7.2) and (7.5), shows that equations (6)–(9) are integral equations with a weak singularity.

2. The kernels

$$\frac{\partial}{\partial \nu} \frac{1}{r^{m-2}}, \quad \frac{\partial}{\partial n} \frac{1}{r^{m-2}}$$

can be obtained one from the other by interchange of arguments. Since these kernels are also real, they are consequently adjoint (c.f. § 1, Chapter 8). Hence it follows that equations (6) and (9) are mutually adjoint, and so are equations (7) and (8).

3. Every square-integrable solution of any of the equations (6)–(9) is continuous on Γ. It was shown in § 3 that

$$\frac{\partial}{\partial \nu} \frac{1}{r^{m-2}} = \frac{A(x, \xi)}{r^{m-1-\frac{1}{2}\alpha}},$$

where the function $A(x, \xi)$ is continuous on Γ. Interchanging the arguments x and ξ, we obtain

$$\frac{\partial}{\partial n} \frac{1}{r^{m-2}} = \frac{A(\xi, x)}{r^{m-1-\frac{1}{2}\alpha}},$$

and the function $A(\xi, x)$ is also continuous on Γ. We further recall that the functions $\phi(x)$ and $\psi(x)$ are continuous by definition. Our assertion then follows from Theorem 8.6.1.

Equations (6)–(9) are usually called the *integral equations of potential theory*.

§ 9. Dirichlet and Neumann Problems in a Half-Space

Up till now we have not given any definition of functions which are harmonic in a half-space or, more generally, in a domain having an infinite boundary. We now extend the definition for a finite region to this case: a function is said to be harmonic in a domain having an infinite boundary if it has second derivatives continuous at every point, and satisfies the homogeneous Laplace's equation, in this domain.

The integral representations obtained in the previous section enable solutions to the Dirichlet and Neumann problems for the homogeneous Laplace equation in a half-space; it is only necessary to insist that the prescribed functions $\phi(x)$ and $\psi(x)$ should decrease at a definite rate at infinity. We shall discuss this more explicitly a little later.

Under certain natural restrictions, the theorems concerning potentials, stated above, can also be extended to the case where the surface Γ is infinite. Thus, if Γ is the hyperplane $\xi_m = 0$ then it is sufficient to prescribe the additional condition that the density $\sigma(\xi)$ of the double-layer potential should be, at infinity, of order

$$\sigma(\xi) = O(\rho^{-\beta}), \quad \rho^2 = \sum_{k=1}^{m-1} \xi_k^2, \quad \beta = \text{const.} > 0, \tag{1}$$

and that the density $\mu(\xi)$ of the single-layer potential should be

$$\mu(\xi) = O(\rho^{-\beta-1}); \tag{2}$$

with these estimates the integrals (3.1) and (6.1) converge.

In the case of a half-space it is sufficient to discuss only the interior problem, and to consider only the integral equations (8.6) and (8.8). By formula (2.4), the kernel of equation (6) has the form

$$\frac{\partial}{\partial v} \frac{1}{r^{m-2}} = -\frac{(m-2)}{r^{m-1}} \cos(r, v).$$

Here the normal v, outwards with respect to the half-space $\xi_m > 0$, is in the opposite direction to the axis ξ_m; if both the points x and ξ lie on the plane $\xi_m = 0$, then the vector r lies on this same plane. But then $\cos(r, v) \equiv 0$, and the kernel of equation (8.6) is identically zero. So is the adjoint kernel of equation (8.8). These equations then give immediately that

$$\sigma(x) = -\frac{2}{(m-2)|S_1|} \phi(x), \quad \mu(x) = \frac{2}{(m-2)|S_1|} \psi(x).$$

The solution to the Dirichlet problem for the half-space $x_m > 0$ is given by the formula

$$u(x) = \frac{2}{(m-2)|S_1|} \int_{-\infty}^{+\infty} \cdots \int_{-\infty}^{+\infty} \phi(\xi) \frac{\partial}{\partial \xi_m} \frac{1}{r^{m-2}} d\xi_1 \ldots d\xi_{m-1}, \qquad (3)$$

and the solution to the Neumann problem by the formula

$$u(x) = \frac{2}{(m-2)|S_1|} \int_{-\infty}^{+\infty} \cdots \int_{-\infty}^{+\infty} \psi(\xi) \frac{1}{r^{m-2}} d\xi_1 \ldots d\xi_{m-1}. \qquad (4)$$

Equations (3) and (4) are appropriate if, at infinity,

$$\phi(\xi) = O(\rho^{-\beta}), \quad \psi(\xi) = O(\rho^{-\beta-1}), \quad \beta = \text{const.} > 0.$$

Carrying out the differentiation in formula (3), we obtain the form

$$u(x) = \frac{2x_m}{|S_1|} \int_{-\infty}^{+\infty} \cdots \int_{-\infty}^{+\infty} \phi(\xi) \frac{1}{r^m} d\xi_1 \ldots d\xi_{m-1}, \qquad (3')$$

$$r^2 = \sum_{k=1}^{m-1} (\xi_k - x_k)^2 + x_m^2.$$

§ 10. The First Pair of Adjoint Equations

Our subsequent investigations of the integral equations of potential theory will be carried out under the assumption that the surface Γ is closed and regular. We note that a regular surface is necessarily a Lyapunov surface, in which the exponent $\alpha = 1$ (prove!).

In the present section we shall show that the integral equations (8.6) and (8.9), corresponding to the interior Dirichlet problem D_i and the exterior Neumann problem N_e, are soluble, and have unique solutions, for any continuous functions $\phi(x)$ and $\psi(x)$. With this aim we consider the homogeneous integral equation for the exterior Neumann problem; we denote the unknown function in this equation by $\mu_0(x)$

$$\mu_0(x) - \frac{2}{(m-2)|S_1|} \int_\Gamma \mu_0(\xi) \frac{\partial}{\partial n} \frac{1}{r^{m-2}} d\xi = 0. \qquad (1)$$

Let $\mu_0 \in L_2(\Gamma)$ be any solution of this equation. As was shown in § 8, the function μ_0 is continuous on Γ. Construct the single-layer potential with density μ_0

$$V_0(x) = \int_\Gamma \mu_0(\xi) \frac{1}{r^{m-2}} d_\xi \Gamma. \qquad (2)$$

The potential (2) has a regular normal derivative from outside Γ, and equation (1) implies that this normal derivative is equal to zero

$$\frac{\partial V_0(x)}{\partial n_e} \equiv 0. \tag{3}$$

By the uniqueness theorem for the exterior Neumann problem

$$V_0(x) \equiv 0, \qquad x \in \Omega'. \tag{4}$$

But the single-layer potential is a continuous function throughout all space. Therefore

$$V_0(x) \equiv 0, \qquad x \in \Gamma. \tag{5}$$

Consider now the potential $V_0(x)$ in the domain Ω located inside Γ. In this domain the function $V_0(x)$ is harmonic and equation (5) shows that it equals zero on the boundary of the domain. By the uniqueness theorem for the interior Dirichlet problem,

$$V_0(x) \equiv 0, \qquad x \in \Omega. \tag{6}$$

But then we also have in Ω that

$$\frac{\partial V_0(x)}{\partial n_i} \equiv 0.$$

Comparing this with formula (3), and using (7.11), we find that $\mu_0(x) \equiv 0$.

Thus the homogeneous integral equation (1) has only the trivial solution. By virtue of Fredholm's Alternative (§ 5, Chapter 8), the integral equation for the exterior Neumann problem (equation (8.9)) is soluble, and has a unique solution, for any function $\psi \in L_2(\Gamma)$, and, therefore, for any continuous function $\psi(x)$.

Thus, the value of the parameter

$$\lambda = \frac{2}{(m-2)|S_1|}$$

is regular for the kernel $(\partial/\partial n)(1/r^{m-2})$; by Fredholm's Third Theorem it is also regular for the adjoint kernel $(\partial/\partial v)(1/r^{m-2})$. Hence it follows that the integral equation for the interior Dirichlet problem is soluble (and has a unique solution) for any function $\phi \in L_2(\Gamma)$ and, therefore, for any continuous function $\phi(x)$.

If the integral equations for the problems D_i and N_e are soluble, then so are the problems themselves. This leads us to the following conclusions:

1. If Γ is a regular surface, then the interior Dirichlet problem for this surface is soluble for any continuous boundary data, and the solution can be represented in the form of a double-layer potential.

2. If Γ is a regular surface, then the exterior Neumann problem for this surface is continuous for any continuous boundary data, and the solution can be represented in the form of a single-layer potential.

§ 11. The Second Pair of Adjoint Equations

The value of the parameter

$$\lambda = -\frac{2}{(m-2)|S_1|}, \qquad (1)$$

which occurs in the integral equations for problems D_e and N_i (equations (8.7) and (8.8)) is an eigenvalue for each of the kernels

$$\frac{\partial}{\partial \nu} \frac{1}{r^{m-2}}, \qquad \frac{\partial}{\partial n} \frac{1}{r^{m-2}}.$$

The third of equations (4.2) shows that the homogeneous integral equation corresponding to problem D_e

$$\sigma_0(x) + \frac{2}{(m-2)|S_1|} \int_\Gamma \sigma_0(\xi) \frac{\partial}{\partial \nu} \frac{1}{r^{m-2}} d_\xi \Gamma = 0 \qquad (2)$$

has the non-trivial solution $\sigma_0(x) \equiv 1$, and this implies that the quantity (1) is an eigenvalue for the kernel $(\partial/\partial \nu)(1/r^{m-2})$. From Fredholm's Third Theorem, it is also an eigenvalue for the adjoint kernel $(\partial/\partial n)(1/r^{m-2})$. In this case the homogeneous integral equation for problem N_i

$$\mu_0(x) + \frac{2}{(m-2)|S_1|} \int_\Gamma \mu_0(\xi) \frac{\partial}{\partial n} \frac{1}{r^{m-2}} d_\xi \Gamma = 0 \qquad (3)$$

has at least one non-trivial solution; we denote this solution by $\mu_0(x)$.

We now show that equations (2) and (3) have no non-trivial solutions which are linearly-independent of the solutions $\sigma_0(x)$ and $\mu_0(x)$ just given. Because of Fredholm's Third Theorem, it will be sufficient to show that equation (3) has this property. Construct the single-layer potential with density $\mu_0(\xi)$

$$V_0(x) = \int_\Gamma \mu_0(\xi) \frac{1}{r^{m-2}} d_\xi \Gamma. \qquad (4)$$

It follows from equation (3) that

$$\frac{\partial V_0}{\partial n_i} \equiv 0. \tag{5}$$

Since $V_0(x)$ is a harmonic function in the domain Ω which lies inside Γ, the uniqueness theorem for the problem N_i gives

$$V_0(x) \equiv c_0 = \text{const.}, \qquad x \in \Omega. \tag{6}$$

Here $c_0 \neq 0$. For if $c_0 = 0$, then $V_0(x) \equiv 0$, $x \in \Omega$. By virtue of the continuity of the single-layer potential, $V_0(x) \equiv 0$, $x \in \Gamma$; by the uniqueness theorem for the exterior Dirichlet problem $V_0(x) \equiv 0$, $x \in \Omega'$. But then

$$\frac{\partial V_0(x)}{\partial n_e} \equiv 0. \tag{7}$$

From relations (5) and (7), and from formula (7.11), it follows that $\mu_0(x) \equiv 0$, and this contradicts the fact that the solution $\mu_0(x)$ is non-trivial.

We have proved in passing the following assertion: if the single-layer potential inside Γ is identically equal to zero, then its density is also identically equal to zero.

We now suppose that equation (3) has one further solution $\mu_1(x)$. We construct the potential

$$V_1(x) = \int_\Gamma \mu_1(\xi) \frac{1}{r^{m-2}} \, d\xi.$$

Repeating the preceding arguments, we can show that if $x \in \Omega$, then the potential $V_1(x) \equiv c_1 = \text{const.}$ Put

$$\mu_2(x) = c_1 \mu_0(x) - c_0 \mu_1(x). \tag{8}$$

Obviously, $\mu_2(x)$ is a solution of the same equation (3). Construct the potential

$$V_2(x) = \int_\Gamma \mu_2(\xi) \frac{1}{r^{m-2}} \, d_\xi \Gamma = c_1 V_0(x) - c_0 V_1(x).$$

If $x \in \Omega$, then $V_2(x) = c_1 c_0 - c_0 c_1 = 0$. It follows from what we have shown above that $\mu_2(x) \equiv 0$. Hence

$$\mu_1(x) = \frac{c_1}{c_0} \mu_0(x). \tag{9}$$

Thus, any solution of equation (3) differs from $\mu_0(x)$ only by a constant multiple. This is what we set out to prove.

Consider now the inhomogeneous equation of the problem N_i (equation (8.8)). By virtue of Fredholm's Fourth Theorem, this equation has a solution if and only if the function $\psi(x)$ is orthogonal to every solution of the adjoint homogeneous equation (8.7). But this latter equation has only one linearly-independent solution $\sigma_0(x) \equiv 1$. Thus, in order that (8.8) be soluble, it is necessary and sufficient that $(\psi, 1) = 0$, or, in other words,

$$\int_\Gamma \psi(x) \, d_x \Gamma = 0. \tag{10}$$

If equation (8.8) has a solution, then, obviously, the problem N_i also has a solution. Thus, the condition (10) is sufficient for the solubility of problem N_i; on the other hand, if the function $u(x)$ is harmonic inside Γ and has the necessary continuous derivatives, then

$$\int_\Gamma \frac{\partial u}{\partial n} \, d_x \Gamma = 0.$$

But

$$\left. \frac{\partial u}{\partial n} \right|_\Gamma = \psi(x),$$

consequently

$$\int_\Gamma \psi(x) \, d_x \Gamma = 0.$$

Hence the condition for solubility, which was obtained as a sufficient condition, is also necessary.

We can now formulate the following result.

Let Γ be a regular surface, and let $\psi \in C(\Gamma)$. Condition (10) is the necessary and sufficient condition that the interior Neumann problem with boundary condition $(\partial u/\partial n)_\Gamma = \psi(x)$ should have a solution; this solution can be expressed in the form of a single-layer potential.

It remains to consider the integral equation for the exterior Dirichlet problem (equation (8.7)).

It is not difficult to write down the necessary and sufficient condition for the solubility of this equation:

$$(\phi, \mu_0) = \int_\Gamma \phi(x) \mu_0(x) \, d_x \Gamma = 0. \tag{11}$$

If condition (11) is satisfied, the integral equation (8.7) has a solution. Then there exists a solution to the exterior Dirichlet problem, represented in the form of a double-layer potential, and, consequently, decreasing at infinity like $|x|^{1-m}$.

If condition (11) is violated, then there does not exist a solution to equation (8.7). This does not mean, however, that the exterior Dirichlet problem is insoluble; it only means that this problem does not have a solution which can be represented in the form of a double-layer potential.

§ 12. Solution of the Exterior Dirichlet Problem

We situate the coordinate origin inside the surface Γ. The function $1/|x|^{m-2}$ is harmonic in any domain not containing the origin, so in particular it is harmonic in Ω'.

We seek a solution of the problem D_e in the form

$$u(x) = \int_\Gamma \sigma(\xi) \frac{\partial}{\partial \nu} \frac{1}{r^{m-2}} d_\xi \Gamma + \frac{1}{|x|^{m-2}} \int_\Gamma \sigma(\xi) d_\xi \Gamma. \qquad (1)$$

For any arbitrary continuous function $\sigma(\xi)$, the right-hand side of equation (1) is harmonic in Ω'; we therefore only need to find $\sigma(\xi)$ such that the boundary condition (8.1) is satisfied.

Repeating the arguments of § 8, we obtain an integral representation for the unknown function:

$$\sigma(x) + \frac{2}{(m-2)|S_1|} \int_\Gamma \left[\frac{\partial}{\partial \nu} \frac{1}{r^{m-2}} + \frac{1}{|x|^{m-2}} \right] \sigma(\xi) d_\xi \Gamma = \frac{2}{(m-2)|S_1|} \phi(x),$$

$$x \in \Gamma. \qquad (2)$$

The kernel of equation (2),

$$\frac{\partial}{\partial \nu} \frac{1}{r^{m-2}} + \frac{1}{|x|^{m-2}},$$

has a weak singularity, as does the kernel $(\partial/\partial \nu)(1/r^{m-2})$. Therefore we can apply Fredholm theory to this equation.

Consider the homogeneous integral equation which results from equation (2) when $\phi(x) \equiv 0$:

$$\sigma_0(x) + \frac{2}{(m-2)|S_1|} \int_\Gamma \left[\frac{\partial}{\partial \nu} \frac{1}{r^{m-2}} + \frac{1}{|x|^{m-2}} \right] \sigma_0(\xi) d_\xi \Gamma = 0. \qquad (3)$$

Let $\sigma_0 \in L_2(\Gamma)$ be any solution of equation (3). As in § 8, it can be shown that this solution is continuous. We construct the function

$$u_0(x) = \int_\Gamma \sigma_0(\xi) \frac{\partial}{\partial \nu} \frac{1}{r^{m-2}} d_\xi \Gamma + \frac{1}{|x|^{m-2}} \int_\Gamma \sigma_0(\xi) d_\xi \Gamma, \qquad (4)$$

harmonic in Ω'.

It follows from equation (3) that $u_0(x)|_\Gamma \equiv 0$; by the uniqueness theorem for the exterior Dirichlet problem $u_0(x) \equiv 0$, $x \in \Omega'$. Taking this into account, we multiply equation (4) by $|x|^{m-2}$, and let $|x| \to \infty$. At infinity the double-layer potential decreases like $|x|^{1-m}$, so that in the limit the first term vanishes, and we obtain

$$\int_\Gamma \sigma_0(\xi) d_\xi \Gamma = 0. \qquad (5)$$

Thus, any solution of equation (3) satisfies the relation (5). But then equation (3) can be simplified, and takes the form

$$\sigma_0(x) + \frac{2}{(m-2)|S_1|} \int_\Gamma \frac{\partial}{\partial \nu} \frac{1}{r^{m-2}} \sigma_0(\xi) d_\xi \Gamma = 0, \qquad (6)$$

which is identical with equation (11.2). As we have shown in § 11, equation (6) has only one linearly-independent solution, viz. unity, in which case its general solution is $\sigma_0(\xi) \equiv C = \text{const}$. Substituting this in (5), we obtain $C|\Gamma| = 0$ or $C = 0$. Then $\sigma_0(\xi) \equiv 0$ and equation (3) has only the trivial solution. From Fredholm's Alternative it follows that the inhomogeneous equation (2) is soluble for any continuous function $\phi(x)$. Moreover, for any continuous boundary function $\phi(x)$ the exterior Dirichlet problem is also soluble; its solution can be represented in the form (1).

Remark 1. All the results of §§ 10–12 can be extended to arbitrary, closed Lyapunov surfaces, if we insist that the given functions $\phi(x)$ and $\psi(x)$ satisfy *Lipschitz conditions with exponent* λ:

$$|\phi(\xi) - \phi(x)| \leq Ar^\lambda, \qquad |\psi(\xi) - \psi(x)| \leq Ar^\lambda,$$

where A and λ are positive constants.

Remark 2. The investigation of the integral equations of potential theory becomes somewhat complicated if the domain Ω (or Ω') is bounded, not by one, but by several closed Lyapunov surfaces; the results turn out to be somewhat different from those in the case of a single bounding surface. These questions are discussed in detail in the book [4].

Remark 3. The method of potentials can be applied to a number of other boundary-value problems. Suppose, for example, it is required to find a harmonic function $u(x)$ either inside or outside a closed, regular surface Γ and satisfying boundary conditions of the form

$$\left[\frac{\partial u}{\partial n}+\beta(x)u\right]_\Gamma = \omega(x), \tag{7}$$

where $\beta(x)$ and $\omega(x)$ are continuous functions on Γ, and n is the outward normal to Γ at the point $x \in \Gamma$. This type of problem is often called the *third boundary-value problem*. We shall seek its solution in the form of a single-layer potential

$$u(x) = \int_\Gamma \mu(\xi) \frac{1}{r^{m-2}}\, d_\xi \Gamma. \tag{8}$$

Using the theorem concerning the limiting values of the normal derivative of the potential (8), we obtain for $\mu(\xi)$ an integral equation with a weak singularity

$$\mu(x) \pm \frac{2}{(m-2)|S_1|}\int_\Gamma \left[\frac{\partial}{\partial n}\frac{1}{r^{m-2}} + \frac{\beta(x)}{r^{m-2}}\right]\mu(\xi)\,d_\xi\Gamma = \pm \frac{2}{(m-2)|S_1|}\omega(x),$$
$$x \in \Gamma. \tag{9}$$

Here the plus sign corresponds to the interior problem, and the minus to the exterior problem.

If the problem (7) for a harmonic function has not more than one solution, then equation (9) is soluble for any $\omega(x)$, and our problem has a solution; if the solution is not unique (i.e. if the homogeneous problem (7) has k linearly independent eigenfunctions) then it exists if and only if $\omega(x)$ satisfies k orthogonality conditions. By Fredholm's First Theorem, the number k is finite. The interior (exterior) problem (7) for a harmonic function has a unique solution if $\beta(x) \geq 0$ and $\beta(x) > 0$ on a set of positive measure on Γ ($\beta(x) \leq 0$ and $\beta(x) < 0$ on a set of positive measure on Γ). Prove this!

§ 13. Case of Two Independent Variables

In the case $m = 2$, the singular solution of Laplace's equation is the function $\ln(1/r)$, $r = |\xi - x|$. Accordingly, the single- and double-layer potentials in a plane are defined by the formulae

$$V(x) = \int_\Gamma \mu(\xi) \ln \frac{1}{r} \, d_\xi \Gamma, \tag{1}$$

$$W(x) = \int_\Gamma \sigma(\xi) \frac{\partial}{\partial \nu} \ln \frac{1}{r} \, d_\xi \Gamma, \tag{2}$$

where Γ is a curve, which we assume to be a closed Lyapunov curve. Inequality (2.6) here takes the form

$$\int_\Gamma \left| \frac{\partial}{\partial \nu} \ln \frac{1}{r} \right| d_\xi \Gamma \leq C = \text{const.} \tag{3}$$

The densities $\sigma(\xi)$ and $\mu(\xi)$ will be assumed continuous.

The expressions (1) and (2) are usually called *logarithmic potentials*. The logarithmic potential of a double-layer is harmonic both inside and outside Γ; at infinity it has the behaviour $O(|x|^{-1})$. The logarithmic potential of a single layer is harmonic inside Γ; outside, it is in general not harmonic (in contrast with the case $m > 2$): a harmonic function in an infinite domain has to be bounded at infinity, and the single-layer potential generally grows like $\ln|x|$ at infinity.

For the potentials (1) and (2) there are theorems analogous to (but not always identical with) the theorems we have established for the case $m > 2$: the single-layer potential is continuous on the whole plane, except possibly at infinitely distant points; there are limiting relations for the double-layer potential

$$\begin{aligned} W_i(x) &= -\pi\sigma(x) + \overline{W(x)}, \\ W_e(x) &= \pi\sigma(x) + \overline{W(x)} \end{aligned} \tag{4}$$

and for the normal derivative of the single-layer potential

$$\begin{aligned} \frac{\partial V(x)}{\partial n_i} &= \pi\mu(x) + \overline{\frac{\partial V(x)}{\partial n}}, \\ \frac{\partial V(x)}{\partial n_e} &= -\pi\mu(x) + \overline{\frac{\partial V(x)}{\partial n}}. \end{aligned} \tag{5}$$

This normal derivative is regular.

The Gauss integral can be evaluated from the formula

$$\int_\Gamma \frac{\partial}{\partial \nu} \ln \frac{1}{r} \, d_\xi \Gamma = \begin{cases} -2\pi, & x \text{ inside } \Gamma, \\ 0, & x \text{ outside } \Gamma, \\ -\pi, & x \in \Gamma. \end{cases} \tag{6}$$

The Dirichlet and Neumann problems can be posed in the usual way; we can retain the notation $\phi(x)$ and $\psi(x)$ for functions prescribed in these problems. It is important to notice that for $m = 2$ the exterior Neumann problem N_e is different in some respects. The uniqueness theorem for the problem N_e in this case is formulated in exactly the same way as for the problem N_i: two solutions of this problem can differ only by a constant term. The converse is also true: two functions which are harmonic outside and which differ only by a constant term are solutions of one and the same problem N_e.

The other special feature of the problem N_e in a plane is defined by the following lemma.

LEMMA 18.13.1. *In order that the problem N_e should have a solution when $m = 2$, it is necessary that*

$$\int_\Gamma \psi(x) \, dx = 0. \tag{7}$$

Suppose that a solution $u(x)$ to the problem N_e exists. Circumscribe the coordinate origin with a circle S_R of radius R sufficiently large that the curve Γ lies inside S_R. The function $u(x)$ is harmonic outside S_R and continuous in the corresponding closed domain; therefore we can use Poisson's formula ((1.6′), Chap. 13)

$$u(x) = \frac{1}{2\pi} \int_0^{2\pi} \frac{\rho^2 - R^2}{\rho^2 - 2R\rho \cos(\omega - \theta) + R^2} u(R, \omega) \, d\omega, \tag{8}$$

where R, ω are the polar coordinates of the point ξ, $u(R, \omega) = u(\xi)$, and ρ, θ are polar coordinates of the point x, with $\rho > R$.

We differentiate formula (8) with respect to the cartesian coordinates of the point x. This can be done most simply as follows. Put $z = x_1 + ix_2$, $\zeta = \xi_1 + i\xi_2$ where x_1, x_2 and ξ_1, ξ_2 are the cartesian coordinates of the points x and ξ. Then $z = \rho e^{i\theta}$, $\zeta = Re^{i\omega}$ and (cf. § 1, Chap. 13)

$$\frac{\rho^2 - R^2}{\rho^2 - 2R\rho \cos(\omega - \theta) + R^2} = \operatorname{Re} \frac{z + \zeta}{z - \zeta}.$$

Hence

$$\frac{\partial}{\partial x_1} \frac{\rho^2 - R^2}{\rho^2 - 2R\rho \cos(\omega - \theta) + R^2} = \frac{\partial}{\partial x_1} \operatorname{Re} \frac{z+\zeta}{z-\zeta}$$

$$= \operatorname{Re} \frac{\partial}{\partial z} \frac{z+\zeta}{z-\zeta} = -\operatorname{Re} \frac{2\zeta}{(z-\zeta)^2}.$$

We now have

$$\frac{\partial u}{\partial x_1} = -\frac{1}{\pi} \int_0^{2\pi} u(R, \omega) \operatorname{Re} \frac{\zeta}{(z-\zeta)^2} d\omega = O\left(\frac{1}{|x|^2}\right).$$

Similarly we find that $\partial u/\partial x_2 = O(1/|x|^2)$ also. These estimates are valid for sufficiently large $|x|$.

We now write formula (2.7), Chap. 12, for the domain $\Omega'_{\bar{R}}$ bounded by the curves Γ and $S_{\bar{R}}$, where $\bar{R} > R$:

$$-\int_\Gamma \frac{\partial u}{\partial n} d_x \Gamma + \int_{S_{\bar{R}}} \frac{\partial u}{\partial n} d_x S_{\bar{R}} = 0;$$

here n is the outward normal to Γ and to $S_{\bar{R}}$ respectively, at x; the minus sign in front of the first integral is because the normal to Γ is interior with respect to the domain $\Omega'_{\bar{R}}$.

The last formula can be put in the form

$$-\int_\Gamma \psi(x) d_x \Gamma + \int_{S_{\bar{R}}} \frac{\partial u}{\partial n} d_x S_{\bar{R}} = 0. \tag{9}$$

From the estimate obtained above it follows that when $\bar{R}$ is sufficiently large

$$\left|\frac{\partial u}{\partial n}\right| = \left|\frac{\partial u}{\partial x_1} \cos(n, x_1) + \frac{\partial u}{\partial x_2} \cos(n, x_2)\right| \leq \frac{c}{\bar{R}^2}, \qquad c = \text{const.}$$

on the circumference $S_{\bar{R}}$.

But then

$$\left|\int_{S_{\bar{R}}} \frac{\partial u}{\partial n} d_x S_{\bar{R}}\right| \leq \frac{c}{\bar{R}^2} 2\pi \bar{R} = \frac{2\pi c}{\bar{R}} \xrightarrow[\bar{R}\to\infty]{} 0.$$

Letting $\bar{R} \to \infty$ in formula (9), we obtain equation (7). The lemma is proved.

As in the general case, we shall seek a solution of the Dirichlet problem in the form of a double-layer potential (2), and the solution of the Neumann problem in the form of a single-layer potential (1). This leads us to the integral equations

$$\sigma(x) \mp \frac{1}{\pi}\int_\Gamma \sigma(\xi) \frac{\partial}{\partial v} \ln \frac{1}{r} d_\xi \Gamma = \mp \frac{1}{\pi} \phi(x) \qquad (D)$$

for the Dirichlet problem, and

$$\mu(x) \pm \frac{1}{\pi}\int_\Gamma \mu(\xi) \frac{\partial}{\partial n} \ln \frac{1}{r} d_\xi \Gamma = \pm \frac{1}{\pi} \psi(x) \qquad (N)$$

for the Neumann problem. The kernels of these equations have a weak singularity. In these circumstances it is easy to prove that they are bounded if the Lyapunov exponent $\alpha = 1$, and continuous if the curve Γ has a continuous curvature. As before, the equations for the problems D_i and N_e constitute an adjoint pair, as do the equations for the problems D_e and N_i. We shall assume henceforth that the curvature of the curve Γ is continuous.

We now proceed to investigate the integral equations (D) and (N), basing ourselves on the following lemma.

LEMMA 18.13.2. *If the integral equation of the problem* N_e

$$\mu(x) - \frac{1}{\pi}\int_\Gamma \mu(\xi) \frac{\partial}{\partial n} \ln \frac{1}{r} d_\xi \Gamma = -\frac{1}{\pi} \psi(x) \qquad (10)$$

is soluble, and $\psi(x)$ satisfies equation (7), *then the single-layer potential* (1) *solves the problem* N_e.

Let equation (10) be soluble. Taking its solution as the density of the potential (1), we obtain a function satisfying the boundary condition for the problem N_e, and harmonic outside Γ everywhere except possibly at infinitely distant points, where this function can turn out to be unbounded. It only remains to show that, if equation (7) is satisfied, then the potential (1) is bounded at infinity. Multiply both sides of equation (10) by $d_x \Gamma$ and integrate over Γ. Using condition (7), we obtain

$$\int_\Gamma \mu(x) d_x \Gamma - \frac{1}{\pi} \int_\Gamma \int_\Gamma \mu(\xi) \frac{\partial}{\partial n} \ln \frac{1}{r} d_\xi \Gamma d_x \Gamma = 0. \qquad (11)$$

In the second integral we interchange the symbols x and ξ; then n is replaced by v. Utilizing the third of equations (6), we find

$$\frac{1}{\pi}\int_{\Gamma}\int_{\Gamma}\mu(\xi)\frac{\partial}{\partial n}\ln\frac{1}{r}\,d_{\xi}\Gamma\,d_{x}\Gamma = \frac{1}{\pi}\int_{\Gamma}\int_{\Gamma}\mu(x)\frac{\partial}{\partial v}\ln\frac{1}{r}\,d_{\xi}\Gamma\,d_{x}\Gamma$$

$$= \frac{1}{\pi}\int_{\Gamma}\mu(x)\left\{\int_{\Gamma}\frac{\partial}{\partial v}\ln\frac{1}{r}\,d_{\xi}\Gamma\right\}d_{x}\Gamma = -\int_{\Gamma}\mu(x)\,d_{x}\Gamma.$$

It now follows from equation (11) that

$$\int_{\Gamma}\mu(x)\,d_{x}\Gamma = 0. \tag{12}$$

In equation (12) we replace the symbol x by ξ, then multiply this equation by $\ln|x|$, and add to equation (1). The result is a new expression for the potential (1):

$$V(x) = \int_{\Gamma}\mu(\xi)\ln\frac{|x|}{r}\,d_{\xi}\Gamma. \tag{13}$$

As $|x| \to \infty$ the integral (13) is bounded (it even tends to zero). Hence the lemma is proved.

Analysis of the integral equations (D) and (N) proceeds essentially in the same way as in §§10–12; we outline this analysis very briefly.

First of all we show that equation (10) for the problem N_e is soluble for any continuous function $\psi(x)$. In accordance with the Fredholm Alternative, we consider the homogeneous equation

$$\mu_0(x) - \frac{1}{\pi}\int_{\Gamma}\mu_0(\xi)\frac{\partial}{\partial n}\ln\frac{1}{r}\,d_{\xi}\Gamma = 0. \tag{14}$$

Let $\mu_0(x)$ be any solution of it. The free term of equation (14), being equal to zero, obviously satisfies condition (7), and because of Lemma 18.13.2, the potential

$$V_0(x) = \int_{\Gamma}\mu_0(\xi)\ln\frac{1}{r}\,d_{\xi}\Gamma$$

solves the homogeneous problem N_e. By the uniqueness theorem for the problem N_e, $V_0(x) \equiv C = \text{const.}$ outside Γ. Being continuous on the whole plane, the potential $V_0(x) \equiv C$ on Γ also. Finally, from the uniqueness theorem for the problem D_i, it follows that $V_0(x) \equiv C$ inside Γ also. But then

$$\frac{\partial V_0(x)}{\partial n_i} = \frac{\partial V_0(x)}{\partial n_e} \equiv 0.$$

It follows from equation (5) that

$$\mu_0(x) = \frac{1}{2\pi} \left[\frac{\partial V_0(x)}{\partial n_i} - \frac{\partial V_0(x)}{\partial n_e} \right] \equiv 0.$$

From Lemma 18.13.2 we now conclude that *condition* (7) *is not only necessary, but also sufficient, for the solubility of the problem* N_e *in the plane.*

Equation (10) and its adjoint, the integral equation for the problem D_i, are always soluble together. Hence it follows that the problem D_i always has a solution in the plane.

The investigation of the integral equations for the problems D_e and N_i also proceeds exactly as in the general case, and leads to the same results: condition (7) is necessary and sufficient for the solubility of the problem N_i; the only solution to the homogeneous integral equation of the problem D_e

$$\sigma_0(x) + \frac{1}{\pi} \int_\Gamma \sigma_0(\xi) \frac{\partial}{\partial \nu} \ln \frac{1}{r} d_\xi \Gamma = 0 \tag{15}$$

is a constant.

The solution to the problem D_e in the plane can be constructed by taking it to be of the form

$$u(x) = \int_\Gamma \sigma(\xi) \frac{\partial}{\partial \nu} \ln \frac{1}{r} d_\xi \Gamma + \int_\Gamma \sigma(\xi) d_\xi \Gamma. \tag{16}$$

This leads to the integral equation

$$\sigma(x) + \frac{1}{\pi} \int_\Gamma \left[\frac{\partial}{\partial \nu} \ln \frac{1}{r} + 1 \right] \sigma(\xi) d_\xi \Gamma = \frac{1}{\pi} \phi(x). \tag{17}$$

As in § 12, it can be shown that equation (17) always has a solution.

§ 14. Equations of Potential Theory for the Circle

Let Γ be the circle $x_1^2 + x_2^2 = R^2$. We calculate the kernel of the double-layer potential under the assumption that both the points $x(x_1, x_2)$ and $\xi(\xi_1, \xi_2)$ lie on this circle. We have

$$\frac{\partial}{\partial \nu} \ln \frac{1}{r} = -\frac{1}{r} \frac{\partial r}{\partial \xi_1} \cos(\nu, \xi_1) - \frac{1}{r} \frac{\partial r}{\partial \xi_2} \cos(\nu, \xi_2) = -\frac{\cos(\nu, r)}{r}.$$

On the circle the direction of ν coincides with the direction of the radius passing through the point ξ. It is clear from Fig. 36 that (we are denoting the angle (v, r) by β)

$$r = 2R \sin \tfrac{1}{2}(\pi - 2\beta) = 2R \cos \beta.$$

Hence

$$\frac{\partial}{\partial v} \ln \frac{1}{r} = -\frac{1}{2R}; \qquad x, \xi \in \Gamma. \tag{1}$$

Interchanging x and ξ, we obtain

$$\frac{\partial}{\partial n} \ln \frac{1}{r} = -\frac{1}{2R}; \qquad x, \xi \in \Gamma. \tag{2}$$

The kernels of equations (D) and (N) have become degenerate, and these equations can be solved by elementary means. We consider here the equations for the interior problems; the other cases are left to the reader.

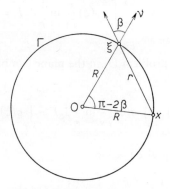

Fig. 36.

Denote by θ and ω the angles which the radius vectors Ox and $O\xi$ make with the x_1 axis. Then $d_\xi \Gamma = R d\omega$. Thus a function of the point $x \in \Gamma$ can be regarded as a function of θ. In accordance with this we shall write $\sigma(\theta)$ in place of $\sigma(x)$, etc.

The equation of the interior Dirichlet problem takes the form

$$\sigma(\theta) + \frac{1}{2\pi} \int_{-\pi}^{\pi} \sigma(\omega) \, d\omega = -\frac{1}{\pi} \phi(\theta). \tag{3}$$

Put $(1/2\pi)\int_{-\pi}^{\pi} \sigma(\omega) d\omega = c$. Then

$$\sigma(\theta) + c = -\frac{1}{\pi} \phi(\theta).$$

Integrating, we obtain

$$c = -\frac{1}{4\pi^2}\int_{-\pi}^{\pi}\phi(\omega)\,d\omega.$$

Now $\sigma(\theta) = -\phi(\theta)/\pi - c$ and

$$u(x) = -\int_{\Gamma}\left[\frac{1}{\pi}\phi(\omega)+c\right]\frac{\partial}{\partial v}\ln\frac{1}{r}\,d_{\xi}\Gamma$$

$$= -\frac{R}{\pi}\int_{-\pi}^{\pi}\phi(\omega)\frac{\partial}{\partial v}\ln\frac{1}{r}\,d\omega - c\int_{\Gamma}\frac{\partial}{\partial v}\ln\frac{1}{r}\,d_{\xi}\Gamma.$$

The point x now lies inside the circle; by formula (13.6) we have

$$u(x) = -\frac{R}{\pi}\int_{-\pi}^{\pi}\phi(\omega)\frac{\partial}{\partial v}\ln\frac{1}{r}\,d\omega + 2\pi c$$

$$= -\frac{1}{2\pi}\int_{-\pi}^{\pi}\left[2R\frac{\partial}{\partial v}\ln\frac{1}{r}-1\right]\phi(\omega)\,d\omega.$$

Moreover,

$$\frac{\partial}{\partial v}\ln\frac{1}{r} = -\frac{1}{r^2}[(\xi_1-x_1)\cos(v,\xi_1)+(\xi_2-x_2)\cos(v,\xi_2)]$$

$$= -\frac{1}{r^2 R}[(\xi_1-x_1)\xi_1+(\xi_2-x_2)\xi_2] = -\frac{R^2-(\xi_1 x_1+\xi_2 x_2)}{r^2 R}.$$

Using the fact that

$$r^2 = (\xi_1-x_1)^2+(\xi_2-x_2)^2 = R^2+\rho^2-2(\xi_1 x_1+\xi_2 x_2),$$
$$\rho^2 = x_1^2+x_2^2,$$

we obtain

$$-\left[2R\frac{\partial}{\partial v}\ln\frac{1}{r}-1\right] = \frac{R^2-\rho^2}{r^2},$$

and, consequently,

$$u(x) = \frac{1}{2\pi}\int_{-\pi}^{\pi}\phi(\omega)\frac{R^2-\rho^2}{r^2}\,d\omega. \qquad (4)$$

We have arrived at the Poisson integral for the circle.

The equation of the problem N_i for the circle takes the form

$$\mu(\theta) - \frac{1}{2\pi}\int_{-\pi}^{\pi} \mu(\omega)\,d\omega = \frac{1}{\pi}\psi(\theta). \qquad (5)$$

Putting $(1/2\pi)\int_{-\pi}^{\pi}\mu(\omega)\,d\omega = c_1$, we have

$$\mu(\theta) - c_1 = \frac{1}{\pi}\psi(\theta).$$

Integration of this equation leads to the same necessary condition which has been obtained previously:

$$\int_{-\pi}^{\pi} \psi(\theta)\,d\theta = 0; \qquad (6)$$

where the constant c_1 remains arbitrary. If equation (6) is satisfied, then the solution to equation (5) takes the form

$$\mu(\theta) = \frac{1}{\pi}\psi(\theta) + c_1;$$

the solution to the problem N_i is then given by

$$u(x) = \frac{R}{\pi}\int_{-\pi}^{\pi} \psi(\omega) \ln\frac{1}{r}\,d\omega + c_1 R \int_{-\pi}^{\pi} \ln\frac{1}{r}\,d\omega.$$

It is not difficult to show that the second integral is a constant, and so we arrive at Dini's formula

$$u(x) = \frac{R}{\pi}\int_{-\pi}^{\pi} \psi(\omega) \ln\frac{1}{r}\,d\omega + C, \qquad C = \text{const.}$$

Chapter 19.
The Oblique-Derivative Problem

§ 1. Formulation of the Problem

We have been considering in the preceding chapters boundary-value problems which have "Fredholm-type" properties: either these problems admit one solution (the problems D_i and D_e, and the problem N_e for $m > 2$), or the uniqueness theorem is violated and the homogeneous problem has linearly-independent solutions. In the latter case the number of such solutions turns out to be finite, and the inhomogeneous problem is soluble if and only if the given boundary function satisfies the same number of orthogonality conditions as there are linearly-independent solutions of the homogeneous problem (the problem N_i, and the problem N_e when $m = 2$). In the present chapter we shall consider a new boundary-value problem which in general is not of Fredholm type – this is called the oblique-derivative (or sometimes the skew-derivative) problem.

Consider a domain Ω in m-dimensional Euclidean space E_m. For the sake of definiteness we shall take this domain to be finite, and shall suppose that its boundary Γ is a regular surface.

Consider some neighbourhood of the surface Γ. We associate with each point x' of this neighbourhood some direction $\lambda = \lambda(x')$; we shall assume that $\lambda(x')$ is a continuous function of x'. The problem is: find a solution in the domain Ω of the elliptic equation

$$-\frac{\partial}{\partial x_j}\left(A_{jk}\frac{\partial u}{\partial x_k}\right) + B_k \frac{\partial u}{\partial x_k} + Cu = f(x) \tag{1}$$

with boundary conditions

$$\lim_{x' \to x} \frac{\partial u(x')}{\partial \lambda} = \psi(x), \qquad x \in \Gamma. \tag{2}$$

The problem (1)–(2) is the *oblique-derivative problem*.

If on the surface Γ the direction cosines $\cos(\lambda, x_k)$ are proportional to the quantities

$$A_{jk} \cos(v, x_j),$$

where v is the normal to Γ, then the oblique-derivative problem reduces to the Neumann problem.

We shall solve the oblique-derivative problem under the following very special circumstances: we restrict ourselves to two dimensions, and take the domain Ω to be the circle

$$x_1^2 + x_2^2 < 1 \tag{3}$$

in the plane, and equation (1) will be the homogeneous Laplace equation

$$\Delta u = \frac{\partial^2 u}{\partial x_1^2} + \frac{\partial^2 u}{\partial x_2^2} = 0. \tag{4}$$

We shall use the notation

$$x_1 + ix_2 = z = \rho e^{i\theta}, \quad i = \sqrt{-1};$$

if $\xi(\xi_1, \xi_2)$ is a point on the circumference Γ of the circle (3), then we shall also write

$$\xi_1 + i\xi_2 = \zeta = e^{i\omega}.$$

Obviously $d_\xi \Gamma = d\omega$, $d_x \Gamma = d\theta$.

The boundary condition (2) can be put in somewhat different form: write $\cos(\lambda, x_1) = a(\theta)$, $\cos(\lambda, x_2) = b(\theta)$ and $\psi(\theta)$ in place of $\psi(x)$. Then the boundary condition takes the form (we omit the limit sign)

$$a(\theta) \frac{\partial u}{\partial x_1}\bigg|_\Gamma + b(\theta) \frac{\partial u}{\partial x_2}\bigg|_\Gamma = \psi(\theta). \tag{5}$$

The functions $a(\theta)$ and $b(\theta)$ are related through the equation

$$a^2(\theta) + b^2(\theta) = 1, \tag{6}$$

and are assumed to be periodic with period 2π and continuous. We also suppose these functions to be continuously differentiable with respect to θ.

§ 2. Hilbert Operator

Consider a set M of functions which have period 2π, are absolutely continuous on $[-\pi, \pi]$, and have square-integrable derivatives on this interval.

Let $\phi(\theta) \in M$. Expand the function $\phi(\theta)$ in a Fourier series

$$\phi(\theta) = \sum_{n=-\infty}^{+\infty} a_n e^{in\theta}. \qquad (1)$$

It is easy to see that the series

$$\sum_{n=-\infty}^{+\infty} |a_n| \qquad (2)$$

converges, and, consequently, the series (1) converges absolutely and uniformly. For $n \neq 0$ we have

$$a_n = \frac{1}{2\pi} \int_{-\pi}^{\pi} \phi(\omega) e^{-in\omega} d\omega$$

$$= -\frac{1}{2\pi} \left[\phi(\omega) \frac{e^{-in\omega}}{in} \right]_{-\pi}^{\pi} + \frac{1}{2\pi i n} \int_{-\pi}^{\pi} \phi'(\omega) e^{-in\omega} d\omega = \frac{\alpha_n}{in}, \qquad (3)$$

where

$$\alpha_n = \frac{1}{2\pi} \int_{-\pi}^{\pi} \phi'(\omega) e^{-in\omega} d\omega$$

is the n-th Fourier coefficient for the derivative $\phi'(\omega)$. The series

$$\sum_{n=-\infty}^{+\infty} |\alpha_n|^2 \qquad (4)$$

converges because of Bessel's inequality; from the inequality

$$|a_n| = \left|\frac{\alpha_n}{n}\right| \leq \frac{1}{2}\left(\frac{1}{n^2} + |\alpha_n|^2\right)$$

it follows that the series (2) converges.

On the set M we prescribe a linear operator P, which acts in the following way: if the function $\phi \in M$ is expanded in the series (1), then

$$(P\phi)(\theta) = i \sum_{n=1}^{\infty} a_n e^{in\theta} - i \sum_{n=-1}^{-\infty} a_n e^{in\theta}. \qquad (5)$$

The operator P is called a *Hilbert operator*.

If the function $\phi(\theta)$ is real, then the function $(P\phi)(\theta)$ is also real. In this case, $a_{-n} = \bar{a}_n$, and consequently,

$$-ia_{-n} e^{-in\theta} = \overline{ia_n e^{in\theta}}.$$

LEMMA 19.2.1. *If $\phi \in M$, then $P\phi \in M$ also.*

Construct the series

$$i \sum_{n=1}^{\infty} \alpha_n e^{in\theta} - i \sum_{n=-1}^{-\infty} \alpha_n e^{in\theta}, \qquad (6)$$

where the coefficients α_n are defined by relations (3). Because of the convergence of series (4), series (6) converges in the metric of $L_2(-\pi, \pi)$, and its sum, which we denote by $\sigma(\theta)$, is square-integrable on the interval $[-\pi, \pi]$. Integrating series (6) term by term, we recover series (5) to within a constant term.

Thus,

$$(P\phi)(\theta) = \int_0^\theta \sigma(\omega)\,d\omega + C, \qquad C = \text{const.},$$

and the function $(P\phi)(\theta)$ is seen to be absolutely continuous, and its derivative $(d/d\theta)(P\phi)(\theta) = \sigma(\theta)$ to be square-integrable. Finally, the function $(P\phi)(\theta)$ is periodic with period 2π – this follows from the fact that the individual terms of series (5) have this property.

LEMMA 19.2.2. *The formula*

$$(P^2\phi)(\theta) = -\phi(\theta) + \frac{1}{2\pi}\int_{-\pi}^{\pi} \phi(\omega)\,d\omega \qquad (7)$$

holds.

If $\phi \in M$, then, as we have just shown, $P\phi \in M$, and we can apply the operator P to $P\phi$ once again. By formula (5)

$$(P^2\phi)(\theta) = -\sum_{n=1}^{\infty} a_n e^{in\theta} - \sum_{n=-1}^{-\infty} a_n e^{in\theta}$$

$$= -\phi(\theta) + a_0 = -\phi(\theta) + \frac{1}{2\pi}\int_{-\pi}^{\pi} \phi(\omega)\,d\omega.$$

Now let $F(z) = U(x_1, x_2) + iV(x_1, x_2)$ be a function holomorphic in the circle $|z| < 1$ and continuous in the closed circle $|z| \leq 1$. Suppose that the value of this function on the circumference $|z| = 1$ is an element of the set M

$$F(e^{i\theta}) \subset M \qquad (8)$$

and that $F(0)$ is a real quantity

$$\text{Im}\, F(0) = V(0, 0) = 0. \qquad (9)$$

From the inclusion (8) it follows that the Taylor series for the function

$F(z)$ converges absolutely and uniformly in the closed circle $|z| \leq 1$. Thus, let

$$F(z) = \sum_{n=0}^{\infty} A_n z^n. \qquad (10)$$

The coefficients A_n can be evaluated from the well-known formula

$$A_n = \frac{1}{2\pi i} \int_{\Gamma_\rho} \frac{F(z)}{z^{n+1}} dz = \frac{1}{2\pi} \int_{-\pi}^{\pi} F(\rho e^{-i\omega}) \rho^{-n} e^{-in\omega} d\omega,$$

where Γ_ρ is the circle $|z| = \rho < 1$.

When $0 < \rho \leq 1$ the integrand is continuous with respect to both the variables ρ and ω. It is therefore possible to let $\rho \to 1$, and to take the limit under the integral sign. We then obtain

$$A_n = \frac{1}{2\pi} \int_{-\pi}^{\pi} F(e^{i\omega}) e^{-in\omega} d\omega.$$

Repeating the arguments set out at the beginning of this section, we see that the series $\sum_{n=0}^{\infty} |A_n|$ converges, and, consequently, the series (10) converges absolutely and uniformly for $|z| \leq 1$.

Note that the coefficient A_0 is real. We write $F(e^{i\theta}) = \phi(\theta) + i\chi(\theta)$, so that

$$\phi(\theta) = U(x_1, x_2)|_{z=\exp(i\theta)}, \qquad \chi(\theta) = V(x_1, x_2)|_{z=\exp(i\theta)}.$$

Obviously $\phi \in M$ and $\chi \in M$; here

$$\phi(\theta) = \tfrac{1}{2}[F(e^{i\theta}) + \overline{F(e^{i\theta})}] = \sum_{n=-\infty}^{+\infty} a_n e^{in\theta}, \qquad (11)$$

where

$$a_n = \begin{cases} \tfrac{1}{2} A_n, & n > 0, \\ A_0, & n = 0, \\ \tfrac{1}{2} \overline{A}_{-n}, & n < 0, \end{cases} \qquad (12)$$

and

$$\chi(\theta) = \frac{1}{2i}[F(e^{in\theta}) - \overline{F(e^{in\theta})}]$$
$$= -i \sum_{n=1}^{\infty} a_n e^{in\theta} + i \sum_{n=-1}^{-\infty} a_n e^{in\theta} = -(P\phi)(\theta). \qquad (13)$$

Equation (13) describes a very important property of the Hilbert operator, which we formulate in the following theorem.

THEOREM 19.2.1. *Let the function $U(x_1, x_2)$, harmonic in the circle $|z| < 1$, take the value $\phi(\theta)$ on the circumference $z = e^{i\theta}$, $-\pi \leq \theta \leq \pi$, where $\phi \in M$. Further, let $V(x_1, x_2)$ be, of all the harmonic functions which are conjugate to $U(x_1, x_2)$, the one which is equal to zero when $z = 0$. Then*

$$V(x_1, x_2)|_{z = \exp(i\theta)} = -(P\phi)(\theta). \tag{14}$$

As consequences of this theorem, we can state the following two properties of the Hilbert operator.

1. If the function $\phi \in M$, then the function

$$\phi(\theta) - i(P\phi)(\theta) \tag{15}$$

represents the value, on the circumference of the unit circle $z = e^{i\theta}$, of some function $f^+(z)$, holomorphic in the circle $|z| < 1$. If the function $\phi(\theta)$ is real, then the quantity $f^+(0)$ is also real.

If $\phi(\theta)$ is represented by the series (1), then

$$\phi(\theta) - i(P\phi)(\theta) = a_0 + 2\sum_{n=1}^{\infty} a_n e^{in\theta}.$$

The sum of this series is the value when $z = e^{i\theta}$ of the function

$$f^+(z) = a_0 + 2\sum_{n=1}^{\infty} a_n z^n, \tag{16}$$

which is holomorphic in the circle $|z| < 1$. If the function $\phi(\theta)$ is real, then the quantity

$$f^+(0) = a_0 = \frac{1}{2\pi}\int_{-\pi}^{\pi} \phi(\omega)\,d\omega$$

is also real.

2. Under the same condition $\phi \in M$, the function

$$\phi(\theta) + i(P\phi)(\theta) \tag{17}$$

is the value on $z = e^{i\theta}$ of the function $f^-(z)$, holomorphic in the exterior of the circle $|z| > 1$; here

$$f^-(\infty) = f^+(0). \tag{18}$$

Now the function (17) can be expanded in the series

$$\phi(\theta) + i(P\phi)(\theta) = a_0 + 2\sum_{n=-1}^{-\infty} a_n e^{in\theta}$$

whose sum is the value on the circumference of the unit circle of the function

$$f^-(z) = a_0 + 2 \sum_{n=1}^{\infty} \frac{a_{-n}}{z^n} \qquad (19)$$

which is holomorphic in the domain $|z| > 1$. Here

$$f^-(\infty) = a_0 = f^+(0).$$

Note that the above results lead to the relations

$$\begin{aligned} \phi(\theta) &= \tfrac{1}{2}[f^+(z) + f^-(z)], \\ (P\phi)(\theta) &= \tfrac{1}{2}\mathrm{i}[f^+(z) - f^-(z)], \end{aligned} \qquad (20)$$

where $z = e^{i\theta}$.

In general, superscripts $+$ and $-$ will henceforth indicate functions which are holomorphic inside and outside the circle respectively.

Remark 1. The set M can easily be turned into a complete, normed space; if the function $\phi \in M$ is represented by the series (1), then we put

$$\|\phi\| = |a_0| + [\sum_{n=-\infty}^{+\infty} |\alpha_n|^2]^{\frac{1}{2}},$$

where the α_n are defined by formula (3).

Obviously the norm of the Hilbert operator is equal to unity in this space.

Remark 2. The norm of the Hilbert operator is also equal to one in the space $L_2(-\pi, \pi)$, on which this operator is defined by the formulae (1) and (5). It can also be shown that the Hilbert operator is bounded in the space $L_p(-\pi, \pi)$, if $1 < p < \infty$.

Remark 3. The Hilbert operator can be represented in the form of the so-called *singular integral*

$$(P\phi)(\theta) = \frac{1}{2\pi} \int_{-\pi}^{\pi} \phi(\omega) \cot \tfrac{1}{2}(\omega - \theta) \, d\omega$$

$$= \lim_{\varepsilon \to 0} \left\{ \frac{1}{2\pi} \int_{-\pi}^{\theta - \varepsilon} \phi(\omega) \cot \tfrac{1}{2}(\omega - \theta) \, d\omega + \frac{1}{2\pi} \int_{\theta + \varepsilon}^{\pi} \phi(\omega) \cot \tfrac{1}{2}(\omega - \theta) \, d\omega \right\}. \qquad (21)$$

Formula (21) makes clear the connection between the Hilbert operator and the theory of *singular integral equations,* which play an important role in the modern theory of partial differential equations. An extensive discussion of singular integral equations can be found in references [14] and [12].

§ 3. Equations with the Hilbert Operator

Consider the equation

$$a(\theta)\phi(\theta)+b(\theta)(P\phi)(\theta) = g(\theta). \tag{1}$$

Here $a(\theta)$, $b(\theta)$, $g(\theta)$ are given functions, and $\phi(\theta)$ is an unknown function of class M; the function $g(\theta)$ also belongs to this class. It is further assumed that the functions $a(\theta)$ and $b(\theta)$ are periodic with period 2π, continuous and continuously differentiable.

We shall solve equation (1) * subject to the condition that the coefficients $a(\theta)$ and $b(\theta)$ satisfy the inequality

$$a^2(\theta)+b^2(\theta) \neq 0, \quad -\pi \leq \theta \leq \pi. \tag{2}$$

We use the formulae (2.20), which enable equation (1) to be expressed in the form

$$f^+(z)+\frac{a(\theta)-ib(\theta)}{a(\theta)+ib(\theta)}f^-(z) = g_1(\theta), \quad z = e^{i\theta}, \tag{3}$$

where $g_1(\theta) = 2g(\theta)/\{a(\theta)+ib(\theta)\}$. The properties of the functions $f^+(z)$ and $f^-(z)$ have been described in the previous section.

The function $\{a(\theta)-ib(\theta)\}/\{a(\theta)+ib(\theta)\}$ is continuous and has period 2π; therefore as θ varies from $-\pi$ to π, the argument of this function changes by a multiple of 2π:

$$\int_{-\pi}^{\pi} d\arg\frac{a(\theta)-ib(\theta)}{a(\theta)+ib(\theta)} = 2\kappa\pi, \tag{4}$$

where κ is some integer.

Put

$$\alpha(\theta) = \ln\left[e^{-i\kappa\theta}\frac{a(\theta)-ib(\theta)}{a(\theta)+ib(\theta)}\right]. \tag{5}$$

It follows from the definition of the number κ that the function $\alpha(\theta)$ is periodic with period 2π: $\alpha(-\pi) = \alpha(\pi)$; also, it is clear that this function is continuous and continuously differentiable. Consequently $\alpha \in M$ and $P\alpha \in M$.

Construct the functions $\beta^+(z)$ and $\beta^-(z)$, holomorphic respectively in the domains $|z| < 1$ and $|z| > 1$, and taking the values

---

* Equation (1) belongs to the class of singular integral equations mentioned at the end of the preceding section.

$$\beta^+(e^{i\theta}) = \tfrac{1}{2}[\alpha(\theta)-i(P\alpha)(\theta)],$$
$$\beta^-(e^{i\theta}) = -\tfrac{1}{2}[\alpha(\theta)+i(P\alpha)(\theta)] \tag{6}$$

on the circumference $|z| = 1$.

Formulae (2.16) and (2.19) show that the Taylor series for the function $\beta^+(z)$ converges absolutely and uniformly in the closed circle $|z| \leq 1$, and the Laurent series for the function $\beta^-(z)$ is absolutely and uniformly convergent in the closed domain $|z| \geq 1$. Hence it follows that both these functions are bounded, and, consequently, the functions $\exp\{\pm\beta^+(z)\}$ and $\exp\{\pm\beta^-(z)\}$ are bounded and not equal to zero in the respective closed domains.

From formulae (5) and (6) it follows that

$$\frac{a(\theta)-ib(\theta)}{a(\theta)+ib(\theta)} = z^\kappa \exp\{\beta^+(z)-\beta^-(z)\}, \qquad z = e^{i\theta}.$$

Substitute this expression in (3), and multiply both sides by $\exp\{-\beta^-(z)\}$. Introducing the notation

$$f^+(z)\exp\{-\beta^+(z)\} = \Phi^+(z),$$
$$f^-(z)\exp\{-\beta^-(z)\} = \Phi^-(z), \tag{7}$$
$$g_1(\theta)\exp\{-\beta^+(e^{i\theta})\} = g_2(\theta),$$

we transform equation (3) to the following, simpler equation:

$$\Phi^+(z)+z^\kappa\Phi^-(z) = g_2(\theta), \qquad z = e^{i\theta}. \tag{8}$$

Further development depends on the value of the number κ.

1. $\kappa = 0$. Equation (8) takes the form

$$\Phi^+(z)+\Phi^-(z) = g_2(\theta), \qquad z = e^{i\theta}. \tag{9}$$

One solution of this equation is defined by the formulae

$$\Phi_1^+(e^{i\theta}) = \tfrac{1}{2}[g_2(\theta)-i(Pg_2)(\theta)]$$
$$\Phi_1^-(e^{i\theta}) = \tfrac{1}{2}[g_2(\theta)+i(Pg_2)(\theta)], \tag{10}$$

together with subsequent analytic continuation of the functions $\Phi_1^+(z)$ and $\Phi_1^-(z)$ with the aid of Taylor and Laurent series. This solution can be described more explicitly as follows: expand $g_2(\theta)$ in a Fourier series

$$g_2(\theta) = \sum_{n=-\infty}^{+\infty} b_n e^{in\theta}. \tag{11}$$

Then

$$\Phi_1^+(z) = \tfrac{1}{2}b_0 + \sum_{n=1}^{\infty} b_n z^n, \qquad \Phi_1^-(z) = \tfrac{1}{2}b_0 + \sum_{n=1}^{\infty} \frac{b_{-n}}{z^n}. \qquad (12)$$

In order to find the general solution of equation (9), we consider the corresponding homogeneous equation

$$\Phi_0^+(z) + \Phi_0^-(z) = 0, \qquad z = e^{i\theta}. \qquad (13)$$

By virtue of Riemann's theorem on analytic continuation through a curve *), the functions $\Phi_0^+(z)$ and $-\Phi_0^-(z)$ can be analytically continued, together, over the whole of the z-plane. By Liouville's theorem, $\Phi_0^+(z) \equiv C$, $\Phi_0^-(z) \equiv -C$, where C is a constant.

Thus, the general solution of equation (9) has the form

$$\Phi^+(z) = \Phi_1^+(z) + C, \qquad \Phi^-(z) = \Phi_1^-(z) - C, \qquad (14)$$

where C is an arbitrary constant, and $\Phi_1^+(z)$ and $\Phi_1^-(z)$ are defined by formulae (12).

Knowing $\Phi^+(z)$ and $\Phi^-(z)$, we recover $f^+(z)$ and $f^-(z)$ from formulae (7). Here the constant C should be chosen such that equation (2.18) is satisfied. It is easy to see from formulae (6) that $\beta^-(\infty) = -\beta^+(0)$. Writing $\exp\{\beta^+(0)\} = \delta$, we obtain for C the equation

$$C\left(\delta + \frac{1}{\delta}\right) = -\tfrac{1}{2}b_0\left(\delta - \frac{1}{\delta}\right). \qquad (15)$$

If $\delta \neq \pm i$, then the constant C and with it the functions $f^+(z)$ and $f^-(z)$ are defined in a unique way. Equation (1) then has one and only one solution, which is given by the first of formulae (2.20)

$$\phi(\theta) = \tfrac{1}{2}[f^+(e^{i\theta}) + f^-(e^{i\theta})].$$

If $\delta = \pm i$, then it is necessary and sufficient for the solubility of equation (1) that

$$b_0 = \frac{1}{2\pi}\int_{-\pi}^{\pi} g_2(\omega)\,d\omega = \frac{1}{\pi}\int_{-\pi}^{\pi} \frac{g(\omega)\exp\{-\beta^+(e^{i\omega})\}}{a(\omega) + ib(\omega)}\,d\omega = 0. \qquad (16)$$

It is also clear that when $\delta = \pm i$, the homogeneous equation

$$a(\theta)\phi_0(\theta) + b(\theta)(P\phi_0)(\theta) = 0 \qquad (17)$$

has one linearly-independent solution.

* Cf., for example, V. I. Smirnov, *A Course of Higher Mathematics*, Vol. III, Part 2, p. 96.

Thus, the case $\kappa = 0$ turns out to be compatible with the Fredholm Alternative: either (if $\delta \neq \pm i$) the homogeneous equation has only the trivial solution and the inhomogeneous equation is always soluble, or (if $\delta = \pm i$) the homogeneous equation has one linearly-independent solution and the inhomogeneous equation has a solution if and only if its free term $g(\theta)$ satisfies the one orthogonality condition (16).

2. $\kappa > 0$. We shall look for a solution of equation (8) in which the Laurent series of the function $\Phi^-(z)$ begins with a term proportional to $z^{-\kappa}$. Then the function $\Psi^-(z) = z^\kappa \Phi^-(z)$ will be holomorphic in the exterior of the circle $|z| > 1$. Putting $\Phi^+(z) = \Psi^+(z)$, we arrive at the equation

$$\Psi^+(z) + \Psi^-(z) = g_2(\theta), \quad z = e^{i\theta},$$

which is identical with equation (9). Clearly one of the required solutions of equation (8) has on the circumference $z = e^{i\theta}$ the form

$$\begin{aligned}\Phi_1^+(e^{i\theta}) &= \tfrac{1}{2}[g_2(\theta) - i(Pg_2)(\theta)], \\ \Phi_1^-(e^{i\theta}) &= \tfrac{1}{2}[g_2(\theta) + i(Pg_2)(\theta)] e^{-i\kappa\theta};\end{aligned} \quad (18)$$

inside and outside the circle $|z| = 1$ it is defined respectively by the series

$$\Phi_1^+(z) = \tfrac{1}{2}b_0 + \sum_{n=1}^\infty b_n z^n,$$

$$\Phi_1^-(z) = \frac{b_0}{2z^\kappa} + \sum_{n=1}^\infty \frac{b_{-n}}{z^{n+\kappa}}.$$

The solution (18) is not unique. To see this, consider the homogeneous equation

$$\Phi_0^+(z) + z^\kappa \Phi_0^-(z) = 0, \quad z = e^{i\theta}. \qquad (19)$$

From Liouville's theorem $z^\kappa \Phi_0^-(z)$ is a polynomial of degree $\leq \kappa$; let

$$z^\kappa \Phi_0^-(z) = -(c_0 + c_1 z + \ldots + c_\kappa z^\kappa).$$

Then

$$\begin{aligned}\Phi_0^+(z) &= c_0 + c_1 z + \ldots + c_\kappa z^\kappa, \\ \Phi_0^-(z) &= -\left(\frac{c_0}{z^\kappa} + \frac{c_1}{z^{\kappa-1}} + \ldots + c_\kappa\right).\end{aligned} \qquad (20)$$

Formulae (20) determine the general solution of equation (19). The general solution of the inhomogeneous equation (8) is given by the formulae

$$\Phi^+(z) = \Phi_1^+(z) + c_0 + c_1 z + \ldots + c_\kappa z^\kappa,$$
$$\Phi^-(z) = \Phi_1^-(z) - \frac{c_0}{z^\kappa} - \frac{c_1}{z^{\kappa-1}} - \ldots - c_\kappa, \quad (21)$$

where $\Phi_1^+(z)$ and $\Phi_1^-(z)$ are given by the equations (18).

From formulae (7) we can find the functions $f^+(z)$ and $f^-(z)$; in this connection, the constants $c_0, c_1, \ldots, c_\kappa$ are subject to condition (2.18). It is not difficult to see that this condition leads to the equation

$$c_0 \delta + \frac{c_\kappa}{\delta} = -\tfrac{1}{2} b_0 \delta, \qquad \delta = \exp\{\beta^+(0)\}. \quad (22)$$

Equation (22) allows c_0 to be expressed in terms of c_κ; the formulae (21), and accordingly the solution of equation (1), contain κ arbitrary constants. Hence it follows that in the case $\kappa > 0$ the inhomogeneous equation (1) has a solution for any free term, and, moreover, has an infinite set of solutions, depending on the arbitrary constants $c_1, c_2, \ldots, c_\kappa$. The homogeneous equation (17) has exactly κ linearly-independent solutions.

Thus we see that the case $\kappa > 0$ is incompatible with the Fredholm Alternative, which does not hold in this case.

3. $\kappa < 0$. Write $k = -\kappa$, so that $k > 0$. We express equation (8) in the form

$$\Phi^+(z) + z^{-k} \Phi^-(z) = g_2(\theta), \qquad z = e^{i\theta}. \quad (23)$$

The function $\Psi^-(z) = z^{-k} \Phi^-(z)$ is holomorphic for $|z| > 1$. Putting $\Phi^+(z) = \Psi^+(z)$ again, we obtain the equation

$$\Psi^+(z) + \Psi^-(z) = g_2(\theta), \qquad z = e^{i\theta},$$

whose general solution, as we have seen, is given by the formulae

$$\Psi^+(z) = \tfrac{1}{2} b_0 + C + \sum_{n=1}^\infty b_n z^n, \qquad \Psi^-(z) = \tfrac{1}{2} b_0 - C + \sum_{n=1}^\infty \frac{b_{-n}}{z^n},$$

derived from relations (11), (12) and (14).

Obviously $\Psi^-(\infty) = 0$, so that necessarily $C = \tfrac{1}{2} b_0$. This gives

$$\Phi^+(z) = \sum_{n=0}^\infty b_n z^n, \qquad \Phi^-(z) = \sum_{n=1}^\infty \frac{b_{-n}}{z^{n-k}}. \quad (24)$$

But the function $\Phi^-(z)$ must be bounded at infinity. Therefore formulae

(24) give the solution to equation (23) if and only if the expansion of the function $\Phi^-(z)$ in (24) contains no positive powers of z. This imposes on the free term in equation (1) $k-1$ orthogonality conditions

$$b_{-n} = 0, \qquad n = 1, 2, \ldots, k-1,$$

or, more explicitly,

$$\int_{-\pi}^{\pi} \frac{g(\omega) \exp\{in\omega - \beta^+(e^{i\omega})\}}{a(\omega) + ib(\omega)} d\omega = 0, \qquad n = 1, 2, \ldots, k-1. \qquad (25)$$

Equation (2.18) gives one further orthogonality condition

$$b_0 \delta - \frac{b_{-k}}{\delta} = 0$$

or

$$\int_{-\pi}^{\pi} \left(\delta - \frac{e^{ik\omega}}{\delta}\right) g(\omega) d\omega = 0. \qquad (26)$$

If the orthogonality conditions (25) and (26) are satisfied, then equation (1) has a unique solution which can be constructed from formulae (24), (7) and (2.20).

The Fredholm Alternative is obviously invalid in the case $\kappa < 0$ also.

We introduce the following definition. Suppose that some problem consists in finding a solution to the equation

$$Au = f, \qquad (27)$$

where $u \in X, f \in Y$; X, Y are Banach spaces, and A is a closed linear operator acting from X into Y. This problem is said to be *normally soluble* if it is necessary and sufficient for its solubility that the free term satisfies some orthogonality conditions, i.e. conditions of the form

$$(F_j, f) = 0, \qquad j = 1, 2, \ldots, \qquad (28)$$

where the F_j are linear functionals bounded in the metric of the space Y. We do not exclude here the case when the number of orthogonality conditions is equal to zero – in this case a normally soluble problem has a solution for any free term. We shall speak of the normal solubility either of equation (27) or of the operator A.

Almost all the problems considered in previous chapters have been normally soluble in respect of the chosen pair of spaces; this is true of equations with completely continuous operators, and of the Dirichlet and Neumann problems for a non-degenerate elliptic equation in a finite domain; also of the problems relating to the homogeneous Laplace equation in a finite or infinite domain. In the present section we have shown that equation (1), containing a Hilbert operator, is normally soluble if condition (2) is satisfied.

We remark, without proof, that equation (1) is not normally soluble if equation (2) is violated at a finite number of points. Another example of a problem which is not normally soluble is the problem $-\Delta u = f(x)$, $u|_\Gamma = 0$, in the case when a finite surface Γ bounds an infinite domain Ω, and the problem is considered in $L_2(\Omega)$.

Let the operator A be normally soluble, and let $\alpha(A)$ denote the number of linearly-independent solutions of the homogeneous equation $Au = 0$; let $\beta(A)$ denote the number of orthogonality conditions (28) necessary and sufficient for the solubility of equation (27). Suppose that at least one of the numbers $\alpha(A)$ and $\beta(A)$ is finite. Then the difference $\alpha(A) - \beta(A)$ is called the *index* of the operator A, or of the corresponding linear problem, and is denoted by Ind A

$$\text{Ind } A = \alpha(A) - \beta(A).$$

If T is completely continuous and I is the identity operator, then $\text{Ind}(I+T) = 0$ – this follows from the Fredholm Alternative. Generally speaking, if the Fredholm Alternative is valid for a particular problem, then the index of this problem is zero. It is not difficult to show that the index of equation (1) is κ; it being assumed that condition (2) is fulfilled. The indices of the other normally soluble problems listed above are equal to zero.

§ 4. Number of Solutions and Index for the Oblique-Derivative Problem in a Plane

We return now to the oblique-derivative problem posed in § 1: to find a function $u(x) = u(x_1, x_2)$, harmonic in the circle $|z| < 1$, $z = x_1 + ix_2$ and satisfying the boundary condition

$$a(\theta) \frac{\partial u}{\partial x_1} \bigg|_{z=\exp(i\theta)} + b(\theta) \frac{\partial u}{\partial x_2} \bigg|_{z=\exp(i\theta)} = \psi(\theta), \tag{1}$$

$$a^2(\theta) + b^2(\theta) = 1. \tag{2}$$

We recall that $a(\theta) = \cos(\lambda, x_1)$, $b(\theta) = \cos(\lambda, x_2)$. As in § 1, we shall assume that $a(\theta)$ and $b(\theta)$ are continuously differentiable, and that $\psi \in M$; the set M has been defined in § 2. We shall look for a solution such that $\phi \in M$, where

$$\phi(\theta) = \frac{\partial u}{\partial x_1}\bigg|_{z=\exp(i\theta)}. \tag{3}$$

If $u = \operatorname{Re}(w(z))$, where $w(z)$ is a harmonic function, and $v = \operatorname{Im}(w(z))$, then

$$w'(z) = \frac{\partial u}{\partial x_1} + i\frac{\partial v}{\partial x_1} = \frac{\partial u}{\partial x_1} - i\frac{\partial u}{\partial x_2}.$$

Hence it follows that the harmonic functions $\partial u/\partial x_1$ and $-\partial u/\partial x_2$ are conjugate. Therefore, if we use the notation

$$\chi(\theta) = \frac{\partial u}{\partial x_2}\bigg|_{z=\exp(i\theta)}, \quad l = \frac{\partial u}{\partial x_2}\bigg|_{z=0}, \tag{4}$$

then by Theorem 19.2.1,

$$\chi(\theta) = (P\phi)(\theta) + l. \tag{5}$$

Substituting expressions (3) and (5) into equation (1), we reduce it to the form

$$a(\theta)\phi(\theta) + b(\theta)(P\phi)(\theta) = \psi(\theta) - lb(\theta). \tag{6}$$

We evaluate the index of equation (6); it is

$$\kappa = \frac{1}{2\pi}\int_{-\pi}^{\pi} d\arg\frac{a(\theta) - ib(\theta)}{a(\theta) + ib(\theta)}$$

or, if we make use of relation (2),

$$\kappa = \frac{1}{\pi}\int_{-\pi}^{\pi} d\arg[a(\theta) - ib(\theta)] = \frac{1}{\pi}\int_{-\pi}^{\pi} d\arg e^{-i(\lambda, x_1)}$$

$$= -\frac{1}{\pi}\int_{-\pi}^{\pi} d(\lambda, x_1) = -\frac{[(\lambda, x_1)]_\Gamma}{\pi}, \tag{7}$$

where $[(\lambda, x_1)]_\Gamma$ denotes the increase in the angle (λ, x_1) in going round the circumference Γ in a positive direction. We observe that the index of equation (6) is even.

We now find the number of solutions and the index of the problem (1)–(2). It will be necessary to consider several cases.

1. $\kappa = 0$, $\delta \neq \pm i$. The homogeneous equation (6) has only trivial solutions, and the homogeneous problem (1)–(2) has two linearly-independent solutions: one arises from the quantity l, which remains arbitrary, and the other is due to the fact that the solution $u(x)$ can be altered by an arbitrary constant term, without thereby violating either Laplace's equation or the boundary condition (1). This latter solution has to be taken into account in the subsequent cases as well. There are no orthogonality conditions in this case, and the index of the problem is equal to 2.

2. $\kappa = 0$, $\delta = \pm i$. There is one orthogonality condition – (3.16) – and one linearly-independent solution of the homogeneous equation (6). If the function $b(\theta)$ satisfies condition (3.16), then the quantity l remains arbitrary; the homogeneous problem (1)–(2) has three linearly-independent solutions, and for the solubility of the inhomogeneous problem (1)–(2) it is necessary and sufficient that one orthogonality condition be fulfilled. The index of this problem is thus 2. If $b(\theta)$ does not satisfy (3.16), then the quantity l is determined by this condition; the orthogonality condition is satisfied, and no longer has to be taken into account. The homogeneous problem (1)–(2) has two non-trivial solutions, and the index of the problem (1)–(2) is 2.

3. $\kappa > 0$. There are no orthogonality conditions; the homogeneous equation (6) has κ linearly-independent solutions. The homogeneous problem (1)–(2) has $\kappa+2$ solutions – two additional solutions appearing just as in case 1. The index of the problem (1)–(2) is equal to $\kappa+2$.

4. $\kappa < 0$. The homogeneous equation (6) has only trivial solutions; there are $-\kappa$ orthogonality conditions (3.25) and (3.26). If $b(\theta)$ satisfies these conditions, then l is arbitrary, and the homogeneous problem has two linearly-independent solutions. The index of the problem (1)–(2) is $\kappa+2$. If, however, $b(\theta)$ does not satisfy all the orthogonality conditions, then the constant l can be eliminated; the number of orthogonality conditions is then decreased by one, and is equal to $-\kappa-1$. The homogeneous problem (1)–(2) has one non-trivial solution – a constant. The index of the problem is $\kappa+2$.

The discussion of paragraphs 1–4 enables us to formulate the following theorem.

THEOREM 19.4.1. *The index of the oblique-derivative problem for a circle in the plane is equal to*

$$\kappa+2 = -\frac{[(\lambda, x_1)]_\Gamma}{\pi} + 2.$$

Example. In the Neumann problem for the circle, the direction λ coincides with the direction of the radius vector. Therefore $(\lambda, x_1) = \theta$ and $[(\lambda, x_1)]_\Gamma = 2\pi$; the index of the Neumann problem is equal to zero.

Remark. Theorem 19.4.1 holds not only for a circle but also for any domain bounded by a finite number of closed, non-intersecting regular curves.

In a space of m dimensions, where $m > 2$, the following statement is true: Let a domain be bounded by a finite number of closed, non-intersecting regular surfaces, and let $\cos(\lambda, \xi_j)$, where λ is the direction of differentiation, be sufficiently smooth functions. If the direction λ is nowhere tangent to the boundary of the domain, then the index of the oblique-derivative problem is equal to zero.

PART VI

Time-Dependent Equations

In this part we study equations of parabolic and hyperbolic type. More specifically, we study two classes of equations which can be regarded as generalized heat-conduction and wave equations for the case of a medium with more complex physical properties (i.e. an inhomogeneous and anisotropic medium) in space of arbitrary dimensionality.

In both the heat-conduction equation and the wave equation one of the independent variables represents time, while the others are spatial coordinates. Because of this, we shall employ the following notation, somewhat different from that of Parts IV and V. The total number of independent variables will be denoted by $m+1$, instead of by m; the first m variables will be $x_1, x_2, \ldots, x_m$, and the last will be t. The quantities $x_1, x_2, \ldots, x_m$ will be regarded as the cartesian coordinates of some point x belonging to an m-dimensional Euclidean space E_m. We shall adhere to the summation rule adopted in preceding chapters: if an index which can vary over values between 1 and m occurs twice in some expression, then it is understood that summation is being performed with respect to this index, over the range 1 to m.

For simplicity we may occasionally write x_{m+1} in place of t.

Chapter 20.
The Heat-Conduction Equation

§ 1. The Heat-Conduction Equation and its Characteristics

Consider the following second-order differential equation with $m+1$ independent variables

$$\frac{\partial u}{\partial t} - A_{jk}(x,t) \frac{\partial^2 u}{\partial x_j \partial x_k} + A_k(x,t) \frac{\partial u}{\partial x_k} + A_0(x,t)u = f(x,t), \quad A_{jk} = A_{kj}. \quad (1)$$

Its matrix of highest coefficients has the form

$$\begin{Vmatrix} -A_{11} & -A_{12} & \ldots & -A_{1n} & 0 \\ -A_{21} & -A_{22} & \ldots & -A_{2n} & 0 \\ \cdot & \cdot & \cdot & \cdot & \cdot \\ -A_{n1} & -A_{n2} & \ldots & -A_{nn} & 0 \\ 0 & 0 & \ldots & 0 & 0 \end{Vmatrix}.$$

One of the eigenvalues of this matrix is equal to zero, and the rest only differ in sign from the eigenvalues of the matrix A_{jk}. If these eigenvalues are all of the same sign, then equation (1) belongs to type $(m, 0, 1)$ and, consequently, is parabolic; when this happens we shall call (1) the *heat-conduction equation*.

It is important to observe that the differential expression

$$-A_{jk} \frac{\partial^2 u}{\partial x_j \partial x_k} + A_k \frac{\partial u}{\partial x_k} + A_0 u, \quad (2)$$

which occurs in the heat-conduction equation, is elliptic in the variables $x_1, x_2, \ldots, x_m$.

In what follows we shall assume that the matrix of coefficients A_{jk} has positive eigenvalues, i.e. that this matrix is positive definite.

We now find the characteristics of equation (1). If $\omega(x, t) = $ const. is the equation of its characteristic surface, then (cf. § 2, Chapter 10) the function ω satisfies the equation

$$A_{jk}\frac{\partial \omega}{\partial x_j}\frac{\partial \omega}{\partial x_k} = 0.$$

But since the matrix $||A_{jk}||_{j,k=1}^{j,k=m}$ is positive definite, it follows that necessarily $\partial\omega/\partial x_k = 0$, $k = 1, 2, \ldots, m$. The function ω thus depends only on t, and the equation of the characteristic surface takes the form $\omega(t) = $ const. If $\omega'(t) \equiv 0$ in a certain interval, then ω is identically constant in this interval and the equation $\omega = $ const. does not define any surface. If, on the other hand, $\omega'(t) \not\equiv 0$, then in the neighbourhood of any value of t for which $\omega'(t) \neq 0$ the equation $\omega(t) = $ const. can be solved for t. The result is, obviously,

$$t = \text{const.} \tag{3}$$

Thus, *the characteristics of the heat-conduction equation are m-dimensional planes, normal to the t-axis.*

In what follows we shall consider the heat-conduction equation subject to the following, more particular assumptions.

1. The elliptic differential expression (2) has the form

$$-\frac{\partial}{\partial x_j}\left(A_{jk}\frac{\partial u}{\partial x_k}\right), \tag{4}$$

so that the heat-conduction equation itself has the form

$$\frac{\partial u}{\partial t} - \frac{\partial}{\partial x_j}\left(A_{jk}\frac{\partial u}{\partial x_k}\right) = f(x, t). \tag{5}$$

In some cases we shall take $f(x, t) \equiv 0$, and consider the homogeneous heat-conduction equation:

$$\frac{\partial u}{\partial t} - \frac{\partial}{\partial x_j}\left(A_{jk}\frac{\partial u}{\partial x_k}\right) = 0. \tag{6}$$

2. The coefficients A_{jk} are independent of t, and are continuously differentiable with respect to $x_1, x_2, \ldots, x_m$.
3. The elliptic expression (4) is non-degenerate.

§ 2. Maximum Principle

Consider a finite domain Ω, with boundary Γ, lying in the plane $t = 0$ (i.e. in the m-dimensional Euclidean space E_m). Construct a cylindrical surface with directrix Γ and generators parallel to the t-axis; denote by B_T the portion of this surface enclosed between the planes $t = 0$ and $t = T$, where T is a positive constant. Further, denote by Ω_T the projection of the domain Ω on to the plane $t = T$, and by Q_T the domain in the space $(x_1, x_2, \ldots, x_m, t)$, whose boundary is $\Omega \cup B_T \cup \Omega_T$ (Fig. 37).

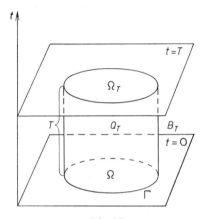

Fig. 37.

We introduce the following notation: if D is some set in the space of the variables $(x_1, x_2, \ldots, x_m, t)$, then $C^{(p,q)}(D)$ will denote the class of functions which have continuous derivatives with respect to $x_1, x_2, \ldots, x_m$ of order $\leq p$, and continuous derivatives with respect to t of order $\leq q$, on the set D.

THEOREM 20.2.1. *Let the function $u(x, t)$ belong to the intersection*

$$C(\bar{Q}_T) \cap C^{(2,1)}(Q_T \cup \Omega_T) \tag{1}$$

and satisfy in Q_T the homogeneous heat-conduction equation (1.6). *Then the function $u(x, t)$ attains both its largest and its smallest value in the closed domain $\bar{Q}_T$ on $\Omega \cup B_T$.*

Theorem 20.2.1 is called the *Maximum Principle for the heat-conduction equation*.

It is sufficient to prove the theorem for the case of a maximum: if the function $u(x, t)$ of the theorem attains a minimum at some point, then the function $-u(x, t)$ attains a maximum at the same point, and also satisfies the conditions of the theorem.

Let
$$M = \max_{(x,t)\in \bar{Q}_T} u(x, t), \qquad \mu = \max_{(x,t)\in \Omega \cup B_T} u(x, t).$$

Obviously, $\mu \leq M$, and the theorem states that $\mu = M$. Suppose the contrary; let $\mu < M$. Then the function $u(x, t)$, continuous in $\bar{Q}_T$, attains its maximum at some point (x_0, t_0) which lies either in Q_T or in Ω_T,
$$u(x_0, t_0) = M, \qquad (x_0, t_0) \in Q_T \cup \Omega_T.$$

We construct an auxiliary function
$$v(x, t) = u(x, t) + \frac{M-\mu}{2T}(t_0 - t). \tag{2}$$

If $(x, t) \in \Omega \cup B_T$, then $t_0 - t \leq t_0 \leq T$, and, consequently,
$$v(x, t)|_{(x,t)\in \Omega \cup B_T} \leq \mu + \tfrac{1}{2}(M-\mu) = \tfrac{1}{2}(M+\mu) < M.$$

On the other hand
$$v(x_0, t_0) = u(x_0, t_0) = M.$$

Thus there is a point outside $\Omega \cup B_T$ where the function v attains the value M, and so on $\Omega \cup B_T$ the value of v is strictly less than M. Hence it follows that $v(x, t)$ achieves a maximum in $\bar{Q}_T$ at a point belonging either to Ω_T or to Q_T.

Denote by (x^0, t^0) the point at which $v(x, t)$ attains its maximum. Suppose first that $(x^0, t^0) \in Q_T$. For arbitrarily chosen orthogonal axes $Ox_1, Ox_2, \ldots, Ox_m$ at the point (x^0, t^0), the conditions for a maximum must be satisfied:
$$\frac{\partial v}{\partial t} = 0, \quad \frac{\partial v}{\partial x_k} = 0, \quad \frac{\partial^2 v}{\partial x_k^2} \leq 0, \qquad k = 1, 2, \ldots, m. \tag{3}$$

For brevity, denote by L the differential expression on the left-hand side of equation (1.6). We calculate the quantity Lv at the point (x^0, t^0):
$$Lv|_{(x^0, t^0)} = \left[\frac{\partial v}{\partial t} - \frac{\partial A_{jk}}{\partial x_j}\frac{\partial v}{\partial x_k} - A_{jk}\frac{\partial^2 v}{\partial x_j \partial x_k}\right]_{(x^0, t^0)} = -A_{jk}\frac{\partial^2 v}{\partial x_j \partial x_k}\bigg|_{(x^0, t^0)}.$$

Choose the direction of the coordinate axis Ox_k in such a way that the matrix $\|A_{jk}\|_{j,k=1}^{j,k=m}$ becomes diagonal at the point x_1; this is possible because the matrix is symmetric. But this matrix is still positive definite, and so in the chosen coordinate system

$$A_{jj}(x^0) > 0, \quad A_{jk}(x^0) = 0, \quad j \neq k,$$

and

$$Lv|_{(x^0, t^0)} = -\sum_{j=1}^{m} A_{jj} \frac{\partial^2 v}{\partial x_j^2}\bigg|_{(x^0, t^0)} \geq 0.$$

On the other hand,

$$Lv = Lu + \frac{M-\mu}{2T} L(t_0 - t) = -\frac{M-\mu}{2T} < 0$$

and we have a contradiction, which means that $(x^0, t^0) \bar\in Q_T$.

Now suppose $(x^0, t^0) \in \Omega_T$. This means that $t^0 = T$, $x^0 \in \Omega$. Then t^0 is an end-point of the interval $(0, T)$, while x^0 is an interior point of the domain Ω; the necessary conditions for maximum at the point (x^0, t^0) take the form

$$\frac{\partial v}{\partial t} \geq 0, \quad \frac{\partial v}{\partial x_k} = 0, \quad \frac{\partial^2 v}{\partial x_k^2} \leq 0, \quad k = 1, 2, \ldots m.$$

As before

$$Lv|_{(x_1, t_1)} = \frac{\partial v}{\partial t}\bigg|_{(x^0, t^0)} - \left[\sum_{j=1}^{m} A_{jj} \frac{\partial^2 v}{\partial x_j^2}\right]_{(x^0, t^0)} \geq 0.$$

On the other hand, $Lv < 0$, also as before; this is another contradiction, which shows that $(x^0, t^0) \bar\in \Omega_T$.

Thus the point (x^0, t^0), which must belong to the union $Q_T \cup \Omega_T$, cannot belong to either Q_T or Ω_T. From this contradiction it follows that the assumption $\mu < M$ is invalid, and, consequently, $\mu = M$. The theorem is proved.

It is clear from the nature of the proof that, in the case of a maximum, the argument remains valid if the equation $Lu = 0$ is replaced by the inequality $Lu \leq 0$. We can therefore formulate the following, strengthened Maximum Principle:

THEOREM 20.2.2. *Let the function $u(x, t)$ belong to the intersection* (1). *If $Lu \leq 0$ everywhere in Q, then the function $u(x, t)$ attains a maximum on $\Omega \cup B_T$. If $Lu \geq 0$ everywhere in Q, then the function $u(x, t)$ attains a minimum on $\Omega \cup B_T$.*

§ 3. Cauchy's Problem and the Mixed Boundary-Value Problem

It was shown in § 2, Chapter 10, that, in the case of the heat-conduction equation, it is possible to prescribe only one of the Cauchy data. The Cauchy problem for equation (1.5) therefore comprises the following: to determine the solution to this equation for any $x \in E_m$, and for any $t > 0$, given the value of this solution when $t = 0$

$$u|_{t=0} = \phi(x), \qquad x \in E_m. \tag{1}$$

Cauchy's problem for the heat-conduction equation admits a simple physical interpretation. Let the heat-conducting medium (which may be inhomogeneous and anisotropic) occupy all space. Suppose that in this medium there are distributed heat sources whose strength (relative to some suitably-chosen unit of measure) $f(x, t)$ is known. Assume finally that the temperature at any point of the medium is known at the initial moment of time. Cauchy's problem is to determine the temperature at any point of the medium at any subsequent time.

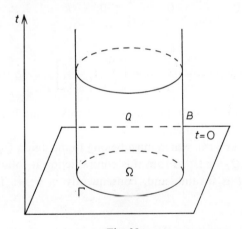

Fig. 38.

An important role is played by so-called mixed boundary-value problems (or, briefly, mixed problems), which can be formulated in the following way. Let Ω be a domain of Euclidean space E_m (Fig. 38), Γ its boundary, and B a cylindrical surface having Γ as directrix and generators parallel to the t-axis; more particularly, we take as B that portion of this surface for which $t > 0$. The mixed problem for equation (1.5) is the following: it is required to find a solution of this equation in a semi-infinite domain of the space of the variables $x_1, x_2, \ldots, x_m, t$ whose boundary is $\Omega \cup B$; this solution must satisfy the Cauchy condition

$$u|_{t=0} = \phi(x), \quad x \in \Omega \tag{2}$$

when $t = 0$, and some boundary condition on the cylindrical surface B.

Different types of boundary conditions lead to different mixed problems. The three most interesting types of boundary conditions are the following:

1) the condition of the first boundary-value problem

$$u|_B = \psi(x, t); \tag{3}$$

2) the condition of the second boundary-value problem

$$\left[A_{jk} \frac{\partial u}{\partial x_k} \cos(n, x_j) \right]_B = \chi(x, t); \tag{4}$$

3) the condition of the third boundary-value problem

$$\left[A_{jk} \frac{\partial u}{\partial x_k} \cos(n, x_j) + \sigma(x, t) u \right]_B = \omega(x, t). \tag{5}$$

In conditions (3)–(5) the functions $\psi, \chi, \omega, \sigma$ are given on B, and are supposed continuous; n is the normal to Γ at the point x.

The mixed problems listed here allow a physical interpretation. In all three cases we are concerned with a heat-conducting medium occupying a domain Ω in space. Heat sources of given strength $f(x, t)$ are distributed through the medium; in addition, the temperature at any point is prescribed at the initial instant of time. Further, it is assumed that some information is given regarding interaction between the heat-conducting medium and the medium surrounding it; this interaction takes place through the surface Γ, so the information about it is expressed in the form of a boundary condition.

Condition (3) implies that we know the temperature at all points of the boundary Γ at any time after the initial instant. In principle, this information can be obtained if the boundary is accessible to observation.

The left-hand side of equation (4) is proportional to the intensity of heat flux through a small neighbourhood of the point $x \in \Gamma$ at time t. Thus, the condition of the second boundary-value problem implies that the heat flux through the boundary of the medium is known at any time after the initial instant.

In condition (5) for the third boundary-value problem, the function $\omega(x, t)$ is proportional to the temperature of the surrounding medium at the point x and time t; this condition describes a process of heat-exchange with the surrounding medium, where the temperature of the latter is supposed known.

Physical aspects of problems connected with the heat-conduction equation are discussed more extensively in reference [8].

We shall consider below only one of the above mixed problems, namely, the first (boundary condition (3)).

§ 4. Uniqueness Theorems

THEOREM 20.4.1. *The mixed problem for the equation*

$$\frac{\partial u}{\partial t} - \frac{\partial}{\partial x_j}\left(A_{jk}\frac{\partial u}{\partial x_k}\right) = f(x, t) \tag{1}$$

subject to initial and boundary conditions

$$u|_{t=0} = \phi(x), \quad x \in \Omega; \quad u|_B = \psi(x, t) \tag{2}$$

has not more than one solution in the class

$$C(\bar{\Omega} \times [0, \infty)) \cap C^{(2, 1)}(\Omega \times (0, \infty)). \tag{3}$$

Suppose the problem (1)–(2) has two solutions, whose difference is $w(x, t)$. Then $w(x, t)$ satisfies the homogeneous heat-conduction equation (1.6), and belongs to the class (3). By virtue of the Maximum Principle, both the greatest and the least values of $w(x, t)$ are attained either in Ω or on the cylinder B. But the function w also satisfies homogeneous initial and boundary conditions

$$w|_{t=0} = 0, \quad x \in \Omega; \quad w|_B = 0.$$

From this it follows that both the greatest and the least values of $w(x, t)$ are zero. Hence $w(x, t) \equiv 0$, and the two solutions of problem (1)–(2) coincide.

For simplicity we investigate the uniqueness of the solution to Cauchy's problem only in the case when $A_{jk} = \delta_{jk}$, so that the elliptic expression contained in the heat-conduction equation reduces to the Laplace operator.

THEOREM 20.4.2. *The equation*

$$Lu = \frac{\partial u}{\partial t} - \Delta u = f(x, t) \tag{4}$$

has, in the class
$$C(E_m \times [0, \infty)) \cap C^{(2,1)}(E_m \times (0, \infty)) \tag{5}$$
not more than one bounded solution satisfying the Cauchy condition
$$u|_{t=0} = \phi(x)$$
for a given function $\phi(x)$.

If there are two such solutions, their difference $w(x, t)$ solves the homogeneous Cauchy problem
$$Lw = \frac{\partial w}{\partial t} - \Delta w = 0, \tag{6}$$
$$w|_{t=0} = 0 \tag{7}$$
and belongs to the class (5). Being the difference of two bounded functions, it is itself bounded; let $|w(x, t)| \leq M$.

Consider in the plane $t = 0$ (i.e. in the Euclidean space E_m) a sphere Σ_R of radius R and with centre at the coordinate origin; denote the bounding surface of this sphere by S_R. Construct a cylindrical surface having generators parallel to the t-axis and directrix S_R; denote by B the portion of this surface for which $t > 0$. Finally, denote by Q the domain of the space $(x_1, x_2, \ldots, x_m, t)$ whose boundary is $\Sigma_R \cup B$.*

Consider the auxiliary function
$$v_R(x, t) = \frac{2Mm}{R^2}\left(\frac{x^2}{2m} + t\right), \qquad x^2 = \sum_{k=1}^{m} x_k^2. \tag{8}$$

It is easy to see that the function v_R satisfies the homogeneous heat-conduction equation. Moreover
$$v_R|_{t=0} = \frac{Mx^2}{R^2} \geq 0;$$
because of inequality (7),
$$v_R|_{t=0} \geq |w||_{t=0}.$$
Finally,
$$v_R|_B = v_R|_{x^2 = R^2} \geq M \geq |w||_B.$$

The last two relations imply that
$$v_R|_{\Sigma_R \cup B} \geq |w||_{\Sigma_R \cup B},$$

* The construction described here corresponds to Fig. 38 (page 416) with $\Omega = \Sigma_R$.

and it is clear that each of the quantities $v_R + w$ and $v_R - w$ is non-negative on $\Sigma_R \cup B$. In addition, both of these quantities satisfy equation (6). But then by the Maximum Principle applied to the closed domain $\bar{Q}_T$, in which $x \in \bar{\Omega}, 0 \leq t \leq T, T = \text{const.}$ (Fig. 37), both the sum $v_R + w$ and the difference $v_R - w$ attain a minimum on $\Sigma_R \cup B$, where these minima are non-negative. Hence it follows that

$$v_R + w \geq 0, \quad v_R - w \geq 0, \quad x^2 \leq R^2, \quad t \geq 0.$$

Thus, when $x^2 \leq R^2, t \geq 0$, the inequality $-v_R \leq w \leq v_R$ is fulfilled, or, equivalently,

$$|w(x, t)| \leq \frac{2Mm}{R^2}\left(\frac{x^2}{2m} + t\right).$$

We fix x and t arbitrarily, and take the limit $R \to \infty$. From the latter inequality it then follows that $|w(x, t)| \leq 0$, i.e. that $w(x, t) = 0$. The theorem is proved.

§ 5. Abstract Functions of a Real Variable

Given a set E on the real number axis, we shall say that an *abstract function* $u(t)$ with values in the space X is defined on E if to any number $t \in E$ there corresponds, through some rule, one and only one element $u(t) \in X$. We shall assume below that the space X is a Banach space.

In a Banach space there exist two types of convergence: strong, or convergence in the norm, and weak. In accordance with these it is possible to introduce for abstract functions of a real variable the concepts of strong and weak continuity, strong and weak derivative, etc. Taking into account subsequent applications, we restrict ourselves to consideration of strong continuity and strong derivative; the word "strong" will henceforth be omitted.

An abstract function is continuous at the point $(t = t_0)$ if

$$\lim_{t \to t_0} \|u(t) - u(t_0)\|_X = 0;$$

it is continuous on some set of values of t if it is continuous at each point of this set.

An abstract function $u(t)$ has a derivative $u'(t)$ at the point t if

$$\lim_{h \to 0} \left\|\frac{u(t+h) - u(t)}{h} - u'(t)\right\|_X = 0.$$

As usual, a function having a derivative at some point is said to be differentiable at that point. It is obvious that a function differentiable at a point is continuous there. Higher derivatives of abstract functions can also be defined in the usual way.

In our subsequent work the following formula for differentiation of a scalar product will play an important role: if $u(t)$ and $v(t)$ are abstract functions with values in a Hilbert space, and if these functions are differentiable at a point t, then

$$\frac{d}{dt}(u(t), v(t)) = (u'(t), v(t)) + (u(t), v'(t)). \tag{1}$$

We have

$$\frac{d}{dt}(u(t), v(t)) = \lim_{h \to 0} \frac{1}{h}[(u(t+h), v(t+h)) - (u(t), v(t))]$$

$$= \lim_{h \to 0} \left[\left(\frac{u(t+h) - u(t)}{h}, v(t+h)\right) + \left(u(t), \frac{v(t+h) - v(t)}{h}\right)\right];$$

and proceeding to the limit within the scalar product, we obtain formula (1).

The concept of the integral of an abstract function can also be introduced in a natural way.

We shall use below the following notation.

Consider abstract functions whose values belong to some class of objects $\mathfrak{K}$, and let these functions be continuous on the set E of values of the variable t. We denote the set of these functions by $C(E; \mathfrak{K})$. If the functions in question are k times continuously differentiable on E, then we shall denote their set by $C^{(k)}(E; \mathfrak{K})$.

§ 6. Generalized Solution of the Mixed Problem

Consider a mixed problem for the heat-conduction equation

$$\frac{\partial u}{\partial t} - \frac{\partial}{\partial x_j}\left(A_{jk}\frac{\partial u}{\partial x_k}\right) = f(x, t), \quad x \in \Omega, \quad t > 0, \tag{1}$$

with a homogeneous boundary condition

$$u|_B = 0 \tag{2}$$

and, in general, an inhomogeneous initial condition

$$u|_{t=0} = \phi(x), \quad x \in \Omega. \tag{3}$$

We assume that the domain Ω is finite, and that its boundary Γ is piecewise-smooth.

We shall suppose that the required solution $u(x, t)$ belongs to the class $C(\bar{\Omega} \times [0, \infty)) \cap C^{(1,2)}(\bar{\Omega} \times (0, \infty))$. For fixed $t \geq 0$, condition (2) implies that

$$u|_\Gamma = 0, \qquad (4)$$

and, for fixed t, $u(x, t)$ can be treated as an element in the domain of definition $D(\mathfrak{A})$ of the operator $\mathfrak{A}$ of the Dirichlet problem for the elliptic differential expression

$$-\frac{\partial}{\partial x_j}\left(A_{jk}\frac{\partial u}{\partial x_k}\right), \qquad x \in \Omega.$$

Moreover, it can be treated as an element of the corresponding energy space $H_{\mathfrak{A}}$.

If we use the concept of abstract function, the mixed boundary-value problem posed above can be formulated in a different way.

We shall regard the function $f(x, t)$ as an abstract function $f(t)$ with values in $L_2(\Omega)$, and the function $\phi(x)$ as an element of the space $L_2(\Omega)$. Finally, the unknown function $u(x, t)$ will be taken to be the abstract function $u(t)$ with values in the domain $D(\mathfrak{A})$; the values of this function are, consequently, elements of both the spaces $L_2(\Omega)$ and $H_{\mathfrak{A}}$ at the same time.

The problem (1)–(3) now reduces to the following abstract Cauchy problem: to integrate the abstract, ordinary, first-order differential equation

$$\frac{du}{dt} + \mathfrak{A}u = f(t), \qquad t > 0, \qquad (5)$$

subject to the initial condition

$$u|_{t=0} = \phi. \qquad (6)$$

We assume that problem (5)–(6) has a solution.

Take an arbitrary abstract function $\eta(t)$ with values in $H_{\mathfrak{A}}$, and scalar multiply both sides of equation (5) (in the sense of the metric of $L_2(\Omega)$) by $\eta(t)$. Using the definition of energy product, we obtain

$$\left(\frac{du}{dt}, \eta\right) + [u, \eta] = (f, \eta). \qquad (7)$$

We omit here, and later, the subscript $\mathfrak{A}$ for energy product and norm.

Conversely, if $u \in C^{(1)}((0, \infty); D(\mathfrak{A}))$, and if this function satisfies equation (7), then it also satisfies equation (5). In fact, if $u \in D(\mathfrak{A})$, then $[u, \eta] = (\mathfrak{A}u, \eta)$ and equation (7) can be written in the form

$$\left(\frac{du}{dt} + \mathfrak{A}u - f, \eta\right) = 0, \qquad \forall \eta \in H_{\mathfrak{A}},$$

and since the elements of the space $H_{\mathfrak{A}}$ generate a dense set in $L_2(\Omega)$, therefore

$$\frac{du}{dt} + \mathfrak{A}u - f = 0.$$

We shall call the abstract function $u(t)$ the *generalized solution of the mixed problem* (1)–(3), if it satisfies the following requirements: 1) $u(t)$ belongs simultaneously to the classes

$$C([0, \infty); L_2(\Omega)), \quad C((0, \infty); H_{\mathfrak{A}}), \quad C^{(1)}((0, \infty); L_2(\Omega)),$$

i.e. this function is continuous for $t \geq 0$, and continuously differentiable for $t > 0$, as an abstract function with values in $L_2(\Omega)$; at the same time it is continuous for $t > 0$ as an abstract function with values in $H_{\mathfrak{A}}$; 2) $u(t)$ satisfies equation (7) for any $t > 0$ and any abstract function $\eta(t)$ with values in $H_{\mathfrak{A}}$; 3) $u(t)$ satisfies the initial condition (6). This last requirement is understood in the sense that

$$\lim_{t \to 0} \|u(t) - \phi\|_{L_2(\Omega)} = 0.$$

From the above discussion it follows that the generalized solution $u(t)$ is also an ordinary solution, if $u(t) \in D(\mathfrak{A})$ for any $t > 0$, and if $u(x, t) \xrightarrow[t \to 0]{} \phi(x)$, not only in the metric of $L_2(\Omega)$, but also uniformly.

THEOREM 20.6.1. *The generalized solution of the mixed problem for the heat-conduction equation is unique.*

Suppose two functions exist which satisfy equation (7) and the boundary condition (6). Their difference, which we denote by $w(t)$, satisfies the equation

$$\left(\frac{dw}{dt}, \eta\right) + [w, \eta] = 0 \tag{8}$$

and initial condition

$$w|_{t=0} = 0. \tag{9}$$

Putting $\eta = w$ in equation (8) and utilizing formula (5.1), we obtain

$$\frac{d}{dt}\|w(t)\|^2 + 2\||w|\|^2 = 0.$$

Hence it is clear that $(d/dt)\|w(t)\|^2 \leq 0$, and, consequently, the numerical function $\|w(t)\|^2$ does not increase with increasing t. But from relation (9), $\|w(0)\|^2 = 0$. Hence $\|w(t)\|^2 = 0$, $t > 0$, and the theorem is proved.

Chapter 21. The Wave Equation

§ 1. Basic Concepts

The second-order equation of the form

$$\frac{\partial^2 u}{\partial t^2} - A_{jk}(x, t) \frac{\partial^2 u}{\partial x_j \partial x_k} + A_k(x, t) \frac{\partial u}{\partial x_k} + A_0(x, t)u = f(x, t), \qquad (1)$$

in which the matrix of the coefficients A_{jk} is positive definite, is called the *wave equation*. The matrix of highest coefficients in equation (1) has the form

$$\begin{Vmatrix} -A_{11} & -A_{12} & \cdots & -A_{1m} & 0 \\ -A_{21} & -A_{22} & \cdots & -A_{2m} & 0 \\ \cdots & \cdots & \cdots & \cdots & \cdots \\ -A_{m1} & -A_{m2} & \cdots & -A_{mm} & 0 \\ 0 & 0 & \cdots & 0 & 1 \end{Vmatrix}. \qquad (2)$$

One of the eigenvalues of the matrix (2) is equal to unity, while the others coincide with the eigenvalues of the matrix $-\|A_{jk}\|$, and, consequently, are negative. Hence it follows that the wave equation belongs to the type $(m, 1, 0)$, i.e. it is hyperbolic.

The characteristic equation of the wave equation has the form

$$\left(\frac{\partial \omega}{\partial t}\right)^2 - A_{jk} \frac{\partial \omega}{\partial x_j} \frac{\partial \omega}{\partial x_k} = 0. \qquad (3)$$

Like all hyperbolic equations, equation (1) has real characteristics. We observe that the function $\omega(x, t) \equiv t$ does not satisfy equation (3); therefore the plane $t = $ const. is not a characteristic surface for the wave equation, and so both Cauchy data can be prescribed for $t = $ const.

We shall consider below a less general form of the wave equation, namely,

$$\frac{\partial^2 u}{\partial t^2} - \frac{\partial}{\partial x_j}\left(A_{jk} \frac{\partial u}{\partial x_k}\right) = f(x, t). \qquad (4)$$

From a physical point of view, equation (4) describes small oscillations of a medium under the action of continuously distributed sources whose strength is proportional to the quantity $f(x, t)$. In the general case, the vibrating medium can be inhomogeneous and anisotropic, and its physical properties can be changing with time. If the properties of the medium are invariant with time, the coefficients A_{jk} are independent of t; *this is the case we shall investigate henceforth*. If the medium is homogeneous, $A_{jk} = $ const.; in this case it is possible to transform the matrix $||A_{jk}||$ into a unit matrix by subjecting the coordinates $x_1, x_2, \ldots, x_m$ to an affine transformation. We then arrive at the simplest form of the wave equation

$$\frac{\partial^2 u}{\partial t^2} - \Delta u = f(x, t). \tag{5}$$

From the point of view of physical application, great interest attaches to a slightly more complex equation

$$\frac{\partial^2 u}{\partial t^2} - a^2 \Delta u = f(x, t), \qquad a^2 = \text{const.} \tag{6}$$

We observe that equation (6) reduces to the form (5) if we change the unit of time, i.e., if we make the substitution $t' = at$.

As we have already said, we shall suppose the coefficients of equation (4) to be independent of time; we further assume that these coefficients are continuously differentiable, and that the matrix is not degenerate.

§ 2. The Mixed Problem and its Generalized Solution

The formulation of the mixed problem for the wave equation is very similar to that for the heat-conduction equation.

We formulate the problem as follows. Let Ω be a given finite domain with a piecewise-smooth boundary Γ in the plane $t = 0$ (i.e. in the space E_m). It is required to find a solution, in the domain Q of Fig. 38 (page 416), of the wave equation

$$\frac{\partial^2 u}{\partial t^2} - \frac{\partial}{\partial x_j}\left(A_{jk}(x)\frac{\partial u}{\partial x_k}\right) = f(x, t), \tag{1}$$

satisfying the initial conditions

$$u|_{t=0} = \phi_0(x), \qquad \frac{\partial u}{\partial t}\bigg|_{t=0} = \phi_1(x) \tag{2}$$

and one of the following boundary conditions:

$$u|_B = \psi(x, t) \tag{3}$$

(first problem);

$$\left[A_{jk}\frac{\partial u}{\partial x_k}\cos(n, x_j)\right]_B = \chi(x, t) \tag{4}$$

(second problem);

$$\left[A_{jk}\frac{\partial u}{\partial x_k}\cos(n, x_j) + \sigma(x, t)u\right]_B = \omega(x, t) \tag{5}$$

(third problem). Other types of boundary conditions are, of course, possible.

In the following discussion we shall restrict ourselves to the case of a homogeneous boundary condition of the first problem

$$u|_B = 0. \tag{6}$$

As with the heat-conduction equation, the mixed problem for the wave equation can be formulated in operational terms. We shall assume for the time being that the solution $u(x, t)$ of the mixed problem belongs to the class

$$C(\bar{\Omega} \times [0, \infty)) \cap C^{(2, 2)}(\bar{\Omega} \times (0, \infty)).$$

Then this solution can be treated as an abstract function $u(t)$ with values in $D(\mathfrak{A})$, where $\mathfrak{A}$ is the operator of the Dirichlet problem; this function has two continuous derivatives in the interval $(0, \infty)$. The function $f(x, t)$ will be regarded as an abstract function with values in $L_2(\Omega)$. Finally, we shall suppose that the function $\phi_0(x)$ is an element ϕ_0 of the energy space $H_{\mathfrak{A}}$, and the function $\phi_1(x)$ is an element ϕ_1 of the space $L_2(\Omega)$. Then the mixed problem for equation (1) with initial conditions (2) and boundary condition (6) can be treated as a Cauchy problem for an abstract ordinary, second-order differential equation

$$\frac{d^2u}{dt^2} + \mathfrak{A}u = f(t) \tag{7}$$

with initial conditions

$$u(0) = \phi_0, \quad u'(0) = \phi_1. \tag{8}$$

Take an arbitrary function $\eta(t)$, belonging to the intersection

$$K = C([0, \infty); H_{\mathfrak{A}}) \cap C^{(1)}([0, \infty); L_2(\Omega)). \tag{9}$$

Scalar multiply (in the metric of $L_2(\Omega)$) both sides of equation (7) by $\eta(t)$

$$\left(\frac{d^2u(t)}{dt^2}, \eta(t)\right) + (\mathfrak{A}u(t), \eta(t)) = (f(t), \eta(t)).$$

This leads to the expression *

$$\left(\frac{d^2u(t)}{dt^2}, \eta(t)\right) + [u(t), \eta(t)] = (f(t), \eta(t)), \qquad \eta \in K. \tag{10}$$

Equation (10) could be used to define the generalized solution. But this would be inexpedient, because the generalized solution so defined would have to have a second derivative. Therefore, we proceed in the following way. We choose an arbitrary instant of time $T > 0$, and require that $\eta(T) = 0$. Next integrate both sides of equation (10) with respect to t over the interval $(0, T)$. By formula (5.1), Chapter 20, we then have

$$\left(\frac{d^2u(t)}{dt^2}, \eta(t)\right) = \frac{d}{dt}\left(\frac{du(t)}{dt}, \eta(t)\right) - \left(\frac{du(t)}{dt}, \frac{d\eta(t)}{dt}\right).$$

Using the values $\eta(T) = 0$, $u'(0) = \phi_1$, we obtain

$$-\int_0^T \left(\frac{du(t)}{dt}, \frac{d\eta(t)}{dt}\right) dt + \int_0^T [u(t), \eta(t)] \, dt - (\phi_1, \eta(0)) =$$
$$= \int_0^T (f(t), \eta(t)) \, dt, \qquad \eta \in K_T. \tag{11}$$

Here K_T denotes the class of functions $\eta(t)$ which have the properties $\eta \in K$ and $\eta(T) = 0$. We use equation (11) to introduce the concept of generalized solution.

We shall say that the abstract function $u(t)$ is a *generalized solution* of the problem (1), (2) and (6), if 1) $u \in K$; 2) $u(0) = \phi_0$; this is to be understood in the sense

$$\lim_{t \to 0} |||u(t) - \phi_0||| = 0;$$

3) $u(t)$ satisfies equation (11), in which $\eta(t)$ is an arbitrary function of class K_T.

It is easy to show that the generalized solution $u(t)$, belonging to the intersection

$$C^{(1)}([0, \infty); D(\mathfrak{A})) \cap C^{(2)}((0, \infty); D(\mathfrak{A})),$$

---

* Here, and subsequently, the subscript $\mathfrak{A}$ in the designation of energy product and norm is omitted.

is also an ordinary solution of the mixed problem for the wave equation.

THEOREM 21.2.1. *The mixed problem for the wave equation has not more than one generalized solution.*

The difference $w(t)$ of two generalized solutions of the same problem is an element of the class K, satisfying the equation

$$-\int_0^T \left(\frac{dw(t)}{dt}, \frac{d\eta(t)}{dt}\right) dt + \int_0^T [w(t), \eta(t)] dt = 0, \qquad \forall \eta \in K_T, \quad (12)$$

and the initial condition

$$w(0) = 0. \tag{13}$$

In equation (12) put

$$\eta(t) = \int_t^T w(\tau) d\tau. \tag{14}$$

This can be done because, obviously, $\eta \in K_T$ here. From formula (14) it follows that $w(t) = -d\eta(t)/dt$, and we obtain

$$\int_0^T \left(\frac{d^2\eta(t)}{dt^2}, \frac{d\eta(t)}{dt}\right) dt - \int_0^T \left[\frac{d\eta(t)}{dt}, \eta(t)\right] dt = 0. \tag{15}$$

Furthermore

$$\left(\frac{d^2\eta(t)}{dt^2}, \frac{d\eta(t)}{dt}\right) = \frac{1}{2}\frac{d}{dt}\left(\frac{d\eta(t)}{dt}, \frac{d\eta(t)}{dt}\right) = \frac{1}{2}\frac{d}{dt}\left\|\frac{d\eta(t)}{dt}\right\|^2 = \frac{1}{2}\frac{d}{dt}\|w(t)\|^2,$$

$$\left[\frac{d\eta(t)}{dt}, \eta(t)\right] = \frac{1}{2}\frac{d}{dt}[\eta(t), \eta(t)] = \frac{1}{2}\frac{d}{dt}|||\eta(t)|||^2.$$

Substituting this into equation (15), we obtain

$$\|w(T)\|^2 - \|w(0)\|^2 - |||\eta(T)|||^2 + |||\eta(0)|||^2 = 0.$$

But $w(0) = 0$ and $\eta(T) = 0$, therefore

$$\|w(T)\|^2 + |||\eta(0)|||^2 = 0.$$

Hence $w(T) = 0$ which proves the result, since T is arbitrary.

§ 3. Wave Equation with Constant Coefficients. Cauchy's Problem. Characteristic Cone

Henceforth in the present chapter we shall consider the wave equation in its simplest form

$$\frac{\partial^2 u}{\partial t^2} - \Delta u = f(x, t). \tag{1}$$

As we explained in § 1, the surface $t = 0$ is not characteristic for equation (1). Cauchy's problem for this equation can be posed in the following way: for any $x \in E_m$ and any $t > 0$, find a solution to equation (1), satisfying the initial conditions

$$u|_{t=0} = \phi_0(x), \qquad \frac{\partial u}{\partial t}\bigg|_{t=0} = \phi_1(x). \tag{2}$$

An important tool for investigating the solutions of Cauchy's problem for the wave equation is the so-called *characteristic cone*.

Take some point (x_0, t_0), and consider the surface defined by the equation

$$t_0 - t = r, \qquad r = |x - x_0|. \tag{3}$$

This is the lower sheet of the cone with vertex at the point (x_0, t_0) and axis parallel to the t-axis. It is easy to see that the surface (3) is characteristic for the wave equation (1). Putting $\omega(x, t) = t_0 - t - r$, we can write equation (3) in the form $\omega(x, t) = 0$. The characteristic equation for equation (1) has the form

$$\left(\frac{\partial \omega}{\partial t}\right)^2 - \sum_{k=1}^{m} \left(\frac{\partial \omega}{\partial x_k}\right)^2 = 0. \tag{4}$$

Here

$$\frac{\partial \omega}{\partial t} = -1, \qquad \frac{\partial \omega}{\partial x_k} = -\frac{\partial r}{\partial x_k} = -\frac{x_k - x_{k0}}{r},$$

where x_{k0} is the k-th coordinate of the point x_0. Now

$$\left(\frac{\partial \omega}{\partial t}\right)^2 - \sum_{k=1}^{m} \left(\frac{\partial \omega}{\partial x_k}\right)^2 = 1 - \frac{1}{r^2} \sum_{k=1}^{m} (x_k - x_{k0})^2 = 0.$$

The surface (3) is called the *characteristic cone* for the wave equation.

We find the direction of the outward normal n to the characteristic cone (Fig. 39). It makes an acute angle with the t-axis, so the cosine of this angle is positive; from a well-known formula of differential geometry

$$\cos(n, t) = -\frac{\partial \omega/\partial t}{\sqrt{\{(\partial \omega/\partial t)^2 + \sum_{k=1}^{m}(\partial \omega/\partial x_k)^2\}}} = \frac{1}{\sqrt{2}}. \qquad (5)$$

From this there follows one further relation

$$\sum_{k=1}^{m} \cos^2(n, x_k) = 1 - \cos^2(n, t) = \tfrac{1}{2}. \qquad (6)$$

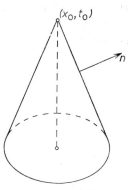

Fig. 39.

§ 4. Uniqueness Theorem for Cauchy's Problem. Domain of Dependence

THEOREM 21.4.1. *Let $|x-x_0|^2 \leq t_0^2$ be some closed sphere in the space E_m. Let the initial functions for two Cauchy problems for the wave equation (3.1) coincide in this sphere. If both problems have solutions which are continuous, together with their first and second derivatives, then these solutions coincide for $t > 0$ inside and on the boundary of the characteristic cone whose vertex is (x_0, t_0).*

Let $w(x, t)$ be the difference between the solutions of the above two problems. Then this difference satisfies the homogeneous wave equation

$$\frac{\partial^2 w}{\partial t^2} - \Delta w = 0 \qquad (1)$$

and initial conditions of the form

$$w|_{t=0} = 0, \quad \frac{\partial w}{\partial t}\bigg|_{t=0} = 0, \qquad (2)$$

$$|x - x_0| \leq t_0.$$

The values of $w|_{t=0}$ and $(\partial w/\partial t)|_{t=0}$ outside the sphere $|x-x_0|^2 \leq t_0^2$ are irrelevant.

Consider the domain D of the space $(x_1, x_2, \ldots, x_m, t)$ bounded by the plane $t = 0$ and the characteristic cone $t_0 - t = |x - x_0|$ (Fig. 40). Take an arbitrary point $(\tilde{x}, \tilde{t})$ inside or on the boundary of this domain, and construct a new characteristic cone $\tilde{t} - t = |\tilde{x} - x|$. Let $\tilde{D}$ denote the domain bounded by the plane $t = 0$ and the new cone. It is important to note that in the plane $t = 0$ the domain $\tilde{D}$ is bounded by the sphere $|x - \tilde{x}|^2 \leq \tilde{t}^2$, which constitutes part of the original sphere $|x-x_0|^2 \leq t_0^2$; hence it follows that the relations (2) hold in the new sphere.

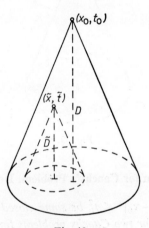

Fig. 40.

Multiply both sides of equation (1) by $\partial w/\partial t$, and integrate over the domain $\tilde{D}$. Making use of the obvious relations

$$\frac{\partial w}{\partial t}\frac{\partial^2 w}{\partial t^2} = \frac{1}{2}\frac{\partial}{\partial t}\left(\frac{\partial w}{\partial t}\right)^2,$$

$$\frac{\partial w}{\partial t}\frac{\partial^2 w}{\partial x_k^2} = \frac{\partial}{\partial x_k}\left(\frac{\partial w}{\partial t}\frac{\partial w}{\partial x_k}\right) - \frac{1}{2}\frac{\partial}{\partial t}\left(\frac{\partial w}{\partial x_k}\right)^2$$

§ 4] UNIQUENESS THEOREM FOR CAUCHY'S PROBLEM

and applying Gauss's theorem, we obtain

$$\int_{\tilde{D}} \frac{\partial w}{\partial t} \left(\frac{\partial^2 w}{\partial t^2} - \Delta w \right) dx\, dt = \frac{1}{2} \int_{\Sigma \cup K} \left\{ \left[\left(\frac{\partial w}{\partial t} \right)^2 + \sum_{k=1}^{m} \left(\frac{\partial w}{\partial x_k} \right)^2 \right] \cos(n, t) \right.$$

$$\left. - 2 \sum_{k=1}^{m} \frac{\partial w}{\partial t} \frac{\partial w}{\partial x_k} \cos(n, x_k) \right\} dS = 0. \qquad (3)$$

Here Σ denotes the sphere $|x - \tilde{x}| \leq \tilde{t}$, K the characteristic cone $\tilde{t} - t = |x - \tilde{x}|$, and dS an element of measure on the boundary $\Sigma \cup K$ of the domain $\tilde{D}$. By virtue of conditions (2), we have that $\partial w/\partial t \equiv 0$ and $w \equiv 0$ identically in the sphere Σ. Differentiating the latter identity with respect to the coordinate x_k, we also obtain $\partial w/\partial x_k \equiv 0$, $k = 1, 2, \ldots, m$. In the middle expression in equation (3) the integral over Σ vanishes, and we obtain the simpler equation

$$\int_K \left\{ \left[\left(\frac{\partial w}{\partial t} \right)^2 + \sum_{k=1}^{m} \left(\frac{\partial w}{\partial x_k} \right)^2 \right] \cos(n, t) - 2 \sum_{k=1}^{m} \frac{\partial w}{\partial t} \frac{\partial w}{\partial x_k} \cos(n, x_k) \right\} dS = 0.$$

Multiply both sides of this equation by the constant $1/\sqrt{2} = \cos(n, t)$, which we include under the integral sign. Taking into account equation (3.6), we obtain

$$\int_K \sum_{k=1}^{m} \left[\frac{\partial w}{\partial t} \cos(n, x_k) - \frac{\partial w}{\partial x_k} \cos(n, t) \right]^2 dS = 0,$$

from which it follows that on the cone K

$$\frac{\partial w}{\partial t} \cos(n, x_k) - \frac{\partial w}{\partial x_k} \cos(n, t) \equiv 0, \qquad k = 1, 2, \ldots, m,$$

and, consequently,

$$\frac{\partial w}{\partial x_1} : \cos(n, x_1) = \ldots = \frac{\partial w}{\partial x_m} : \cos(n, x_m) = \frac{\partial w}{\partial t} : \cos(n, t).$$

These equations imply that on the cone K the vector grad w is parallel to the normal.

Take an arbitrary point (x, t) on the cone K, and through this point draw a generator l of the cone. Obviously the vector grad w is orthogonal to l.

Therefore

$$\frac{\partial w}{\partial l} = \prod p_l \operatorname{grad} w = 0.$$

Hence it follows that $w = \text{const.}$ along any generator of the cone K. In particular, the value of w at the vertex $(\tilde{x}, \tilde{t})$ coincides with the value of w at that point of the generator l which lies in the plane $t = 0$. But condition (2) gives that $w = 0$ at this point. Hence $w(\tilde{x}, \tilde{t}) = 0$, and since the point $(\tilde{x}, \tilde{t})$ was chosen arbitrarily in $\bar{D}$, therefore $w(x, t) \equiv 0$, $(x, t) \in \bar{D}$. The theorem is proved.

We observe that Theorem 21.4.1 is also true in the case of two wave equations of the form (1) whose right-hand sides coincide in the domain D.

Let $u(x, t)$ be a solution of Cauchy's problem for equation (1); let the right-hand side of this equation, $f(x, t)$, be fixed. It follows from the theorem of the present section that the value of the function u at any point (x_0, t_0) is determined only by the values of the initial functions in the sphere $|x - x_0| \leq t_0$. This sphere is called the *domain of dependence* for the point (x_0, t_0).

If instead of (1) we consider the equation

$$\frac{\partial^2 u}{\partial t^2} - a^2 \Delta u = f(x, t),$$

then the domain of dependence for the point (x_0, t_0) will be the sphere $|x - x_0| \leq at_0$.

§ 5. The Propagation of Waves

The uniqueness theorem proved in the preceding section leads to a number of consequences of a physical kind, which we shall now discuss briefly.

Consider the homogeneous wave equation

$$\frac{\partial^2 u}{\partial t^2} - a^2 \Delta u = 0, \qquad a = \text{const.}, \tag{1}$$

with initial conditions

$$u|_{t=0} = \phi_0(x), \qquad \frac{\partial u}{\partial t}\bigg|_{t=0} = \phi_1(x), \qquad x \in E_m. \tag{2}$$

Suppose that the initial functions $\phi_0(x)$ and $\phi_1(x)$ are identically equal to zero outside some finite domain $D \subset E_m$ (Fig. 41); inside this domain these initial functions will, in general, be taken to be different from zero. We shall assume that the Cauchy problem here posed has a solution.

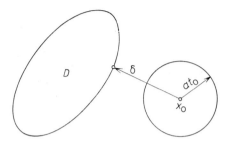

Fig. 41.

Take any point $x_0 \in E_m$ lying outside the domain D. At the initial instant the value of u at the point x_0 is equal to zero – this is obvious from the initial conditions. At this instant the point x_0 is in a stationary state. Consider the instant t_0, sufficiently close to the initial instant, namely, let

$$t_0 < \delta/a,$$

where δ is the shortest distance from the point x_0 to the boundary of the domain D. The domain of dependence for the point x_0 at time t_0 is a sphere of radius at_0 and with centre at x_0 – this sphere does not intersect the domain D. This means that in this domain of dependence the initial functions are zero; from the theorem of the previous section, $u(x_0, t_0) = 0$, and the point x_0 remains in a state of rest at the time t_0, so long as $t_0 < \delta/a$.

Now let $t_0 > \delta/a$. The domain of dependence intersects the domain D (in Fig. 42 this intersection is shaded), and then the initial functions are not identically zero, so that, in general, $u(x_0, t_0) \neq 0$.

Thus we see that the instant $t_0 = \delta/a$ can be regarded as the instant when a disturbance arrives at the point x_0; up to this moment the point is in a state of rest, after it is in a disturbed state.

We can also easily resolve the following question: let t_0 be a given moment of time; which domain is at rest and which is disturbed at this moment?

Let Γ be the boundary of the domain D of initial disturbance. From every point of the boundary Γ as centre, draw a sphere of radius at_0. The envelope Γ_{t_0} of these spheres (more precisely, the locus of points which lie outside

D and are at a distance at_0 from Γ) separates the stationary domain from the domain whose points are, in general, undergoing disturbances. The surface Γ_{t_0} is called the *leading wave front*.

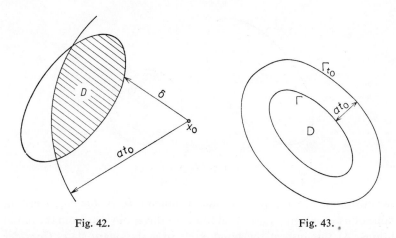

Fig. 42.   Fig. 43.

The process of propagation of a disturbance is called *wave propagation*. Obviously the disturbance propagates with velocity a in a direction normal to Γ.

Remark. If the dimensionality of space is odd and greater than one, then in a homogeneous medium there can appear under certain conditions a so-called *rear wave front*: at each point the disturbance vanishes after some moment of time. We return to this question in Chapter 24.

§ 6. Generalized Solution of the Cauchy Problem

The formulation of the Cauchy problem for the wave equation given in § 3 assumes that the required function possesses at least those derivatives which occur in the differential equation. The conditions under which this is in fact the case are known. S. L. Sobolev has shown that the solution of Cauchy's problem for the homogeneous wave equation has continuous second derivatives if the initial function $\phi_0(x)$ has all the generalized derivatives of order up to $[\frac{1}{2}m]+3$, and the function $\phi_1(x)$ up to order $[\frac{1}{2}m]+2$, and these derivatives are square-integrable. If $m \geq 5$ then these numbers are, in general, necessary; for $m = 1, 2, 3$, we can make do with a smaller number of derivatives of the initial functions.

Thus, it is not difficult to verify that the function *

$$u(x, t) = \tfrac{1}{2}[\phi_0(x+t)+\phi_0(x-t)] \qquad (1)$$

solves Cauchy's problem for the string equation

$$\frac{\partial^2 u}{\partial t^2} - \frac{\partial^2 u}{\partial x^2} = 0 \qquad (2)$$

with initial conditions

$$u|_{t=0} = \phi_0(x), \qquad \frac{\partial u}{\partial t}\bigg|_{t=0} = 0. \qquad (3)$$

It is clear that the solution $u(x, t)$ in this case has second derivatives, if the second derivative $\phi_0''(x)$ exists. But the requirement that the second derivative $\phi''(x)$ should exist does not follow from physical considerations. In fact, conditions (3) imply that at the initial instant the string was disturbed from its equilibrium state by the application of an initial displacement $\phi_0(x)$ and without the application of an initial velocity. The graph of the equation $u = \phi_0(x)$ gives the form of the string at the initial instant.

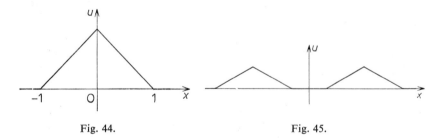

Fig. 44.   Fig. 45.

Suppose that the graph of the function $\phi_0(x)$ is the open polygon of Fig. 44. At three points, $x = -1, 0, 1$, this function does not even have a first derivative, yet the problem of vibrations of a string having this initial shape is completely meaningful, and the solution to this problem is given by the same equation (1). The form of the string at moments of time not too close to the initial instant is shown in Fig. 45. It is clear that the function (1) has not even first, much less second, derivatives, when the variables have values which are connected by the relations $x \pm t = -1, 0, 1$.

---

* Further discussion of this is to be found below, in § 7, Chap. 24.

It becomes necessary to consider generalized solutions of Cauchy's problem; this involves a kind of generalization which we have not encountered at any stage in the course of this book. We first introduce the definition of a *generalized solution of a differential equation*.

Let L be a linear differential expression of, say, second order, and let M be a differential expression formally adjoint to L.

Let the function $u \in C^{(2)}(\Omega)$ satisfy in the finite domain Ω the equation

$$Lu = f(x), \tag{4}$$

where x here denotes the set of all independent variables. Further, let $\Phi(x) \in \mathfrak{M}^{(2)}(\Omega)$ (§ 1, Chapter 2). Apply Green's formula to the functions $u(x)$ and $\Phi(x)$ (formula (6.4), Chapter 10). Then the surface integral vanishes, because the function Φ and its derivatives are equal to zero on the boundary of the domain Ω, and we arrive at the relation

$$\int_\Omega u M\Phi \, dx + \int_\Omega f\Phi \, dx, \qquad \forall \Phi \in \mathfrak{M}^{(2)}(\Omega). \tag{5}$$

It is not difficult to show that the function $u \in C^{(2)}(\Omega)$, satisfying (5), also satisfies equation (4).

Introduce the following definition: *a function $u(x)$ which is integrable in Ω and which satisfies relation (5) is called the generalized solution of equation (4)*.

It is easy to see that the concepts of generalized solutions to various problems, which we introduced earlier in this book, are compatible with the definition just given. In accordance with this definition, we shall say that the function $u(x, t)$ is a generalized solution of the wave equation

$$\Box u = \frac{\partial^2 u}{\partial t^2} - \Delta u = f(x, t) \tag{6}$$

in the half-space $t > 0$ if in any finite domain D of variation of the variables $x_1, x_2, \ldots, x_m, t$, this function is integrable and satisfies the relation

$$\int_D u \, \Box \, \Phi \, dx \, dt = \int_D f\Phi \, dx \, dt, \qquad \Phi \in \mathfrak{M}^{(2)}(D). \tag{7}$$

We now introduce the concept of the generalized solution of Cauchy's problem. Let equation (6) be supplemented by the initial conditions

$$u|_{t=0} = \phi_0(x), \qquad \left.\frac{\partial u}{\partial t}\right|_{t=0} = \psi(x). \tag{8}$$

The function $u(x, t)$ will be called the generalized solution to Cauchy's problem (6), (8) if this function 1) is a generalized solution of equation (6); 2) is square-integrable and has square-integrable generalized first derivatives in any finite domain of variation of the variables x_1, $x_2, \ldots, x_m, t$; 3) satisfies in any finite domain Ω of variation of the variables $x_1, x_2, \ldots, x_m$ the limiting relation

$$\lim_{t \to \infty} \int_\Omega \left\{ [u(x, t) - \phi_0(x)]^2 + \sum_{k=1}^m \left[\frac{\partial u(x, t)}{\partial x_k} - \frac{\partial \phi_0}{\partial x_k} \right]^2 \right. $$
$$\left. + \left[\frac{\partial u(x, t)}{\partial t} - \phi_1(x) \right]^2 \right\} dx = 0. \qquad (9)$$

The definition given here requires that in any finite domain $\Omega \subset E_m$ the initial functions $\phi_0(x)$ and $\phi_1(x)$ should be square-integrable, and the function $\phi_0(x)$ should also have square-integrable generalized first derivatives.

For a more extensive discussion of generalized solutions of the wave equation, see S. L. Sobolev [7]. The definition of a generalized solution to a boundary-value problem is given in the book by O. A. Ladyzhenskaya [3].

Chapter 22. Fourier's Method

In essence, Fourier's method for solving mixed problems and Cauchy's problem is a method based on the utilization of spectral properties of the elliptic operator which occurs in the equation. In the classical works of Fourier himself and his successors, the method was associated with separation of variables in the differential equation; this latter technique was used in § 3, Chapter 15. In the present chapter, Fourier's method is used as a basis for obtaining solutions to mixed problems for the heat-conduction equation and wave equation.

§ 1. Fourier's Method for the Heat-Conduction Equation

In this section we shall give a method for constructing the generalized solution to the mixed heat-conduction problem; in the process of doing so, we shall obtain a proof of the existence of this solution. The concept of a generalized solution and a proof of its uniqueness have been presented in § 6, Chapter 20. We recall that, in the present problem, the generalized solution is an abstract function of t of the class

$$C([0, \infty); L_2(\Omega)) \cap C((0, \infty), H_\mathfrak{A}) \cap C^{(1)}((0, \infty); L_2(\Omega)),$$

satisfying the relation

$$\left(\frac{du(t)}{dt}, \eta(t)\right) + [u(t), \eta(t)]_\mathfrak{A} = (f(t), \eta(t)) \tag{1}$$

and the initial condition

$$u(0) = \phi. \tag{2}$$

Here $\mathfrak{A}$ is the operator of the Dirichlet problem (cf. § 2, Chapter 14) for a finite domain $\Omega \subset E_m$ with a piecewise-smooth boundary Γ; $\eta(t)$ is an arbitrary, abstract function of t with values in the energy space $H_\mathfrak{A}$, and

ϕ is an element of the space $L_2(\Omega)$. Finally, $f(t)$ is an abstract function of t with values in $L_2(\Omega)$; we suppose that $f \in C^{(1)}([0, \infty); L_2(\Omega))$.

Let us assume that a solution $u(t) = u(x, t)$ of the problem (1)–(2) exists. For any $t \geq 0$ this solution is an element of the space $L_2(\Omega)$ and can be expanded in a series with respect to any complete, orthonormal system in $L_2(\Omega)$; in particular, with respect to the system of eigenelements of the operator $\mathfrak{A}$. Denote these elements by $u_n = u_n(x)$, and the corresponding eigenvalues by λ_n. Writing

$$(u(t), u_n) = c_n(t), \tag{3}$$

we have

$$u(t) = \sum_{n=1}^{\infty} c_n(t) u_n. \tag{4}$$

The problem reduces to the evaluation of the coefficients $c_n(t)$. To do this, we put $\eta(t) = u_k$ in equation (1). The element u_k is independent of t, and by formula (5.1), Chapter 20,

$$\left(\frac{du(t)}{dt}, u_k\right) = \frac{d}{dt}(u(t), u_k) = c_k'(t).$$

From the definition of a (generalized) eigenfunction (cf. (3.2), Chapter 6) we have

$$[u(t), u_k] = \lambda_k(u(t), u_k) = \lambda_k c_k(t).$$

Writing

$$(f(t), u_k) = f_k(t), \tag{5}$$

we finally obtain

$$c_k'(t) + \lambda_k c_k(t) = f_k(t). \tag{6}$$

This is an ordinary, first-order differential equation containing numerical functions: $f_k(t)$ given and $c_k(t)$ unknown. The general integral of this equation is

$$c_k(t) = \exp(-\lambda_k t)\left[C_k + \int_0^t \exp(\lambda_k \tau) f_k(\tau) d\tau\right], \quad C_k = \text{const}.$$

The initial condition for equation (6) follows from formulae (2) and (3):

$$c_k(0) = (\phi, u_k).$$

Hence $C_k = (\phi, u_k)$ and

$$c_k(t) = (\phi, u_k) \exp(-\lambda_k t) + \int_0^t \exp\{-\lambda_k(t-\tau)\} f_k(\tau) d\tau. \tag{7}$$

We have only to substitute (7) into formula (4), and we arrive at the following conclusion.

If the mixed heat-conduction problem has a generalized solution, then this solution can be represented by the series

$$u(x, t) = \sum_{n=1}^{\infty} (\phi, u_n) \exp(-\lambda_n t) u_n(x)$$
$$+ \sum_{n=1}^{\infty} u_n(x) \int_0^t \exp\{-\lambda_n(t-\tau)\} f_n(\tau) \, d\tau. \qquad (8)$$

From this there follows, among other things, the uniqueness of the solution, proved earlier in § 6, Chapter 20.

§ 2. Justification of the Method

We now show that the series (1.8) does in fact give the generalized solution to the heat-conduction problem. The proof reduces to the verification of the statements formulated below.

a) In the metric of $L_2(\Omega)$, the series (1.8) converges uniformly with respect to t on any segment $[0, T]$.

The series (1.8) is orthogonal in $L_2(\Omega)$, and it is sufficient to prove that, on the segment $[0, T]$, the series of squares of the coefficients converges uniformly:

$$\sum_{n=1}^{\infty} \left\{ (\phi, u_n) \exp(-\lambda_n t) + \int_0^t \exp\{-\lambda_n(t-\tau)\} f_n(\tau) \, d\tau \right\}^2$$
$$\leq 2 \sum_{n=1}^{\infty} (\phi, u_n)^2 \exp(-2\lambda_n t) + 2 \sum_{n=1}^{\infty} \left\{ \int_0^t \exp\{-\lambda_n(t-\tau)\} f_n(\tau) \, d\tau \right\}^2. \qquad (1)$$

The abstract function $f(t)$ is continuous for $t \geq 0$; hence it follows that the quantity $\|f(t)\|$ is continuous on the segment $[0, T]$.

It is not difficult now to show that the second series on the right of (1) converges uniformly. By the Schwartz-Bunyakovskii inequality,

$$\left\{ \int_0^t \exp\{-\lambda_n(t-\tau)\} f_n(\tau) \, d\tau \right\}^2 \leq \int_0^t \exp\{-2\lambda_n(t-\tau)\} \, d\tau \int_0^t f_n^2(\tau) \, d\tau$$
$$= \frac{1 - \exp(-2\lambda_n t)}{2\lambda_n} \int_0^t f_n^2(\tau) \, d\tau < \frac{1}{2\lambda_n} \int_0^t f_n^2(\tau) \, d\tau. \qquad (2)$$

Replacing λ_n by the smallest eigenvalue λ_1, we find that the general term of the above series has the bound $(1/2\lambda_1)\int_0^t f_n^2(\tau)d\tau$. The equation

$$\sum_{n=1}^{\infty} f_n^2(\tau) = \|f(\tau)\|^2 \tag{3}$$

shows that the series (3), with non-negative continuous terms, converges and has a continuous sum. From Dini's theorem we know that this series converges on the segment $[0, T]$ for any $T > 0$. But then the second series on the right-hand side of (1) also converges uniformly.

The convergence of the first series in (1) can be established quite simply:

$$2(\phi, u_n)^2 \exp(-2\lambda_k t) \leq 2(\phi, u_n)^2,$$

and the series $2\sum_{n=1}^{\infty}(\phi, u_n)^2$ converges because of Bessel's inequality.

From what we have shown it follows that the sum of the series (1.8)

$$u(x, t) = u(t) \in C([0, \infty); L_2(\Omega)).$$

b) In the metric of $H_\mathfrak{A}$, the series (1.8) converges uniformly with respect to t on any segment $[\bar{t}, T]$ where $0 < \bar{t} < T < \infty$.

In the metric of $H_\mathfrak{A}$, the functions $u_n(x)/\sqrt{\lambda_n}$ are orthonormal; series (1.8) can be represented in the form

$$u(x, t) = \sum_{n=1}^{\infty} \sqrt{\lambda_n} \left\{(\phi, u_n)\exp(-\lambda_n t) + \int_0^t \exp\{-\lambda_n(t-\tau)\}f_n(\tau)d\tau\right\}\frac{u_n(x)}{\sqrt{\lambda_n}}. \tag{4}$$

It is sufficient that the series of squares of the coefficients should converge; we estimate these squares. We have

$$\sqrt{\lambda_n}\exp(-\lambda_n t) = \frac{1}{t\sqrt{\lambda_n}}\lambda_n t\exp(-\lambda_n t) \leq \frac{1}{\bar{t}\sqrt{\lambda_n}}\max z e^{-z} = \frac{1}{\bar{t}e\sqrt{\lambda_n}}. \tag{5}$$

Now

$$\lambda_n c_n^2(t) \leq 2(\sqrt{\lambda_n}\exp(-\lambda_n t))^2(\phi, u_n)^2 + 2\lambda_n\left(\int_0^t \exp\{-\lambda_n(t-\tau)\}f_n(\tau)d\tau\right)^2$$

$$\leq \frac{2}{\bar{t}^2 e^2 \lambda_n}(\phi, u_n)^2 + \int_0^t f_n^2(\tau)d\tau, \tag{6}$$

where we have here used the estimate (2). The series whose general term is (6) converges uniformly, and so statement b) is proven. From this statement it follows that

$$u(x, t) = u(t) \in C((0, \infty); H_\mathfrak{A})).$$

c) The series obtained by differentiating (1.8) with respect to t converges uniformly, in the metric of $L_2(\Omega)$, with respect to t in any segment $[\bar{t}, T]$, where $0 < \bar{t} < T < \infty$.

After differentiating (1.8) with respect to t, we obtain the following series:

$$\sum_{n=1}^{\infty} \left\{ -\lambda_n(\phi, u_n) \exp(-\lambda_n t) + f_n(t) - \lambda_n \int_0^t \exp\{-\lambda_n(t-\tau)\} f_n(\tau) d\tau \right\} u_n(x)$$

or, on integrating by parts,

$$\sum_{n=1}^{\infty} \left\{ -\lambda_n(\phi, u_n) \exp(-\lambda_n t) + f_n(0) \exp(-\lambda_n t) + \right.$$
$$\left. + \int_0^t \exp\{-\lambda_n(t-\tau)\} f_n'(\tau) d\tau \right\} u_n(x). \quad (7)$$

As before, this is a series with respect to a complete, orthonormal system $\{u_n(x)\}$ in $L_2(\Omega)$. We estimate its coefficients. From Cauchy's inequality, the square of the coefficient of $u_n(x)$ in (7) does not exceed the value

$$3(\lambda_n \exp(-\lambda_n t))^2 (\phi, u_n)^2 + 3 f_n^2(0) \exp(-2\lambda_n t)$$
$$+ 3 \left(\int_0^t \exp\{-\lambda_n(t-\tau)\} f_n'(\tau) d\tau \right)^2$$
$$\leq \frac{3}{(\bar{t}e)^2} (\phi, u_n)^2 + 3 f_n^2(0) + \frac{3}{2\lambda_n} \int_0^T [f_n'(\tau)]^2 d\tau. \quad (8)$$

From our assumptions regarding the functions $\phi(x)$ and $f(x, t)$, it follows that the series with general term (8) converges. Then the series (7) converges uniformly, in the metric of $L_2(\Omega)$, on the segment $[\bar{t}, T]$. Statement c) is proved. From this statement it follows that the sum of the series (1.8) is

$$u(x, t) = u(t) \in C^{(1)}((0, \infty); L_2(\Omega)).$$

d) The sum of the series (1.8) satisfies the initial condition (1.2).

The properties established in a) imply that in this series it is permissible to take the limit $t \to 0$ term by term. Therefore

$$\lim_{t \to 0} \|u(t) - \phi\|_{L_2} = \left\| \sum_{n=1}^{\infty} (\phi, u_n) u_n - \phi \right\| = 0.$$

e) The sum of the series (1.8) satisfies the integral relation (1.1).

Let $\eta(t)$ be an arbitrary abstract function of t, $0 < t < \infty$, with values in $H_\mathfrak{A}$. First differentiate both sides of (1.8) with respect to t, then scalar multiply (in the metric of $L_2(\Omega)$) by $\eta(t)$

$$\left(\frac{du(t)}{dt}, \eta(t)\right) = \sum_{n=1}^{\infty} c'_n(t)(u_n, \eta(t)).$$

Substituting for $c'_n(t)$ from formula (1.6), we obtain

$$\left(\frac{du(t)}{dt}, \eta(t)\right) = \sum_{n=1}^{\infty} f_n(t)(u_n, \eta(t)) - \sum_{n=1}^{\infty} \lambda_n c_n(t)(u_n, \eta(t))$$

$$= \left(\sum_{n=1}^{\infty} f_n(t)u_n, \eta(t)\right) - \sum_{n=1}^{\infty} c_n(t)[u_n, \eta(t)]$$

$$= (f(t), \eta(t)) - \left[\sum_{n=1}^{\infty} c_n(t)u_n, \eta(t)\right] = (f(t), \eta(t)) - [u(t), \eta(t)].$$

Our statement is proved.

§ 3. On the Existence of the Classical Solution. A Special Case

The existence, under quite wide conditions, of the generalized solution to the mixed heat-conduction problem having been proved, it is now appropriate to pose the following question: what restrictions have to be imposed on the data in order that the solution should be "classical"? By this we mean that this solution should be continuous in $Q \cup B \cup \Omega$ (Fig. 38, page 416), and have in Q a continuous first derivative with respect to t and continuous second derivatives with respect to the coordinates.

We shall give the answer to this question in the simplest case, when $m = 1$, the domain Ω is the interval $(0, 1)$ of the Ox-axis, and heat-conduction equation has the form

$$\frac{\partial u}{\partial t} - \frac{\partial^2 u}{\partial x^2} = 0. \tag{1}$$

The initial and boundary conditions are

$$u(x, 0) = \phi(x), \qquad 0 \leq x \leq 1, \tag{2}$$

$$u(0, t) = u(1, t) = 0; \tag{3}$$

the solution is determined in a semi-infinite strip (Fig. 46).

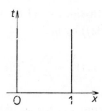

Fig. 46.

If the solution $u(x, t)$ of our problem is continuous in the semi-infinite strip $0 \leq x \leq 1, t \geq 0$, then, first of all, it is necessary that $\phi(x)$ should be continuous. The function $u(x, t)$ must be continuous at the points $x = 0$, $t = 0$ and $x = 1$, $t = 0$. At both of these points it is possible to calculate a value of $u(x, t)$ using either the initial condition or the boundary conditions (3). But the two processes must lead to the same result. Hence it follows that the function $\phi(x)$ must necessarily satisfy *compatibility conditions*

$$\phi(0) = \phi(1) = 0. \tag{4}$$

It is now possible to formulate the answer to the question posed above: *the generalized solution to problem* (1)–(3) *will also be a classical solution if the initial function* $\phi(x)$ *is absolutely continuous on the segment* [0, 1], *its derivative* $\phi' \in L_2(0, 1)$, *and the compatibility conditions* (4) *are fulfilled.* We now prove this statement.

In the present situation, the operator $\mathfrak{A}$ acts according to the formula $\mathfrak{A}u = -d^2u/dx^2$, and is defined on functions of the class $C^{(2)}(0, 1)$, which are equal to zero when $x = 0$ and $x = 1$. Formulae (8.3) and (8.4a) of Chapter 6 give the generalized eigenvalues and eigenfunctions of the operator $\mathfrak{A}$:

$$\lambda_n = n^2\pi^2, \qquad u_n(x) = \sqrt{2} \sin n\pi x.$$

From (1.8) we find

$$u(x, t) = \sum_{n=1}^{\infty} b_n e^{-n^2\pi^2 t} \sin n\pi x, \tag{5}$$

where for brevity we have written

$$b_n = 2 \int_0^1 \phi(x) \sin n\pi x \, dx. \tag{6}$$

We first show that the series (5) is uniformly convergent for $0 \leq x \leq 1$,

§ 3] THE EXISTENCE OF THE CLASSICAL SOLUTION

$t \geq 0$. To do this, integrate (6) by parts; making use of condition (4), we obtain $b_n = \beta_n/n$, where

$$\beta_n = \frac{2}{\pi} \int_0^1 \phi'(x) \cos n\pi x \, dx$$

is the n-th Fourier coefficient of the function $(\sqrt{2}/\pi)\phi'(x)$ in its expansion with respect to the orthonormal system $\{\sqrt{2} \cos n\pi x\}$. The series $\sum_{n=1}^{\infty} \beta_n^2$ converges, and then, because of the inequality $|b_n| \leq \frac{1}{2}(\beta_n^2 + 1/n^2)$, the series

$$\sum_{n=1}^{\infty} |b_n| \tag{7}$$

also converges. But the series (7) is majorant for the series (5) which therefore converges absolutely and uniformly; the terms of (5) are continuous for $0 \leq x \leq 1$, $t \geq 0$, and their sum is also continuous.

We now show that the function $u(x, t)$ – the sum of the series (5) – has derivatives of all orders with respect to x and t for $t > 0$. To do this it is sufficient to prove that after differentiating the series (5) any number of times with respect to x and t, we obtain a series which converges uniformly for $0 \leq x \leq 1$, $t \geq \bar{t}$, where $\bar{t}$ is an arbitrary positive constant. Differentiating the series (5) p times with respect to x and q times with respect to t we obtain

$$(-1)^q \sum_{n=1}^{\infty} n^{p+2q} \pi^{p+2q} b_n e^{-n^2\pi^2 t} \sin(n\pi x + \tfrac{1}{2}p\pi). \tag{8}$$

The Fourier coefficients b_n are obviously bounded, and series (8) has as majorant the following series:

$$C \sum_{n=1}^{\infty} n^{2s} e^{-n^2\pi^2 \bar{t}}, \quad C = \text{const.}, \, 2s \geq p + 2q, \quad s = \text{const.}$$

Moreover

$$n^{2s} e^{-n^2\pi^2 \bar{t}} < \frac{n^{2s}}{1 + n^2\pi^2\bar{t} + \ldots + \dfrac{n^{2s+2}}{(2s+2)!}(\pi^2\bar{t})^{2s+2}} < \frac{(2s+2)!}{(\pi^2\bar{t})^{2s+2} n^2},$$

and we can construct a stronger majorant series

$$C_1 \sum_{n=1}^{\infty} \frac{1}{n^2}, \quad C_1 = \text{const.},$$

which obviously converges. Our assertion is proved.

As is obvious from the nature of the proof, the existence of an infinite number of derivatives for $t > 0$ can be established with the condition $\phi \in L(0, 1)$ alone; the additional restrictions on the function $\phi(x)$ were necessary only to prove the continuity of the solution for $t = 0$.

§ 4. Fourier's Method for the Wave Equation

The generalized solution $u(x, t) = u(t)$ of the mixed problem for the wave equation (cf. § 2, Chapter 21) is an abstract function of t belonging to the class K (formula (2.9), Chapter 21) and satisfying the equation

$$-\int_0^T \left(\frac{du(t)}{dt}, \frac{d\eta(t)}{dt}\right) dt + \int_0^T [u(t), \eta(t)] dt - (\phi_1, \eta(0)) = \int_0^T (f(t), \eta(t)) dt,$$

$$\eta \in K_T, \qquad (1)$$

and the initial condition

$$u(0) = \phi_0. \qquad (2)$$

Here $\phi_0 \in H_{\mathfrak{A}}$, $\phi_1 \in L_2(\Omega)$ we assume that $f(t) = f(x, t)$ is an abstract function of class $C([0, \infty); L_2(\Omega))$. Condition (2) is understood in the sense of a limit process in the energy metric:

$$\lim_{t \to 0} |||u(t) - \phi_0||| = 0.$$

Suppose that the solution $u(t)$ of problem (1)–(2) exists. Expand it, in the metric of $L_2(\Omega)$, in a series of the system of eigenelements of the operator $\mathfrak{A}$:

$$u(t) = \sum_{n=1}^\infty c_n(t) u_n, \qquad c_n(t) = (u(t), u_n). \qquad (3)$$

When written in the form

$$u(t) = \sum_{n=1}^\infty \sqrt{\lambda_n} c_n(t) \frac{u_n}{\sqrt{\lambda_n}},$$

this series also gives the expansion of the solution $u(t)$, in the metric of $H_{\mathfrak{A}}$, with respect to the orthonormal system $u_n/\sqrt{\lambda_n}$.

In equation (1) put $\eta(t) = (T-t)u_n$. Recalling that

$$[u(t), u_n] = \lambda_n(u(t), u_n) = \lambda_n c_n(t),$$

we obtain the following equation for the unknown coefficients $c_n(t)$:

$$\int_0^T c_n'(t)\,dt - T(\phi_1, u_n) + \lambda_n \int_0^T (T-t)c_n(t)\,dt = \int_0^T (T-t)f_n(t)\,dt, \quad (4)$$

where

$$f_n(t) = (f(t), u_n). \quad (5)$$

Differentiate equation (4) with respect to T, and substitute t and τ for T and t respectively:

$$c_n'(t) - (\phi_1, u_n) + \lambda_n \int_0^t c_n(\tau)\,d\tau = \int_0^t f_n(\tau)\,d\tau. \quad (6)$$

Equation (6) shows that the second derivative $c_n''(t)$ exists. Differentiating, we see that $c(t)$ satisfies a second-order, ordinary differential equation with constant coefficients:

$$\frac{d^2 c_n(t)}{dt^2} + \lambda_n c_n(t) = f_n(t); \quad (7)$$

the initial conditions for this equation are obtained from relations (2), (3) and (6):

$$c_n(0) = (\phi_0, u_n), \quad c_n'(0) = (\phi_1, u_n). \quad (8)$$

The solution of the problem has the form

$$c_n(t) = (\phi_0, u_n)\cos\sqrt{\lambda_n}\,t + \frac{(\phi_1, u_n)}{\sqrt{\lambda_n}}\sin\sqrt{\lambda_n}\,t$$

$$+ \frac{1}{\sqrt{\lambda_n}}\int_0^t \sin\sqrt{\lambda_n}(t-\tau)f_n(\tau)\,d\tau. \quad (9)$$

Hence

$$u(x, t) = \sum_{n=1}^\infty \left\{ (\phi_0, u_n)\cos\sqrt{\lambda_n}\,t + \frac{(\phi_1, u_n)}{\sqrt{\lambda_n}}\sin\sqrt{\lambda_n}\,t \right.$$

$$\left. + \frac{1}{\sqrt{\lambda_n}}\int_0^t \sin\sqrt{\lambda_n}(t-\tau)f_n(\tau)\,d\tau \right\} u_n(x). \quad (10)$$

Thus, if the generalized solution of the mixed problem for the wave equation exists, then it necessarily has the form (10). As with the heat-conduction equation, this implies the uniqueness of the generalized solution.

Formula (10) is rather cumbersome, and therefore we shall give its justification in the following way.

Let the functions $v(x, t)$ and $w(x, t)$ be generalized solutions of the following problems: the homogeneous wave equation with inhomogeneous initial conditions

$$\frac{d^2 v}{dt^2} + \mathfrak{A} v = 0, \qquad v|_{t=0} = \phi_0, \quad \frac{dv}{dt}\bigg|_{t=0} = \phi_1, \qquad (11)$$

and the inhomogeneous wave equation with homogeneous initial conditions

$$\frac{d^2 w}{dt^2} + \mathfrak{A} w = f(t), \qquad w|_{t=0} = 0, \quad \frac{dw}{dt}\bigg|_{t=0} = 0. \qquad (12)$$

Then obviously $u = v + w$. From the general formula (10) we deduce the following formulae for v and w:

$$v(x, t) = \sum_{n=1}^{\infty} \left\{ (\phi_0, u_n) \cos \sqrt{\lambda_n}\, t + \frac{(\phi_1, u_n)}{\sqrt{\lambda_n}} \sin \sqrt{\lambda_n}\, t \right\} u_n(x), \qquad (13)$$

$$w(x, t) = \sum_{n=1}^{\infty} \frac{u_n(x)}{\sqrt{\lambda_n}} \int_0^t \sin \sqrt{\lambda_n}(t-\tau) f_n(\tau)\, d\tau. \qquad (14)$$

In the next two sections we shall present justifications for Fourier's method in respect of each of the problems (11) and (12) separately.

§ 5. Justification of the Method for the Homogeneous Equation

As in the case of the heat-conduction equation (§ 3), the justification of Fourier's method reduces to the verification of a number of assertions.

a) The series (4.13) converges in the metric of $H_\mathfrak{A}$, uniformly with respect to t on the whole axis.

Write the series (4.13) in the form

$$v(x, t) = \sum_{n=1}^{\infty} \left\{ \sqrt{\lambda_n}(\phi_0, u_n) \cos \sqrt{\lambda_n}\, t + (\phi_1, u_n) \sin \sqrt{\lambda_n}\, t \right\} \frac{u_n(x)}{\sqrt{\lambda_n}}$$

$$= \sum_{n=1}^{\infty} \left\{ \left[\phi_0, \frac{u_n}{\sqrt{\lambda_n}}\right] \cos \sqrt{\lambda_n}\, t + (\phi_1, u_n) \sin \sqrt{\lambda_n}\, t \right\} \frac{u_n(x)}{\sqrt{\lambda_n}}. \qquad (1)$$

The latter is a series with respect to the system of functions $\{u_n/\sqrt{\lambda_n}\}$, orthonormal in the metric of $H_\mathfrak{A}$, and it is sufficient to show that the series of squares of the coefficients

$$\sum_{n=1}^{\infty} \left\{ \left[\phi_0, \frac{u_n}{\sqrt{\lambda_n}}\right] \cos \sqrt{\lambda_n} t + (\phi_1, u_n) \sin \sqrt{\lambda_n} t \right\}^2 \qquad (2)$$

converges uniformly with respect to t. The sum of this series does not exceed the value

$$2 \sum_{n=1}^{\infty} \left[\phi_0, \frac{u_n}{\sqrt{\lambda_n}}\right]^2 + 2 \sum_{n=1}^{\infty} (\phi_1, u_n)^2.$$

By Bessel's inequality, both of these series converge. At the same time, their terms are independent of t. Therefore, by Weierstrass's theorem, the series (2) converges uniformly.

The series (4.13) also converges uniformly with respect to t in the metric of $L_2(\Omega)$ – this follows immediately from the positive definiteness inequality (inequality (3.3), Chapter 5).

From the assertion made above it also follows that

$$v(x, t) = v(t) \in C([0, \infty), H_{\mathfrak{A}}).$$

b) The series obtained by differentiating (4.13) with respect to t,

$$\sum_{n=1}^{\infty} \{-\sqrt{\lambda_n}(\phi_0, u_n) \sin \sqrt{\lambda_n} t + (\phi_1, u_n) \cos \sqrt{\lambda_n} t\} u_n(x) =$$

$$= \sum_{n=1}^{\infty} \left\{ -\left[\phi_0, \frac{u_n}{\sqrt{\lambda_n}}\right] \sin \sqrt{\lambda_n} t + (\phi_1, u_n) \cos \sqrt{\lambda_n} t \right\} u_n(x), \qquad (3)$$

converges uniformly with respect to t in the metric of $L_2(\Omega)$.

It is sufficient to write down the inequality

$$\left\{ -\left[\phi_0, \frac{u_n}{\sqrt{\lambda_n}}\right] \sin \sqrt{\lambda_n} t + (\phi_1, u_n) \cos \sqrt{\lambda_n} t \right\}^2 \leq 2 \left[\phi_0, \frac{u_n}{\sqrt{\lambda_n}}\right]^2 + 2(\phi_1, u_n)^2,$$

and to apply Bessel's inequality and Weierstrass's theorem, as in a).

From a) and b) it follows that

$$v(x, t) = v(t) \in C^{(1)}([0, \infty); L_2 \Omega)),$$

and, consequently, $v(t) \in K$.

c) The sum of the series (4.13) satisfies the initial conditions (4.11). Because of the results in a), it is possible to take the limit as $t \to 0$ (in the metric of $H_{\mathfrak{A}}$) term by term in series (1). Hence

$$\lim_{t\to 0} |||v(t)-\phi_0||| = \left\|\sum_{n=1}^{\infty}\left[\phi_0, \frac{u_n}{\sqrt{\lambda_n}}\right]\frac{u_n}{\sqrt{\lambda_n}} - \phi_0\right\| = 0.$$

Further, from b) we have that the sum of the series (3) is equal to $dv(t)/dt$, and term by term passage to the limit is permissible in this series also:

$$\lim_{t\to 0}\left\|\frac{dv(t)}{dt}-\phi_1\right\| = \left\|\sum_{n=1}^{\infty}(\phi_1, u_n)u_n - \phi_1\right\| = 0.$$

d) The sum of the series (4.13) satisfies the equation

$$-\int_0^T \left(\frac{dv(t)}{dt}, \frac{d\eta(t)}{dt}\right)dt + \int_0^T [v(t), \eta(t)]\,dt - (\phi_1, \eta(0)) = 0, \quad \forall \eta \in K_T, \quad (4)$$

which is obtained by putting $f \equiv 0$ in equation (4.1).

For brevity, let $\gamma_n(t)$ denote the coefficients of the series (4.13):

$$\gamma_n(t) = (\phi_0, u_n)\cos\sqrt{\lambda_n}\,t + \frac{(\phi_1, u_n)}{\sqrt{\lambda_n}}\sin\sqrt{\lambda_n}\,t.$$

Then

$$v(x, t) = v(t) = \sum_{n=1}^{\infty}\gamma_n(t)u_n.$$

The coefficients $\gamma_n(t)$ satisfy the differential equation (4.7) with $f_n(t) \equiv 0$:

$$\gamma_n''(t) + \lambda_n \gamma_n(t) = 0. \tag{5}$$

We have

$$-\int_0^T \left(\frac{dv}{dt}, \frac{d\eta}{dt}\right)dt = -\sum_{n=1}^{\infty}\int_0^T \gamma_n'(t)\left(u_n, \frac{d\eta(t)}{dt}\right)dt.$$

Term by term integration is permissible because the series (3) converges uniformly with respect to t. Integrating by parts, we obtain

$$-\int_0^T \gamma_n'(t)\left(u_n, \frac{d\eta(t)}{dt}\right)dt = \int_0^T \gamma_n''(t)(u_n, \eta(t))\,dt - \gamma_n'(t)(u_n, \eta(t))\big|_{t=0}^{t=T}$$

$$= \int_0^T \gamma_n''(t)(u_n, \eta(t))\,dt + (\phi_1, u_n)(u_n, \eta(0)).$$

Hence

$$-\int_0^T \left(\frac{dv(t)}{dt}, \frac{d\eta(t)}{dt}\right)dt = \sum_{n=1}^{\infty}\int_0^T \gamma_n''(t)(u_n, \eta(t))\,dt + \sum_{n=1}^{\infty}(\phi_1, u_n)(u_n, \eta(0)).$$

But by Parseval's theorem

$$\sum_{n=1}^{\infty} (\phi_1, u_n)(u_n, \eta(0)) = (\phi_1, \eta(0)),$$

and, consequently,

$$-\int_0^T \left(\frac{dv(t)}{dt}, \frac{d\eta(t)}{dt}\right) dt = \sum_{n=1}^{\infty} \int_0^T \gamma_n''(t)(u_n, \eta(t)) dt + (\phi_1, \eta(0)). \quad (6)$$

Moreover,

$$\int_0^T [u(t), \eta(t)] dt = \sum_{n=1}^{\infty} \int_0^T \gamma_n(t)[u_n, \eta(t)] dt = \sum_{n=1}^{\infty} \int_0^T \lambda_n \gamma_n(t)(u_n, \eta(t)) dt. \quad (7)$$

If now we add equations (6) and (7), and make use of equation (5), we obtain equation (4).

§ 6. Justification of the Method for Homogeneous Initial Conditions

We now show that the sum of the series (4.14) is a generalized solution of the problem (4.12). With this as our aim, we show that all the statements a) – d) of the preceding section hold for the series (4.14), with one exception: uniform convergence takes place, not on the whole of the t-axis, but only on some segment of the form $[0, \bar{t}]$, $\bar{t} = \text{const.} > 0$.

The proof of statement a) reduces to the verification of the fact that the series

$$\sum_{n=1}^{\infty} \left\{\int_0^t \sin\sqrt{\lambda_n}(t-\tau) f_n(\tau) d\tau\right\}^2 \quad (1)$$

converges uniformly on the segment $[0, \bar{t}]$. By the Schwartz-Bunyakovskii inequality,

$$\left\{\int_0^t \sin\sqrt{\lambda_n}(t-\tau) f_n(\tau) d\tau\right\}^2 \leq \left\{\int_0^t |f_n(t)| dt\right\}^2 \leq \bar{t}\int_0^t f_n^2(\tau) d\tau. \quad (2)$$

From Parseval's theorem,

$$\sum_{n=1}^{\infty} f_n^2(\tau) = \|f(\tau)\|^2. \quad (3)$$

The series (3), having continuous, non-negative terms, converges to a continuous function. By Dini's theorem, this series converges uniformly. But then the integrated series also converges uniformly. From inequality (2) it now follows that series (1) also converges uniformly.

From this it follows that the series (4.14) converges uniformly on the segment $[0, \tilde{t}]$, in the metric of $L_2(\Omega)$, and also that $w(x, 0) = 0$.

We turn to statement b). Formally differentiating the series (4.14) with respect to t, we obtain the series

$$\sum_{n=1}^{\infty} u_n(x) \int_0^t \cos \sqrt{\lambda_n}(t-\tau) f_n(\tau) \, d\tau.$$

We can prove the uniform convergence of this series in the metric of $L_2(\Omega)$, if we can establish that the series

$$\sum_{n=1}^{\infty} \left\{ \int_0^t \cos \sqrt{\lambda_n}(t-\tau) f_n(\tau) \, d\tau \right\}^2$$

converges uniformly on the segment $[0, \tilde{t}]$. But this can be shown exactly as in a). It is now clear that

$$\frac{dw(t)}{dt} = \sum_{n=1}^{\infty} u_n(x) \int_0^t \cos \sqrt{\lambda_n}(t-\tau) f_n(\tau) \, d\tau$$

and that $(dw(t)/dt)|_{t=0} = 0$.

Nothing further need be said regarding statement c); we have just proved it, in passing.

We therefore now turn to d). We show that $w(x, t)$ satisfies equation (4.1), which in this case takes the form

$$-\int_0^T \left(\frac{dw(t)}{dt}, \frac{d\eta(t)}{dt} \right) dt + \int_0^T [w(t), \eta(t)] \, dt = \int_0^T (f(t), \eta(t)) \, dt,$$

$$\eta \in K_T. \quad (4)$$

We have

$$\int_0^T \left(\frac{dw(t)}{dt}, \frac{d\eta(t)}{dt} \right) dt = \sum_{n=1}^{\infty} \int_0^T \left(u_n, \frac{d\eta(t)}{dt} \right) \left\{ \int_0^t \cos \sqrt{\lambda_n}(t-\tau) f_n(\tau) \, d\tau \right\} dt.$$

Integrate by parts the outer integral

$$\int_0^T \left(u_n, \frac{d\eta(t)}{dt} \right) \left\{ \int_0^t \cos \sqrt{\lambda_n}(t-\tau) f_n(\tau) \, d\tau \right\} dt =$$

$$= \left[(u_n, \eta(t)) \int_0^t \cos \sqrt{\lambda_n}(t-\tau) f_n(\tau) \, d\tau \right]_{t=0}^{t=T}$$

$$- \int_0^T (u_n, \eta(t)) \left\{ f_n(t) - \int_0^t \sqrt{\lambda_n} \sin \sqrt{\lambda_n}(t-\tau) f_n(\tau) \, d\tau \right\} dt.$$

The integrated term vanishes, since $\eta(T) = 0$. Using the fact that

$$\sum_{n=1}^{\infty} f_n(t) u_n = f(t)$$

and that

$$\sqrt{\lambda_n}(u_n, \eta(t)) = \frac{1}{\sqrt{\lambda_n}}[u_n, \eta(t)],$$

we find that

$$\int_0^T \left(\frac{dw(t)}{dt}, \frac{d\eta(t)}{dt}\right) dt = -\int_0^T (f(t), \eta(t)) dt + \int_0^T [w(t), \eta(t)] dt,$$

so that relation (4) is proved.

§ 7. Equation of the Vibrating String. Conditions for Existence of the Classical Solution

We now return to the question, under what conditions is the generalized solution of the mixed problem for the wave equation at the same time the classical solution? We consider this question only for the simplest case of the string equation *

$$\frac{\partial^2 u}{\partial t^2} - \frac{\partial^2 u}{\partial x^2} = 0, \qquad 0 \leq x \leq 1, t \geq 0. \tag{1}$$

We shall solve this equation with the boundary conditions

$$u(0, t) = u(1, t) = 0 \tag{2}$$

and initial conditions

$$u\bigg|_{t=0} = \phi_0(x), \qquad \frac{\partial u}{\partial t}\bigg|_{t=0} = \phi_1(x). \tag{3}$$

In the present case $\lambda_n = n^2 \pi^2$, $u_n(x) = \sqrt{2} \sin n\pi x$; from the general formula (4.13) the generalized solution has the form

$$u(x, t) = \sum_{n=1}^{\infty} \left(a_n \cos n\pi t + \frac{b_n}{n\pi} \sin n\pi t\right) \sin n\pi x, \tag{4}$$

---

* Concerning the general case, see ref. [3].

where

$$a_n = 2\int_0^1 \phi_0(x)\sin n\pi x\,dx,$$
$$b_n = 2\int_0^1 \phi_1(x)\sin n\pi x\,dx.$$
(5)

The solution (4) is said to be classical if for $0 \leq x \leq 1$, $t > 0$, it is continuous, together with its first and second derivatives. This will be the case if the series (4), and the series obtained by differentiating (4) once and twice, are uniformly convergent.

We now show that this uniform convergence is the case if the following conditions are fulfilled: 1) the functions

$$\phi_0^{(k)}(x), \ k = 0, 1, 2; \qquad \phi_1^{(k)}(x), \ k = 0, 1$$

are absolutely continuous on the segment $[0, 1]$;

2) $\phi_0''' \in L_2(0, 1)$, $\qquad \phi_1'' \in L_2(0, 1)$;

3) the compatibility conditions are satisfied:

$$\phi_0(0) = \phi_0(1) = 0, \ \phi_1(0) = \phi_1(1) = 0, \ \phi_0''(0) = \phi_0''(1) = 0. \quad (6)$$

Note that the compatibility conditions are necessary if the solution (4) is to be classical. The first two of conditions (6) follow from the continuity of the function $u(x, t)$ at the points $x = 0, t = 0$ and $x = 1, t = 0$; the second two from the continuity of the derivative $\partial u/\partial t$ at the same points. The third pair are obtained in the following way. Putting $t = 0$ in equation (1), we find

$$\left.\frac{\partial^2 u}{\partial t^2}\right|_{t=0} - \phi_0''(x) = 0.$$

Differentiating equation (2), we obtain

$$\left.\frac{\partial^2 u}{\partial t^2}\right|_{x=0} = \left.\frac{\partial^2 u}{\partial t^2}\right|_{x=1} = 0.$$

Putting $t = 0$ here, and $x = 0$ and $x = 1$ in the preceding relation, we arrive at the third pair of (6).

Once conditions (1)–(3) have been formulated, the rest follows simply.

The formulae for the coefficients a_n and b_n are transformed through integration by parts:

$$\begin{aligned}
a_n &= -\left.\frac{2\phi_0(x)\cos n\pi x}{n\pi}\right|_0^1 + \frac{2}{n\pi}\int_0^1 \phi_0'(x)\cos n\pi x\,dx \\
&= \frac{2}{n\pi}\int_0^1 \phi_0'(x)\cos n\pi x\,dx \\
&= \left.\frac{2\phi_0'(x)\sin n\pi x}{n^2\pi^2}\right|_0^1 - \frac{2}{n^2\pi^2}\int_0^1 \phi_0''(x)\sin n\pi x\,dx \\
&= -\frac{2}{n^2\pi^2}\int_0^1 \phi_0''(x)\sin n\pi x\,dx \\
&= \left.\frac{2\phi_0''(x)\cos n\pi x}{n^3\pi^3}\right|_0^1 - \frac{2}{n^3\pi^3}\int_0^1 \phi_0'''(x)\cos n\pi x\,dx \\
&= -\frac{2}{n^3\pi^3}\int_0^1 \phi_0'''(x)\cos n\pi x\,dx.
\end{aligned}$$

In exactly the same way

$$b_n = -\frac{2}{n^2\pi^2}\int_0^1 \phi_1''(x)\sin n\pi x\,dx.$$

Write

$$\alpha_n = -\frac{2}{\pi^3}\int_0^1 \phi_0'''(x)\cos n\pi x\,dx, \qquad \beta_n = -\frac{2}{\pi^3}\int_0^1 \phi_1''(x)\sin n\pi x\,dx.$$

Then

$$a_n = \frac{\alpha_n}{n^3}, \qquad b_n = \frac{\beta_n \pi}{n^2}.$$

The quantities α_n and β_n are the Fourier coefficients of the functions $-(\sqrt{2}/\pi^3)\phi_0'''(x)$ and $-(\sqrt{2}/\pi^3)\phi_1''(x)$, belonging to the space $L_2(0,1)$. Hence it follows that the series $\sum_{n=1}^\infty \alpha_n^2$, $\sum_{n=1}^\infty \beta_n^2$ converge. Formula (4) takes the form

$$u(x,t) = \sum_{n=1}^\infty \frac{1}{n^3}(\alpha_n \cos n\pi t + \beta_n \sin n\pi t)\sin n\pi x. \tag{7}$$

Majorant series for (7) and its first and second derivatives are respectively the series

$$\sum_{n=1}^{\infty} \frac{|\alpha_n|+|\beta_n|}{n^3}, \quad C'\sum_{n=1}^{\infty} \frac{|\alpha_n|+|\beta_n|}{n^2}, \quad C''\sum_{n=1}^{\infty} \frac{|\alpha_n|+|\beta_n|}{n},$$

$$C', C'' = \text{const.},$$

all of which converge. Hence it follows that the sum of the series (4) – the function $u(x, t)$ – is continuous together with its first and second derivatives. This is the required result.

Chapter 23. Cauchy's Problem for the Heat-Conduction Equation

§ 1. Some Properties of the Fourier Transform

In this section we shall outline the simplest properties of the multi-dimensional Fourier transform. It is assumed that the concept of the one-dimensional Fourier transform, and its basic properties, are familiar to the reader.

We shall consider the function $u \in L(E_m)$, such that

$$\int_{E_m} |u(x)|\,dx < \infty.$$

The Fourier transform of this function is defined by the formula

$$(Fu)(x) = \tilde{u}(x) = (2\pi)^{-\frac{1}{2}m} \int_{E_m} e^{-i(x,y)} u(y)\,dy. \qquad (1)$$

Here (x, y) denotes a scalar product in the space E_m:

$$(x, y) = x_k y_k.$$

Because of the absolute and uniform convergence of the integral (1), the function $\tilde{u}$ is defined and continuous at every point $x \in E_m$.

A multiple Fourier transform can be obtained by consecutive applications of the one-dimensional transform; if we construct consecutively the functions

$$u_1(x_1, x_2, \ldots, x_m) = (2\pi)^{-\frac{1}{2}} \int_{-\infty}^{+\infty} u(x_1, x_2, \ldots, x_{m-1}, y_m) \exp(-ix_m y_m)\,dy_m,$$

$$u_2(x_1, x_2, \ldots, x_m) =$$

$$= (2\pi)^{-\frac{1}{2}} \int_{-\infty}^{+\infty} u_1(x_1, x_2, \ldots, y_{m-1}, x_m) \exp(-ix_{m-1} y_{m-1})\,dy_{m-1}$$

$$= (2\pi)^{-1} \int_{-\infty}^{+\infty} \int_{-\infty}^{+\infty} u(x_1, x_2, \ldots, y_{m-1}, y_m) \times$$
$$\times \exp\{-i(x_{m-1} y_{m-1} + x_m y_m)\} dy_{m-1} dy_m,$$

. . . .

$$u_m(x_1, x_2, \ldots, x_m) = (2\pi)^{-\frac{1}{2}} \int_{-\infty}^{+\infty} u_{m-1}(y_1, x_2, \ldots, x_m) \exp(-ix_1 y_1) dy_1$$
$$= (2\pi)^{-\frac{1}{2}m} \int_{-\infty}^{+\infty} \ldots \int_{-\infty}^{+\infty} u(y_1, y_2, \ldots, y_m) \times$$
$$\times \exp\{-i(x_1 y_1 + x_2 y_2 + \ldots + x_m y_m)\} dy_1 dy_2 \ldots dy_m, \quad (2)$$

then, obviously, $u_m(x_1, x_2, \ldots, x_m) = \tilde{u}(x)$.

THEOREM 23.1.1. *If k is a positive integer, and if the product $(1+|x|^k)u(x)$ is integrable in E_m, then the Fourier transform of the function $u(x)$ has continuous derivatives of order not exceeding k everywhere in E_m.*

Since $|u(x)| \leq (1+|x|^k)|u(x)|$, the function $u \in L(E_m)$, and the transform $\tilde{u}(x)$ exists and is continuous in E_m.

We formally differentiate the integral (1) l_1 times with respect to $x_1, \ldots, l_m$ times with respect to x_m; and let

$$l_1 + l_2 + \ldots + l_m = l \leq k.$$

This leads us to the integral

$$(-i)^l (2\pi)^{-\frac{1}{2}m} \int_{E_m} y_1^{l_1} y_2^{l_2} \ldots y_m^{l_m} u(y) e^{-i(x,y)} dy,$$

which converges absolutely and uniformly, since

$$|y_1^{l_1} y_2^{l_2} \ldots y_m^{l_m}| \leq |y|^l,$$

and, consequently,

$$|e^{-i(x,y)} y_1^{l_1} y_2^{l_2} \ldots y_m^{l_m} u(y)| \leq |y|^l |u(y)| \leq (1+|y|^k)|u(y)|.$$

The majorant does not depend on x and is integrable, so that differentiation under the integral sign is permissible:

$$\frac{\partial^l \tilde{u}}{\partial x_1^{l_1} \partial x_2^{l_2} \ldots \partial x_m^{l_m}} = (-i)^l (2\pi)^{-\frac{1}{2}m} \int_{E_m} y_1^{l_1} y_2^{l_2} \ldots y_m^{l_m} e^{-i(x,y)} u(y) dy. \quad (3)$$

Thus, derivatives of order not exceeding k are continuous.

THEOREM 23.1.2. *Let the function $u(x)$ be k-times continuously differentiable at any point $x \in E_m$. Further, let the function itself and all its derivatives of order not exceeding k be integrable in E_m, and tend to zero at infinity. Then, for sufficiently large values of $|x|$,*

$$\tilde{u}(x) = (Fu)(x) = O(|x|^{-k}).$$

Consider some point $x \in E_m$, and denote by j the number of its coordinate having the largest absolute value. Integrate (1) by parts k times with respect to the variable y_j. The integrated terms vanish, and we obtain

$$\tilde{u}(x) = (2\pi)^{-\frac{1}{2}m}(ix_j)^{-k} \int_{E_m} \frac{\partial^k u}{\partial y_j^k} e^{-i(x,y)} dy.$$

We estimate the modulus of $\tilde{u}(x)$:

$$|\tilde{u}(x)| \leq (2\pi)^{-\frac{1}{2}m} |x_j|^{-k} \int_{E_m} \left| \frac{\partial^k u}{\partial y_j^k} \right| dy.$$

The coordinate x_j is the largest in modulus, and so

$$|x| = \sqrt{\sum_{k=1}^{m} x_k^2} \leq \sqrt{m}\,|x_j|, \quad \frac{1}{|x_j|} \leq \frac{\sqrt{m}}{|x|},$$

and, finally

$$|\tilde{u}(x)| \leq \frac{c}{|x|^k},$$

where we can put, for example,

$$c = (2\pi)^{-\frac{1}{2}m} m^{\frac{1}{2}k} \int_{E_m} \sum \left| \frac{\partial^k u(y)}{\partial y_1^{k_1} \partial y_2^{k_2} \ldots \partial y_m^{k_m}} \right| dy;$$

the summation extends over all sets of non-negative indices $k_1, k_2, \ldots, k_m$ such that $k_1 + k_2 + \ldots + k_m = k$.

When the function $u(x)$ satisfies certain conditions we have the *inversion formula for the Fourier transform*:

$$u(x) = (2\pi)^{-\frac{1}{2}m} \int_{E_m} \tilde{u}(y)\, e^{i(x,y)} dy. \tag{4}$$

In general the function $\tilde{u}$ is not integrable in E_m, and therefore it is necessary to state, in each case, in what sense formula (4) is to be understood. Obviously, if $\tilde{u}$ is integrable in E_m, then this integral can be understood in the ordinary sense. For example, the following theorem holds.

THEOREM 23.1.3. *Let the function $u(x)$ be different from zero only in some finite domain, and let it have continuous first derivatives over all space. Then the inversion formula* (4) *holds, with the integral in it understood in the following sense*

$$\int_{E_m} \tilde{u}(y)\, e^{i(x,y)}\, dy = \lim_{N_m \to \infty} \int_{-N_m}^{N_m} \left\{ \ldots \lim_{N_2 \to \infty} \int_{-N_2}^{N_2} \left\{ \lim_{N_1 \to \infty} \int_{-N_1}^{N_1} \tilde{u}(y)\, e^{ix_1 y_1}\, dy_1 \right\} \times e^{ix_2 y_2}\, dy_2 \ldots \right\} e^{ix_m y_m}\, dy_m. \quad (5)$$

Proof. The finite domain in which the function $u(x)$ is different from zero can be placed inside some cube. Let this be the cube

$$-a \leqq x_k \leqq a, \quad k = 1, 2, \ldots, m.$$

Obviously, the function $u(x)$ is integrable with respect to x_m for fixed values of the remaining arguments, and has a derivative with respect to x_m at every point of space; applying the inversion theorem for the one-dimensional Fourier integral to the function u_1 (formula (2)), we obtain

$$u(x) = \lim_{N_m \to \infty} (2\pi)^{-\frac{1}{2}} \int_{-N_m}^{N_m} u_1(x_1, x_2, \ldots, x_{m-1}, y_m)\, e^{ix_m y_m}\, dy_m.$$

We have, moreover,

$$\int_{-\infty}^{+\infty} |u_1(x_1, x_2, \ldots, x_{m-1}, x_m)|\, dx_{m-1}$$

$$= \int_{-\infty}^{+\infty} (2\pi)^{-\frac{1}{2}} \left| \int_{-\infty}^{+\infty} u(x_1, x_2, \ldots, x_{m-1}, y_m)\, e^{-ix_m y_m}\, dy_m \right| dx_{m-1}$$

$$\leqq (2\pi)^{-\frac{1}{2}} \int_{-\infty}^{+\infty} \int_{-\infty}^{+\infty} |u(x_1, x_2, \ldots, x_{m-1}, y_m)|\, dx_{m-1}\, dy_m$$

$$= (2\pi)^{-\frac{1}{2}} \int_{-a}^{a} \int_{-a}^{a} |u(x_1, x_2, \ldots, x_{m-1}, y_m)|\, dx_{m-1}\, dy_m \leqq \frac{4Ma^2}{\sqrt{2\pi}},$$

where M denotes the upper bound of the values of u. Thus, the function $u_1(x_1, x_2, \ldots, x_{m-1}, x_m)$ is integrable with respect to x_{m-1} for all values of the remaining arguments.

We next show that the derivative $\partial u_1/\partial x_{m-1}$ exists at any point. We represent u_1 in the form

$$u_1(x_1, x_2, \ldots, x_{m-1}, x_m) = (2\pi)^{-\frac{1}{2}} \int_{-a}^{a} u(x_1, x_2, \ldots, x_{m-1}, y_m) e^{-ix_m y_m} dy_m.$$

The right-hand side is the integral of a continuously differentiable function over a finite interval. This integral has continuous derivatives with respect to all arguments on which it depends, and in particular with respect to x_{m-1}.

The same inversion theorem for the one-dimensional Fourier integral now gives

$$u_1(x_1, x_2, \ldots, x_{m-1}, x_m) =$$
$$= (2\pi)^{-\frac{1}{2}} \lim_{N_{m-1} \to \infty} \int_{-N_{m-1}}^{N_{m-1}} u_2(x_1, x_2, \ldots, y_{m-1}, x_m) \exp(ix_{m-1} y_{m-1}) dy_{m-1}$$

and, consequently,

$$u(x) = (2\pi)^{-1} \lim_{N_m \to \infty} \int_{-N_m}^{N_m} \left\{ \lim_{N_{m-1} \to \infty} \int_{-N_{m-1}}^{N_{m-1}} u_2(x_1, x_2, \ldots, y_{m-1}, y_m) \times \right.$$
$$\left. \times \exp(ix_{m-1} y_{m-1}) dy_{m-1} \right\} \exp(-ix_m y_m) dy_m.$$

Continuing this procedure, we eventually arrive at formula (5). *

§ 2. Derivation of Poisson's Formula

Consider the homogeneous heat-conduction equation

$$\frac{\partial u}{\partial t} - \Delta u = 0 \tag{1}$$

with the boundary condition

$$u|_{t=0} = \phi(x). \tag{2}$$

We shall assume that all the steps taken below are valid, and with this assumption we shall derive a formula for the solution of the problem (1)–(2).

---

* A more general theorem is proved in ref. [9]; in the same ref. can be found other theorems concerning the validity of the inversion formula (4) under various conditions.

Take the Fourier transform with respect to x of both sides of equation (1)

$$(2\pi)^{-\frac{1}{2}m}\int_{E_m}\frac{\partial u(y,t)}{\partial t}e^{-i(x,y)}dy-(2\pi)^{-\frac{1}{2}m}\int_{E_m}\sum_{k=1}^{m}\frac{\partial^2 u(y,t)}{\partial y_k^2}e^{-i(x,y)}dy=0. \tag{3}$$

Integration with respect to y and differentiation with respect to t are independent, so that in the first term we can take the differentiation with respect to t outside the integral sign. As a result, we obtain

$$(2\pi)^{-\frac{1}{2}m}\int_{E_m}\frac{\partial u(y,t)}{\partial t}e^{-i(x,y)}dy=(2\pi)^{-\frac{1}{2}m}\frac{\partial}{\partial t}\int_{E_m}u(y,t)e^{-i(x,y)}dy=\frac{\partial \tilde{u}(x,t)}{\partial t},$$

where $\tilde{u}(x,t)$ denotes the Fourier transform of the function $u(x,t)$:

$$\tilde{u}(x,t)=(2\pi)^{-\frac{1}{2}m}\int_{E_m}u(y,t)\,e^{-i(x,y)}dy.$$

In equation (3) we now integrate by parts each of the integrals in the second term

$$(2\pi)^{-\frac{1}{2}m}\int_{E_m}\frac{\partial^2 u(y,t)}{\partial y_k^2}e^{-i(x,y)}dy = -(2\pi)^{-\frac{1}{2}m}x_k^2\int_{E_m}u(y,t)\,e^{-i(x,y)}dy$$

$$= -x_k^2\tilde{u}(x,t).$$

Equation (3) takes the form

$$\frac{\partial \tilde{u}}{\partial t}+|x^2|\tilde{u}=0. \tag{4}$$

This is an ordinary, first-order differential equation with independent variable t; the coordinates $x_1, x_2, \ldots, x_m$ play the role of parameters.

Integrating equation (4), we obtain

$$\tilde{u}(x,t)=C(x)\exp(-|x|^2 t).$$

Putting $t=0$ here, we find

$$C(x)=\tilde{u}(x,0).$$

Thus, the function $C(x)$ is the Fourier transform of the initial value of the function $u(x,t)$. Because of condition (2), $u(x,0)=\phi(x)$, and consequently,

$$C(x)=\tilde{\phi}(x)=(2\pi)^{-\frac{1}{2}m}\int_{E_m}\phi(y)\,e^{-i(x,y)}dy.$$

§ 2] DERIVATION OF POISSON'S FORMULA

and
$$\tilde{u}(x, t) = \tilde{\phi}(x) \exp(-|x|^2 t).$$

We now use the inversion formula (1.4):
$$u(x, t) = (2\pi)^{-\frac{1}{2}m} \int_{E_m} \tilde{\phi}(y) \exp\{-|y|^2 t + i(x, y)\} dy.$$

Substituting here for $\tilde{\phi}(y)$ the expression obtained above, and changing the order of integration, we have
$$u(x, t) = (2\pi)^{-m} \int_{E_m} \phi(z) \left\{ \int_{E_m} \exp\{-|y|^2 t + i(x-z, y)\} dy \right\} dz. \quad (5)$$

We evaluate the inner integral in (5). We have
$$\int_{E_m} \exp\{-|y|^2 t + i(x-z, y)\} dy$$
$$= \int_{-\infty}^{+\infty} \cdots \int_{-\infty}^{+\infty} \exp\left\{ \sum_{k=1}^{m} [-y_k^2 t + i(x_k - z_k) y_k] \right\} dy_1 dy_2 \ldots dy_m$$
$$= \prod_{k=1}^{m} \int_{-\infty}^{+\infty} \exp\{-y^2 t + i(x_k - z_k) y\} dy. \quad (6)$$

In the integral on the right, y is a real variable, which ranges between the limits $-\infty < y < \infty$.

Consider the k-th term in the product (6), and for brevity put $x_k - z_k = \alpha$. Then the problem reduces to the evaluation of the integral
$$I(\alpha) = \int_{-\infty}^{+\infty} \exp(-y^2 t + i\alpha y) dy = \exp\left(-\frac{\alpha^2}{4t}\right) \int_{-\infty}^{+\infty} \exp\left\{-t\left(y - \frac{i\alpha}{2t}\right)^2\right\} dy.$$

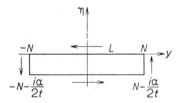

Fig. 47.

Consider the complex plane of the variable $\zeta = y + i\eta$. Suppose for definiteness that $\alpha > 0$. Construct a rectangle with perimeter L, as shown in

Fig. 47. By Cauchy's theorem,

$$\int_L \exp(-t\zeta^2)\,d\zeta = 0$$

or, when written out in full,

$$\int_{-N}^{+N} \exp\left\{-t\left(y-\frac{i\alpha}{2t}\right)^2\right\}dy + \int_{-\alpha/2t}^{0} \exp\{-t(N+i\eta)^2\}i\,d\eta$$

$$-\int_{-N}^{N} \exp(-ty^2)\,dy - \int_{-\alpha/2t}^{0} \exp\{-t(-N+i\eta)^2\}i\,d\eta = 0.$$

Now let $N \to \infty$. Then the second and fourth integrals tend to zero, since

$$\left|\int_{-\alpha/2t}^{0} \exp\{-t(N\pm i\eta)^2\}i\,d\eta\right| \leq \int_{-\alpha/2t}^{0} \exp\{-t(N^2-\eta^2)\}\,d\eta$$

$$= \exp(-tN^2)\int_{-\alpha/2t}^{0} \exp(t\eta^2)\,d\eta \xrightarrow[N\to\infty]{} 0.$$

Hence it follows that

$$\int_{-\infty}^{+\infty} \exp\left\{-t\left(y-\frac{i\alpha}{2t}\right)^2\right\}dy = \int_{-\infty}^{+\infty} \exp(-ty^2)\,dy.$$

It is easy to see that the case $\alpha < 0$ reduces to the same result. The substitution $y\sqrt{t} = s$ gives, moreover,

$$\int_{-\infty}^{+\infty} e^{-ty^2}\,dy = \frac{1}{\sqrt{t}}\int_{-\infty}^{+\infty} e^{-s^2}\,ds = \sqrt{\frac{\pi}{t}}.$$

Now

$$\exp\left(-\frac{\alpha^2}{4t}\right)\int_{-\infty}^{+\infty} \exp\left\{-t\left(y-\frac{i\alpha}{2t}\right)^2\right\}dy = \sqrt{\frac{\pi}{t}}\exp\left\{-\frac{(x_k-z_k)^2}{4t}\right\},$$

and the integral (6) turns out to be equal to the quantity

$$\left(\frac{\pi}{t}\right)^{\frac{1}{2}m} \exp\left\{-\frac{1}{4t}\sum_{k=1}^{m}(x_k-z_k)^2\right\} = \left(\frac{\pi}{t}\right)^{\frac{1}{2}m}\exp\left(-\frac{r^2}{4t}\right), \qquad r = |x-z|.$$

Substituting this result in formula (5), we obtain the so-called *Poisson's formula*

$$u(x,t) = \frac{1}{(2\sqrt{\pi t})^m}\int_{E_m} \phi(z)\exp\left(-\frac{r^2}{4t}\right)dz \qquad (7)$$

§ 3. Justification of Poisson's Formula

We shall not attempt to prove the validity of the steps taken in the previous section. Instead, we shall establish directly that Poisson's formula gives a bounded solution to Cauchy's problem for the heat-conduction equation (2.1), under the single assumption that the initial function $\phi(x)$ is continuous and bounded in the space E_m.

First of all we show that Poisson's formula defines a function which is continuous for $t > 0$.

In the space of the $m+1$ variables $x_1, x_2, \ldots, x_m, t$, consider the domain defined by the inequalities

$$|x^2| \leq a^2, \quad 0 \leq t \leq T, \tag{1}$$

where a and T are positive constants. We show that the integral

$$\int_{E_m} \phi(z) \exp\left(-\frac{r^2}{4t}\right) dz, \tag{2}$$

which occurs in Poisson's formula, converges uniformly with respect to x and t in the domain (1). Take a sufficiently large number R, and estimate the integral

$$\int_{|z|>R} \phi(z) \exp\left(-\frac{r^2}{4t}\right) dz.$$

The function $\phi(z)$ is bounded; let $|\phi(z)| \leq M = \text{const}$. Further, $r = |x-z| \geq |z|-|x| \geq |z|-a$. We shall suppose that $R > 2a$. Then $a < \frac{1}{2}R \leq \frac{1}{2}|z|$ and $r > \frac{1}{2}|z|$. Now

$$\exp\left(-\frac{r^2}{4t}\right) < \exp\left(-\frac{|z|^2}{16T}\right)$$

and, consequently,

$$\left|\int_{|z|>R} \phi(z) \exp\left(-\frac{r^2}{4t}\right) dz\right| < M \int_{|z|>R} \exp\left(-\frac{|z|^2}{16T}\right) dz$$

$$= M|S_1| \int_R^\infty \exp\left(-\frac{\rho^2}{16T}\right) \rho^{m-1} d\rho. \tag{3}$$

The integral

$$\int_0^\infty \rho^{m-1} \exp\left(-\frac{\rho^2}{16T}\right) d\rho$$

converges, so that for R sufficiently large the integral on the right-hand side in (3) is as small as we please; since it depends neither on x nor on t, the integral (2) converges uniformly. Hence it follows that the function defined by Poisson's formula is continuous for $t > 0$.

We next show that for $t > 0$ the function $u(x, t)$ is infinitely continuously differentiable with respect to t and with respect to the coordinates x, and that all the derivatives can be obtained by differentiating Poisson's formula under the integral sign.

Consider, for example, the derivative $\partial u/\partial t$. If we differentiate formally with respect to t the right-hand side of Poisson's formula, we obtain the expression

$$-\frac{m}{2^{m+1}\pi^{\frac{1}{2}m}t^{\frac{1}{2}(m+2)}}\int_{E_m} \phi(z) \exp\left(-\frac{r^2}{4t}\right) dz +$$
$$+ \frac{1}{2^{m+2}\pi^{\frac{1}{2}m}t^{\frac{1}{2}(m+4)}}\int_{E_m} r^2\phi(z) \exp\left(-\frac{r^2}{4t}\right) dz. \quad (4)$$

As we have seen, the first integral converges uniformly in the domain (1). In exactly the same way it can be verified that the second integral also converges uniformly in this domain. Hence, as usual, it follows that the derivative exists, is continuous, and is identical with expression (4). The existence of other derivatives is established analogously.

Direct differentiation shows that the function defined by Poisson's formula satisfies the heat-conduction equation (2.1).

It remains to prove that the function $u(x, t)$ is bounded and satisfies the initial condition (2.2):

$$\lim_{t\to 0} u(x, t) = \lim_{t\to 0} \frac{1}{(2\sqrt{\pi t})^m}\int_{E_m} \phi(z) \exp\left(-\frac{r^2}{4t}\right) dz = \phi(x).$$

Make the substitution $z = x + 2\sqrt{t}\xi$; then Poisson's formula takes the form

$$u(x, t) = \pi^{-\frac{1}{2}m}\int_{E_m} \phi(x+2\sqrt{t}\xi) e^{-|\xi|^2} d\xi. \quad (5)$$

Using the well-known result

$$\int_{-\infty}^{+\infty} e^{-\rho^2} d\rho = \sqrt{\pi}$$

we immediately obtain

$$\pi^{-\frac{1}{2}m}\int_{E_m} e^{-|\xi|^2}\,d\xi = 1. \tag{6}$$

It now follows from formula (5) that

$$|u(x,t)| \le M\pi^{-\frac{1}{2}m}\int_{E_m} e^{-|\xi|^2}\,d\xi = M,$$

and the function $u(x,t)$ is bounded. Further, from (5) and (6),

$$u(x,t)-\phi(x) = \pi^{-\frac{1}{2}m}\int_{E_m} [\phi(x+2\sqrt{t}\xi)-\phi(x)]\,e^{-|\xi|^2}\,d\xi. \tag{7}$$

The integral in (7) can be divided into two integrals, taken respectively over the domains $|\xi| > R$ and $|\xi| < R$, where R is some constant. We have

$$\left|\pi^{-\frac{1}{2}m}\int_{|\xi|>R} [\phi(x+2\sqrt{t}\xi)-\phi(x)]\,e^{-|\xi|^2}\,d\xi\right| \le 2M\pi^{-\frac{1}{2}m}\int_{|\xi|>R} e^{-|\xi|^2}\,d\xi$$

$$= 2M\pi^{-\frac{1}{2}m}|S_1|\int_R^\infty \rho^{m-1} e^{-\rho^2}\,d\rho.$$

The integral

$$\int_0^\infty \rho^{m-1} e^{-\rho^2}\,d\rho$$

converges, and it is possible to choose $R_0(\varepsilon)$, such that for $R > R_0(\varepsilon)$, we have

$$2M\pi^{-\frac{1}{2}m}|S_1|\int_R^\infty \rho^{m-1} e^{-\rho^2}\,d\rho < \tfrac{1}{2}\varepsilon.$$

We fix some value of $R > R_0(\varepsilon)$. Then it is possible to find a $t_0(\varepsilon)$ such that for $0 < t < t_0(\varepsilon)$, and for any ξ, $|\xi| \le R$, we obtain

$$|\phi(x+2\sqrt{t}\xi)-\phi(x)| < \tfrac{1}{2}\varepsilon.$$

Now

$$\left|\pi^{-\frac{1}{2}m}\int_{|\xi|<R} [\phi(x+2\sqrt{t}\xi)-\phi(x)]\,e^{-|\xi|^2}\,d\xi\right| < \tfrac{1}{2}\pi^{-\frac{1}{2}m}\varepsilon\int_{E_m} e^{-|\xi|^2}\,d\xi = \tfrac{1}{2}\varepsilon,$$

and finally

$$|u(x,t)-\phi(x)| < \varepsilon, \qquad 0 < t < t_0(\varepsilon).$$

This completes the justification of Poisson's formula.

If we are interested in integrating the inhomogeneous heat-conduction equation

$$\frac{\partial u}{\partial t} - \Delta u = f(x, t) \tag{8}$$

with the Cauchy conditions (2), then we can apply a Fourier transform with respect to the coordinates, and obtain the differential equation

$$\frac{d\tilde{u}}{dt} + x^2 \tilde{u} = \tilde{f}(x, t) \tag{9}$$

and initial condition

$$\tilde{u}(x, 0) = \tilde{\phi}(x). \tag{10}$$

In equation (9)

$$\tilde{f}(x, t) = \frac{1}{(2\pi)^{\frac{1}{2}m}} \int_{E_m} f(y, t) e^{-i(x, y)} dy.$$

Equations (9) and (10) are satisfied by the function

$$\tilde{u}(x, t) = \exp(-|x|^2 t)\tilde{\phi}(x) + \int_0^t \exp(-x^2(t-\tau))\tilde{f}(x, \tau) d\tau.$$

Applying the inverse Fourier transform, we eventually obtain the following expression for the required function:

$$u(x, t) = \frac{1}{(2\sqrt{\pi t})^m} \int_{E_m} \phi(z) \exp\left(-\frac{r^2}{4t}\right) dz +$$

$$+ \int_0^t \int_{E_m} f(z, \tau) \frac{1}{(2\sqrt{\pi(t-\tau)})^m} \exp\left(-\frac{r^2}{4(t-\tau)}\right) dz\, d\tau, \quad r = |z-x|.$$

§ 4. Infinite Velocity of Heat-Conduction

A consequence of Poisson's formula is that heat propagates with infinite velocity. Suppose that the heat-conducting medium occupies the whole of the space E_m. At the initial instant, let the whole medium, except for some finite domain D, have zero temperature ($\phi(x) \equiv 0$), and let the points of the domain D be heated to some temperature $\phi(x) > 0$. At any point $x \in E_m$ and at any moment of time $t > 0$, the temperature of the medium $u(x, t)$ is defined by Poisson's formula

$$u(x, t) = \frac{1}{(2\sqrt{\pi t})^m} \int_D \phi(z) \exp\left(-\frac{r^2}{4t}\right) dz; \tag{1}$$

the integral over $E_m \setminus D$ vanishing, since in that region $\phi(x) = 0$. But it is clear from formula (1) that $u(x, t) > 0$. Thus, however small the value of t, and however far the point x from the domain D, the heat emerging from the domain D succeeds in reaching the point x in the interval t. This implies that heat is propagated with infinite velocity.

This physically contradictory conclusion does not in practice cause any complication. If $|x|$ is large and t small, then the negative exponent $-r^2/4t$ in formula (1) is large in absolute value, and the value of the temperature is negligibly small. In practice, therefore, Poisson's formula gives (to within a negligibly small quantity) a finite velocity of heat propagation.

Chapter 24.
Cauchy's Problem for the Wave Equation

§ 1. Application of the Fourier Transform

We shall look for a solution of the wave equation

$$\frac{\partial^2 u}{\partial t^2} - \Delta u = 0 \tag{1}$$

throughout the whole of the space E_m, and for times $t > 0$. Let the required function satisfy the initial conditions

$$u|_{t=0} = \phi_0(x), \qquad \frac{\partial u}{\partial t}\bigg|_{t=0} = \phi_1(x). \tag{2}$$

We shall assume that all the operations carried out below are permissible. Apply the Fourier transform to both sides of equations (1) and (2). Proceeding just as in the case of the heat-conduction equation, we arrive at the following Cauchy problem for an ordinary, linear, second-order differential equation with constant coefficients

$$\frac{d^2 \tilde{u}}{dt^2} + |x|^2 \tilde{u} = 0, \tag{3}$$

$$\tilde{u}|_{t=0} = \tilde{\phi}_0(x), \qquad \frac{d\tilde{u}}{dt}\bigg|_{t=0} = \tilde{\phi}_1(x); \tag{4}$$

where the symbol $\sim$ denotes a Fourier transform.

The general solution of equation (3) has the form

$$\tilde{u}(x, t) = A(x) \cos |x|t + B(x) \sin |x|t.$$

For $t = 0$ we find

$$\tilde{\phi}_0(x) = A(x), \qquad \tilde{\phi}_1(x) = |x| B(x),$$

and we obtain the solution to the problem (3)–(4):

$$\tilde{u}(x, t) = \tilde{\phi}_0(x) \cos |x|t + \tilde{\phi}_1(x) \frac{\sin |x|t}{|x|}. \tag{5}$$

Using the inverse Fourier transform, we arrive at the following formula for the required function:

$$u(x, t) = (2\pi)^{-\frac{1}{2}m} \int_{E_m} \left[\tilde{\phi}_0(y) \cos |y|t + \tilde{\phi}_1(y) \frac{\sin |y|t}{|y|} \right] e^{i(x, y)} dy. \tag{6}$$

A justification of formula (6) can be given under the following circumstances. We suppose that the function $\phi_0(x)$ has continuous derivatives up to order $m+3$ inclusive throughout all the space E_m, and that the function $\phi_1(x)$ has derivatives up to and including order $m+2$. We further suppose that the initial functions themselves, and their derivatives of the above-mentioned orders, are different from zero only in some finite domain of the space E_m.

It follows from Theorem 23.1.2 that for sufficiently large $|x|$ we have the estimates

$$\tilde{\phi}_0(x) = O(|x|^{-m-3}), \qquad \tilde{\phi}_1(x) = O(|x|^{-m-2}).$$

In this case the integrand in (6) has order of magnitude $O(|y|^{-m-3})$ at infinity, while the first and second derivatives with respect to $x_1, x_2, \ldots, x_m, t$ of the integrand function have orders of magnitude $O(|y|^{-m-2})$ and $O(|y|^{-m-1})$ respectively. In all cases these order estimates are uniform with respect to x and t. Hence it follows that both the integral (6) and the integrals obtained by differentiating it either once or twice, converge uniformly with respect to x and t. But in this case the function (6) is continuous, and twice continuously differentiable with respect to spatial coordinates and time, and these derivatives can be obtained by differentiating under the integral sign.

It is now easy to show that the function (6) satisfies the initial conditions (2) and the wave equation (1). Putting $t = 0$ in (6), we find

$$u(x, 0) = (2\pi)^{-\frac{1}{2}m} \int_{E_m} \tilde{\phi}_0(y) e^{i(x, y)} dy = \phi_0(x);$$

it is not difficult to see that the conditions of Theorem 23.1.3 are fulfilled in this case. Differentiating (6) with respect to t, and putting $t = 0$, we also

find
$$\left.\frac{\partial u(x,t)}{\partial t}\right|_{t=0} = (2\pi)^{-\frac{1}{2}m}\int_{E_m}\tilde{\phi}_1(y)e^{i(x,y)}dy = \phi_1(x).$$

Moreover
$$\frac{\partial^2 u}{\partial t^2} = -(2\pi)^{-\frac{1}{2}m}\int_{E_m}|y|^2\left[\tilde{\phi}_0(y)\cos|y|t+\tilde{\phi}_1(y)\frac{\sin|y|t}{|y|}\right]e^{i(x,y)}dy,$$

$$\Delta u = -(2\pi)^{-\frac{1}{2}m}\int_{E_m}|y|^2\left[\tilde{\phi}_0(y)\cos|y|t+\tilde{\phi}_1(y)\frac{\sin|y|t}{|y|}\right]e^{i(x,y)}dy = \frac{\partial^2 u}{\partial t^2};$$

the function (6) therefore satisfies the wave equation.

The same method – reduction to an ordinary differential equation through the use of a Fourier transform – can be applied to the inhomogeneous wave equation
$$\frac{\partial^2 u}{\partial t^2} - \Delta u = f(x,t). \tag{7}$$

Let Cauchy's problem with initial conditions (2) be posed for this equation. Taking the Fourier transform with respect to the coordinates, and putting
$$\tilde{f}(x,t) = \frac{1}{(2\pi)^{\frac{1}{2}m}}\int_{E_m} f(y,t)e^{-i(x,y)}dy,$$

we obtain
$$\frac{d^2\tilde{u}}{dt^2} + |x|^2\tilde{u} = \tilde{f}(x,t). \tag{8}$$

The solution of this equation which satisfies conditions (4) is the function
$$\tilde{u}(x,t) = \tilde{\phi}_0(x)\cos|x|t + \tilde{\phi}_1(x)\frac{\sin|x|t}{|x|} + \int_0^t \tilde{f}(x,\tau)\sin\frac{|x|(t-\tau)}{|x|}d\tau. \tag{9}$$

If the functions $\phi_0(x)$, $\phi_1(x)$ and $f(x,t)$ have a sufficient number of derivatives, and are different from zero only in a finite domain of variation of the variables *, then application of the inverse Fourier transform to equation (9) gives the solution to Cauchy's problem for the inhomogeneous wave equation.

* This requirement can be substantially relaxed.

§ 2. Transformation of the Solution

It is possible to improve the formula (1.6), which gives the solution to Cauchy's problem for the wave equation. First of all, if we replace the functions $\tilde{\phi}_0(y)$ and $\tilde{\phi}_1(y)$ by their Fourier-integral representations through ϕ_0 and ϕ_1, then we obtain integrals of multiplicity $2m$; but in fact it is possible to treat integrals of much smaller multiplicity. Further, in the justification of formula (1.6), we required the existence of derivatives of the initial functions of unnecessarily high order; the requirement that these functions should be different from zero only in some finite domain is also unnecessary. Finally, it is not immediately clear from the form of (1.6) that the value of $u(x, t)$ is defined solely through the values of the initial functions in the domain of dependence (§ 4, Chapter 21).

We now construct a transformation of formula (1.6) which renders it more accessible to analysis. In the next section we shall derive the results for the case $m = 3$; here we shall carry out some preliminary transformations in the general case.

Introduce the notation

$$T_\omega(x, t) = (2\pi)^{-\frac{1}{2}m} \int_{E_m} \tilde{\omega}(y) \frac{\sin |y|t}{|y|} e^{i(x, y)} dy. \tag{1}$$

If differentiation under the integral sign is permissible, then

$$\frac{\partial}{\partial t} T_\omega(x, t) = (2\pi)^{-\frac{1}{2}m} \int_{E_m} \tilde{\omega}(y) e^{i(x, y)} \cos |y| t \, dy,$$

and formula (1.6) can be written in the more convenient form

$$u(x, t) = \frac{\partial}{\partial t} T_{\phi_0}(x, t) + T_{\phi_1}(x, t). \tag{2}$$

Our aim is to give the expression $T_\omega(x, t)$ in as simple a form as possible. Substituting for $\tilde{\omega}(y)$ the expression

$$\tilde{\omega}(y) = (2\pi)^{-\frac{1}{2}m} \int_{E_m} \omega(z) e^{-i(y, z)} dz,$$

we obtain

$$T_\omega(x, t) = (2\pi)^{-\frac{1}{2}m} \int_{E_m} \frac{\sin |y|t}{|y|} e^{i(x, y)} \left\{ \int_{E_m} \omega(z) e^{-i(y, z)} dz \right\} dy.$$

It is not possible here to change the order of integration – this would lead to a divergent integral. To overcome this difficulty, we introduce a new quantity

$$T_\omega(x, t, \lambda) = (2\pi)^{-\frac{1}{2}m} \int_{E_m} \tilde{\omega}(y) e^{-\lambda|y|} \frac{\sin |y|t}{|y|} e^{i(x, y)} dy, \qquad \lambda > 0. \qquad (3)$$

Let the function $\omega(x)$ satisfy the same conditions as does $\phi_1(x)$ (cf. § 1). It follows from these conditions and from Theorem 23.1.2 that $\tilde{\omega} \in L(E_m)$. Let t vary over the segment $[0, \bar{t}]$, $\bar{t} = $ const. > 0. From the inequality $|\sin \alpha| \leq |\alpha|$ it follows that the integrand function in (3) has an integrable majorant $\bar{t}|\tilde{\omega}(y)|$, which is independent of x, t, λ. Hence it follows that

$$T_\omega(x, t, \lambda) \underset{\lambda \to 0}{\to} T_\omega(x, t) \qquad (4)$$

uniformly with respect to x and t, when x varies through E_m, and t through any finite interval.

In the integral

$$T_\omega(x, t, \lambda) = (2\pi)^{-m} \int_{E_m} e^{-\lambda|y| + i(x, y)} \frac{\sin |y|t}{|y|} \left\{ \int_{E_m} \omega(z) e^{-i(z, y)} dz \right\} dy \qquad (5)$$

it is possible to change the order of integration:

$$T_\omega(x, t, \lambda) = (2\pi)^{-m} \int_{E_m} \omega(z) \left\{ \int_{E_m} e^{-\lambda|y| + i(x-z, y)} \frac{\sin |y|t}{|y|} dy \right\} dz.$$

The inner integral converges; we evaluate it.

Put $x - z = p$, and

$$\Phi(p, t, \lambda) = \int_{E_m} e^{-\lambda|y| + i(p, y)} \frac{\sin |y|t}{|y|} dy. \qquad (6)$$

Then

$$\frac{\partial \Phi(p, t, \lambda)}{\partial t} = \int_{E_m} e^{-\lambda|y| + i(p, y)} \cos |y|t \, dy, \qquad (7)$$

where from formula (6) we see that $\Phi|_{t=0} = 0$. Note that

$$\frac{\partial \Phi}{\partial t} = \tfrac{1}{2}\Phi_1(p, t, \lambda) + \tfrac{1}{2}\Phi_1(p, -t, \lambda), \qquad (8)$$

where

$$\Phi_1(p, t, \lambda) = \int_{E_m} \exp\{-(\lambda - it)|y| + i(p, y)\} dy. \qquad (9)$$

Introduce spherical coordinates with centre at the origin. Denote these coordinates by $\rho, v_1, \ldots, v_{m-2}, v_{m-1}$ so that $|y| = \rho$. Here $dy = \rho^{m-1} d\rho \, dS_1$. Denote also by r the distance between the points x and z: $r = |x-z| = |p|$ and by γ the angle between the vectors p and y. Then

$$(p, y) = r\rho \cos \gamma.$$

Now

$$\Phi_1(p, t, \lambda) = \int_{S_1} \left\{ \int_0^\infty \exp\{-(\lambda - it - ir \cos \gamma)\rho\}\rho^{m-1} d\rho \right\} dS_1$$

$$= (m-1)! \int_{S_1} \frac{dS_1}{(\lambda - it - ir \cos \gamma)^m}. \tag{10}$$

We choose cartesian coordinates such that the axis Oy_1 is directed along the vector p; the other axes can be chosen arbitrarily, so long as the coordinate system remains rectangular. With this choice of coordinates $v_1 = \gamma$; the general formula

$$dS_1 = \sin^{m-2} v_1 \sin^{m-3} v_2 \ldots \sin v_{m-2} dv_1 dv_2 \ldots dv_{m-1}$$

can be written in the form

$$dS_1 = \sin^{m-2} \gamma \, d\gamma \, d\sigma_1,$$

where

$$d\sigma_1 = \sin^{m-3} v_2 \ldots \sin v_{m-2} dv_2 \ldots dv_{m-1}$$

is an element of surface area on the unit sphere in $(m-1)$-dimensional space. Substituting this into formula (10), we obtain

$$\Phi_1(p, t, \lambda) = (m-1)!|\sigma_1| \int_0^\pi \frac{\sin^{m-2} \gamma \, d\gamma}{(\lambda - it - ir \cos \gamma)^m},$$

or, since the surface area of the unit sphere in $(m-1)$-dimensional space is equal to

$$|\sigma_1| = \frac{2\pi^{\frac{1}{2}(m-1)}}{\Gamma(\frac{1}{2}(m-1))},$$

we have

$$\Phi_1(p, t, \lambda) = \frac{2(m-1)!\pi^{\frac{1}{2}(m-1)}}{\Gamma(\frac{1}{2}(m-1))} \int_0^\pi \frac{\sin^{m-2} \gamma \, d\gamma}{(\lambda - it - ir \cos \gamma)^m}. \tag{11}$$

This integral is elementary and can be easily evaluated. In the general case, however, the result turns out to be somewhat cumbersome, so that we henceforth restrict ourselves to the case $m = 3$.

§ 3. Case of Three-Dimensional Space

If $m = 3$, then $\Gamma(\frac{1}{2}(m-1)) = \Gamma(1) = 1$, and formula (2.11) gives

$$\Phi_1(p, t, \lambda) = 4\pi \int_0^\pi \frac{\sin \gamma \, d\gamma}{(\lambda - it - ir \cos \gamma)^3} = \frac{2\pi}{ir} \left[\frac{1}{(\lambda - it - ir)^2} - \frac{1}{(\lambda - it + ir)^2} \right].$$

By formula (2.8)

$$\frac{\partial \Phi}{\partial t} = \frac{\pi}{ir} \left[\frac{1}{(\lambda - it - ir)^2} - \frac{1}{(\lambda - it + ir)^2} + \frac{1}{(\lambda + it - ir)^2} - \frac{1}{(\lambda + it + ir)^2} \right],$$

and since $\Phi|_{t=0} = 0$, we obtain

$$\Phi(p, t, \lambda) = \frac{8\pi t \lambda}{[\lambda^2 + (t-r)^2][\lambda^2 + (t+r)^2]}.$$

It follows from (2.5) that

$$T_\omega(x, t, \lambda) = \frac{t}{\pi^2} \int_{E_3} \frac{\lambda \omega(z) \, dz}{[\lambda^2 + (t-r)^2][\lambda^2 + (t+r)^2]}$$

and finally

$$T_\omega(x, t) = \lim_{\lambda \to 0} \frac{t}{\pi^2} \int_{E_3} \frac{\lambda \omega(z) \, dz}{[\lambda^2 + (t-r)^2][\lambda^2 + (t+r)^2]}. \tag{1}$$

Note that if $r \neq t$ then the integrand function tends to zero as $\lambda \to 0$.

Introduce spherical coordinates with centre at the point x; let one of the coordinates be r; we have $dz = r^2 dr \, dS_1$ and

$$T_\omega(x, t) = \lim_{\lambda \to 0} \frac{t}{\pi^2} \int_{S_1} \left\{ \int_0^\infty \frac{\lambda \omega(z) r^2 \, dr}{[\lambda^2 + (t-r)^2][\lambda^2 + (t+r)^2]} \right\} dS_1. \tag{2}$$

We shall now take the limit in integral (2), although it is not strictly justified. However, this does not matter, since the resultant formulae (the so-called Kirchhoff formula) will be established rigorously later.

In the inner integral we divide the interval of integration into three parts

$$(0, \infty) = (0, t-\delta) \cup [t-\delta, t+\delta] \cup (t+\delta, \infty),$$

where δ is a constant such that $0 < \delta < t$.

CASE OF THREE-DIMENSIONAL SPACE

Formula (2) takes the form

$$T_\omega(x,t) = \lim_{\lambda \to 0} \frac{t}{\pi^2} \left\{ \int_{S_1} \left[\int_0^{t-\delta} \frac{\lambda \omega(z) r^2 \, dr}{[\lambda^2+(t-r)^2][\lambda^2+(t+r)^2]} \right] dS_1 \right.$$

$$+ \int_{S_1} \left[\int_{t-\delta}^{t+\delta} \frac{\lambda \omega(z) r^2 \, dr}{[\lambda^2+(t-r)^2][\lambda^2+(t+r)^2]} \right] dS_1$$

$$\left. + \int_{S_1} \left[\int_{t+\delta}^\infty \frac{\lambda \omega(z) r^2 \, dr}{[\lambda^2+(t-r)^2][\lambda^2+(t+r)^2]} \right] dS_1 \right\}. \quad (2a)$$

In the intervals $(0, t-\delta)$ and $(t+\delta, \infty)$ the quantity $|r-t| > \delta$, and in these intervals the integrand function tends to zero as $\lambda \to 0$. We suppose, without proof, that the first and third terms in (2a) also tend to zero together with λ. This leads to a simpler expression for $T_\omega(x,t)$:

$$T_\omega(x,t) = \lim_{\lambda \to 0} \frac{t}{\pi^2} \int_{S_1} \left\{ \int_{t-\delta}^{t+\delta} \frac{\lambda \omega(z) r^2 \, dr}{[\lambda^2+(r-t)^2][\lambda^2+(r+t)^2]} \right\} dS_1. \quad (3)$$

Let the spherical coordinates (defined above) of the point z be r, ϑ, ϕ, and write

$$\omega(z) = \omega(x+r\theta),$$

where θ is a point on the surface of the unit sphere with angular coordinates ϑ and ϕ. The integrand function in (3) can be represented as the product of two factors

$$\frac{\lambda}{\lambda^2+(t-r)^2} \frac{\omega(z) r^2}{\lambda^2+(t+r)^2}.$$

If δ is sufficiently small, then the difference between r and t is small in (3); for sufficiently small λ and δ, the second of the above factors is not very different from $\tfrac{1}{4}\omega(x+\theta t)$. We therefore replace the second factor by this latter quantity, so that

$$T_\omega(x,t) = \frac{t}{4\pi^2} \lim_{\lambda \to 0} \int_{S_1} \omega(x+t\theta) \left\{ \int_{t-\delta}^{t+\delta} \frac{\lambda \, dr}{\lambda^2+(t-r)^2} \right\} dS_1. \quad (4)$$

Further,

$$\int_{t-\delta}^{t+\delta} \frac{\lambda \, dr}{\lambda^2+(t-r)^2} = 2\arctan \frac{\delta}{\lambda} \xrightarrow[\lambda \to 0]{} \pi$$

and, consequently,

$$T_\omega(x,t) = \frac{t}{4\pi} \int_{S_1} \omega(x+\theta t) \, dS_1 = \frac{t}{4\pi} \int_0^\pi \int_0^{2\pi} \omega(x+\theta t) \sin \vartheta \, d\vartheta \, d\phi. \quad (5)$$

This integral can be written in a slightly different form. The equation $r = t$ is the equation of the sphere S_t of radius t with centre at the point x; here $dS_t = t^2 dS_1$, and formula (5) takes the form

$$T_\omega(x, t) = \frac{1}{4\pi t} \int_{S_t} \omega(z) \, dS_t. \tag{6}$$

From formula (2.2) we now obtain the solution to the Cauchy problem for the three-dimensional wave equation in the form

$$u(x, t) = \frac{\partial}{\partial t} \frac{1}{4\pi t} \int_{S_t} \phi_0(z) \, dS_t + \frac{1}{4\pi t} \int_{S_t} \phi_1(z) \, dS_t, \tag{7}$$

where, as we said before, S_t is the spherical surface $|z - x| = t$. Formula (7) is called *Kirchhoff's formula*.

§ 4. Justification of Kirchhoff's Formula

We now show that Kirchhoff's formula solves Cauchy's problem for the wave equation, if at any point of the space E_3 the function $\phi_0(z)$ has continuous derivatives up to third order inclusive, and the function $\phi_1(z)$ has continuous derivatives of first and second order; it is not necessary to impose any restrictions on the behaviour of the initial functions and their derivatives at infinity.

Let $\omega(z)$ be a function having continuous first and second derivatives at any point of space. Consider the function

$$u_1(x, t) = T_\omega(x, t) = \frac{1}{4\pi t} \int_{S_t} \omega(z) \, dS_t. \tag{1}$$

If formula (3.5) is used, then the function $u_1(x, t)$ can be expressed in a different form:

$$u_1(x, t) = \frac{t}{4\pi} \int_0^\pi \int_0^{2\pi} \omega(x + \theta t) \sin \vartheta \, d\vartheta \, d\phi. \tag{2}$$

Obviously,

$$u_1(x, 0) = u_1(x, t)|_{t=0} = 0. \tag{3}$$

Construct the derivative $\partial u_1 / \partial t$:

$$\frac{\partial u_1}{\partial t} = \frac{1}{4\pi} \int_0^\pi \int_0^{2\pi} \omega(x + t\theta) \sin \vartheta \, d\vartheta \, d\phi + \frac{t}{4\pi} \int_0^\pi \int_0^{2\pi} \frac{\partial \omega}{\partial x_k} \theta_k \sin \vartheta \, d\vartheta \, d\phi. \tag{4}$$

§ 4]   JUSTIFICATION OF KIRCHHOFF'S FORMULA

As usual, repetition of the subscript k means that summation is performed – on this occasion, between the limits 1 to 3; θ_k denotes the components of the vector θ. Their values are

$$\theta_1 = \cos\vartheta, \qquad \theta_2 = \sin\vartheta\cos\phi, \qquad \theta_3 = \sin\vartheta\sin\phi.$$

Putting $t = 0$ in equation (4), we find

$$\left.\frac{\partial u_1}{\partial t}\right|_{t=0} = \frac{\omega(x)}{4\pi}\int_0^\pi\int_0^{2\pi}\sin\vartheta\,d\vartheta\,d\phi = \omega(x). \tag{5}$$

Thus, the function $u_1(x, t)$ satisfies the initial conditions (3) and (5). We also show that $u_1(x, t)$ satisfies the wave equation.

We transform formula (4) as follows:

$$\frac{\partial u_1(x,t)}{\partial t} = \frac{u_1(x,t)}{t} + \frac{1}{4\pi t}\int_{S_t}\frac{\partial\omega(z)}{\partial z_k}\theta_k\,dS_t, \qquad z = x+t\theta.$$

Note that $\theta_k = \cos(r, x_k) = \cos(\nu, x_k)$, where ν is the outward normal to the sphere S_t. By Gauss's theorem,

$$\frac{\partial u_1(x,t)}{\partial t} = \frac{u_1(x,t)}{t} + \frac{1}{4\pi t}\int_{\Sigma_t}\Delta_z\omega\,dz = \frac{u_1(x,t)}{t} + \frac{I(x,t)}{4\pi t}. \tag{6}$$

Here Σ_t is the sphere of radius t with centre at the point x, and

$$I(x, t) = \int_{\Sigma_t}\Delta_z\omega\,dz. \tag{7}$$

Differentiating once again with respect to t and using formula (6), we obtain

$$\frac{\partial^2 u_1(x,t)}{\partial t^2} = \frac{1}{4\pi t}\frac{\partial I(x,t)}{\partial t}. \tag{8}$$

Writing (7) in the form

$$I(x, t) = \int_0^t\int_0^\pi\int_0^{2\pi}r^2\Delta_z\omega\sin\vartheta\,dr\,d\vartheta\,d\phi$$

and differentiating with respect to t, we find

$$\frac{\partial I(x,t)}{\partial t} = \int_0^\pi\int_0^{2\pi}t^2\Delta_z\omega\sin\vartheta\,d\vartheta\,d\phi = \int_{S_t}\Delta_z\omega\,dS_t.$$

On the other hand, differentiation of equation (2) with respect to the coordinates gives

$$\Delta u_1(x,t) = \frac{t}{4\pi} \int_0^\pi \int_0^{2\pi} \Delta_z \omega \sin\vartheta\, d\vartheta\, d\phi = \frac{1}{4\pi t} \int_{S_t} \Delta_z \omega\, dS_t,$$
$$z = x + t\theta, \tag{9}$$

and from formulae (8) and (9) it follows that

$$\frac{\partial^2 u_1}{\partial t^2} - \Delta u_1 = 0. \tag{10}$$

We now suppose that the function $\omega(z)$ has continuous derivatives up to and including the third order, and consider the function

$$u_2(x,t) = \frac{\partial}{\partial t} T_\omega(x,t) = \frac{\partial u_1}{\partial t}. \tag{11}$$

It is easy to see that the function u_2 has continuous second derivatives. It satisfies the wave equation

$$\frac{\partial^2 u_2}{\partial t^2} - \Delta u_2 = \frac{\partial^3 u_1}{\partial t^3} - \Delta \frac{\partial u_1}{\partial t} = \frac{\partial}{\partial t}\left(\frac{\partial^2 u_1}{\partial t^2} - \Delta u_1\right) = 0. \tag{12}$$

Formula (5) shows that

$$u_2(x,t)|_{t=0} = \omega(x). \tag{13}$$

Furthermore,

$$\left.\frac{\partial u_2}{\partial t}\right|_{t=0} = \left.\frac{\partial^2 u_1}{\partial t^2}\right|_{t=0} = \Delta u_1|_{t=0} = \Delta u_1(x,0),$$

which is equal to zero because of (3). Thus,

$$\left.\frac{\partial u_2}{\partial t}\right|_{t=0} = 0. \tag{14}$$

If we now substitute first $\omega = \phi_0$, and then $\omega = \phi_1$, and use the relations (3), (5), (13) and (14), and also equations (10) and (12), we see that the function defined by Kirchhoff's formula satisfies both the wave equation and the initial conditions

$$u|_{t=0} = \phi_0(x), \quad \left.\frac{\partial u}{\partial t}\right|_{t=0} = \phi_1(x).$$

§ 5. Rear Wave Front

Kirchhoff's formula permits the elucidation of an interesting aspect of the phenomenon of wave-propagation in three-dimensional space – the appearance of a so-called *rear front* of the wave. This feature is easily explained on the basis of the fact that Kirchhoff's formula involves integration only over a spherical surface of variable radius.

It is clear in Kirchhoff's formula that $u(x, t) = 0$ if $\phi_0(z) \equiv 0$ and $\phi_1(z) \equiv 0$ for $z \in S_t$. Suppose now that the initial functions are different from zero only in some domain D of the space E_3 (Fig. 48). Let the point x be situated, for example, outside the domain D. Denote by δ and δ_1 the smallest and greatest distances respectively of the point x from the points of the boundary of the domain D.

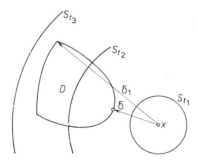

Fig. 48.

At times near to the initial instant, namely, so long as $t < \delta$, the sphere S_t does not intersect the domain D (the sphere S_{t_1} in Fig. 48), so on this sphere the initial functions are equal to zero and $u(x, t) = 0$. If $\delta < t < \delta_1$, then the sphere S_t intersects the domain D (the sphere S_{t_2} in Fig. 48); on that portion of the sphere S_t which lies inside D the initial functions are not identically zero, and, in general, $u(x, t) \neq 0$. These two cases both occurred in § 5, Chapter 21, in connection with the wave equation in space of an arbitrary number of dimensions.

Now let $t > \delta_1$. The sphere S_t does not intersect the domain D (sphere S_{t_3} of Fig. 48) and $u(x, t) = 0$ again; the point x was subject to disturbance in the course of the time-interval $\delta < t < \delta_1$, and afterwards returned to a state of rest.

If we look at the vibrating medium at a time not too close to the initial instant, we observe in the medium points of three types: the first are at rest because the disturbance has not yet reached them, the second are in a disturbed state, and the third are again at rest, because the disturbance has already passed through them. The leading wave front (cf. § 5, Chapter 21) defines the domain which the disturbance has not yet reached, separating it from the domain of disturbance. The surface which separates the domain of the disturbance from the domain through which the disturbance has already passed is called the rear wave front. If, as before, we denote by δ_1 the greatest distance from the point x to the boundary of the domain D, then the rear wave front passes through the point x at the moment of time $t = \delta_1$.

If the wave equation has the form

$$\frac{\partial^2 u}{\partial t^2} - a^2 \Delta u = 0, \quad a = \text{const.},$$

then all we have said in the present section remains true if only we replace t by at.

§ 6. The Case $m = 2$ (Vibrating-Membrane Equation)

The solution to Cauchy's problem for the vibrating-membrane equation,

$$\frac{\partial^2 u}{\partial t^2} - \Delta u = \frac{\partial^2 u}{\partial t^2} - \frac{\partial^2 u}{\partial x_1^2} - \frac{\partial^2 u}{\partial x_2^2} = 0, \tag{1}$$

$$u|_{t=0} = \phi_0(x) = \phi_0(x_1, x_2),$$

$$\left.\frac{\partial u}{\partial t}\right|_{t=0} = \phi_1(x) = \phi_1(x_1, x_2), \tag{2}$$

can be obtained from Kirchhoff's formula if we assume that the initial functions ϕ_0 and ϕ_1 in that formula are independent of the coordinate x_3.

Consider the function

$$T_\omega(x, t) = \frac{1}{4\pi t} \int_{S_t} \omega(z)\, dS_t$$

and suppose that $\omega(z)$ is independent of the third coordinate z_3, i.e., $\omega(z) = \omega(z_1, z_2)$.

Let the sphere S_t be divided into two hemispheres by a plane which passes through the point x and is parallel to the (x_1, x_2)-plane; the projection of both hemispheres on to this latter plane is a circle of radius t and with centre at the point (x_1, x_2). Denote this circle by C_t. Since $\omega(z)$ is independent of z_3, the volume integrals over the hemispheres are equal to each other, and each of them can be replaced by an integral over the circle C_t if we use the fact that

$$dS_t = \frac{dz_1\, dz_2}{|\cos(v, z_3)|},$$

where v is the outward normal to the sphere S_t. Thus,

$$T_\omega(x, t) = \frac{1}{2\pi t} \int\int_{C_t} \frac{\omega(z_1, z_2)}{|\cos(v, z_3)|}\, dz_1\, dz_2.$$

The normal to the sphere is directed along the radius, so that

$$|\cos(v, z_3)| = \frac{|z_3 - x_3|}{t} = \frac{1}{t}\sqrt{\{t^2 - (z_1 - x_1)^2 - (z_2 - x_2)^2\}}$$

and

$$T_\omega(x, t) = \frac{1}{2\pi} \int\int_{C_t} \frac{\omega(z_1, z_2)\, dz_1\, dz_2}{\sqrt{\{t^2 - (z_1 - x_1)^2 - (z_2 - x_2)^2\}}}.$$

Kirchhoff's formula now reduces to the form

$$u(x, t) = u(x_1, x_2, t)$$
$$= \frac{\partial}{\partial t} \frac{1}{2\pi} \int\int_{C_t} \frac{\phi_0(z_1, z_2)\, dz_1\, dz_2}{\sqrt{\{t^2 - (z_1 - x_1)^2 - (z_2 - x_2)^2\}}}$$
$$+ \frac{1}{2\pi} \int\int_{C_t} \frac{\phi_1(z_1, z_2)\, dz_1\, dz_2}{\sqrt{\{t^2 - (z_1 - x_1)^2 - (z_2 - x_2)^2\}}}, \qquad (3)$$

which solves Cauchy's problem for the vibrating-string equation; here C_t is the two-dimensional circle defined by the inequality $|z - x| \leq t$.

§ 7. Equation of the Vibrating String

The solution to Cauchy's problem for the vibrating-string equation

$$\frac{\partial^2 u}{\partial t^2} - \frac{\partial^2 u}{\partial x^2} = 0 \qquad (1)$$

can be obtained from formula (6.3), but it is simpler to derive this solution directly.

It is not difficult to obtain an expression which contains all the solutions to the string equation. To do this, introduce new variables $\xi = x+t$, $\eta = x-t$. Then equation (1) transforms to the following:

$$\frac{\partial^2 u}{\partial \xi \, \partial \eta} = 0.$$

Putting this equation in the form

$$\frac{\partial}{\partial \eta}\left(\frac{\partial u}{\partial \xi}\right) = 0,$$

we then find $\partial u/\partial \xi = \vartheta(\xi)$, where $\vartheta(\xi)$ is an arbitrary function. Integrating with respect to ξ, we obtain

$$u = \theta_1(\xi) + \theta_2(\eta), \quad \theta_1(\xi) = \int \vartheta(\xi)\,d\xi,$$

where θ_1 and θ_2 are arbitrary differentiable functions. Returning to the original variables, we obtain the general solution of equation (1):

$$u(x, t) = \theta_1(x+t) + \theta_2(x-t). \tag{2}$$

Formula (2) is called *D'Alembert's solution*.

We now find the solution to equation (1) satisfying the Cauchy conditions:

$$u|_{t=0} = \phi_0(x), \quad \frac{\partial u}{\partial t}\bigg|_{t=0} = \phi_1(x).$$

Putting $t = 0$ in (2), we obtain

$$\theta_1(x) + \theta_2(x) = \phi_0(x). \tag{3}$$

Differentiating (2) with respect to t, and then putting $t = 0$, we again obtain

$$\theta_1'(x) - \theta_2'(x) = \phi_1(x).$$

Integrate this equation:

$$\theta_1(x) - \theta_2(x) = \int_0^x \phi_1(z)\,dz + C. \tag{4}$$

From equations (3) and (4) we find

$$\theta_1(x) = \tfrac{1}{2}\left[\phi_0(x) + \int_0^x \phi_1(z)\,dz + C\right],$$

$$\theta_2(x) = \tfrac{1}{2}\left[\phi_0(x) - \int_0^x \phi_1(z)\,dz - C\right].$$

Formula (2) now gives the required solution

$$u(x,t) = \tfrac{1}{2}\{\phi_0(x+t)+\phi_0(x-t)\}+\tfrac{1}{2}\int_{x-t}^{x+t}\phi_1(z)\,dz. \tag{5}$$

Equation (5) is called *D'Alembert's formula*.

§ 8. Wave Equation with Variable Coefficients *

Consider the equation

$$\frac{\partial^2 u}{\partial t^2} - \frac{\partial}{\partial x_j}\left(A_{jk}(x)\frac{\partial u}{\partial x_k}\right) + C(x)u = 0. \tag{1}$$

We make the following assumptions: 1) the coefficients A_{jk} are continuously differentiable and the coefficient C is continuous in the whole of the space E_m; 2) these coefficients are bounded in the space; 3) $C(x) \geq 0$; 4) the matrix $||A_{jk}||_{j,k=1}^{j,k=m}$ is positive definite for any $x \in E_m$.

Let $\hat{A}$ be an operator given in the space $L_2(E_m)$, and let the domain of definition $D(\hat{A})$ of this operator be generated by functions which are twice continuously differentiable in E_m and which are equal to zero outside some sphere (a particular one for each function). Let the operator itself act according to the formula

$$\hat{A}u = -\frac{\partial}{\partial x_j}\left(A_{jk}\frac{\partial u}{\partial x_k}\right) + Cu. \tag{2}$$

It is easy to see that the operator $\hat{A}$ is positive; it will be positive definite if $C(x) \geq C_0$, where C_0 is a positive constant. The operator $\hat{A}$ can be extended by Friedrichs's method (§ 7, Chapter 5) into self-adjoint form **. Denote this self-adjoint extension by A, and in place of equation (1) consider the abstract, ordinary, second-order differential equation

$$\frac{d^2 u}{dt^2} + Au = 0; \tag{3}$$

---

* For an understanding of this section it is necessary to know the theorem concerning the spectral decomposition of a function of a self-adjoint operator. The arguments presented below can also be applied to the heat-conduction equation with variable coefficients.

** If $\hat{A}$ is a positive definite operator, then, as we saw in § 7, Chapter 5, its Friedrichs extension is a self-adjoint extension. If it is merely positive, then we consider the positive definite operator $\hat{B} = \hat{A}+I$ (where I is the identity operator). If B is a self-adjoint extension of the operator $\hat{B}$, then $A = B-I$ is a self-adjoint extension of the operator $\hat{A}$.

we shall look for a solution of this equation satisfying the initial conditions

$$u(0) = \phi_0, \qquad u'(0) = \left.\frac{du(t)}{dt}\right|_{t=0} = \phi_1. \tag{3'}$$

We shall assume that

$$\phi_0 \in D(A), \qquad \phi_1 \in D(A^{\frac{1}{2}}). \tag{4}$$

Under these circumstances it is not difficult to show that the solution of problem (1)–(2) is the generalized solution of Cauchy's problem (§ 6, Chapter 21) for the wave equation (1) with initial conditions (3'). If A were a constant numerical coefficient, then the solution of the problem (3)–(3') would be given by the formula

$$u(t) = (\cos\sqrt{A}\,t)\phi_0 + \frac{\sin\sqrt{A}\,t}{\sqrt{A}}\phi_1. \tag{5}$$

We now show that this formula remains true in our case as well, so long as the symbols $\cos\sqrt{A}\,t$ and $(\sin\sqrt{A}\,t)/\sqrt{A}$ are understood to mean the respective functions of the operator A.

The numerical functions $\cos\sqrt{\lambda}\,t$ and $(\sin\sqrt{\lambda}\,t)/\sqrt{\lambda}$ are bounded, so that for any t the operators $\cos\sqrt{A}\,t$ and $(\sin\sqrt{A}\,t)/\sqrt{A}$ are also bounded; formula (5) defines a function $u(t)$ with values in $L_2(E_m)$. It is easy to see that $u(t) \in D(A)$ for any t. Since a self-adjoint operator commutes with any bounded function of itself, we have

$$A\left(\cos\sqrt{A}\,t + \frac{\sin\sqrt{A}\,t}{\sqrt{A}}\right) = (\cos\sqrt{A}\,t)A + (\sin\sqrt{A}\,t)\sqrt{A};$$

because of condition (4) we can thus attach a meaning to the expression

$$Au = A\left((\cos\sqrt{A}\,t)\phi_0 + \frac{\sin\sqrt{A}\,t}{\sqrt{A}}\phi_1\right) = (\cos\sqrt{A}\,t)A\phi_0 + (\sin\sqrt{A}\,t)\sqrt{A}\phi_1. \tag{6}$$

We next show that the function (5) has a second derivative with respect to t, and evaluate this derivative.

Let $\mathscr{E}_\lambda$ be the spectral function of the operator A. This operator is positive, and so its spectrum is contained on the semi-axis $0 \leq \lambda < \infty$. We have

$$u(t) = \int_0^\infty \cos\sqrt{\lambda}\,t\,d\mathscr{E}_\lambda \phi_0 + \int_0^\infty \frac{\sin\sqrt{\lambda}\,t}{\sqrt{\lambda}}\,d\mathscr{E}_\lambda \phi_1. \tag{7}$$

§ 8] WAVE EQUATION WITH VARIABLE COEFFICIENTS

Put
$$v(t) = \int_0^\infty \cos\sqrt{\lambda}t\, d\mathscr{E}_\lambda \phi_0, \qquad w(t) = \int_0^\infty \frac{\sin\sqrt{\lambda}t}{\sqrt{\lambda}}\, d\mathscr{E}_\lambda \phi_1. \qquad (8)$$

We assert that each of the functions $v(t)$ and $w(t)$ has first and second derivatives with respect to t, and that these derivatives can be obtained by formal differentiation of the integrals (8). We give the proof in the case of the function $v(t)$; for $w(t)$ it is the same apart from some obvious changes. Put
$$v_1(t) = -\int_0^\infty \sqrt{\lambda}\sin\sqrt{\lambda}t\, d\mathscr{E}_\lambda \phi_0. \qquad (9)$$

The integral (9) converges uniformly with respect to t, since
$$\left\| \int_{\lambda_1}^{\lambda_2} \sqrt{\lambda}\sin\sqrt{\lambda}t\, d\mathscr{E}_\lambda \phi_0 \right\|^2 = \int_{\lambda_1}^{\lambda_2} \lambda\sin^2\sqrt{\lambda}t \|d\mathscr{E}_\lambda \phi_0\|^2 \leq \int_{\lambda_1}^{\lambda_2} \lambda \|d\mathscr{E}_\lambda \phi_0\|^2.$$

The last integral here tends to zero as $\lambda_1, \lambda_2 \to \infty$, since the integral
$$\int_0^\infty \lambda^2 \|d\mathscr{E}_\lambda \phi_0\|^2 = \|A\phi_0\|^2$$

converges. It is now easily shown by the usual means that
$$\left\| \frac{v(t+h)-v(t)}{h} - v_1(t) \right\|^2 \underset{h\to 0}{\to} 0,$$

and, consequently,
$$\frac{dv}{dt} = v_1(t) = -\int_0^\infty \sqrt{\lambda}\sin\sqrt{\lambda}t\, d\mathscr{E}_\lambda \phi_0.$$

Similarly it can be shown that
$$\frac{d^2 v}{dt^2} = -\int_0^\infty \lambda \cos\sqrt{\lambda}t\, d\mathscr{E}_\lambda \phi_0,$$
$$\frac{d^2 w}{dt^2} = -\int_0^\infty \sqrt{\lambda}\sin\sqrt{\lambda}t\, d\mathscr{E}_\lambda \phi_1.$$

But then the second derivative $d^2 u/dt^2$ exists, and
$$\frac{d^2 u}{dt^2} = -\int_0^\infty \lambda \left[\cos\sqrt{\lambda}t\, d\mathscr{E}_\lambda \phi_0 + \frac{\sin\sqrt{\lambda}t}{\sqrt{\lambda}}\, d\mathscr{E}_\lambda \phi_1 \right].$$

The right-hand side of this expression is equal to $-Au$, and, consequently, the function u satisfies equation (3).

We now show that the function (6) satisfies the initial conditions (4), understood in the following sense:

$$\lim_{t \to 0} \|u(t) - \phi_0\|_{L_2(E_m)} = 0$$

$$\lim_{t \to 0} \left\| \frac{du(t)}{dt} - \phi_1 \right\|_{L_2(E_m)} = 0. \tag{10}$$

We give the proof for the second of equations (10); the other is similar. From the preceding discussion we have

$$\frac{du(t)}{dt} = -\int_0^\infty \sqrt{\lambda} \sin \sqrt{\lambda} t \, d\mathscr{E}_\lambda \phi_0 + \int_0^\infty \cos \sqrt{\lambda} t \, d\mathscr{E}_\lambda \phi_1,$$

and, consequently,

$$\frac{du}{dt} - \phi_1 = -\int_0^\infty \sqrt{\lambda} \sin \sqrt{\lambda} t \, d\mathscr{E}_\lambda \phi_0 + \int_0^\infty (\cos \sqrt{\lambda} t - 1) \, d\mathscr{E}_\lambda \phi_1.$$

Hence

$$\left\| \frac{du(t)}{dt} - \phi_1 \right\|^2 \leq 2 \int_0^\infty \lambda \sin^2 \sqrt{\lambda} t \|d\mathscr{E}_\lambda \phi_0\|^2 +$$

$$+ 2 \int_0^\infty (1 - \cos \sqrt{\lambda} t)^2 \|d\mathscr{E}_\lambda \phi_1\|^2. \tag{11}$$

In particular, it follows from equation (5) that

$$\int_0^\infty \lambda \|d\mathscr{E}_\lambda \phi_0\|^2 < \infty, \qquad \int_0^\infty \|d\mathscr{E}_\lambda \phi_1\|^2 < \infty,$$

and therefore, for sufficiently large $a > 0$,

$$\int_a^\infty \lambda \|d\mathscr{E}_\lambda \phi_0\|^2 < \tfrac{1}{16}\varepsilon^2, \qquad \int_a^\infty \|d\mathscr{E}_\lambda \phi_1\|^2 < \tfrac{1}{16}\varepsilon^2,$$

where ε is an arbitrary, prescribed, positive number. It now follows from inequality (11) that

$$\left\| \frac{du(t)}{dt} - \phi_1 \right\|^2 \leq 2 \int_0^a \lambda \sin^2 \sqrt{\lambda} t \|d\mathscr{E}_\lambda \phi_0\|^2 +$$

$$+ 2 \int_0^a (1 - \cos \sqrt{\lambda} t)^2 \|d\mathscr{E}_\lambda \phi_1\|^2 + \tfrac{1}{4}\varepsilon^2;$$

it only remains to choose t_0 small enough so that for $t < t_0$ the sum of the integrals on the right-hand side is less than $\frac{3}{4}\varepsilon^2$.

Remark 1. It can be shown in a similar way that the function (6) also satisfies the initial conditions in the following sense:

$$\lim_{t \to 0} \|Au(t) - A\phi_0\| = 0$$

$$\lim_{t \to 0} \left\| \sqrt{A}\frac{du(t)}{dt} - \sqrt{A}\phi_1 \right\| = 0. \qquad (12)$$

Remark 2. It is possible to form the generalized solution of the problem (1), (4) as a function which satisfies the initial conditions (10) and the equation

$$-\int_0^T \left(\frac{du}{dt}, \frac{d\eta}{dt}\right) dt + \int_0^T [u(t), \eta(t)] dt - (\phi_1, \eta(0)) = 0; \qquad (13)$$

in this equation T is an arbitrary positive number, $\eta(t)$ is an arbitrary function of the class

$$C^{(1)}([0, \infty); L_2(E_m)) \cap C([0, \infty); H_A), \qquad (14)$$

equal to zero when $t = T$. The generalized solution belonging to the class (14) is given by the same formula (6); here it is sufficient to require that $\phi_0 \in D(\sqrt{A})$, $\phi_1 \in L_2(E_m)$.

Remark 3. The method and results of the present section can easily be extended to the case of the inhomogeneous wave equation.

PART VII

Properly and Improperly Posed Problems

Chapter 25. The Proper Posing of Problems in Mathematical Physics

§ 1. The Fundamental Theorem

In Chapter 9 (§ 5) we introduced the concept of proper posing in mathematical physics. In essence, this concept can be reduced to the following. Denote by Φ the totality of data of a particular problem, by U the totality of unknowns, and by A the operator which transforms U into Φ, so that

$$AU = \Phi. \tag{1}$$

The problem is to find U for given A and Φ. Suppose that U and Φ can be regarded as elements of the metric spaces B_1 and B_2 respectively. We say that the problem (1) is properly posed in the pair of spaces B_1 and B_2 if, for any $\Phi \in B_2$, it has one and only one solution, and if to a sufficiently small (in the metric of B_2) change in Φ there corresponds an arbitrarily small (in the metric of B_1) change in U. We shall henceforth only interest ourselves in the case when the operator A is linear and B_1, B_2 are Banach spaces. In this case the following theorem holds.

THEOREM 25.1.1. *In order that the linear problem* (1) *be properly posed in the pair of Banach spaces* (B_1, B_2), *it is necessary and sufficient that there exists the operator* $R = A^{-1}$, *acting from* B_2 *into* B_1, *for which* $D(R) = B_2$ *and R is bounded as an operator from* B_2 *into* B_1.

Necessity. If the problem (1) is properly posed, then, first of all, its solution exists for any $\Phi \in B_2$ and is unique. The uniqueness of the solution implies the existence of the operator $R = A^{-1}$, while the existence of the solution for any $\Phi \in B_2$ implies that the operator R is defined in the whole of the space B_2.

Now replace the element Φ by $\Phi + \phi$, where $\phi \in B_2$, and let the required solution of problem (1) be $U + u$. Then $A(U+u) = \Phi + \phi$, and since A is a linear operator, so $Au = \phi$. Problem (1) is properly posed, so that, given a number $\varepsilon > 0$, we can find a number $\delta > 0$, such that $\|u\| = \|R\phi\| < \delta$ when $\|\phi\| < \varepsilon$. We fix ε and the δ which corresponds to it. If

$\psi \in B_2$ and $\|\psi\| = 1$, then $\|\tfrac{1}{2}\varepsilon\psi\| = \tfrac{1}{2}\varepsilon < \varepsilon$ and, consequently, $\|R\tfrac{1}{2}\varepsilon\psi\| = \tfrac{1}{2}\varepsilon\|R\psi\| < \delta$. Hence

$$\|R\psi\| < \frac{2\delta}{\varepsilon}, \qquad \|\psi\| = 1,$$

and this means that $\|R\| < 2\delta/\varepsilon$. The operator R is bounded.

Sufficiency. If the operator R exists, then problem (1) has not more than one solution. If $D(R) = B_2$, then problem (1) is soluble for any $\Phi \in B_2$. Finally, if R is a bounded operator, and $\|\phi\|_{B_2} < \varepsilon$, then $\|u\|_{B_1} = \|R\phi\|_{B_1} < \delta$, where $\delta = \varepsilon\|R\|$.

It is important to emphasize that proper or improper posing of a problem depends on what spaces the given and unknown quantities are embedded in; the same problem can turn out to be properly posed in one pair of spaces and improperly posed in another. We shall look into this question more thoroughly in § 8.

In mathematical physics (as in analysis generally) improperly posed problems play quite an important role. Thus, for example, it can be shown that in the pair of spaces $\bigl(C^{(1)}(\Omega), C(\Gamma)\bigr)$ the Dirichlet problem for the homogeneous Laplace equation is improperly posed, even though this problem turns up in the theory of mechanics of deformable media (particularly in elasticity theory). One of the simplest improperly posed problems is that of finding a solution to the equation

$$Tu = f, \qquad (2)$$

where T is a completely continuous operator acting from an infinite-dimensional Banach space X into a similar space Y. A special case of equation (2) is the so-called Fredholm integral equation of the first kind

$$\int_\Omega K(x, y) u(y)\,dy = f(x),$$

where $K(x, y)$ is a Fredholm kernel.

The fact that problem (2) is improperly posed follows at once from the following considerations. If it were properly posed, there would exist a bounded operator T^{-1}, and then the identity operator $I = T^{-1}T$ would be completely continuous in the infinite-dimensional space X (which it is not).

In recent times a considerable amount of work has been devoted to finding approximate solutions of improperly posed problems (on the condition, of course, that an exact solution exists). One of the first contributions along these lines was the work of A. N. Tikhonov [3], who proposed a method based on the idea that the solution of an improperly posed problem could be regarded as the limit of the solutions of a certain special sequence of properly posed problems.

§ 2. Positive Definite Problems

1. Let A be a positive definite operator in a Hilbert space H. Consider the equation

$$Au = f. \tag{1}$$

We construct the energy space H_A of the operator A, and seek a generalized solution of equation (1), i.e. an element of the space H_A, satisfying the identity

$$[u, \eta] = (f, \eta), \quad \forall \eta \in H_A. \tag{2}$$

This solution exists and is unique; thus, there exists the operator $R = A^{-1}$, acting from H into H_A, and defined in the whole of the space H. Put $B_1 = H_A$, $B_2 = H$.

We now show that the operator R is bounded. Let u be the generalized solution of problem (1). Putting $\eta = u$ in equation (2), we obtain

$$|||u|||_A^2 = (f, u) \leq \|f\| \cdot \|u\|.$$

Let γ^2 be the lower bound of the operator A. Then $\|u\| \leq |||u|||_A/\gamma$. Substituting this into the preceding inequality, we obtain

$$|||u|||_A = |||Rf|||_A \leq \frac{1}{\gamma} \|f\|.$$

This means that

$$\|R\|_{H \to H_A} \leq \frac{1}{\gamma},$$

and so problem (1) is properly posed in the pair of spaces (H_A, H) when the operator A is positive definite.

2. We consider some examples.

1. Let $\Omega \subset E_m$ be a finite domain with a piecewise-smooth boundary Γ.

Consider the Dirichlet problem

$$-\frac{\partial}{\partial x_j}\left(A_{jk}(x)\frac{\partial u}{\partial x_k}\right) + C(x)u = f(x), \qquad u|_\Gamma = 0, \qquad (3)$$

where the coefficients $A_{jk}(x)$ and $C(x)$ are subject to the usual conditions (cf. Chapter 14). The operator $\mathfrak{A}$ for problem (3) is positive definite in the space $L_2(\Omega)$, and so this problem is properly posed in the pair of spaces $(H_\mathfrak{A}, L_2(\Omega))$. We recall that the metric in $H_\mathfrak{A}$ is given by the formula

$$|||u|||_\mathfrak{A}^2 = \int_\Omega \left(A_{jk}\frac{\partial u}{\partial x_j}\frac{\partial u}{\partial x_k} + Cu^2\right) dx. \qquad (4)$$

2. In equation (3), let $C(x) \geqq C_0 = \text{const.} > 0$, and impose the Neumann boundary condition

$$\left[A_{jk}\frac{\partial u}{\partial x_j}\cos(v, x_k)\right]_\Gamma = 0. \qquad (5)$$

The operator $\mathfrak{N}$ for this problem is positive definite in $L_2(\Omega)$, and the Neumann problem is properly posed in the pair of spaces $(H_\mathfrak{N}, L_2(\Omega))$. The metric in $H_\mathfrak{N}$ is also defined by formula (4).

3. Now let $C(x) \equiv 0$. In this case the operator $\mathfrak{N}_0$ of the Neumann problem is positive definite in the space $\tilde{L}_2(\Omega)$ – the subspace of $L_2(\Omega)$ which is orthogonal to unity – and the Neumann problem is properly posed in the pair of spaces $(H_{\mathfrak{N}_0}, \tilde{L}_2(\Omega))$.

§ 3. Dirichlet Problem for the Homogeneous Laplace Equation

Let Γ be a regular surface and Ω a domain either inside or outside Γ. We pose the Dirichlet problem for the homogeneous Laplace equation

$$\Delta u = 0, \qquad u|_\Gamma = \phi(x). \qquad (1)$$

It is easy to suggest a pair of spaces in which this problem is properly posed. One such pair is $B_1 = G(\bar{\Omega})$, $B_2 = C(\Gamma)$, where $G(\bar{\Omega})$ denotes the subspace of the space $C(\bar{\Omega})$ generated by functions which are harmonic in Ω and continuous in $\bar{\Omega}$ (c.f. Corollary 11.9.1). We see that, for any function $\phi \in C(\Gamma)$, problem (1) has a solution which is unique and continuous in $\bar{\Omega}$. Let us now concentrate on the interior Dirichlet problem. It follows from the Maximum Principle that

$$\max_{x\in\bar{\Omega}}|u(x)| = \max_{x\in\Gamma}|u(x)| = \max_{x\in\Gamma}|\phi(x)|,$$

or
$$\|u\|_{B_1} = \|\phi\|_{B_2}. \tag{2}$$

If we denote by R the operator which transforms the given function $\phi(x)$ into the required function $u(x)$: $u = R\phi$, then it follows from equation (2) that $\|R\| = 1$; consequently problem (1) is properly posed.

We now pass to the exterior Dirichlet problem. First let $m > 2$. Then from the condition $u(x) = O(|x|^{2-m})$, $|x| \to \infty$, it follows that $u(x) \to 0$. Construct a sphere S_R of sufficiently large radius R (Fig. 21, page 256) that the surface Γ lies inside the sphere. In the finite domain Ω_R, bounded by the surfaces Γ and S_R, the Maximum Principle holds:

$$\max_{x \in \Omega_R} |u(x)| = \max \{\max_{x \in \Gamma} |u(x)|, \max_{x \in S_R} |u(x)|\} = \max \{\max_{x \in \Gamma} |\phi(x)|, \max_{x \in S_R} |u(x)|\}.$$

Take the limit $R \to \infty$. This leads us to the relation

$$\max_{x \in \bar{\Omega}} |u(x)| = \max \{\max_{x \in \Gamma} |\phi(x)|, 0\} = \max_{x \in \Gamma} |\phi(x)|.$$

The rest proceeds as for the interior problem.

If $m = 2$ and the domain Ω is infinite, then the exterior Dirichlet problem can be transformed into an interior problem by conformally mapping Ω into a finite domain. In this way the exterior Dirichlet problem can be shown to be properly posed.

§ 4. Exterior Neumann Problem

Let the infinite domain Ω be bounded by a regular surface Γ. Take the space to have dimension $m > 2$. If the function $\psi(x)$ is continuous on Γ, then the exterior Neumann problem

$$\Delta u = 0, \quad \frac{\partial u}{\partial n}\bigg|_\Gamma = \psi(x) \tag{1}$$

has one and only one solution which is continuous in $\bar{\Omega}$, and can be represented in the form of a single-layer potential

$$u(x) = \int_\Gamma \mu(\xi) \frac{1}{r^{m-2}} \, d_\xi(\Gamma) \tag{2}$$

with a continuous density $\mu(\xi)$. This density satisfies the integral equation

(§ 8, Chapter 18)

$$\mu(x) - \frac{2}{(m-2)|S_1|} \int_\Gamma \mu(\xi) \frac{\partial}{\partial n} \frac{1}{r^{m-2}} d_\xi \Gamma = -\frac{2}{(m-2)|S_1|} \psi(x). \quad (3)$$

As we explained in Chapter 18, this equation has one and only one continuous solution on Γ, whenever the function $\psi(x)$ is continuous on Γ. This solution can be represented in the form

$$\mu = Q\psi, \quad (4)$$

where Q is the inverse of the operator on the left-hand side of equation (3), correct to a constant multiplier. From this it follows that the operator Q acts in the space $C(\Gamma)$ of continuous functions on Γ, and is defined over the whole space. We now show that the operator Q is bounded in $C(\Gamma)$.

Suppose the contrary. Then there exists a sequence of functions $\psi_n \in C(\Gamma)$ such that

$$\|\mu_n\| \geq n\|\psi_n\|, \qquad \mu_n = Q\psi_n.$$

Put

$$\frac{\psi_n(x)}{\|\mu_n\|} = \psi_n^*(x), \qquad \frac{\mu_n(x)}{\|\mu_n\|} = \mu_n^*(x).$$

Then $\mu_n^* = Q\psi_n^*$, $\|\mu_n^*\| = 1$, $\|\psi_n^*\| \to 0$. The first of these shows that μ_n^* satisfies the equation

$$\mu_n^*(x) - \frac{2}{(m-2)|S_1|} \int_\Gamma \mu_n^*(\xi) \frac{\partial}{\partial n} \frac{1}{r^{m-2}} d_\xi \Gamma = -\frac{2}{(m-2)|S_1|} \psi_n^*(x).$$

If for brevity we put

$$\frac{2}{(m-2)|S_1|} \int_\Gamma \mu(\xi) \frac{\partial}{\partial n} \frac{1}{r^{m-2}} d_\xi \Gamma = (K\mu)(x),$$

then the last equation can be written in the concise form

$$\mu_n^* - K\mu_n^* = -\frac{2}{(m-2)|S_1|} \psi_n^*. \quad (5)$$

K is an operator with a weak singularity and is completely continuous in $C(\Gamma)$ (cf. § 4, Chapter 7), and the sequence $\{\mu_n^*\}$ is bounded ($\|\mu_n^*\| = 1$). It is then possible to pick out a subsequence $\mu_{n_k}^*$, such that the limit

$$\lim K\mu_{n_k}^* = \mu^*$$

exists. At the same time $\psi_{n_k}^* \to 0$, and it follows from equation (5) that $\mu_{n_k}^* \to \mu^*$. If now we proceed to the limit in (5) under the operator K, we find that μ^* satisfies the homogeneous integral equation

$$\mu^* - K\mu^* = 0. \tag{6}$$

Equation (3) has a unique solution, and therefore the corresponding homogeneous equation (6) has only the trivial solution. Hence $\mu^* = 0$. On the other hand,

$$\|\mu^*\| = \lim \|\mu_{n_k}^*\| = 1.$$

We have thus obtained a contradiction, which shows that the operator Q is bounded *. It follows that there exists a constant α such that

$$\|\mu\| \leq \alpha \|\psi\|, \tag{7}$$

where μ is the solution to equation (3) and the norm is taken in the metric of the space $C(\Gamma)$.

From formula (2) we have

$$|u(x)| \leq \max_{\eta \in \Gamma} |\mu(\eta)| \int_\Gamma \frac{d_\xi \Gamma}{r^{m-2}}, \qquad x \in \bar{\Omega}.$$

This latter integral is the potential of a single layer with density equal to one. This potential is continuous over the whole of the space E_m, and, moreover, is equal to zero at infinity. Hence it follows that it is bounded:

$$\int_\Gamma \frac{d_\xi \Gamma}{r^{m-2}} \leq \beta = \text{const.}$$

In this case

$$|u(x)| \leq \beta \max_{\xi \in \Gamma} |\mu(\xi)| = \beta \|\mu\| \leq \alpha\beta \|\psi\|.$$

Taking the maximum of the left-hand side, we finally obtain

$$\|u\|_{G(\bar{\Omega})} \leq \alpha\beta \|\psi\|_{C(\Gamma)}. \tag{8}$$

It is now clear that when $m > 2$ the exterior Neumann problem is properly posed in the pair of spaces $(G(\bar{\Omega}), C(\Gamma))$. In fact, denote by R the operator which transforms the function $\psi(x)$ into the function $u(x)$

---

* The boundedness of the operator Q can also be shown in the following way. It follows from Theorem 7.4.1 that the operator on the left-hand side of equation (3) is bounded in $C(\Gamma)$. Since it is defined on the whole of this space, its boundedness implies that it is closed; the inverse operator Q is therefore also closed. From Banach's well-known theorem, an operator which is defined and closed in the whole space $C(\Gamma)$ is bounded in this space.

— the solution of the exterior Neumann problem. Then 1) the operator R exists (the solution is unique); 2) it is defined on the whole of the space $C(\Gamma)$ (the solution exists for any continuous function $\psi(x)$); 3) it is bounded as an operator from $C(\Gamma)$ into $G(\bar{\Omega})$ (inequality (8)).

§ 5. Interior Neumann Problem

The results of this section will also be true for the exterior Neumann problem when $m = 2$, since this can be conformally transformed into an interior Neumann problem.

As before, let Γ be a regular surface, but now let Ω be a domain lying inside Γ. The problem consists in determining a harmonic function in Ω with the boundary condition

$$\left.\frac{\partial u}{\partial n}\right|_\Gamma = \psi(x). \tag{1}$$

In the pair of spaces $(G(\bar{\Omega}), C(\Gamma))$, the interior Neumann problem is improperly posed because it is not always soluble, and the solution (if it exists) is not unique.

We introduce the space $C^\perp(\Gamma)$ which is a subspace of the space $C(\Gamma)$, and which is defined by the additional relation

$$\int_\Gamma \psi(\xi)\,d_\xi\Gamma = 0. \tag{2}$$

If $\psi \in C^\perp(\Gamma)$ in the boundary condition (1), then the interior Neumann problem is soluble, but not in a unique way; we therefore have to concern ourselves with the question of selecting one solution from an infinite set. The solution of the interior Neumann problem can be represented (cf. § 11, Chapter 18) in the form of a single-layer potential

$$u(x) = \int_\Gamma \mu(\xi)\frac{d_\xi\Gamma}{r^{m-2}} \tag{3}$$

with continuous density $\mu(\xi)$; for $\mu(\xi)$ we can take any solution of the integral equation

$$\mu(x) + \frac{2}{(m-2)|S_1|}\int_\Gamma \mu(\xi)\frac{\partial}{\partial n}\frac{1}{r^{m-2}}\,d_\xi\Gamma = \frac{2}{(m-2)|S_1|}\psi(x). \tag{4}$$

If $m = 2$, then $u(x)$ can be represented in the form of a logarithmic potential, and equation (4) is correspondingly changed. This case is also included in the following discussion.

Equation (4) is soluble if $\psi \in C^\perp(\Gamma)$, and its solutions are continuous. There are infinitely-many of them: the homogeneous equation has one linearly-independent solution $\mu_0(x)$. If we denote any solution of equation (4) by $\hat{\mu}(x)$, then the general solution has the form

$$\mu(x) = \hat{\mu}(x) + c\mu_0(x), \tag{5}$$

where c is an arbitrary constant. Different solutions of the Neumann problem correspond to different values of c, and all these solutions are continuous in $\bar{\Omega}$.

We shall use the following method of choosing the constant c in (5): we require that, in the metric of $L_2(\Gamma)$, the function $\mu(x)$ should be orthogonal to $\mu_0(x)$. This gives the following values for c and $\mu(x)$:

$$c = -\frac{(\hat{\mu}, \mu_0)_{L_2}}{\|\mu_0\|_{L_2}^2} = -\frac{\int_\Gamma \hat{\mu}(x)\mu_0(x)\,d\Gamma}{\int_\Gamma \mu_0^2(x)\,d\Gamma},$$

$$\mu(x) = \hat{\mu}(x) - \frac{(\hat{\mu}, \mu_0)_{L_2}}{\|\mu_0\|_{L_2}^2}\mu_0(x). \tag{6}$$

It is not difficult to see that the function $\mu(x)$ defined by formula (6) is independent of the choice of the particular solution $\hat{\mu}(x)$, and, consequently, is uniquely defined.

Henceforth we shall understand by $\mu(x)$ the function (6), and by the solution to the interior Neumann problem we shall mean the potential (3) whose density coincides with the function (6). This solution is unique.

We now show that with this choice of solution the interior Neumann problem is properly posed in the pair of spaces $(G(\bar{\Omega}), C^\perp(\Gamma))$. We have already seen that in this pair of spaces the interior Neumann problem has one and only one solution. It therefore only remains to show that the operator R, which transforms the function $\psi(x)$ into the solution $u(x)$ of the problem, is bounded.

We first show that there exists a constant α such that

$$\|\mu\|_{C(\Gamma)} \leq \alpha\|\psi\|_{C(\Gamma)}. \tag{7}$$

Suppose this is not so. Then, as in the previous section, we can see that there exist two sequences $\{\psi_n^*\}$ and $\{\mu_n^*\}$, associated with equation (4), such that

$$\|\mu_n^*\| = 1, \qquad \|\psi_n^*\| \underset{n\to\infty}{\to} 0.$$

Retaining the symbol K for the integral operator introduced in the preceding section, we can write equation (4) for the functions μ_n^* and ψ_n^* in the form

$$\mu_n^* + K\mu_n^* = \frac{2}{(m-2)|S_1|}\psi_n^*.$$

The rest proceeds as in § 4: we select a sequence $\mu_{n_k}^*$ such that the limit

$$\mu^* = \lim K\mu_{n_k}^* = -\lim \mu_{n_k}^*$$

exists. This limit satisfies the homogeneous equation

$$\mu^* + K\mu^* = 0.$$

Hence we must have

$$\mu^*(x) = C\mu_0(x).$$

Moreover,

$$(\mu^*, \mu_0)_{L_2(\Gamma)} = \lim (\mu_{n_k}^*, \mu_0)_{L_2(\Gamma)} = 0.$$

But then $C = 0$ and $\mu^*(x) \equiv 0$, which is impossible because $\|\mu^*\| = \lim \|\mu_{n_k}^*\| = 1$.

The inequality (7) is thus established; as in § 4, it follows from this that

$$\|u\|_{G(\bar{\Omega})} \leqq \alpha\beta\|\psi\|_{C^1(\Gamma)}.$$

This latter inequality shows that the operator R is bounded, and so the interior Neumann problem as formulated above is properly posed.

§ 6. Heat-Conduction Problems

1. The mixed problem. Consider the problem discussed in §§ 1, 2, Chapter 20: find in the domain Q (Fig. 37, page 413) the generalized solution of the equation

$$\frac{\partial u}{\partial t} - \frac{\partial}{\partial x_j}\left(A_{jk}(x)\frac{\partial u}{\partial x_k}\right) = f(x,t) \tag{1'}$$

with boundary and initial conditions

$$u|_B = 0, \qquad u|_{t=0} = \phi(x). \tag{1''}$$

This problem has a solution, which is unique, in the class $C^{(1)}((0, \infty);$ $H_{\mathfrak{A}}) \cap C([0, \infty); L_2(\Omega))$ if $\phi \in L_2(\Omega)$ and $f \in C^{(1)}([0, \infty); L_2(\Omega))$.

We shall consider the solution in the domain Q_T of Fig. 37, page 413, i.e. in the domain $\Omega \times [0, T]$. The arguments of §§ 1, 2, Chapter 20, can be adapted to the present case without any difficulty, with t varying only on the finite interval $[0, T]$. It is sufficient to assume here that $f \in C^{(1)}([0; T];$ $L_2(\Omega))$, and then we can show that a solution to problem (1) exists and is unique in the class $C^{(1)}([0, T]; H_{\mathfrak{A}}) \cap C([0, T]; L_2(\Omega))$.

Introduce now the spaces B_1 and B_2 which figure in the definition of proper posing. For B_1 we take the space $C([0, T]; L_2(\Omega))$ in which we introduce a norm in the following way:

$$\|u\|_1 = \max_{0 \le t \le T} \|u(t)\|_{L_2(\Omega)}. \tag{2}$$

For B_2 we take a space of pairs of elements of the form

$$\Phi = (\phi(x); f(t, x)), \tag{3}$$

where $\phi \in L_2(\Omega)$ and $f \in C^{(1)}([0, T]; L_2(\Omega))$; the norm in this new space is defined by the formula

$$\|\Phi\|_2 = \|\phi\|_{L_2(\Omega)} + \max_{0 \le t \le T} \|f(t)\|_{L_2(\Omega)} + \max_{0 \le t \le T} \|f'(t)\|_{L_2(\Omega)}. \tag{4}$$

We now show that problem (1) is properly posed in the pair of spaces (B_1, B_2). We have already discussed the existence and uniqueness of the solution, and it will be sufficient to convince ourselves of the boundedness of the operator R which acts from B_2 into B_1, and transforms the element Φ into the solution u.

We shall prove a stronger and simpler assertion: the operator R is bounded as an operator from B_3 into B_1, where B_3 is the space of the elements (3) with the metric

$$\|\Phi\|_3 = \|\phi\|_{L_2(\Omega)} + \max_{0 \le t \le T} \|f(t)\|_{L_2(\Omega)}.$$

We appeal to formula (1.8), Chapter 22. The system $\{u_n\}$ is orthonormal in $L_2(\Omega)$, so that

$$\|u(t)\|_{L_2(\Omega)}^2 = \sum_{n=1}^{\infty} \left[(\phi, u_n) \exp(-\lambda_n t) + \int_0^t \exp(-\lambda_n(t-\tau)) f_n(\tau)\, d\tau \right]^2.$$

This series coincides with the series (2.1), Chapter 22: from the estimate calculated in § 2, Chapter 22, it follows at once that the sum of this series does not exceed the quantity

$$2\sum_{n=1}^{\infty}(\phi, u_n)^2 + \frac{1}{\lambda_1}\int_0^t \sum_{n=1}^{\infty} f_n^2(\tau)\,d\tau = 2\|\phi\|_{L_2(\Omega)}^2 + \frac{1}{\lambda_1}\int_0^t \|f(\tau)\|_{L_2(\Omega)}^2\,d\tau.$$

Hence

$$\|u(t)\|_{L_2(\Omega)}^2 \leq 2\|\phi\|_{L_2(\Omega)}^2 + \frac{T}{\lambda_1}\max_{0\leq t\leq T}\|f(t)\|_{L_2(\Omega)}^2 \leq \alpha^2\|\Phi\|_3^2,$$

where $\alpha^2 = \max(2, T/\lambda_1)$. Taking the maximum of the left-hand side, we find

$$\|u\|_1 \leq \alpha\|\Phi\|_3$$

and, consequently,

$$\|R\|_{B_3 \to B_1} \leq \alpha.$$

2. Cauchy's problem. For B_2 take the space of continuous and bounded functions in E_m, with norm

$$\|\phi\|_2 = \sup_{x \in E_m} |\phi(x)|. \tag{5}$$

For B_1 take the space of continuous and bounded functions in $E_m \times [0, \infty)$, with norm

$$\|u\|_1 = \sup_{x \in E_m,\, t \geq 0} |u(x, t)|. \tag{6}$$

If $\phi \in B_2$ then a solution to the problem

$$\frac{\partial u}{\partial t} - \Delta u = 0, \qquad u|_{t=0} = \phi(x) \tag{7}$$

exists (§ 3, Chapter 23) and is unique (§ 4, Chapter 20) in the space B_1. This implies that the operator R which transforms the initial function ϕ into the solution exists and is defined over the whole of the space B_2. Moreover, from Poisson's formula, written in the form (3.4), Chapter 23:

$$u(x, t) = \pi^{-\frac{1}{2}m} \int_{E_m} \phi(x + 2\sqrt{t}\,\xi)\, e^{-\xi^2}\, d\xi,$$

it follows that

$$|u(x, t)| \leq \sup_{z \in E_m} |\phi(z)| \pi^{-\frac{1}{2}m} \int_{E_m} e^{-\xi^2}\, d\xi = \sup_{z \in E_m} |\phi(z)| = \|\phi\|_2.$$

This inequality is not violated if we substitute for the left-hand side its upper bound:

$$\|u\|_1 = \|R\phi\|_1 \leq \|\phi\|_2,$$

and, consequently, $\|R\| < 1$. Thus, Cauchy's problem for the heat-conduction equation is properly posed in the pair of spaces (B_1, B_2), in which the norms are given by formulae (6) and (5).

§ 7. Problems Connected with the Wave Equation

We shall not bother to consider the mixed problem for the wave equation, since its proper posing can be analyzed by the same procedure as for the heat-conduction equation; we leave it to the reader to formulate and prove the corresponding results.

Consider the Cauchy problem for the wave equation

$$\frac{\partial^2 u}{\partial t^2} - \Delta u = 0, \tag{1}$$

$$u|_{t=0} = \phi_0(x), \quad \frac{\partial u}{\partial t}\bigg|_{t=0} = \phi_1(x).$$

In the role of B_1 we take the space $C^{(2)}(E_m \times [0, \infty))$ of functions which are continuous and bounded, together with their first and second derivatives, for any $x \in E_m$ and any $t \geq 0$. For the norm of an element u of this space we take the quantity

$$\|u\|_1 = \max \left\{ |u(x)| + \sum_{k=1}^{m+1} \left|\frac{\partial u}{\partial x_k}\right| + \sum_{j,k=1}^{m+1} \left|\frac{\partial^2 u}{\partial x_j \partial x_k}\right| \right\}, \tag{2}$$

$$x_{m+1} = t.$$

We take the space of pairs $\Phi = (\phi_0, \phi_1)$ for B_2, with norm

$$\|\Phi\|_2 = \max \left\{ \sum_{n=0}^{m+3} \sum \left|\frac{\partial^n \phi_0}{\partial x_{i_1} \partial x_{i_2} \ldots \partial x_{i_n}}\right| + \sum_{n=0}^{m+2} \sum \left|\frac{\partial^n \phi_1}{\partial x_{i_1} \partial x_{i_2} \ldots \partial x_{i_n}}\right| \right\}, \tag{3}$$

the inner summation being carried out over all possible combinations of the indices $i_1, i_2, \ldots, i_n \leq m$ whose sum is equal to n.

As usual we denote by R the operator which transforms the pair Φ of initial functions into the solution $u(x, t)$ of the Cauchy problem.

From the uniqueness theorem for Cauchy's problem (§ 4, Chapter 21), it follows that the operator R exists, and from the results of § 1, Chapter 24, that this operator acts from B_2 into B_1, and is defined on the whole of the space B_2. Moreover, it follows from formula (1.6) of Chapter 24, and from Theorem 23.1.2, that R is bounded as an operator from B_2 into B_1. Hence it follows that Cauchy's problem for the wave equation is properly posed in the pair of spaces B_1 and B_2 defined here.

§ 8. Improperly Posed Problems in Mathematical Physics

We have already remarked in § 1 that one and the same problem can be properly posed in one pair of spaces and improperly posed in another. We shall elucidate this statement with the aid of two simple examples.

1. Dirichlet problem for the homogeneous Laplace equation. This problem has been considered in § 3, with the following result: if Ω is a domain and Γ its contour, then the problem is properly posed in the pair of spaces $(G(\bar{\Omega}), C(\Gamma))$. There are many cases, however, when it is important to know the value of the Dirichlet integral of the required function

$$D(u) = \int_\Omega \sum_{k=1}^m \left(\frac{\partial u}{\partial x_k}\right)^2 dx,$$

since this integral is usually proportional to the energy of the configuration described by the function u. In keeping with this, we consider the question of the proper posing of Dirichlet's problem for the homogeneous Laplace equation in the pair of spaces $(H_2(\Omega), C(\Gamma))$, where $H_2(\Omega)$ is the space of functions which are square-integrable, together with their first derivatives, in Ω; the norm in $H_2(\Omega)$ is defined by the formula

$$\|u\|_{H_2(\Omega)}^2 = \int_\Omega \left[u^2 + \sum_{k=1}^m \left(\frac{\partial u}{\partial x_k}\right)^2\right] dx. \tag{1}$$

It can easily be demonstrated that this problem is not properly posed. Consider the well-known *Hadamard's problem*. Let Ω be the circle $x_1^2 + x_2^2 < 1$. The problem is: find a function $u(x_1, x_2)$ which is harmonic in Ω and satisfies the boundary condition

$$u|_{\rho=1} = \phi(\theta), \qquad \phi(\theta) = \sum_{n=1}^\infty \frac{1}{2^n} \cos(2^{2n}\theta). \tag{2}$$

Here ρ and θ are the polar coordinates of the point (x_1, x_2).

The terms of the series (2) are continuous on the circumference $\Gamma: \rho = 1$, and the series itself converges uniformly, since it has the convergent majorant

$$\sum_{n=1}^{\infty} \frac{1}{2^n}.$$

Hence it follows that $\phi \in C(\Gamma)$. The solution to problem (2) can be represented in the form of a series (cf. § 1, Chapter 13)

$$u(x_1, x_2) = \sum_{n=1}^{\infty} \frac{1}{2^n} \rho^{2^{2n}} \cos(2^{2n}\theta). \qquad (3)$$

Dirichlet's integral for the function (3) diverges. If $0 < \rho_0 < 1$, it is not difficult to establish that

$$\iint_{\rho < \rho_0} \left[\left(\frac{\partial u}{\partial x_1}\right)^2 + \left(\frac{\partial u}{\partial x_2}\right)^2\right] dx_1 dx_2 = \pi \sum_{n=1}^{\infty} \rho_0^{2^{2n+1}}. \qquad (4)$$

The sum of the series (4) tends to infinity as $\rho_0 \to 1$, since if we prescribe a positive integer N and choose $\rho_1 < 1$ such that $\rho_1^{2^{2N+1}} > \frac{1}{2}$, then, for $\rho_0 > \rho_1$,

$$\sum_{n=1}^{\infty} \rho_0^{2^{2n+1}} > \sum_{n=1}^{N} \rho_0^{2^{2n+1}} > \tfrac{1}{2}N \underset{N \to \infty}{\to} \infty.$$

If the Dirichlet integral of the function (3) over the circle of unit radius converged, then the smaller integral (4) would be bounded.

Thus, the solution of problem (2) does not belong to the space $H_2(\Omega)$. Consider now the operator R which transforms the boundary function $\phi(\theta)$ into the solution of the Dirichlet problem $u(x_1, x_2)$. If this operator is regarded as an operator from $C(\Gamma)$ into $H_2(\Gamma)$, then it is not defined on the whole of the space $C(\Gamma)$. From this alone it follows that the Dirichlet problem is improperly posed in the pair of spaces $(H_2(\Gamma), C(\Gamma))$.

2. Cauchy's problem for the string equation. We pose Cauchy's problem

$$\frac{\partial^2 u}{\partial t^2} - \frac{\partial^2 u}{\partial x^2} = 0, \quad u|_{t=0} = \phi_0(x), \quad \left.\frac{\partial u}{\partial t}\right|_{t=0} = \phi_1(x) \qquad (5)$$

subject to the following assumptions: the functions $\phi_0(x)$ and $\phi_1(x)$ are given on the segment $0 \le x \le 1$, where

$$\phi_0 \in C^{(2)}[0, 1], \quad \phi_1 \in C^{(1)}[0, 1].$$

The segment [0, 1] is the domain of dependence for the triangle T (Fig. 49); in this triangle the solution to Cauchy's problem exists, is unique, and can be represented by D'Alembert's formula ((7.5), Chapter 24). It is easy to

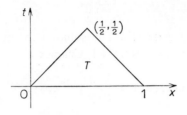

Fig. 49.

see from this formula that the solution to problem (5) $u \in C^{(2)}(\overline{T})$, and that this problem is properly posed in the pair of spaces $(C^{(2)}(\overline{T}), B_2)$, where B_2 is the space of pairs $\Phi = (\phi_0, \phi_1)$, with norm

$$||\Phi|| = ||\phi_0||_{C^{(2)}} + ||\phi_1||_{C^{(1)}}.$$

D'Alembert's formula also shows that problem (5) is improperly posed in, for example, the pair of spaces $(C^{(k)}(\overline{T}), B_2)$ for any $k > 2$.

Appendices

Appendices

Appendix 1. Elliptic Systems

1. Definition of elliptic systems. Consider a system of N linear partial differential equations with N unknown functions $u_1, u_2, \ldots, u_N$ and m independent variables $x_1, x_2, \ldots, x_m$. The system can be written in the form

$$\sum_{k=1}^{N} L_{jk} u_k = f_j(x), \qquad j = 1, 2, \ldots, N, \tag{1}$$

where the L_{jk} are certain linear differential expressions, and x denotes, as usual, a point of m-dimensional Euclidean space with coordinates $x_1, x_2, \ldots, x_m$.

It will be convenient in what follows to write system (1) in somewhat different form.

Let $\alpha_1, \alpha_2, \ldots, \alpha_m$ be non-negative integers. We shall call the ordered set $\alpha = (\alpha_1, \alpha_2, \ldots, \alpha_m)$ a *multi-index*, and we shall denote by $|\alpha|$ the sum of the components of the multi-index α:

$$|\alpha| = \alpha_1 + \alpha_2 + \ldots + \alpha_m.$$

If ξ is an m-component vector, $\xi = (\xi_1, \xi_2, \ldots, \xi_m)$, then we shall write

$$\xi^\alpha = \xi_1^{\alpha_1} \xi_2^{\alpha_2} \ldots \xi_m^{\alpha_m}.$$

We shall also put

$$D_k u = \frac{\partial u}{\partial x_k}, \qquad k = 1, 2, \ldots, m,$$

and write

$$D^\alpha = D_1^{\alpha_1} D_2^{\alpha_2} \ldots D_m^{\alpha_m}.$$

If the order of the differential expression L_{jk} is equal to s_{jk}, then we can obviously write

$$L_{jk} u_k = \sum_{|\alpha| \leq s_{jk}} A_{jk}^{(\alpha)}(x) D^\alpha u_k.$$

Introduce the notation

$$\sum_{|\alpha| \leq s_{jk}} A_{jk}^{(\alpha)}(x)D^\alpha = L_{jk}(x, D). \tag{2}$$

It is clear that $L_{jk}(x, D)$ is a polynomial with respect to $D_1, D_2, \ldots, D_m$. The system (1) can now be represented in the form

$$\sum_{k=1}^{N} L_{jk}(x, D)u_k = f_j(x), \quad j = 1, 2, \ldots, N,$$

or, if we introduce a matrix of order N,

$$L(x, D) = \|L_{jk}(x, D)\|_{j,k=1}^{j,k=N}$$

and N-component vectors

$$u = (u_1, u_2, \ldots, u_N),$$
$$f = (f_1, f_2, \ldots, f_N),$$

in the form

$$L(x, D)u = f(x). \tag{3}$$

Denote by $L_{jk}^0(x, D)$ the so-called *principal part* of the polynomial matrix $L(x, D)$; this is obtained by retaining on the left-hand side of equation (2) only those terms for which $|\alpha| = s_k$, where s_k is the highest order of the derivatives of u_k contained in the system (1):

$$L_{jk}^0(x, D) = \sum_{|\alpha|=s_k} A_{jk}^{(\alpha)}(x)D^\alpha. \tag{4}$$

Let $\xi = (\xi_1, \xi_2, \ldots, \xi_m)$ be an arbitrary point of the space E_m. Put

$$L_{jk}^0(x, \xi) = \sum_{|\alpha|=s_k} A_{jk}^{(\alpha)}(x)\xi^\alpha. \tag{5}$$

The system (1) (or (3)) is said to be *elliptic in the sense of I. G. Petrovskii* [5] at the point x, if at that point the determinant Det $L^0(x, \xi)$, where

$$L^0(x, \xi) = \begin{Vmatrix} L_{11}^0(x, \xi) & L_{12}^0(x, \xi) & \ldots & L_{1N}^0(x, \xi) \\ L_{21}^0(x, \xi) & L_{22}^0(x, \xi) & \ldots & L_{2N}^0(x, \xi) \\ \cdot & \cdot & \cdot & \cdot \\ L_{N1}^0(x, \xi) & L_{N2}^0(x, \xi) & \ldots & L_{NN}^0(x, \xi) \end{Vmatrix}, \tag{6}$$

is equal to zero only when $\xi_1 = \xi_2 = \ldots = \xi_m = 0$. The system (1) is said to be elliptic on some set, if it is elliptic at every point of the set.

It is easy to see that, in the case of a single second-order elliptic equation, the definition of Petrovskii-ellipticity coincides with the definition of § 2, Chapter 9.

A more general definition of an elliptic system is given in ref. [11].

Example 1. The equation
$$\Delta^n u = f(x), \tag{7}$$
where Δ is the Laplace operator, and n is a positive integer, is elliptic in the sense of Petrovskii. In this case $N = 1$, and the matrix (6) reduces to a single element
$$L_{11}^0(x, \xi) = L_{11}(x, \xi) = (\xi_1^2 + \xi_2^2 + \ldots + \xi_m^2)^n;$$
clearly this expression is equal to zero only when $\xi_1 = \xi_2 = \ldots = \xi_m = 0$.

Example 2. The system of equations of static elasticity theory has the form
$$\Delta u + \omega \operatorname{grad} \operatorname{div} u = f(x). \tag{8}$$

Here $u = (u_1, u_2, \ldots, u_m)$ and $f = (f_1, f_2, \ldots, f_m)$ are m-component vectors, and ω is a numerical parameter. In this case, the matrix (6) has the form
$$\begin{Vmatrix} \xi^2 + \omega \xi_1^2 & \omega \xi_1 \xi_2 & \ldots & \omega \xi_1 \xi_m \\ \omega \xi_2 \xi_1 & \xi^2 + \omega \xi_2^2 & \ldots & \omega \xi_2 \xi_m \\ \ldots & \ldots & \ldots & \ldots \\ \omega \xi_m \xi_1 & \omega \xi_m \xi_2 & \ldots & \xi^2 + \omega \xi_m^2 \end{Vmatrix},$$
where $\xi^2 = \xi_1^2 + \xi_2^2 + \ldots + \xi_m^2$.

The determinant of this matrix is simple to evaluate; it is equal to
$$(1 + \omega) \xi^{2m}.$$

Hence it is clear that the system (8) is elliptic for all values of the parameter ω with the exception of $\omega = -1$.

If we replace $f(x)$ by $\omega f(x)$, and put $\omega = \infty$, we obtain the system
$$\operatorname{grad} \operatorname{div} u = f(x).$$

This is not elliptic; the determinant of matrix (6) is identically zero in this case. For this reason we can say that the system of equations of static elasticity theory is elliptic for all values of ω, with the exception of $\omega = -1$ and $\omega = \infty$.

2. Formulation of the boundary-value problem. Complementing condition. If the dimensionality of our space is $m \geq 3$, the determinant of matrix (6) for an elliptic system is a polynomial of even degree with respect to ξ; in this connection, if ξ and ξ' are linearly-independent vectors in the space E_m, then the polynomial Det $L^0(x, \xi + \tau\xi')$ has with respect to τ the same number of zeros with positive imaginary part as with negative imaginary part (cf. [4], [11]).

If $m = 2$, then we shall consider only those elliptic systems for which the determinant of matrix (6) has both the above-mentioned properties.

Let Ω be a finite domain of the space E_m whose boundary Γ is an $(m-1)$-dimensional surface; for simplicity we shall assume it to be infinitely differentiable. Let the boundary conditions have the form

$$\left[\sum_{k=1}^{N} B_{jk}(x, D) u_k\right]_{\Gamma} = \phi_j(x), \qquad j = 1, 2, \ldots, N_1, \tag{9}$$

where the B_{jk} are polynomials with respect to D, and, consequently, the left-hand sides of (9) are the values of certain differential expressions of the functions u_k, evaluated on the surface Γ.

We shall suppose that the coefficients of the polynomials $L_{jk}(x, D)$ and $B_{jk}(x, D)$ are infinitely differentiable functions of x in $\bar{\Omega}$.

We subject the expressions B_{jk} to the following condition, which in refs. [10] and [11] is called the *complementing condition*. It was formulated earlier, in somewhat different form, in refs. [4] and [9].

Denote by $B^0_{jk}(x, D)$ the principal part of the polynomial $B_{jk}(x, D)$. Let $x \in \Gamma$, ξ be an arbitrary non-zero vector, tangent to Γ at the point x, and v a unit vector normal to Γ at this point. Denote by $\tau_k(x, \xi)$, $k = 1, 2, \ldots, p$, the roots of the equation Det $L^0(x, \xi + \tau v) = 0$, Im $\tau_k > 0$, and by $M(x, \xi, \tau)$ the polynomial (with respect to the variable τ)

$$M(x, \xi, \tau) = \prod_{k=1}^{p} (\tau - \tau_k(x, \xi)).$$

Finally, let $C^0(x, \xi)$ denote the matrix consisting of the algebraic complements of the elements of matrix (6).

The complementing condition is the following:

The rows of the matrix $B^0(x, \xi + \tau v) C^0(x, \xi + \tau v)$ must be linearly-independent modulo $M(x, \xi, \tau)$.

Example 3. Let the system (1) represent one non-degenerate elliptic equation of the second order. For this equation the complementing condition is satisfied in the Dirichlet and Neumann problems; if the dimensionality of the space is $m \geq 3$, then the complementing condition is satisfied in the oblique derivative problem at those, and only those, points of the boundary where the direction of differentiation is not tangent to the boundary. If $m = 2$, then the complementing condition in the oblique derivative problem is satisfied on the whole of the boundary.

Example 4. We pose the Dirichlet problem for the system (8) of equations for static elasticity theory: $u|_\Gamma = \phi(x)$, where $\phi(x)$ is a vector function prescribed on the boundary of the elastic body. The complementing condition is satisfied for $\omega \neq -2$ and violated for $\omega = -2$; it is understood, of course, that we are excluding the values $\omega = \infty$ and $\omega = -1$, for which the system ceases to be elliptic.

It is possible to arrive at the complementing condition in the following way. We fix a point $x_0 \in \Gamma$, and make it the origin of a local coordinate system $y_1, y_2, \ldots, y_m$; as usual, we direct the y_m axis along the normal to Γ, and the other axes then lie in the tangent plane. In the system (1) and boundary conditions (9) we retain only the principal terms, and "freeze" their coefficients, replacing the point x by x_0. Also, we make the system (1) homogeneous, by putting the free terms $f_j(x)$ equal to zero. Finally, we replace the domain Ω by the half-space $y_m > 0$. We thus arrive at a greatly simplified boundary-value problem: the equations and boundary conditions are homogeneous with respect to the order of differentiation, and their coefficients are constants, the differential equations of the problem are also homogeneous in the ordinary sense, and the solution is required in the simplest domain – a half-space. If we effect a Fourier transform with respect to $y_1, y_2, \ldots, y_{m-1}$, then we obtain a boundary-value problem for a system of ordinary, linear differential equations with constant coefficients on the semi-axis $y_m > 0$. The complementing condition is necessary and sufficient in order that this latter problem should have one and only one solution which tends to zero as $y_m \to \infty$.

3. **Sobolev spaces and Slobodetskii spaces.** We have already mentioned Sobolev spaces, $W_p^{(l)}(\Omega)$, in § 5, Chapter 2 (see also refs. [7], [8]). The norm in $W_p^{(l)}(\Omega)$ can be obtained from the formula

$$||u|| = \int_\Omega |u(x)|\, dx + \left\{ \int_\Omega \sum_{|\alpha|=l} |D^\alpha u(x)|^p\, dx \right\}^{1/p}, \qquad (10)$$

which differs only in notation from formula (5.1) of Chapter 2. The norm

$$\left\{\int_\Omega \sum_{|\alpha|=0}^{l} |D^\alpha u(x)|^p \, dx\right\}^{1/p} \tag{11}$$

is equivalent to the norm (10); if $p = 2$, then the norm (11) makes $W_2^{(l)}(\Omega)$ a Hilbert space.

The functions generating the space $W_p^{(l)}(\Omega)$ can be vector functions. In that case, the expressions $|u(x)|$ and $|D^\alpha u(x)|$ in formulae (10) and (11) are understood to mean the lengths of the respective vectors.

When $l = 0$, the Sobolev space becomes the space $L_2(\Omega)$.

Now let l be positive, but not an integer. Put $l_0 = [l]$, $\lambda = l - [l_0]$. A Slobodetskii space [6], $W_p^{(l)}(\Omega)$, is generated by functions $u(x)$ which have all possible generalized derivatives of order l_0 in the domain Ω, which are integrable to degree p, and where these derivatives are such that

$$\sum_{|\alpha|=l_0} \int_\Omega \int_\Omega \frac{|D^\alpha u(x) - D^\alpha u(y)|^p}{|x-y|^{\lambda p + m}} \, dx \, dy < \infty.$$

The norm in a Slobodetskii space can be found from the formula

$$\|u\|^p = \sum_{|\alpha|=0}^{l_0} \int_\Omega |D^\alpha u(x)|^p \, dx$$
$$+ \sum_{|\alpha|=l_0} \int_\Omega \int_\Omega \frac{|D^\alpha u(x) - D^\alpha u(y)|^p}{|x-y|^{\lambda p + m}} \, dx \, dy. \tag{12}$$

Henceforth we shall restrict ourselves to the case $p = 2$. Then the space $W_2^{(l)}$ with norm (12) will be a Hilbert space.

It is possible to consider spaces $W_p^{(l)}$ which are generated by functions given, not in a domain, but on some manifold, in particular on some surface.

4. Coercive boundary-value problems. The boundary-value problem (1), (9) is called *elliptic* or *coercive* in the domain Ω if 1) the system (1) is elliptic in the closed domain $\bar\Omega$; 2) the complementing condition is satisfied at all points of the boundary of Ω.

Denote by L the operator generated by the left-hand side of the matrix equation (3), by B the operator generated by the matrix of operators B_{jk} which occur in the boundary conditions (9). Finally, let A denote the pair (L, B). We require that all the differential expressions L_{jk} should have one and the same order s.

Let m_j be the highest order of differentiation in the jth condition (9), and let n be any integer, not less than $1 + \max m_j$. Denote by $V^{(n)}(\Gamma)$ the orthogonal sum of the Slobodetskii spaces $W_2^{(n-m_j-\frac{1}{2})}(\Gamma), j = 1, 2, \ldots, N_1$. Further, if l is a sufficiently large integer, then denote by $H_l(\Omega, \Gamma)$ the orthogonal sum of the vector Sobolev space $W_2^{(l)}(\Omega)$ and the space $V^{(l)}(\Gamma)$. We shall assume that $l \geq l_0 = \max(s, m_j+1)$.

Obviously, the operator A introduced above acts from $W_2^{(l)}(\Omega)$ into $H_l(\Omega, \Gamma)$.

The following theorem holds.

THEOREM. *Coerciveness of the problem* (1), (9) *is equivalent to each of the following conditions*:

1. *If* $u \in W_2^{(l_0)}(\Omega)$, $Lu \in W_2^{(l-s)}(\Omega)$ *and* $Bu \in V^{(l)}(\Gamma)$ *then* $u \in W_2^{(l)}(\Omega)$, *and the estimate*

$$\|u\|_{W_2^{(l)}} \leq C\{\|Lu\|_{W_2^{(l-s)}} + \|Bu\|_{V^{(l)}} + \|u\|_{L_2}\} \qquad (13)$$

holds, where the constant C is independent of the function u.

2. *The operator A is normally soluble and has a finite index.*

The theorem stated here has been proved in ref. [1] for the wider class of so-called *singular integro-differential* boundary conditions, with respect to which our boundary condition (9) represents a special case.

5. The index of a coercive problem. It follows from the above theorem that every coercive problem has a finite index. Considerable interest attaches to the problem of calculating this index and investigating its properties.

A very general formula for the index, useful in a wide class of problems and, in particular, in the coercive problem (1), (9), was given by M. F. Atiyah and I. M. Singer [12]. But the actual evaluation of an index from this formula is very difficult. It is therefore of interest to look at cases where some statement about the index can be formulated comparatively simply. We give one such statement.

Let two coercive problems of the type (1), (9) *differ only in their boundary conditions. Then the difference between the indices of these problems is equal to the index of a system of singular integral equations which is wholly defined by the data of the two problems* [1].

The index of a system of singular integral equations can be calculated comparatively easily. Therefore, if the index is known for a given elliptic system (1) subject to some boundary conditions, it is not difficult to calculate it for any other boundary conditions which satisfy the complementing requirement.

6. Non-coercive problems for elliptic systems. These problems are extremely difficult. More extensively developed is the oblique derivative problem for one non-degenerate, second-order elliptic equation in the case when the direction of differentiation is tangent to the surface of the boundary, on a manifold of lower dimensionality. An investigation of the basis results relating to the oblique derivative problem in the non-coercive case can be found in refs. [2], [3], [13].

Appendix 2. Cauchy's Problem for Hyperbolic Equations

V. M. BABICH

1. In the course of this book, the solution to Cauchy's problem for the wave equation

$$u_{tt} - a^2 \Delta u = 0$$

has been obtained for the case where the initial data are prescribed at $t = 0$.

It is possible to derive a formula which solves the far more general Cauchy problem

$$u_{tt} - a^2 \Delta u = f(x, t),$$
$$u|_S = u_0(t, x), \tag{1}$$
$$\frac{\partial u}{\partial n}\bigg|_S = u_1(t, x).$$

Here f, u_0, u_1 are sufficiently smooth functions, u_0 and u_1 being prescribed on a sufficiently smooth surface S, and n is the direction of the normal to S. The surface S must fulfil one further important condition, namely, it must satisfy the inequality

$$\cos^2(n, t) - a^2 \sum_{i=1}^{m} \cos^2(n, x_i) > 0. \tag{2}$$

A surface S which satisfies condition (2) is called *spatially-orientated* or *space-like*.

We outline a method whereby it is possible to construct a solution to problem (1).

Let m be an even number and (x_0, t_0) be the point at which we wish to evaluate the function u (we assume that the solution of (1) exists). Let the characteristic cone with vertex at the point (x_0, t_0),

$$a^2(t - t_0)^2 - \sum_{i=1}^{m}(x_i - x_i^0) = 0 \tag{3}$$

cut out from S a finite portion Σ with sufficiently smooth $(m-1)$-dimensional boundary σ. Using condition (2), we can easily show that if the point (x_0, t_0) is close to S then the previous assumption always holds.

We apply Green's formula

$$\int_\Xi uLv\,d\Xi = \int_F (uPv - vPu)\,dF + \int_\Xi vLu\,d\Xi,$$

$$L = \frac{\partial^2}{\partial t^2} - a^2\Delta, \quad P = \cos(n,t)\frac{\partial}{\partial t} - \sum_{i=1}^m \cos(n, x_i)\frac{\partial}{\partial x_i} \quad (4)$$

to the required solution of problem (1), and to the function

$$v = v_\lambda(x, t, x_0, t_0) =$$
$$= \frac{(-1)^{\frac{1}{2}(m-2)}\Gamma(\frac{1}{2}(m-1))}{2\pi^{\frac{1}{2}(m+1)}a}\left[a^2(t-t_0)^2 - \sum_{i=1}^m (x_i - x_i^0)^2\right]_+^\lambda. \quad (5)$$

Here Γ is the Gamma function, λ is a complex number; we use the notation $\zeta_+^\lambda = \zeta^\lambda$ for $\zeta > 0$ and $\zeta_+^\lambda = 0$ for $\zeta \leq 0$. The domain of integration Ξ is here bounded by the surface of the cone (3) and the portion Σ which this cone cuts out of S. Denote the boundary of this domain by F. Both sides of equation (4) are regular functions of λ for Re $\lambda > 2$. Continuing analytically both sides of equation (5) to the point $\lambda = -\frac{1}{2}(m-1)$ (it can be shown that this analytic continuation is possible), we obtain on the left-hand side of equation (4) the function $u(x_0, t_0)$, and on the right we obtain an expression which contains no unknown functions.

In fact, when Re $\lambda > 2$, v and Pv are zero on the surface of the cone, and $Lu = f$, $u|_S = u_0$; $P(u)$ can easily be expressed in terms of $u|_S = u_0$ and $(\partial u/\partial n)|_S = u_1$.

It can be shown by direct verification that the formula obtained by this method solves problem (1).

When $\lambda = -\frac{1}{2}(m-1)$, the function (5), continued to be equal to zero in the domain $a^2(t-t_0)^2 - \sum_{i=1}^m(x_i - x_i^0)^2 < 0$ and in the half-space $t > t_0$ (or $t_0 < t$), is called *the fundamental solution of Cauchy's problem* for the wave equation. In particular, when $m = 2$ this fundamental solution takes the form

$$v = \frac{1}{2\pi a}\left[a^2(t-t_0)^2 - (x_1 - x_1^0)^2 - (x_2 - x_2^0)^2\right]_+^{-\frac{1}{2}}. \quad (6)$$

Note that the formula which solves Cauchy's problem for the wave equation in the plane case has the form (cf. § 6, Chapter 24)

$$u(x_0, t_0) = \iint_{t=0} u_1 v\,dx_1\,dx_2 + \frac{\partial}{\partial t}\iint_{t=0} u_0 v\,dx_1\,dx_2.$$

If the number m of space variables is odd and greater than one, then the solution to problem (1) can be obtained in the same way, except that instead of the function (5) we have to substitute in Green's formula (4) the function

$$v = v_\lambda = \frac{1}{2\pi^{\frac{1}{2}(m-1)}a} \frac{\left[a^2(t-t_0)^2 - \sum_{i=1}^{m}(x_i-x_i^0)^2\right]_+^\lambda}{\Gamma(\lambda+1)},$$

$$\mathrm{Re}\,\lambda > 2$$

and both sides of the resulting equation have to be analytically continued to the point $\lambda = -\frac{1}{2}(m-1)$.

The fundamental solution of Cauchy's problem is the generalized function v,* which solves the problem

$$Lv = \delta(x_1-x_1^0,\ldots,x_m-x_m^0,t-t_0),$$
$$v|_{t>t_0} = 0 \ (\text{or } v|_{t<t_0} = 0). \tag{7}$$

Here δ is the Dirac delta function.

In the case when m is odd, the fundamental solution of Cauchy's problem has the form

$$v = \frac{1}{2\pi^{\frac{1}{2}(m-1)}a} \delta^{\frac{1}{2}(m-3)}\left[a^2(t-t_0)^2 - \sum_{i=1}^{m}(x_i-x_i^0)^2\right], \tag{8}$$

$$v|_{t>t_0} = 0 \ (\text{or } v|_{t<t_0} = 0).$$

When m is even, the fundamental solution of Cauchy's problem is given by the function (5) if we put $\lambda = \frac{1}{2}(1-m)$.

2. For the hyperbolic equation

$$Lu = \sum_{i,j=0}^{m} a_{ij}(x)\frac{\partial^2 u}{\partial x_i \partial x_j} + \sum_{i=0}^{m} b_i(x)\frac{\partial u}{\partial x_i} + c(x)u = f(x),$$
$$x = (x_0, x_1, \ldots, x_m), \tag{9}$$

the Cauchy problem

$$Lu = f, \quad u|_S = u_0, \quad \frac{\partial u}{\partial n}\bigg|_S = u_1, \tag{10}$$

is well understood, if S is a space-like surface.

---

* For generalized functions see ref. [5] (cf. also Appendix 3, pages 532–534).

At every point x_0 we construct the cone

$$\sum_{i,j=0}^{m} A_{ij}(x_0)(x_i-x_i^0)(x_j-x_j^0) = 0. \tag{11}$$

The matrix $A_{ij}(x_0)$ is the inverse of the matrix $a_{ij}(x_0)$. The cone (11) can be obtained by an affine transformation of the cone (3). The cone (11) divides the space into three domains: two "interior" and one "exterior". The surface S is space-like if and only if for every point $x_0 \in S$ the normal at this point has no common point either with the cone (11) (except, of course, x_0 itself), or with its exterior domain.

We can introduce the concept of a fundamental solution to Cauchy's problem for the differential operator (9).

In this case, the fundamental solution of Cauchy's problem is the generalized function v satisfying zero initial conditions on any (space-like) surface and satisfying close to S the equation

$$Lv = \delta(x_0-x_0^0, \ldots, x_m-x_m^0) \quad (x_0 = (x_0^0, \ldots, x_m^0) \bar\in S).$$

Singularities of the fundamental solution lie on the characteristic conoid corresponding to the point x_0 – this is called the characteristic surface having a conical singular point at x_0.

The singularities of the fundamental solutions in the case of the wave operator and the operator (10) have a similar analytic character for the same m, if the characteristic conoid of operator (10) has no singularities other than the point x_0.

If the characteristic conoid has singular points, then the fundamental solution in their neighbourhood has a very complex structure. Note that the fundamental solution for operator (10) in the case $m = 1$ is called the *Riemann function*, while for even m it is called *Hadamard's elementary solution*.

3. An extensive study has been made of discontinuous solutions (solutions in a generalized sense or in the sense of generalized functions) of equation (10). It can be shown that, under normal conditions, the solution of (10) can have discontinuities only on characteristics, if the a_{ij}, b_i, c, f are smooth.

Suppose equation (10) ($m = 3$) describes the physical process of wave propagation.

The characteristic $F(x_1, x_2, x_3, t) = 0, t = x_0$, on which u has a discontinuity, is, from the viewpoint of an observer in the coordinate system x_1, x_2, x_3, a moving surface along which the function u varies especially quickly. These surfaces are called *wave fronts*. In the case of the wave equation with variable velocity

$$\frac{1}{c^2(x)} u_{tt} - \Delta u = 0, \qquad x = (x_1, x_2, x_3), \tag{12}$$

the orthogonal trajectories to the wave fronts are called *rays*; rays satisfy Fermat's Principle: along a ray, the variation of the *Fermat functional*

$$\int \frac{ds}{c(x)} = \int \frac{\sqrt{(dx_1^2 + dx_2^2 + dx_3^2)}}{c(x_1, x_2, x_3)} \tag{13}$$

is equal to zero.

A formal method has been developed for finding the singularities in the solutions to problems for equation (12), in which an important constituent part is the construction of rays, i.e. the extremals of equation (13). This method, known as the *ray method* is widely used (even though without rigorous justification) in mathematical problems of wave theory.

A rigorous justification of the ray method in the general case is difficult, and the problem has not yet been solved.

4. We turn to the wave equation for $c(x) = a = \text{const}$. Let the initial functions $u_0(x_1, x_2, \ldots, x_m) = u|_{t=0}$ and $u_1(x_1, \ldots, x_m) = u_t|_{t=0}$ be equal to zero outside some domain Ξ. Construct the characteristic cones (cf. formula (3)) with vertices at each point $t = 0$, $x_1^0, \ldots, x_m^0$, where $x_1^0, \ldots, x_m^0 \in \Xi$.

We define a set F by the formula

$$F = \bigcup_{(x^0{}_1, \ldots, x^0{}_m) \in \Xi} \left\{ (t, x_1, \ldots, x_m) : a^2 t^2 = \sum_{i=1}^{m} (x_i - x_i^0)^2 \right\}.$$

This set F is the union of points belonging to the characteristic cones whose vertices lie on the plane $t = 0$ at points of the domain Ξ. It is obvious that if Ξ collapses to the point x_0, then the set F contracts to the characteristic cone

$$a^2 t^2 = (x - x_0)^2.$$

It follows from Kirchhoff's formula for $m = 3$ (cf. § 3, Chapter 24) that the solution $u(t, x_1, x_2, x_3)$ to Cauchy's problem is zero outside the set F. This fact is usually expressed by saying "there is no wave diffusion for the wave equation with $m = 3$". (For more details, cf. [9].)

The concept of an *equation without wave diffusion* can easily be extended to the general case of equation (9). The role of the characteristic cones in this case will be taken by the characteristic conoids.

It is not difficult to show in the case of the wave equation that when $m = 5, 7, 9, \ldots$ wave-diffusion will be absent, while when $m = 1, 2, 4, 6, 8, \ldots$ the function u is in general not zero outside the set F. More precisely, u will be equal to zero outside the sets

$$F_1 = \bigcup_{(x^0_1, \ldots, x^0_m) \in \Xi} \left\{ (t, x_1, \ldots, x_m) : a^2 t^2 \geq \sum_{i=1}^{m} (x_i - x_i^0)^2 \right\}$$

and will be identically non-zero on the set F_1.

The fact that u is equal to zero outside F_1 is a consequence of the finiteness of the domain of dependence on the initial data of the solutions of Cauchy's problem (cf. par. 7 below). Precise analogues of this fact hold for the most general hyperbolic equations, whereas the absence of wave diffusion is in a certain sense exceptional.

In the case of hyperbolic equations with variable coefficients, diffusion always occurs when m is even. When m is odd, it is obvious that if the equation reduces to the wave equation through a change of variables, then there is no wave diffusion. For $m = 5, 7, 9, \ldots$ there exist equations, not reducing to the wave equation through a change of variables, for which there is nevertheless no diffusion [13]. When $m = 3$, * it is not known whether or not there exist equations without diffusion, which do not reduce to the wave equation. If it could be shown that "diffusionless" equations not reducing to the wave equation do not exist, this would point to the exceptional character of the space-time continuum (x_1, x_2, x_3, t).

The questions which we have so briefly discussed here are considered in detail in the classical work of Hadamard [10], in the article by M. Riesz [12], and in the book by Courant [7]. The ray method and its applications are also examined extensively in [3].

5. Consider the partial differential equation with constant coefficients

$$Lu = \sum_{\Sigma \alpha_i = p} a_{\alpha_0 \alpha_1 \ldots \alpha_m} \frac{\partial^p u}{\partial t^{\alpha_0} \partial x_1^{\alpha_1} \ldots \partial x_m^{\alpha_m}},$$

$$a_{p0\ldots 0} = 1, \qquad a_{\alpha_0 \ldots \alpha_m} = \text{const.} \tag{14}$$

---

* The Polish mathematician M. Mathison has shown that if in the operator (10), $m = 3$, $a_{ij} = $ const., and the equation $Lu = 0$ is an equation without diffusion, then this equation can be reduced to the wave equation by a change of variables [11].

Equation (14) is said to be *hyperbolic in the sense of Petrovskii*, if for any numbers $\omega_1, \ldots, \omega_m$, not all zero, the polynomial with respect to λ (with coefficients depending on $\omega = (\omega_1, \ldots, \omega_m)$)

$$\Delta(\lambda, \omega) = \sum_{\Sigma \alpha_i = p} a_{\alpha_0 \ldots \alpha_m} \lambda^{\alpha_0} \omega_1^{\alpha_1} \ldots \omega_m^{\alpha_m}$$

has exactly p real roots, and these roots are all distinct. We shall look for a solution of equation (14) in the form

$$u = f(vt + \sum_{i=1}^{m} \omega_i x_i), \quad \sum_{i=1}^{m} \omega_i^2 = 1, \tag{15}$$

where f is an arbitrary function.

The solution of equation (14) having the form (15) is called a solution of *plane-wave type*, propagating with velocity v in the direction $-\omega$. It is obvious that

$$u = f(vt + \sum_{i=1}^{m} \omega_i x_i) = \text{const.},$$

if

$$vt + \sum_{i=1}^{m} \omega_i x_i = C = \text{const.}, \tag{16}$$

i.e. the plane-wave solution of the form (15) has a constant value on surfaces which are orthogonal to the vector ω and which move in the direction $-\omega$ with velocity v.

Substituting (15) into equation (14), we obtain

$$Lf(vt + \sum_{i=1}^{m} \omega_i x_i) = \Delta(v, \omega) f^{(p)}(vt + \sum_{i=1}^{m} \omega_i x_i).$$

In order that equation (14) should be satisfied for any function f it is necessary and sufficient that

$$\Delta(v, \omega) = 0.$$

Thus, the condition that equation (14) is hyperbolic in the sense of Petrovskii is equivalent to the statement that for any direction ω there exist exactly p plane waves with different velocities, propagating in the direction $-\omega$.

The generalized function h is called a fundamental solution of Cauchy's problem for equation (14) if it satisfies the conditions

$$Lh = \delta(x - x_0, t - t_0),$$
$$h|_{t > t_0} = 0.$$

When $p \geq 2$, the fundamental solution h enables us to solve by quadratures the problem

$$Lv = F(x), \quad v|_{t=0} = v_0(x), \ldots, \left.\frac{\partial^{p-1}v}{\partial t^{p-1}}\right|_{t=0} = v_{p-1}(x)$$

(F, v_i are sufficiently smooth functions).

The fundamental solution can be constructed through the superposition of plane waves for any $m = 1, 2, 3, \ldots$ and $p = 2, 3, 4, \ldots$. For example, for odd m and $p \geq m+1$

$$h = -\frac{(-1)^{\frac{1}{2}(m-1)}}{2(2\pi)^{m-1}(p-m-1)!} \times$$

$$\times \int_\Xi \left(\sum_{k=1}^m X_k \xi_k + t - t_0\right)^{p-m-1} \operatorname{sgn}\left(\sum_{k=1}^m X_k \xi_k + t - t_0\right) d\Xi,$$

$$d\Xi = \frac{d\sigma}{|\operatorname{grad} H| \operatorname{sgn} \sum_{k=1}^m \xi_k H_{\xi_k}}.$$

Here $H(\xi_1, \xi_2, \ldots, \xi_m) \equiv \Delta(1, \xi_1, \ldots, \xi_m)$, $X_k = x_k - x_k^0$, $d\sigma$ is an element of the surface $H = 0$.

The derivation of the formula for the fundamental solution can be found in ref. [1].

6. The surface $\gamma(t, x_1, \ldots, x_m) = 0$ is said to be characteristic for equation (14), if at points of this surface

$$\Delta\left(\frac{\partial \gamma}{\partial t}, \frac{\partial \gamma}{\partial x_1}, \ldots, \frac{\partial \gamma}{\partial x_m}\right) = 0.$$

It is interesting to note that fundamental solutions are singular only on the characteristic conoids. A surface in the space $x_1, \ldots, x_m, t$ is called a characteristic conoid * if it is the envelope of the family of planes

---

* A different definition has been given (par. 2) for the characteristic conoid of a second-order hyperbolic equation with, in general, non-constant coefficients. The two definitions are simultaneously applicable only in the case of a single hyperbolic equation with constant coefficients, of the form

$$u_{tt} = a_{ij} \partial^2 u / \partial x_i \partial x_j, \quad a_{ij} = \text{const.},$$

in which case they lead to one and the same characteristic conoid:

$$(t - t_0)^2 = A_{ij}(x_i - x_i^0)(x_j - x_j^0),$$

$$\|A_{ij}\|_{i,j=1}^{i,j=m} = \|a_{ij}\|^{-1}.$$

$$v(t-t_0)+\sum_{i=1}^{m}\omega_i(x_i-x_i^0)=0,$$

where
$$\Delta(v,\boldsymbol{\omega})=0, \quad \boldsymbol{\omega}=(\omega_1,\ldots,\omega_m).$$

In the case
$$m=2 \text{ and } \Delta(v,\boldsymbol{\omega})\equiv v^2-a^2\sum_{i=1}^{m}\omega_i^2$$

equation (14) coincides with the wave equation, and the characteristic conoid coincides with the conoid (3). In the general case, the characteristic conoid is the characteristic surface for equation (14) with the singular point $x_i=x_i^0, t=t_0$. Generally speaking, the characteristic conoid represents $[\frac{1}{2}(p-1)]+1$ cones having the point $x_i=x_i^0, t=t_0$ as their common vertex (the symbol [ ] denotes integral part). The singularities of the fundamental solution in the neighbourhood of a regular point of the characteristic conoid have been studied in [2]. It is found that their character can only be algebraic, logarithmic, or "delta-shaped".

In the neighbourhood of singular points of the characteristic conoid, the singularity of the fundamental solutions has a very complex character; this question has not yet been fully clarified.

7. The fundamental solution is equal to zero outside the "broadest" of the cones which make up the characteristic conoid. This is closely connected with one very important property of hyperbolic equations:

If the initial data for Cauchy's problem for the wave equation $u_{tt}-a^2\Delta u=0$ are given for $t=0$, then the solution at the point $x_i=x_i^0$, $t=t_0, t_0>0$ depends only on the values of the initial data inside the sphere $(x_i-x_i^0)^2=a^2t_0^2$. Thus, in the case of the wave equation, the domain of dependence of the solution to Cauchy's problem is finite at every point. For non-hyperbolic equations the domain of dependence is not finite.

In Cauchy's problem for the heat-conduction equation, for example, i.e. $u_t=\Delta u; u|_{t=0}=\phi(x), x=x_1,\ldots,x_m$, the solution $u(x_0,t_0)$ depends, for any $t_0>0$, on the values of $\phi(x)$ for all x (i.e., for $-\infty<x_i<\infty$, $i=1,2,\ldots,m$. Cf. Poisson's formula, § 2, Chapter 23).

In the case of equation (14), with data prescribed at $t=0$, the domain of dependence at the point $x=x_0, t=t_0$ can be found in the following way: construct the characteristic conoid with vertex at the point $x=x_0, t=t_0$. The domain which cuts out from the plane $t=0$ the "broadest" cone of the conoid is the domain of dependence for the point (x_0,t_0).

8. The system of equations

$$\frac{\partial^{n_i} u_i}{\partial t^{n_i}} = \sum_{\substack{\Sigma \alpha_j = n_i \\ \alpha_0 < n_i \\ s = 1, 2, \ldots, N}} a_{\alpha_0 \alpha_1 \ldots \alpha_m}^{(s, i)} \frac{\partial^{n_i} u_s}{\partial t^{\alpha_0} \partial x_1^{\alpha_1} \ldots \partial x_m^{\alpha_m}} \tag{17}$$

$$(a_{\alpha_0 \ldots \alpha_m}^{(s, i)} = \text{const.}, s = 1, 2, \ldots, N)$$

is called *hyperbolic in the sense of Petrovskii*, if for any real $\omega_1, \ldots, \omega_m$, not all zero, the polynomial with respect to λ,

$$\det \left\| \lambda^{n_i} \delta_{is} - \sum_{\substack{\Sigma \alpha_j = n_i \\ \alpha_0 < n_i}} a_{\alpha_0 \ldots \alpha_m}^{(s, i)} \lambda^{\alpha_0} \omega_1^{\alpha_1}, \ldots, \omega_m^{\alpha_m} \right\| \tag{18}$$

($s, i = 1, 2, \ldots, N$, δ_{is} is the Kronecker delta) has only real and distinct roots.

The discussion concerning the properties of equation (14), given in pars. 7 and 8, can be repeated with very little change for the system (17).

9. Consider the system (in general, non-linear) of N equations for the N functions $u_1, \cdots, u_N$:

$$\frac{\partial^{n_i} u_i}{\partial t^{n_i}} = F_i \left(t, x, u_s, \ldots, \frac{\partial^r u_s}{\partial t^{\alpha_0}, \partial x_1^{\alpha_1}, \ldots, \partial x_m^{\alpha_m}} \right)$$

$$(\alpha_0 < n_i; \; r \leq n_i; \; i, s = 1, 2, \ldots, N). \tag{19}$$

Let the initial data prescribed for this system be

$$u_i \bigg|_{t=0} = u_{i0}(x), \ldots, \frac{\partial^{n_i - 1} u_i}{\partial t^{n_i - 1}} \bigg|_{t=0} = u_{i, n_i - 1}(x),$$

$$x = (x_1, \ldots, x_m). \tag{20}$$

If $u_{i\alpha}(x)$ and F_i are sufficiently smooth functions, then it is possible to evaluate at $t = 0$ all the derivatives of $u_i(t, x_1, \ldots, x_m)$ ($i = 1, 2, \ldots, N$) which are present in the system (18). The system (19) is called hyperbolic for $t = 0$, for initial data (20), if the determinant (18), where

$$a_{\alpha_0 \alpha_1 \ldots \alpha_m}^{(s, i)} = \frac{\partial F_i}{\partial \left[\dfrac{\partial^p u}{\partial t^{\alpha_0} \partial x_1^{\alpha_1} \ldots \partial x_m^{\alpha_m}} \right]} \bigg|_{t=0}, \tag{21}$$

has only real, distinct roots λ.

Hyperbolicity in the case of a linear system of form (19) does not depend on the initial data. If (19) is hyperbolic and linear, then we can introduce for this system the concept of a fundamental matrix (generalization of the concept of the fundamental solution to Cauchy's problem), as well as the concept of a characteristic conoid for each point (x_0, t_0) (cf. [1]).

In the most general case, when system (18) is Petrovskii-hyperbolic at $t = 0$, Cauchy's problem (19)–(20) is properly posed near $t = 0$; i.e. Cauchy's problem is uniquely soluble for sufficiently smooth $u_{is}(x)$, with the solution depending continuously on the initial data in the following sense: the vector function $\boldsymbol{u}^{(\kappa)}(x_0, t_0)$ ($t_0 > 0$, t_0 sufficiently small), $\boldsymbol{u}^{(\kappa)} = (u_1^{(\kappa)}, \ldots, u_N^{(\kappa)})$, which solves the Cauchy problem (19)–(20) with initial data $u_{is}^{(\kappa)}$ ($\kappa = 1, 2, 3, \ldots$), tends to the solution $\boldsymbol{u}(x_0, t_0)$ of Cauchy's problem with initial data $u_{is}(x)$, if, when $x \to \infty$, the initial data $u_{is}^{(\kappa)}(x)$, together with a sufficient number of derivatives, converge to u_i on any compactum on the surface.

The proof that this most general problem is properly posed can be obtained with the aid of an estimate of the required solution in the so-called energy norm (cf. [8], [6]).

10. I. M. Gel'fand and G. Ye. Shilov (c.f. [5]) call the system of linear equations with constant coefficients

$$\frac{\partial u_j}{\partial t} = \sum_{k=1}^{m} P_{jk}\left(i\frac{\partial}{\partial x_l}\right) u_k(x, t) \qquad (22)$$

hyperbolic, if the function $\Lambda(s) = \max_j \operatorname{Re} \lambda_j(s)$ (where the $\lambda_j(s)$ are the characteristic roots of the matrix $||P_{jk}(s)H||$) satisfies the following conditions:

1) $\Lambda(s) \leq a|s|+b$, $\quad a, b = \text{const.}$;

2) for real $s = \sigma$

$$\Lambda(s) \leq c, \qquad c = \text{const.}$$

For such systems, the Cauchy problem with data for $t = 0$ is properly posed in the same sense that the Petrovskii-hyperbolic system (19) is properly posed. But here it is not necessary that t be small.

If we confine ourselves to systems of the form (22), then the conditions stated above are not only sufficient for Cauchy's problem to be properly posed (in the sense of par. 9), but also necessary.

Appendix 3. Some Questions in the Theory of General Differential Operators

V. G. MAZ'YA

1. Generalized functions. Until recent times the study of the theory of differential equations was concentrated on operators of three types: elliptic, parabolic and hyperbolic. In this book an extensive account has been given of the Laplace, heat-conduction and wave operators, which are typical and very simple representatives of the above three types respectively. The investigation of these operators and their generalizations, which was begun at the beginning of the nineteenth century, has led to an enormous literature, has enabled the accumulation and analysis of a large number of facts and relations, and has stimulated the development of the theory of functions, of functional analysis and of other branches of mathematics.

Alongside this, even about two decades ago almost nothing was known about equations and systems not belonging to these three classical types. But in recent years there has been a profound development in this direction, so much that we can speak today of a theory of general differential operators. A brief and necessarily incomplete presentation of some aspects of this theory are given in the present outline.

The study of general differential operators was made possible in the first place by the creation of the theory of generalized functions.

In non-rigorous form, generalized functions were first used in physics. For example, the representation of an electric point charge leads to the intuitive definition of the so-called delta function, i.e. the function which is equal to zero everywhere except at one point, where it is infinite, and having an integral equal to unity. The concept of a dipole impels the introduction of the derivative of a δ-function.

These and other "strange" objects also arose in mathematical questions, for example in the theory of hyperbolic equations. All this pointed to the inadequacy of the classical concept of a function.

The first generalized functions were introduced in mathematics and used by S. L. Sobolev, in 1936 [11], in the study of Cauchy's problem for linear hyperbolic equations. The theory of generalized functions in its contemporary form was expounded by L. Schwartz [24], and received further elaboration in a series of monographs by I. M. Gelfand and other authors *.

What are generalized functions from a mathematical standpoint? If we are speaking of the δ-function in this context, then mathematical analysis has long been in a position to give it a meaningful definition. In fact, all that is involved here is a finite measure, concentrated at a point. But when we turn even to the derivative of the δ-function, we find that measure theory breaks down.

Returning to the δ-function, we recall that every measure can be identified with a linear continuous functional on a space of continuous functions, and we observe that for the δ-function this functional has the form

$$(\delta(x), f(x)) = f(0),$$

where $f(x)$ is any function continuous, for example, on the interval $[-1, 1]$. Thus the δ-function is, by definition, that functional on $C[-1, 1]$ which associates with the function $f(x)$ its value at the point 0.

Using the idea of this definition, we can introduce generalized functions as linear continuous functionals on some space or another (including spaces more restricted than C). These are usually called spaces of basic functions. The choice of the space of basic functions depends on the analytic problem under consideration. These spaces are topological, but not necessarily normed or metric; it will often turn out to be convenient to use so-called enumerably-normed spaces. The concept of convergence of a sequence of basic functions naturally generates convergence in the adjoint space – the space of generalized functions.

As examples of basic spaces we take the spaces $D(\Omega)$ and $S(E_m)$. The elements of $D(\Omega)$ are functions which are infinitely differentiable in the domain Ω and each of which is equal to zero outside some closed sub-domain of Ω. The space $S(E_m)$ consists of functions which are infinitely differentiable in E_m and which tend to zero, together with their derivatives, more rapidly than any power of $|x|^{-1}$ as $|x| \to \infty$. The corresponding adjoint spaces are denoted by $D'(\Omega)$ and $S'(E_m)$. Henceforth we shall discuss only these spaces of generalized functions.

The differentiation operator is defined on $D(\Omega)$ and $S(E_m)$, and leaves these spaces invariant, so that it is permissible to introduce differentiation in the functional spaces as an adjoint operator. Thus, all generalized functions of the classes $D'(\Omega)$ and $S'(E_m)$ turn out to be infinitely differentiable, and the differentiation operator is a continuous operator.

* The ideas in this Appendix are closely related to those of ref. [3].

In a similar way we can define multiplication of a generalized function by any function which is a multiplier in the corresponding space of basic functions *. In particular, a generalized function of the class $D'(\Omega)$ can be multiplied by any infinitely differentiable function in Ω, and a generalized function of the class $S'(E_m)$ by any infinitely differentiable function which grows at infinity not faster than a positive power of $|x|$.

Similar considerations enable the introduction, for generalized functions in $D'(E_m)$ and $S'(E_m)$, of a Fourier transform, convolution (generalization of the integral operator with a difference kernel) and other operations of classical analysis. In particular the Fourier transform maps the space $S(E_m)$ into itself, and therefore it is possible to define a Fourier transform of generalized functions within the confines of the space $S'(E_m)$. In this way the Fourier transform is extended over wide classes of functions, even those which increase at infinity.

The flexibility and breadth of the theory of generalized functions made it a natural apparatus for developing the theory of general differential operators, and brought about a rapid evolution of this theory. Henceforth we shall confine ourselves to basic operators with constant coefficients, for which a comparatively complete series of results have been obtained during the past fifteen years.

2. Fundamental solution. Consider a differential expression of order l with constant coefficients:

$$P(D)u = \sum_{|\alpha| \leq l} a^{\alpha_1, \ldots, \alpha_m} \left(\frac{1}{i} \frac{\partial}{\partial x_1}\right)^{\alpha_1} \cdots \left(\frac{1}{i} \frac{\partial}{\partial x_m}\right)^{\alpha_m} u. \tag{1}$$

The first general result with respect to an operator of this type was the proof of the existence of a fundamental solution in various spaces of generalized functions [23], [17].

The generalized function $e(x)$ is called a fundamental solution if it satisfies the equation

$$P(D)e(x) = \delta(x). \tag{2}$$

In the case of the Laplace operator in $E_m (m > 2)$, the fundamental solution has the form

$$e(x) = (2-m)^{-1}|S_1|^{-1} |x|^{2-m}.$$

---

* The function $\lambda(x)$ is called a multiplier in some class of functions A if the product $\lambda u \in A$ for any function $u \in A$.

As we know, this function serves as the kernel of the integral operator (of the Newtonian potential) which represents a particular solution of the equation $\Delta u = f$. The fundamental solution of the operator $P(D)$ plays a similar role. The question of the existence of such a solution can only be posed in terms of generalized functions: it is well-known that in the class of ordinary functions even the simplest equation may not have a fundamental solution. The Fourier transform F converts equation (2) into an equivalent algebraic equation

$$P(\xi)(Fe)(\xi) = 1,$$

where

$$P(\xi) = \sum_{|\alpha| \leq l} a^{\alpha_1, \ldots, \alpha_m} \zeta_1^{\alpha_1} \ldots \zeta_m^{\alpha_m}.$$

Thus, the problem of constructing a fundamental solution $e(x)$ reduces to the construction of a generalized function $1/P(\xi)$ in some space or another. In this way we can not only obtain an existence theorem for the function $e(x)$, but can also investigate its differential properties.

As we have said, if we know the function $e(x)$, we can construct the solution of the inhomogeneous equation

$$P(D)u = f. \tag{3}$$

Moreover, with the aid of this function it is possible to derive a very general analogue of Poisson's formula, i.e. to express the solution of the equation

$$P(D)u = 0 \tag{4}$$

at any point, in terms of the values of this solution in some arbitrarily thin layer surrounding the point. These representations, together with information concerning the smoothness of the fundamental solution and its singularities, provide one way of studying the differential properties of solutions to general partial differential equations.

3. Hypoelliptic equations. It is well-known that any twice continuously differentiable solution of Laplace's equation is in fact an analytic function. An analogous property applies for any elliptic equations and systems with analytic coefficients. It was shown by I. G. Petrovskii [10] that analyticity of all solutions is a characteristic property of operators of elliptic type with constant coefficients. Thus, a simple "superficial" property of the equations – ellipticity – turns out to be equivalent to a profound and considerably less obvious property – the analyticity of its solutions.

L. Schwartz [24] posed the question of describing more general differential expressions $P(D)$ having the property that any solution of equation (3) $u \in C^{(\infty)}$ if $f \in C^{(\infty)}$. Such differential expressions are called *hypoelliptic*. The simplest example of a differential operator which is hypoelliptic, but not elliptic, is the heat-conduction operator.

An exhaustive discussion of hypoelliptic operators was given in 1955 by L. Hörmander [13] (cf. also [12], [16]). It turns out that for the operator P to be hypoelliptic it is necessary and sufficient that for all zeros of the polynomial $P(\xi)$, $|\xi| \to \infty$ should imply $\text{Im }|\xi| \to \infty$.

If the polynomial $P(\xi)$ satisfies this condition, then there exists a positive constant γ such that the inequality

$$|\text{Im }\xi| \geq a|\text{Re }\xi|^\gamma - b \tag{5}$$

is satisfied for the zeros of $P(\xi)$; where a and b are constants. The minimal value of γ is called the *exponent of hypoellipticity*. It is always the case that $\gamma \leq 1$, where $\gamma = 1$ is equivalent to an elliptic operator. The index of hypoellipticity can be defined by the following equation ([7]):

$$\gamma = \varlimsup_{|\xi| \to \infty} \frac{\ln(|\text{grad }P(\xi)|/|P(\xi)|)}{\ln|\xi|}.$$

Hypoelliptic operators are also distinguished by the following feature: the fundamental solution of the operator $P(D)$ is infinitely differentiable everywhere outside the origin if and only if $P(D)$ is hypoelliptic [20].

Through constructing fundamental solutions and examining their properties, we can describe a wider class of so-called partially hypoelliptic equations whose solutions are infinitely differentiable with respect to some of the variables [19] (cf. also [12]).

4. Representation of solutions of equations with constant coefficients. A general representation has been constructed for the solution of equation (4). It is a far-reaching generalization of the elementary representation, in the form of a sum of exponential solutions of an ordinary, homogeneous, linear differential equation with constant coefficients. We recall that the exponents of the exponentials in the latter case are the zeros of a characteristic polynomial. For a partial differential equation the corresponding expression has the form

$$u(x) = \int_{P(\xi)=0} e^{i(x,\xi)} \mu(d\xi), \tag{6}$$

where μ is an arbitrary generalized function of some class (if the zeros of $P(\xi)$ are simple, the μ is a measure) [18], [8].

The representation (6) follows from the fact that equation (4) in E_m is equivalent to the equation $P(\xi)(Fu)\xi = 0$ in spaces of generalized functions. This equation implies that the generalized function $Fu \equiv \mu$ is different from zero only on the set $\{\xi \colon P(\xi) = 0\}$, and this leads eventually to relation (6).

5. Existence of properly posed problems. Equation (6) implies the existence of an infinite set of solutions to any homogeneous equation with constant coefficients. It is therefore natural to consider the question of the additional conditions which select a unique solution from the whole family of solutions. We know that, as regards the equations of mathematical physics, this aim is met by setting various boundary-value problems, i.e. of all the solutions the one chosen is that which satisfies given conditions on the boundary of the domain.

It was stated in Chapter 25 that this problem is said to be properly posed for some pair of spaces (A, B) if it is uniquely soluble in A for all data from B, and the solution depends continuously on this data.

The question arises, does there exist, for a given equation in a fixed domain, even one properly posed problem? It was shown by Hörmander [13] that this question can be answered satisfactorily in the case of an equation with constant coefficients if $A = B = L_2(\Omega)$, and Ω is a bounded domain. In order to present the essence of this result, we introduce some definitions.

In conjunction with the operator $P(D)$, we shall consider the formally adjoint operator $\bar{P}(D)$, whose coefficients are obtained from the coefficients of $P(D)$ by replacing them with complex conjugates. We construct with respect to the operator P the so-called *minimal operator* P_0 which represents the closure in $L_2(\Omega)$ of the operator P given on $D(\Omega)$. The *maximal operator* is the adjoint in $L_2(\Omega)$ of the minimal operator $\bar{P}_0$.

Thus, the set of functions satisfying any homogeneous boundary conditions contains the domain of definition of the minimal operator, and is contained within the domain of definition of the maximal operator. The question of the existence of properly posed problems for $P(D)$ can now be formulated in the following way: is it possible to find an extension of the minimal operator which is at the same time a contraction of the maximal operator, and which has a bounded inverse defined on the whole of the space $L_2(\Omega)$?

From results obtained by M. I. Vishik [1], it follows that the necessary and sufficient condition for the existence of this extension is fulfilled for all functions $u \in D(\Omega)$ by the inequalities

$$\|u\|_{L_2(\Omega)} \leq C\|P(D)u\|_{L_2(\Omega)}, \tag{7}$$
$$\|u\|_{L_2(\Omega)} \leq C\|\bar{P}(D)u\|_{L_2(\Omega)},$$

where C is a constant independent of u. (For the Laplace operator, for example, these inequalities are simple consequences of Friedrichs' inequality.) The above-mentioned result of Hörmander was a proof of the surprising fact that these inequalities are true for any operator with constant coefficients.

This fact is a special case of some general theorems of Hörmander regarding comparison of differential operators. By definition, the operator $P(D)$ is stronger than the operator $Q(D)$ if $D(P_0) \subset D(Q_0)$ where P_0 and Q_0 are the respective minimal operators.

General considerations show that P is stronger than Q if and only if there exists a constant C such that for all functions $u \in D(\Omega)$ the inequality

$$\|Qu\|_{L_2(\Omega)} \leq C(\|Pu\|_{L_2(\Omega)} + \|u\|_{L_2(\Omega)}) \tag{8}$$

is satisfied, or, equivalently because of (7)

$$\|Qu\|_{L_2(\Omega)} \leq C'\|Pu\|_{L_2(\Omega)}. \tag{9}$$

Developing the technique of energy integrals, Hörmander gave in algebraic terms the following necessary and sufficient condition for which the estimate (9) is true:

$$|Q(\xi)| \leq C \sum |P^{(\alpha)}(\xi)|, \quad \forall \xi \in E_m, \tag{10}$$

where $P^{(\alpha)}(\xi)$ are derivatives of the polynomial $P(\xi)$, and the summation extends over all multi-indices α.

Hence it follows in particular that the operator $P(D)$ is elliptic if and only if it is stronger than any operator whose order is not higher than the order of P.

In the same work [13] it was shown that the necessary and sufficient condition for complete continuity of the operator $Q_0 P_0^{-1}$ is

$$\frac{\sum |Q^{(\alpha)}(\xi)|}{\sum |P^{(\alpha)}(\xi)|} \to 0 \text{ for } \xi \to \infty. \tag{11}$$

Hörmander also showed that if P and Q are maximal operators, and if $D(P) \subset D(Q)$, then either $Q = aP + b$, where a and b are constants, or P and Q are ordinary differential operators, with the degree of Q not greater than the degree of P.

It is interesting to note that the answer to the question of the existence of properly posed boundary-value problems turns out to be more complex for a system of equations than for one equation. Even in the simple case of the system $u_x + v_y = f_1$, $v_x = f_2$, the minimal operator does not have solvable extensions [4]. At the present time, the necessary and sufficient condition is known for the existence of a properly posed problem from $L_2(\Omega)$ into $L_2(\Omega)$ for a system with constant coefficients [9], [21].

Thus, the problem of existence of properly posed problems is fully answered in the case of constant coefficients. As we have seen, however, the question of explicit description of all proper boundary conditions for general operators is far from being answered, apart from a few isolated cases of equations and systems which have been analyzed in detail.

An examination of properly posed problems for general equations and systems in a half-space was first undertaken in the work of Gel'fand and Shilov [2]. They found uniqueness classes for solutions of Cauchy's problem for the system

$$\frac{\partial u_i(x,t)}{\partial t} = \sum_{j=1}^{k} P_{ij}(D) u_j(x,t) \quad (1 \leq i \leq k), \tag{12}$$

described by inequalities of the form

$$|u| \leq C e^{\alpha |x|^p}.$$

The exponent p in this inequality is defined by the matrix $\|P_{ij}(\xi)\|$. In ref. [2] the method used was based on a Fourier transform of generalized functions with respect to spatial variables, which reduces the system (12) to a system of ordinary differential equations. In a later development of this approach, Shilov [16] gave a description of classes of solubility for Cauchy's problem for the system (12). More general properly posed problems in a half-space were examined in refs. [5], [6] (cf. also [16]).

6. Operators with variable coefficients. The investigation of these operators involves very great difficulties in comparison with the case of constant coefficients. This is already clear from the fact in the case of equations with variable coefficients, the question of the existence of even one solution in a given domain cannot always be answered satisfactorily. In fact, the first-order equation

$$-i \frac{\partial u}{\partial x_1} + \frac{\partial u}{\partial x_2} - 2(x_1 + ix_2) \frac{\partial u}{\partial x_3} = f$$

for some infinitely differentiable function f, does not have a solution in the space D' of generalized functions in whatever sub-domain of E_3 ([22], [14]). Hörmander (cf. [14]) gave necessary and sufficient conditions for the existence of solutions in a small domain.

This result does not, of course, exhaust the information known at the present time regarding general equations with variable coefficients. Sufficient conditions have been obtained, for example, for the hypoellipticity of such equations, and also conditions for the existence of a properly posed problem "in the small" [14]. But the boundaries of the general theory of equations with variable coefficients have not yet been defined.

Appendix 4. Non-Linear, Second-Order Elliptic Equations

I. YA. BAKEL'MAN

Together with linear partial differential equations, an important place in mathematical physics and its applications is occupied by non-linear equations. In the following, we shall consider non-linear, second-order equations of elliptic type. These equations arise naturally from various problems in mechanics, variational calculus, geometry and other branches of mathematics.

1. Basic concepts. Let Ω be some domain in the m-dimensional Euclidean space E_m. We fix in E_m some cartesian coordinate system $x_1, x_2, \ldots, x_m$. As usual, a point of the space E_m will be denoted by

$$x = (x_1, x_2, \ldots, x_m).$$

Consider the set $C^{(2)}(\Omega)$ of twice continuously differentiable functions in Ω. Introduce the following notation:

$$z_i = \frac{\partial z}{\partial x_i} \quad (i = 1, 2, \ldots, m),$$

$$z_{ik} = \frac{\partial^2 z}{\partial x_i \partial x_k} \quad (i, k = 1, 2, \ldots, m),$$

$$Dz = (z_1, \ldots, z_m),$$

$$D^2 z = (z_{11}, z_{12}, \ldots, z_{mm}).$$

Now let

$$F(x_1, \ldots, x_m, z, p_1, \ldots, p_m, r_{11}, r_{12}, \ldots, r_{mm})$$

be a continuous function of the independent variables x_i, z, p_i, r_{ik} for all $x \in \Omega$ and for all finite values of the other arguments. For brevity we shall write the function F as $F(x, z, p, r)$, where $p = (p_1, \ldots, p_m)$ and $r = (r_{11}, r_{12}, \ldots, r_{mm})$. The quantities p and r can conveniently be interpreted as points of some Euclidean spaces P and R, which have dimensionality

m and $\tfrac{1}{2}m(m+1)$ respectively, and in which we take cartesian coordinates $p_1, \ldots, p_m$ and $r_{11}, r_{12}, \ldots, r_{mm}$.

Every function $F(x, z, p, r)$ on the set $C^{(2)}(\Omega)$ of functions generates an operator
$$\Phi(z) = F(x, z, \mathrm{D}z, \mathrm{D}^2 z),$$
which arises as a result of substituting for z, p, r in F the function $z(x)$ and its first and second derivatives respectively.

We assume now in addition that the function F has continuous first derivatives with respect to the variables r_{ik} ($i, k = 1, 2, \ldots, m$) for all $x \in \Omega$ and for all finite values of the other variables. We shall say that the operator $\Phi(z)$ is elliptical on the function $z_0 \in C^{(2)}(\Omega)$ if the quadratic form
$$T(\Phi, z_0) = \sum_{i,k=1}^{m} F^0_{r_{ik}} \xi_i \xi_k$$
is defined at all points of the domain Ω. The function $F^0_{r_{ik}}$ is the result of substituting in $F_{r_{ik}}(x, z, p, r)$ the function $z_0(x)$, its first and second derivatives. Since the functions $F^0_{r_{ik}}$ are continuous in Ω, the quadratic form $T(\Phi, z_0)$ retains the same sign everywhere in Ω. Therefore, if the operator $\Phi(z)$ is elliptic on the function $z_0 \in C^{(2)}(\Omega)$, then the quadratic form $T(\Phi, z_0)$ is either positive definite or negative definite everywhere in Ω. Note that the operator $\Phi(z)$ can be elliptic on some functions of $C^{(2)}(\Omega)$ and not on others. Respective examples will be given later.

The differential equation
$$F(x, z, \mathrm{D}z, \mathrm{D}^2 z) = 0$$
is called *elliptic* if the operator $\Phi(z)$ is elliptic on all solutions of this equation.

It is obvious that the definition of ellipticity given for linear equations (§ 2, Chapter 9) is a special case of the concept of ellipticity just introduced for the general equation
$$F(x, z, \mathrm{D}z, \mathrm{D}^2 z) = 0.$$

The most important types of non-linear equations are quasi-linear equations and equations of Monge-Ampère type.

a) The name quasi-linear is given to equations of the form
$$A_{ik}(x, z, \mathrm{D}z) \frac{\partial^2 z}{\partial x_i \partial x_k} + B(x, z, \mathrm{D}z) = 0. \tag{1}$$

These equations are linear with respect to the second derivatives of the unknown function $z(x)$. Linear equations are obviously a special case of quasi-linear equations.

Quasi-linear equations are elliptic if for all $x \in \Omega, z \in (-\infty, +\infty), p \in P$ the quadratic form $A_{ik}(x, z, p)\xi_i\xi_k$ is defined for all points of Ω. As we have already observed, the quadratic form

$$A_{ik}(x, z, p)\xi_i\xi_k \tag{2}$$

retains its sign for any $x \in \Omega, z, p \in P$. Therefore the ellipticity of equation (1) follows from the condition that the quadratic form (2) is positive (or negative) definite for any x, z, p. It can be shown that this condition is not only sufficient but also necessary for the ellipticity of equation (1).

b) An equation of the form

$$z_{x_1x_1}z_{x_2x_2} - z_{x_1x_2}^2 + \sum_{i,k=1}^{2} A_{ik}(x, z, Dz)z_{x_ix_k} + B(x, z, Dz) = 0, \quad A_{ik} = A_{ki} \tag{3}$$

is called a *Monge-Ampère equation*. This is an equation for functions with two independent variables. Its distinguishing feature is the presence of the expression

$$z_{x_1x_1}z_{x_2x_2} - z_{x_1x_2}^2$$

as the principal term. The operator

$$H(z) = z_{x_1x_1}z_{x_2x_2} - z_{x_1x_2}^2$$

is called the simplest *Monge-Ampère operator*.

The function

$$F(x, z, p, r) = r_{11}r_{22} - r_{12}r_{21} + \sum_{i,k=1}^{2} A_{ik}(x, z, p)r_{ik} + B(x, z, p)$$

is continuous if and only if the functions $A_{ik}(x, z, p), B(x, z, p)$ are continuous for all $x \in \Omega$ (here Ω is a domain in the plane of the variables x_1, x_2) and for all $z \in (-\infty, \infty), p \in P$. If these conditions are fulfilled, the function $F(x, z, p, r)$ also has continuous derivatives

$$\frac{\partial F}{\partial r_{11}} = r_{22} + A_{11}, \qquad \frac{\partial F}{\partial r_{12}} = -r_{21} + A_{12},$$

$$\frac{\partial F}{\partial r_{21}} = -r_{12} + A_{21}, \qquad \frac{\partial F}{\partial r_{22}} = r_{11} + A_{22}.$$

We now explain under what conditions equation (3) is elliptic. Let

$z(x_1, x_2)$ be an arbitrary function of $C^{(2)}(\Omega)$. Then

$$T(\Phi, z) = (z_{x_2 x_2} + A_{11})\xi_1^2 - 2(z_{x_1 x_2} - A_{12})\xi_1 \xi_2 + (z_{x_1 x_1} + A_{22})\xi_2^2.$$

The quadratic form $T(\Phi, z)$ is defined if and only if the inequality

$$(z_{x_2 x_2} + A_{11})(z_{x_1 x_1} + A_{22}) - (z_{x_1 x_2} - A_{12})^2 > 0$$

is satisfied in Ω.

Let $z(x_1, x_2)$ be a solution of equation (3). Then

$$z_{x_1 x_1} z_{x_2 x_2} - z_{x_1 x_2}^2 + \sum_{i,k=1}^{2} A_{ik} z_{x_i x_k} + B = 0$$

and consequently, in order that (3) should be an elliptic equation, it is enough to require that the inequality

$$A_{11} A_{22} - A_{12}^2 - B > 0$$

be satisfied for all $(x_1, x_2) \in \Omega$ and all finite values of z, p_1, p_2. It can be shown that this condition is also necessary for the ellipticity of equation (3). It is easy to see that the quadratic form

$$T(\Phi, z_0) = \sum_{i,k=1}^{2} F_{rik}^0 \xi_i \xi_k \tag{4}$$

retains the same sign everywhere if the operator $\Phi(z)$ is elliptic on the function z_0. This leads to the fact that there are functions z_0 for which the quadratic form (4) is positive definite in Ω, there are functions for which this form is negative definite, and, finally, there are functions for which it alternates in sign.

We investigate this in the case of the simplest Monge-Ampère equation

$$z_{x_1 x_1} z_{x_2 x_2} - z_{x_1 x_2}^2 = \phi(x_1, x_2), \tag{5}$$

where $\phi(x_1, x_2)$ is a continuous function in the domain Ω. For the operator

$$\Phi(z) = z_{x_1 x_1} z_{x_2 x_2} - z_{x_1 x_2}^2 - \phi(x_1, x_2)$$

$T(\Phi, z)$ has the form

$$T(\Phi, z) = z_{x_2 x_2} \xi_1^2 - 2 z_{x_1 x_2} \xi_1 \xi_2 + z_{x_1 x_1} \xi_2^2.$$

Obviously $T(\Phi, z)$ is positive definite on the functions $z = \frac{1}{2} k^2 (x_1^2 + x_2^2)$, negative definite on the functions $z = -\frac{1}{2} k^2 (x_1^2 + x_2^2)$, and, finally, alternating in sign for the functions $z = \frac{1}{2} k^2 (x_1^2 - x_2^2)$.

The condition for ellipticity of equation (5) is the inequality

$$\phi(x_1, x_2) > 0.$$

Thus, the solutions of the elliptic equation (5) will necessarily be convex functions. In fact, if $u(x_1, x_2)$ is a solution of equation (5), then $u_{x_1 x_1} u_{x_2 x_2} - u_{x_1 x_2}^2 = \phi(x_1, x_2) > 0$ everywhere in the domain Ω, and, consequently $T(\Phi, u)$ is a definite form everywhere in Ω, maintaining one and the same sign.

Together with the solution $u(x_1, x_2)$ of equation (5), another solution of this equation is the function $v(x_1, x_2) = -u$, therefore (5) has two classes of solution: in one of them the form $T(\Phi, u)$ is positive definite for any solution, and in the other the same form is negative for any solution. This phenomenon has a simple geometrical interpretation: the solutions of the former class are convex functions, with their convexity facing downwards, while the solutions of the latter class are convex functions with their convexity facing upwards.

In just the same way any elliptic equation of Monge-Ampère type (3) has two classes of solutions. For solutions of the first class, the form $T(\Phi, z)$ is positive definite, and for solutions of the second class it is negative definite.

c) The analogue of the simplest Monge-Ampère equation for functions of m variables is the equation

$$\begin{vmatrix} z_{x_1 x_1} & z_{x_1 x_2} & \cdots & z_{x_1 x_m} \\ z_{x_2 x_1} & z_{x_2 x_2} & \cdots & z_{x_2 x_m} \\ \cdot & \cdot & \cdot & \cdot \\ z_{x_m x_1} & z_{x_m x_2} & \cdots & z_{x_m x_m} \end{vmatrix} = \phi(x, z, Dz). \tag{6}$$

The analogue of the general Monge-Ampère equation in the m-dimensional case is taken to be the equation

$$\begin{vmatrix} z_{x_1 x_1} + A_{11} & z_{x_1 x_2} + A_{12} & \cdots & z_{x_1 x_m} + A_{1m} \\ \cdot & \cdot & \cdot & \cdot \\ z_{x_m x_1} + A_{m1} & z_{x_m x_2} + A_{m2} & \cdots & z_{x_m x_m} + A_{mm} \end{vmatrix} + B(x, z, Dz) = 0,$$

where $A_{ik} = A_{ki}$, B are continuous functions of x, z, p.

These equations also have two classes of solutions, with all the geometrical properties of the solutions to equation (6) being analogous to the properties of the solutions to the simplest, two-dimensional Monge-Ampère equation.

In the field of non-linear elliptic equations, most attention has been paid to the Dirichlet problem. Just as for linear elliptic equations, this problem consists in finding a function which satisfies a given non-linear elliptic equation inside some domain and which is equal to a given function on the boundary of the domain.

2. Maximum Principle. Uniqueness theorem for the Dirichlet problem. In the theory of elliptic equations, an important part is played by the so-called *maximum principle* which essentially states that under certain definite conditions solutions of elliptic equations cannot achieve either a positive maximum or a negative minimum inside a domain.

We glance briefly at the case of linear equations.

MAXIMUM PRINCIPLE. *Let the equation*

$$L(z) \equiv a_{ik}(x)z_{ik} + b_i(x)z_i + c(x)z = 0 \tag{7}$$

be given in a domain Ω. *It is assumed that the coefficients of this equation are continuous in* Ω, *that for all* $x \in \Omega$ *the quadratic form*

$$a_{ik}(x)\xi_i\xi_k$$

is positive, and that the function $c(x)$ *is non-positive. Further, let* $z(x) \in C^{(2)}(\Omega)$ *be a solution of equation* (7).

Then $z(x)$ *cannot attain either a negative minimum or a positive maximum in* Ω.

In § 10, Chapter 11, the maximum principle for linear equations was proved under the further assumption that $c(x) < 0$. *

The uniqueness theorem for the solution of the Dirichlet problem for a linear, elliptic, second-order equation follows immediately from the maximum principle.

The maximum principle and the uniqueness theorem for linear elliptic equations have been the subject of numerous investigations. It has been explained that the requirement of continuity of the coefficients in (7), and of the twofold continuous differentiability of the solution, can be substantially relaxed. In fact, it is sufficient to regard the coefficients of the equation as elements in a space L_p, and to regard the solutions as belonging to the class of functions having generalized derivatives. These results can be found in refs. [10], [12]. Geometrical methods have been applied to these questions in refs. [1], [2].

---

* The equation discussed in § 10, Chapter 11, differs from equation (7) by the sign of the left-hand side, and therefore the condition on the coefficient $C(x)$ there had the form $C(x) > 0$.

We now turn to the uniqueness theorem for the Dirichlet problem for non-linear elliptic equations.

Let $F(x, z, p, r)$ be a function defined for all $x \in \Omega$, $-\infty < z < +\infty$, $p \in P, r \in R$ and continuously differentiable for all other values of the arguments. As we observed in par. 1, the function F generates the operator

$$\Phi(z) = F(x, z, Dz, D^2z)$$

on the class of functions of $C^{(2)}(\Omega)$.

We shall say that the operator $\Phi(z)$ is *elliptic convex*, if the fact that the quadratic form

$$T(\Phi, z) = \sum_{i,k=1}^{n} \frac{\partial F(x, z, Dz, D^2z)}{\partial r_{ik}} \xi_i \xi_k,$$

is positive (negative) definite on the functions $z^0(x), z^1(x) \in C^{(2)}(\Omega)$, implies that this form is also positive (negative) definite on all the functions $z^\tau(x) = (1-\tau)z^0 + \tau z^1$, when τ varies between zero and one.

UNIQUENESS THEOREM FOR THE DIRICHLET PROBLEM. *Let the function $F(x, z, p, r)$, defined and continuously differentiable for all $x \in \Omega$, $-\infty < z < +\infty, p \in P, r \in R$, generate an elliptic convex operator $\Phi(z)$. Let the operator $\Phi(z)$ be positive (negative) definite on the solutions $z^1(x)$, $z^2(x) \in C^{(2)}(\Omega)$ of the equation*

$$F(x, z, Dz, D^2z) = 0.$$

If the functions $z^1(x)$ and $z^2(x)$ coincide on the boundary of Ω, and if

$$F_z(x, z, p, r) \leq 0 \quad (F_z(x, z, p, r) \geq 0)$$

for all $x \in \Omega$, $-\infty < z < \infty, p \in P, r \in R$, then the functions $z^1(x)$ and $z^2(x)$ coincide in Ω.

This theorem also permits various generalizations along the lines mentioned above.

From the uniqueness theorem for a general non-linear equation of elliptic type, we can deduce as simple corollaries the following uniqueness theorems.

1. *The Dirichlet problem for the quasi-linear equation*

$$\sum_{i,k=1}^{m} a_{ik}(x, Dz)z_{ik} + b(x, z, Dz) = 0,$$

where $a_{ik}(x, p)$ and $b(x, z, p)$ are continuously differentiable functions for all

$x \in \Omega$, $-\infty < z < +\infty$, $p \in P$, has not more than one solution, if the lowing conditions are fulfilled: for any $x \in \Omega$, $-\infty < z < +\infty$, $p \in P$,

a) $$\sum_{i,k=1}^{m} a_{ik}(x,p)\xi_i\xi_k > 0,$$

b) $$b_z(x,z,p) \leq 0.$$

2. *The Dirichlet problem for the Monge-Ampère equation*

$$\Phi(z) = z_{11}z_{22} - z_{12}^2 + \sum_{i,k=1}^{2} A_{ik}(x, \mathrm{D}z)z_{ik} + B(x, z, \mathrm{D}z) = 0,$$

$$A_{ik} \equiv A_{ki}, \tag{7}$$

where $A_{ik}(x, p)$ and $B(x, z, p)$ are continuously differentiable functions for all $x \in \Omega$, $-\infty < z < +\infty$, $p \in P$, has not more than one solution on which the quadratic form $T(\Phi, z)$ is positive (negative), if the following conditions are fulfilled: for any $x \in \Omega$, $z \in (-\infty, \infty)$, $p \in P$

a) $-B + A_{11}A_{22} - A_{12}^2 > 0,$ b) $B_z \leq 0.$

(This concerns uniqueness in the class of solutions of equation (7) on which the form $T(\Phi, z)$ is positive definite. Condition b) has to be replaced by $B_z \geq 0$ if we consider uniqueness in the class of solutions on which the form $T(\Phi, z)$ is negative definite.)

There are corresponding uniqueness theorems for the m-dimensional analogue of the Monge-Ampère equation.

In the formulation of the above uniqueness theorems for the quasi-linear and Monge-Ampère equations, we were concerned with solutions which had continuous second derivatives. But these theorems remain true for weaker assumptions in respect of the differential properties of the solutions. For example, it is sufficient to require that the solution should belong to the class $W_p^{(2)}$ for any $p > m$ (where m is the dimensionality of the space). The uniqueness theorem for the Monge-Ampère equation is of special interest from a geometrical point of view: certain special cases of the Monge-Ampère equation correspond to basic problems of geometry "in the large", connected with questions of the existence and uniqueness of a surface with a given interior metric, with given functions of smooth normal curvatures, etc. In refs. [13] and [14], methods have been developed which enable the application under very complex conditions of the maximum principle to the above-mentioned classes of Monge-Ampère equations, thereby leading to uniqueness theorems for the Dirichlet problem for these equations.

3. Existence theorems. The Dirichlet problem occupies a central position in the analysis of the solubility of non-linear elliptic equations. Proofs of existence theorems for solutions to this problem are extremely complicated, and within the framework of this Appendix we can only give a very brief outline of the basic methods and results.

The basic guide-lines for the investigation of this field were laid down in 1900, in two problems posed by D. Hilbert. The first was concerned with the validity of the hypothesis that all sufficiently smooth solutions of elliptic equations with analytic coefficients are themselves analytic functions; the second consisted in the following: it is required to show that the variational problem of finding a function which takes a given value on the boundary, and which achieves a minimum value for the given functional

$$J(z) = \int_\Omega \phi(x, z, \mathrm{D}z)\,\mathrm{d}x,$$

– where $\phi(x, z, p)$ is a convex function of the variables $p_1, \ldots, p_m$, bounded from below – always has a solution, if this solution is sought in a sufficiently wide class of functions. Since Euler's equation for a function which achieves an extremum for the functional $J(z)$ is the quasi-linear elliptic equation

$$\sum_{i=1}^{m} \frac{\partial}{\partial x_i}\left(\frac{\partial \phi}{\partial p_i}\right) - \frac{\partial \phi}{\partial z} = 0,$$

subsequent investigations of this problem were closely related to analyses of the Dirichlet problem for quasi-linear elliptic equations.

The first fundamental results for non-linear elliptic equations of general form with two variables, which contained the solution to Hilbert's problem, were presented in the classical work of S. N. Bernshtein, in 1908–1912. Bernshtein showed that three-times continuously differentiable solutions of non-linear, analytic equations of elliptic type are analytic functions of their arguments. Further, he obtained the following results concerning the solubility of the Dirichlet problem for the equation

$$F(x_1, x_2, z, z_1, z_2, z_{11}, z_{12}, z_{22}) = 0. \tag{8}$$

Assuming the function F to be analytic with respect to all its arguments, and assuming the analyticity of the boundary conditions, Bernshtein proved that if it were possible, in the Dirichlet problem for equation (8), to introduce in an analytic way a parameter $t \in [0, 1]$ such that (i) the Dirichlet problem had an analytic solution for $t = 0$, (ii) it became identical with the original problem for $t = 1$, and (iii) it were possible for all $t \in [0, 1]$ to obtain uniform estimates in the metric of the space $C^{(2)}$ for the solutions of all the auxiliary Dirichlet problems (assuming only their existence), then the original problem had a solution in the class of analytic functions. For quasi-linear equations, estimates of the solutions to the auxiliary problems in the metric of $C^{(1)}$ would be sufficient.

Estimates for the solutions of differential equations which are obtained only with the assumption that the solutions exist, and whose derivation makes use only of the properties of the coefficients and the boundary conditions of the problem, are commonly called *a priori estimates*. For analytic Dirichlet problems in the case of two variables, the question of solubility of the Dirichlet problem was reduced to the question of obtaining uniform a priori estimates, in the metric of $C^{(2)}$ for general equations and in the metric of $C^{(1)}$ for quasi-linear equations.

For the quasi-linear elliptic equations

$$\sum_{i,k=1}^{2} a_{ik}(x, Dz)z_{ik} + b(x, z, Dz) = 0$$

with analytic coefficients satisfying the condition

$$|b| \leq R_M \sum_{j,k=1}^{2} a_{jk}(x, p) p_j p_k$$

for all $x \in \Omega$ and $|z| \leq M, p_1^2 + p_2^2 \geq 1$, where R is a constant depending only on M, a priori estimates can be obtained for the moduli of the first derivatives in the Dirichlet problem. Here it is assumed that the domain Ω is bounded by an analytic contour with strictly positive curvature, $\partial b/\partial z \leq 0$, and that there is an a priori estimate of the modulus of the solution; the existence of an a priori estimate of the modulus of the solution is guaranteed if $\partial b/\partial z \leq \text{const.} < 0$.

The solubility of the Dirichlet problem has been established for the above class of equations. The variational problem for minimizing the functional

$$\int_{\Omega} \phi(x, z, Dz) \, dx, \tag{9}$$

where ϕ satisfies the inequality

$$\sum_{j,k=1}^{2} \frac{\partial^2 \phi}{\partial p_j \partial p_k} \xi_j \xi_k \geq \alpha_0 \sum_{k=0}^{2} \xi_k^2, \quad \alpha_0 = \text{const.} > 0,$$

leads to the solution of the Dirichlet problem for a quasi-linear equation of the above-mentioned class; this variational problem also turns out to be soluble.

It can be shown that it is possible to carry over Bernshtein's results to spaces of Hölder functions $C^{k,\alpha}$; in these spaces the appropriate results can be formulated much more simply and naturally. Bernshtein's basic theorems regarding general elliptic equations and variational problems have been proved under comparatively simple conditions of smoothness (it is sufficient that the solutions should belong to the space $C^{2,\alpha}$ for general equations, and to the space $C^{1,\alpha}$ for variational problems). These proofs rely on the work of J. Schauder concerning a priori estimates and solubility of boundary-value problems for linear elliptic equations in spaces of Hölder functions.

J. Leray and J. Schauder developed topological methods for solving elliptic and various other functional equations. These methods represent a broad generalization of Bernshtein's method.

Parallel with the series of works which originated with Hilbert, the solution of variational problems also proceeded along other lines. So-called direct methods were developed, that provide a minimizing sequence, converging to a function which achieves an extremum for a given functional. Without further analysis it is only possible to guarantee in this case that the function belongs to a space of the form $W_p^{(1)}$, $p > 1$; this function is naturally assumed to be a generalized solution of the variational problem. In the details of the development, the number of independent variables does not, as a rule, play any role.

In the case of two independent variables it is possible, by imposing various smoothness requirements on the integrand function in the functional (9), to establish that the generalized solution is sufficiently smooth.

The proof of the smoothness of the generalized solution in the case $m > 2$ requires the development of new methods of a priori estimates in $C^{1,\alpha}$, which would involve the specification of a large number of independent variables. A method of this type has been developed for functionals of the form (9), in which the integrand function $\phi(x, z, p)$ has order of magnitude $|p|^\alpha$, $\alpha > 1$ as $|p| \to \infty$. A series of theorems is then obtained regarding the solubility and differential properties of the solutions of the variational problem (9); in many cases it turns out that the requirements imposed on the function ϕ cannot be relaxed further. Theorems have been proved concerning the solubility in the classical sense of quasi-linear elliptic equations of the form

$$a_{jk}(x, z, Dz) \frac{\partial^2 z}{\partial x_j \partial x_k} + a(x, z, Dz) = 0$$

(non-divergent form) and also of equations of the form

$$\frac{\partial}{\partial x_k} a_k(x, z, Dz) + a(x, z, Dz) = 0$$

(divergent form). A more extensive discussion of all the questions mentioned here can be found in refs. [10] and [8].

In recent times a series of works has appeared in which degenerate elliptic problems (the case $\alpha = 1$) are discussed. These problems are closely connected with the classical problem of the minimal surface.

A body of work has been devoted to theorems of existence for equations of the Monge-Ampère type. A. D. Alexandrov has formulated in geometric terms some problems connected with the construction of convex surfaces with respect to their interior metric, Gauss integral curvature and other characteristics, and has developed geometric methods for solving these problems. From an analytical point of view, the question is the construction of generalized solutions for some special types of Monge-Ampère equations. It turns out that when the geometrical characteristics are sufficiently smooth functions, the generalized solutions are also smooth [14].

Recently, there have been studies (cf. [3], [14]) of generalized solution for a wide class of Monge-Ampère equations and their multi-dimensional analogues. The smoothness of the generalized solutions has been established under conditions of sufficient smoothness of the coefficients of the equations and boundary conditions.

The generalized solution of the Dirichlet problem has been investigated (cf., for example, [5], [7]) for an extensive class of quasi-linear, strongly elliptic systems of arbitrary order.

Bibliography

Bibliography to Part I

1. Mikhlin, S. G., Problem of the minimum of a quadratic functional, 1964, Holden-Day.
2. Smirnov, V. I., A course of higher mathematics Vol. V, 1964, Pergamon.
3. Sobolev, S. L., Applications of functional analysis in mathematical physics, 1963, Amer. Math. Soc.

Bibliography to Part II

1. Akhiyezer, N. I., Calculus of variations, 1962, Blaisdell.
2. Bliss, G. A., Lectures on the calculus of variations, 1946, Univ. of Chicago.
3. Courant, R. and Hilbert, D., Methods of mathematical physics Vol. I, 1953; Vol. II, 1962; Interscience.
4. Lavrent'yev, M. A. and Lyusternik, L. A., Foundations of the calculus of variations Vol. I, Part 1, 1935, O.N.T.I. (in Russian).
5. Mikhlin, S. G., Problem of the minimum of a quadratic functional, 1964, Holden-Day.
6. Smirnov, V. I., A course of higher mathematics Vol. IV, 1964, Pergamon.
7. Friedrichs, K. O., Spektraltheorie halbbeschränkter Operatoren und Anwendung der Spektralzerlegung von Differentialoperatoren, 1934, Math. Ann. **109**, 465.

Bibliography to Part III

1. Kantorovich, L. V. and Akilov, G. P., Functional analysis in normed spaces, 1964, Pergamon.
2. Mikhlin, S. G., On the solubility of linear equations in Hilbert spaces, 1947, Dokl. Akad. Nauk. U.S.S.R. **57**, 1 (in Russian).
3. Mikhlin, S. G., Integral equations and their applications, 1964, Pergamon.
4. Petrovskii, I. G., Lectures on the theory of integral equations, 1957, Graylock.
5. Riesz, F. and Sz.-Nagy, B., Functional analysis, 1955, Ungar.
6. Smirnov, V. I., A course of higher mathematics Vol. IV, 1964, Pergamon.
7. Sobolev, S. L., Applications of functional analysis in mathematical physics, 1963, Amer. Math. Soc.
8. Tricomi, F., Integral equations, 1957, Interscience.

Bibliography to Part IV

1. Oleinik, O. A., A boundary-value problem for linear elliptic-parabolic equations, 1965, Univ. of Maryland.
2. Smirnov, M. M., Degenerate elliptic and hyperbolic equations, 1966, Nauka (in Russian).

3. Tricomi, F., Sulle equazioni lineari alle derivate parziali di seconde ordine di tipo misto, 1923, Rend. Reale Accad. Lincei, Ser. 5, **14**, 134.

Bibliography to Part V

1. Vilenkin, N. Ya., Special functions and the theory of representation of groups, 1968, Amer. Math. Soc.
2. Vishik, M. I., Strongly elliptic systems of differential equations, 1951, Matem. Sbor. **29 (71)**, 615 (in Russian).
3. Hobson, E. W., Theory of spherical and ellipsoidal harmonics, 1931, Cambridge Univ.
4. Günter, N. M., Potential theory and its application to basic problems of mathematical physics, 1953, Gostekhizdat (in Russian).
5. Carleman, T., Über die asymptotische Verteilung der Eigenwerte partieller Differentialgleichungen, 1936, Berichte Akad. Leipzig, Bd. 88.
6.–7. Courant, R. and Hilbert, D., Methods of mathematical physics Vol. I, 1953; Vol. II, 1962; Interscience.
8. Ladyzhenskaya, O. A., The mixed problem for a hyperbolic equation, 1953, Gostekhizdat (in Russian).
9. Landkof, N. S., Foundations of modern potential theory, 1966, Nauka (in Russian).
10. Miranda, C., Equazioni alle derivate parziali di tipo ellittico, 1955, Springer.
11. Mikhlin, S. G., Problem of the minimum of a quadratic functional, 1964, Holden-Day.
12. Mikhlin, S. G., Multidimensional singular integrals and integral equations, 1965, Pergamon.
13. Mikhlin, S. G., Numerical implementation of variational methods, 1966, Nauka (in Russian).
14. Muskhelishvili, N. I., Singular integral equations, 1953, Noordhoff.
15. Petrovskii, I. G., Partial differential equations, 1967, Iliffe.
16. Slobodetskii, L. N., Generalized Sobolev spaces and their applications to boundary-value problems for partial differential equations, 1958, Uchen. Zapis. Leningrad. Ped. Int. **197**, 54 (in Russian).
17. Smirnov, V. I., A course of higher mathematics Vol. III, 1964, Pergamon.
18. Sobolev, S. L., On a theorem of functional analysis, 1938, Matem. Sbor. **4 (46)**, 3 (in Russian).
19. Sobolev, S. L., Applications of functional analysis in mathematical physics, 1963, Amer. Math. Soc.

Bibliography to Part VI

1.–2. Courant, R. and Hilbert, D., Methods of mathematical physics Vol. II, 1962, Interscience.
3. Ladyzhenskaya, O. A., The mixed problem for a hyperbolic equation, 1953, Gostekhizdat (in Russian).
4. Petrovskii, I. G., Partial differential equations, 1967, Iliffe.
5. Smirnov, V. I., A course of higher mathematics Vols. II and IV, 1964, Pergamon.
6. Sobolev, S. L., Applications of functional analysis in mathematical physics, 1963, Amer. Math. Soc.
7. Sobolev, S. L., Partial differential equations of mathematical physics, 1965, Pergamon.
8. Tikhonov, A. I. and Samarskii, A. A., Equations of mathematical physics, 1963, Pergamon.
9. Shilov, G. Ye., Mathematical analysis, 1965, Pergamon.

Bibliography to Part VII

1. Slobodetskii, L. N., Generalized Sobolev spaces and their applications to boundary-value problems for partial differential equations, 1958, Uchen. Zapis. Leningrad. Ped. Int. **197**, 54 (in Russian).
2. Sobolev, S. L., Partial differential equations of mathematical physics, 1963, Pergamon.
3. Tikhonov, A. N., On the solution of improperly posed problems and the method of regularization, 1963, Dokl. Akad. Nauk. U.S.S.R. **151**, 3, 501 (in Russian).
4. Hörmander, L., On the theory of general partial differential operators, 1955, Acta Math. **94**, 161.

Bibliography to Appendix 1

1. Agranovich, M. S. and Dynin, A. S., General boundary-value problems for elliptic systems in a multi-dimensional domain, 1962, Dokl. Akad. Nauk. U.S.S.R. **146**, 2, 511 (in Russian).
2. Bitsadze, A. V., Boundary-value problems for second order elliptic equations, 1968, North-Holland.
3. Yegorov, Yu. V. and Kondrat'yev, V. A., The oblique derivative problem, 1966, Dokl. Akad. Nauk. U.S.S.R. **170**, 4, 770 (in Russian).
4. Lopatinskii, Ya. B., On a method of transforming boundary-value problems for a system of elliptic differential equations into regular integral equations, 1953, Ukr. Matem. Zhurn. **5**, 2, 123 (in Russian).
5. Petrovskii, I. G., Some problems in the theory of partial differential equations, 1946, Usp. Matem. Nauk. **1**, 3–4, 44 (in Russian); Amer. Math. Soc. Translation No. 12.
6. Slobodetskii, L. N., Generalized Sobolev spaces and their applications to boundary-value problems for partial differential equations, 1958, Uchen. Zapis. Leningrad. Ped. Int. **197**, 54 (in Russian).
7. Smirnov, V. I., A course of higher mathematics Vol. V, 1964, Pergamon.
8. Sobolev, S. L., Applications of functional analysis in mathematical physics, 1963, Amer. Math. Soc.
9. Shapiro, Z. Ya., On general boundary-value problems for elliptic equations, 1953, Izv. Akad. Nauk. U.S.S.R. Ser. Matem. **17**, 6 (in Russian).
10.–11. Agmon, S., Douglis, A. and Nirenberg, L., Estimates near the boundary for solutions of elliptic partial differential equations satisfying general boundary conditions Part I, 1959, Comm. Pure Appl. Math. **12**, 623; Part II, 1964, ibid **17**, 35.
12. Atiyah, M. F. and Singer, I. M., The index of elliptic operators on compact manifolds, 1963, Bull. Amer. Math. Soc. **69**, 3, 422.
13. Borelli, R. L., The singular second-order oblique derivative problem, 1966, J. Math. Mech. **16**, 1, 51.

Bibliography to Appendix 2

1. Babich, V. M., Fundamental solutions of hyperbolic equations with variable coefficients, 1960, Matem. Sbor. **52 (94)**, 2 (in Russian).
2. Borovikov, V.A., Fundamental solutions of linear equations with constant coefficients, 1949, Trud. Mosk. Matem. Obshch. **8**, 199 (in Russian).
3. Problems in the dynamical theory of seismic wave propagation V, 1961, Leningrad Univ. (in Russian).
4.–5. Gel'fand, I. M. and Shilov, G. Ye., Generalized functions Vol. 1, Properties and operations; Vol. 3, Theory of differential equations, 1964, Academic.

6. Gårding, L., Cauchy's problem for hyperbolic equations, 1958, Univ. of Chicago.
7. Courant, R. and Hilbert, D., Methods of mathematical physics Vol. II, 1962, Interscience.
8. Petrovskii, I. G., Cauchy's problem for a system of partial differential equations, 1937, Matem. Sbor. **2 (44)**, (in Russian).
9. Smirnov, V. I., A course of higher mathematics Vol. II, 1964, Pergamon.
10. Hadamard, J., Le problème de Cauchy et les équations aux dérivées partielles linéaires hyperboliques, 1932, Hermann (Paris).
11. Mathisson, M., Le problème de M. Hadamard, 1939, Acta Math. **71**.
12. Riesz, M., L'intégrale de Riemann-Liouville et le problème de Cauchy, 1949, Acta Math. **81**, 1.
13. Stellmacher, K., Eine Klasse Huyghenschen Differentialgleichungen und ihre Integration, 1955, Math. Ann. **130**, 3, 219.

Bibliography to Appendix 3

1. Vishik, M. I., General boundary-value problems for elliptic differential equations, 1952, Trud. Mosk. Matem. Obshch. **1**, 187 (in Russian).
2. Gel'fand, I. M. and Shilov, G. Ye., Fourier transforms of rapidly increasing functions and questions of uniqueness in Cauchy's problem, 1953, Usp. Matem. Nauk. **8**, 6, 3 (in Russian).
3. Gel'fand, I. M. and Shilov, G. Ye., Generalized functions Vol. I, Properties and operations; Vol. 2, Functions and generalized functions and spaces; Vol. 3, Theory of differential equations, 1964, Academic.
4. Dezin, A. A., Existence and uniqueness theorems for the solutions of boundary-value problems in functional spaces, 1959, Usp. Matem. Nauk. **14**, 3, 21 (in Russian).
5. Dikopolov, G. V. and Shilov, G. Ye., Properly posed problems for partial differential equations in a half-space, 1960, Izv. Akad. Nauk. U.S.S.R. Ser. Matem. **24**, 369 (in Russian).
6. Palamodov, V. P., Properly posed problems for partial differential equations in a half-space, 1960, Izv. Akad. Nauk. U.S.S.R. Ser. Matem. **24**, 381 (in Russian).
7. Palamodov, V. P., Conditions for properly posed solutions in the large, 1960, Sov. Math. Dokl. **1**, 602.
8. Palamodov, V. P., The general form of solutions to homogeneous differential equations, 1961, Dokl. Akad. Nauk. U.S.S.R. **137**, 4, 774 (in Russian).
9. Paneyakh, B. P., General systems of differential equations, 1961, Dokl. Akad. Nauk. U.S.S.R. **138**, 2, 297 (in Russian).
10. Petrovskii, I. G., Sur l'analyticité des solutions des systèmes d'équations différentielles, 1939, Matem. Sbor. **5 (47)**, 3.
11. Sobolev, S. L., Méthode nouvelle à résoudre le problème de Cauchy, 1936, Matem. Sbor. **1 (43)**, 39.
12. Treves, F., Lectures on the theory of partial differential equations.
13. Hörmander, L., On the theory of general partial differential operators, 1955, Acta Math. **94**, 161.
14. Hörmander, L., Linear partial differential operators, 1963, Academic.
15. Shilov, G. Ye., Conditions for the proper posing of Cauchy's problem, 1955, Usp. Matem. Nauk. **10**, 4, 89 (in Russian).
16. Shilov, G. Ye., Mathematical analysis, second special course, 1965, Nauka (in Russian).
17. Ehrenpreis, L., Solution of some problems of division I, 1954, Amer. J. Math. **76**, 883.
18. Ehrenpreis, L., The fundamental principle and some of its applications, 1960, Warsaw Conf. on Funct. Anal.

19. Gårding, L. and Malgrange, B., Opérateurs différentiels partiellement hypoelliptic, 1958, C.R. Acad. Sci. Paris **247**, 2083.
20. Hörmander, L., Local and global properties of fundamental solutions, 1957, Math. Scand. **5**, 27.
21. Fuglede, B., A priori inequalities connected with systems of partial differential equations, 1961, Acta Math. **105**, 177.
22. Levi, H., An example of a smooth linear partial differential equation without solution, 1957, Ann. Math. (2) **66**, 155.
23. Malgrange, B., Equations aux dérivées partielles à coefficients constants. Solutions élémentaires, 1953, C.R. Acad. Sci. Paris, **237**, 1620.
24. Schwartz, L., Théorie des distributions, 1, 2, 1950–51, Hermann, Paris.

Bibliography to Appendix 4

1. Aleksandrov, A. D., The maximum principle, I–VI, Izv. Vyssh. Uch. Zav., Matem., 1958, **5**; 1959, **3**, **5**; 1960, **3**, **5**; 1961, **1** (in Russian).
2. Aleksandrov, A. D., Majorants of solutions and uniqueness conditions for elliptic equations, 1966, Vest. Leningrad. Univ. **7** (in Russian).
3. Bakel'man, I. Ya., Geometric methods of solving elliptic equations, 1965, Nauka (in Russian).
4. Bernshtein, S. N., Colected works Vol. III, Partial differential equations, 1960, Akad. Nauk. U.S.S.R. (in Russian).
5. Browder, F. E., Variational boundary-value problems for quasi-linear elliptic equations of arbitrary order, 1963, Sov.-Amer. Symp. on Part. Diff. Eq., Novosibirsk.
6. Browder, F. E., Non-linear elliptic boundary-value problems, 1963, Sov.-Amer. Symp. on Part. Diff. Eq., Novosibirsk.
7. Vishik, M. I., Quasi-linear strongly elliptic systems, 1963, Trud. Mosk. Matem. Obshch. **12**, 125 (in Russian).
8. De Giorgi, E., Sulla differenziabilità e l'analicità delle estremali degli integrali multipli regolari, 1957, Mem. Acad. Sci. Torino, Cl. Sci. Fis. Mat. Nat., Ser. 3, Vol. 3, **1**, 25.
9. Courant, R. and Hilbert, D., Methods of mathematical physics Vol. II, 1962, Interscience.
10. Ladyzhenskaya, O. A. and Ural'tseva, N. N., Linear and quasilinear equations of elliptic type, 1969, Academic Press.
11. Leray, J. and Schauder, J., Topologie et équations fonctionnelles, 1934, Ann. Ecole Norm. Sup. **51**, 45.
12. Miranda, C., Equazioni alle derivate parziali di tipo ellittico, 1955, Springer.
13. Pogorelov, A. V., Bending of convex surfaces, 1951, Gostekhizdat (in Russian).
14. Pogorelov, A. V., On Monge-Ampère equations of elliptic type, 1964, Stechert.

Subject Index

Abel's theorem, 274
Absolute minimum, 40
Abstract function, 420
Arzela's theorem, 162
A priori estimate, 550
Averaging kernel, 9
Averaging radius, 11

Banach space, 30, 38, 153, 195
Banach's theorem, 168
Bessel functions, 319
Bessel's inequality, 340
Bilinear transformation, 281
Boundary conditions, 37, 192
– –, natural, 76
– –, principal, 79
Boundary-strip, 15
Boundary-value problem, exterior, 254
– – –, first, 254, 307
– – –, interior, 254
– – –, mixed, 416
– – –, second, 254
– – –, third, 381
Brachistochrone problem, 35, 51

Calculus of variations, 37
Canonical form, 205
Cauchy conditions, 193
Cauchy data, 193
– – on characteristics, 202
Cauchy-Riemann equations, 279
Cauchy surface, 193
Cauchy's inequality, 104
Caucy's problem, 192, 416, 430
– –, abstract, 422
– – for heat-conduction equation, 463
– – for hyperbolic equations, 521
– – for wave equation, 472
– –, fundamental solution of, 522
Characteristic cone, 430

– conoid, 524
– curve, 202
– equation, 202
– surface, 202
Characteristics, 202
–, invariance of, 203
Coercive equations, 518
Compatibility conditions, 190, 446
Complementing condition, 516
Completely continuous operator (c.c.o.), 153
Cone condition, 30, 329
Conformal transformation, 219, 280
Conjugate functions, 273
Convergence in the mean, 245
Convex functions, 545

D'Alembert's solution, 486
Density, 226
Differential expression, 184
– –, formally adjoint, 209
– –, formally self-adjoint, 210
Dini's formula, 277
Dirac delta function, 221, 523
Dirichlet integral, 70, 212
Dirichlet problem, 191, 254, 371, 498
– –, eigenvalues of, 316
– –, energy space for, 296
– –, exterior, 255
– – for annulus, 277
– – for circle, 274
– – for half-space, 373
– – for sphere, 260, 285
– –, generalized solution of, 299
– –, interior, 254
– –, operator of, 292
– –, spectrum of, 315
Domain of definition, 37, 86
Domain of dependence, 434
Double-layer potential, 226, 356
– – –, direct value of, 357

Eigenelement, 124
Eigenvalue, 124
-, rank of, 126
-, smallest, 131
Eigenvalues, generalized, 128
-, orthonormal, 127
Elliptic equations, 4, 187
- -, non-degenerate, 293, 306
- -, non-linear, 541
- -, quasi-linear, 542
Elliptic systems, 513
Embedding operator, 30
Embedding theorems, 30
Energy functional, 100
- norm, 93
- product, 93
Energy space, 92, 296
- -, elements of, 93
- -, separable, 107
Euclidean space, 4
Euler's equation, 51
- -, invariance of, 53
Euler's theorem, 56
Existence of solutions, 194
Exponent of hypoellipticity, 536
Extremal, 64
Extremum, 37

Fermat's principle, 36
Finite function, 15
Fourier series, 99
Fourier's method, 440
Fourier transform, 459
- -, inversion theorem for, 461
Fredholm alternative, 176
- integral equation, 167
- kernel, 155
Fredholm operator, 155
- -, adjoint, 168
Fredholm theorems, 166, 340
Friedrichs extension, 111
- -, eigenvalues of, 128
Friedrichs' inequality, 290
- theorem, 111
Fubini's theorem, 86
Functional, 37
-, absolute minimum of, 40
-, bilinear, 84
-, domain of definition of, 37
-, energy, 100
-, extremum of, 38
-, Fermat, 525

-, first variation of, 41
-, gradient of, 40
-, homogeneous quadratic, 84
-, linear, 38
-, minimizing sequence for, 60
-, quadratic, 85
-, relative minimum of, 40
-, second variation of, 54
-, symmetric bilinear, 84
-, value of, 37

Gamma function, 5
Gauss's integral, 357
Gauss's theorem, 18
Generalized derivative, 19
- divergence, 71
- eigenelement, 128
- eigenvalue, 128
- functions, 523, 532
Generalized solution, 104
- - of Cauchy's problem, 438
- - of Dirichlet problem, 299
- - of mixed problem, 423
Gram-Schmidt orthogonalization, 154
Green's formulae, 210
Green's theorem, 18

Hadamard's elementary solution, 524
Hadamard's problem, 508
Harmonic function, 218
- -, integral representation of, 224
Harnack's theorem, 243
Hausdorff's theorem, 154
Heat-conduction equation, 3, 185, 411
- - -, Cauchy's problem for, 459
Hilbert operator, 392
Hilbert space, 44
- -, complex, 124
- -, separable, 84
Hölder condition, 231
Hölder functions, 551
Homogeneous polynomials, 282
Hyperbolic equations, 4, 187
Hypoelliptic equations, 535

Ideal elements, 93
Improperly posed problems, 496
Index of a coercive equation, 519
Index of an operator, 404
Inequality
-, Bessel's, 340
-, Cauchy's, 104

–, Friedrichs', 290
–, Poincaré's, 337
–, positive-definiteness, 88
–, Schwartz-Bunyakovskii, 14
–, triangle, 15
Initial conditions, 418
Integrable function, 11
Integral equations
– –, Fredholm, 167
– – of potential theory, 372
– –, singular, 397
– – with weak singularity, 167
Integral representation, 224, 329
Integration by parts formula, 18
Isoperimetric problem, 56

Kelvin's transformation, 219
Kirchhoff's formula, 480

Lagrange's formula, 64
Laplace operator, 3
Laplace's equation, 3, 185, 217
– –, singular solution of, 219, 248
– –, uniqueness theorems for, 256, 270
Lebesgue-integrable function, 11
Lebesgue measure, 4
Legendre's conditon, 66
L'Hôpital's Rule, 11
Liouville's theorem, 267
Lipschitz condition, 230
Logarithmic potential, 382
Lyapunov
– conditions, 344
– curve, 382
– sphere, 345
– surface, 344
Lyapunov's theorem, 362

Manifold, linear, 38
Matrix of highest coefficients, 184
Maximum principle, 240, 413
Mean function, 11
Mean value theorem, 233, 237
Minimax principle, 145
Minimizing sequence, 60, 132
Mixed problem, 416, 426
Monge-Ampère equations, 543
Multi-index, 513
Multiplier rule, 56

Neumann problem, 192, 370
– –, eigenvalues of, 334

– –, exterior, 255, 499
– – for annulus, 278
– – for circle, 276
– – for half-space, 373
– – for sphere, 286
– –, generalized solution of, 335
– –, interior, 254, 502
– –, operator of, 325
– –, spectrum of, 334
Newtonian potential, 226
Newton-Leibnitz formula, 28
Norm, 50
–, energy, 93
Normal derivative, direct value of, 367
Normally soluble problem, 403

Oblique-derivative problem, 392
Operator, 85
–, adjoint, 154
–, completely continuous, 153
–, contraction of, 175
–, domain of definition of, 86
–, elliptic convex, 547
–, finite-dimensional, 153
–, Fredholm, 155
–, Friedrichs' extension of, 111
–, generalized eigenelement of, 128
–, generalized eigenvalue of, 128
–, Hilbert, 392
–, index of, 404
–, integral, with weak singularity, 158
–, inverse, 109
–, Laplace, 3
–, linear, 86
–, lower bound of, 111, 129
–, maximal, 537
–, merely positive, 121
–, minimal, 537
–, Monge-Ampère, 543
– of boundary-value problem, 195
– of Dirichlet problem, 292
– of Neumann problem, 325
–, positive, 87
–, positive definite, 88, 294
–, self-adjoint, 111
–, singular integral, 165
–, symmetric, 86
Ostrogradskii's formula, 18

Parabolic equations, 4, 187
Parametrix, 248
Parseval's theorem, 453

SUBJECT INDEX

Petrovskii-ellipticity, 514
Petrovskii-hyperbolicity, 527
Poincaré's inequality, 337
Poisson kernel, 263
Poisson's equation, 217
Poisson's formula, 263, 467
– – for circle, 275
Polyharmonic equation, 312
Positive definiteness inequality, 88
Potential, 225
Properly posed problems, 194, 495

Quadratic form, 84

Ray method, 525
Regular normal derivative, 258
Regular surface, 258
Relative minimum, 40
Riemann function, 524
Riesz's theorem, 44

Scalar product, 48
Schwartz-Bunyakovskii inequality, 14
Schwartz-Christoffel transformation, 282
Schwartz's formula, 275
Separation of variables, 318
Single-layer potential, 226, 363
Singular integral, 397
Singularity, weak, 158
Slobodetskii space, 517
Sobolev integral representation, 329
Sobolev space, 30, 517
Solid angle, 349
Space
–, Banach, 30
–, energy, 92
–, Euclidean, 4
–, Hilbert, 44, 124
–, separable, 84
–, Slobodetskii, 517
–, Sobolev, 30, 517
– of basic functions, 533
Space-like surface, 521

Spectral decomposition, 487
Spectral function, 488
Spectrum
–, discrete, 135
–, eigenvalue, 124
–, generalized, 127
Spherical harmonics, 282
Strongly elliptic systems, 309
Sturm-Liouville problem, 138, 148
Sub-domain, 13
Summable function, 11

Test function, 9
Transformation
–, affine, 426
–, bilinear, 281
–, conformal, 219
–, Kelvin's, 219
–, regular, 200
–, Schwartz-Christoffel, 282
Triangle inequality, 15
Tricomi's equation, 187
Type of an equation, 186, 206

Ultrahyperbolic equation, 188
Uniqueness of solutions, 194
Uniqueness theorems, 256, 270, 418, 431

Variation, first, 41
Variation, second, 54
Vibrating-membrane equation, 2, 184, 484
Vibrating-string equation, 2, 184, 455, 485

Wave diffusion, 525
Wave equation, 2, 188, 425
– –, Cauchy's problem for, 472
Wave front, leading, 436
Wave front, rear, 436, 483
Wave, plane, 527
Wave propagation, 436
Weak singularity, 158, 167
Weierstrass' theorem, 451